Machine Learning in Communication Systems and Networks

Machine Learning in Communication Systems and Networks

Editors

Yichuang Sun
Haeyoung Lee
Oluyomi Simpson

MDPI

Basel • Beijing • Wuhan • Barcelona • Belgrade • Novi Sad • Cluj • Manchester

Editors

Yichuang Sun
University of Hertfordshire
Hatfield
UK

Haeyoung Lee
University of Hertfordshire
Hatfield
UK

Oluyomi Simpson
University of Hertfordshire
Hatfield
UK

Editorial Office
MDPI
St. Alban-Anlage 66
4052 Basel, Switzerland

This is a reprint of articles from the Topic published online in the open access journals *Applied Sciences* (ISSN 2076-3417), *Sensors* (ISSN 1424-8220), *Electronics* (ISSN 2079-9292), *Photonics* (ISSN 2304-6732), *Journal of Sensor and Actuator Networks* (ISSN 2224-2708), and *Telecom* (ISSN 2673-4001) (available at: https://www.mdpi.com/topics/ml_communication_networks).

For citation purposes, cite each article independently as indicated on the article page online and as indicated below:

Lastname, A.A.; Lastname, B.B. Article Title. *Journal Name* **Year**, *Volume Number*, Page Range.

ISBN 978-3-7258-0725-3 (Hbk)
ISBN 978-3-7258-0726-0 (PDF)
doi.org/10.3390/books978-3-7258-0726-0

Contents

About the Editors

Yichuang Sun

Yichuang Sun is a Professor of Communications and Electronics, the Head of the Wireless Communications and RF Systems Research Lab, and the Head of Electrical and Electronic Engineering at the University of Hertfordshire, UK. Professor Sun's expertise and interests are in wireless communications, RF systems, microelectronics, neuromorphic computing, and machine learning. He has published four books and 420 papers. He was an Editor of the *IEE Circuits, Devices and Systems* book series, an Associate Editor of *IEEE Transactions on Circuits and Systems I: Regular Papers*, and the lead or sole Guest Editor of 8 *IEEE* and *IET/IEE* Special Issues.

Haeyoung Lee

Haeyoung Lee is a Senior Lecturer in Wireless Communication in the School of Physics, Engineering, and Computer Science, at the University of Hertfordshire. She has contributed to various EU projects focusing on cognitive communication, IoT communication, and dynamic coverage extension. Additionally, she has been actively involved in 5G PPP architecture working groups. Her research interests encompass dynamic spectrum use, radio resource management employing optimization and machine learning, and the performance evaluation of mobile communication systems.

Oluyomi Simpson

Oluyomi Simpson is a Principal Lecturer in Communications and Electronics Engineering, serving as the Academic Lead of the Communications Lab and the Manager of the Wireless Communication and RF Systems Research Lab within the School of Physics, Engineering, and Computer Science at the University of Hertfordshire, UK. His research and technical expertise encompass algorithms, optimization, hardware design, and the implementation of communication systems, including wireless communication, RF systems, and digital systems. Dr. Simpson is an Associate Editor of *IEEE Communications Letters* and has served as a Guest Editor for four Special Issues.

Preface

The communication systems and networks landscape is rapidly transforming due to the widespread use of mobile devices and growing data transmission demands. The ubiquity of mobile devices, coupled with the popularity of their applications, has led to unprecedented data traffic levels, presenting substantial challenges in managing infrastructure complexity and optimizing user experiences.

In addressing these challenges, machine learning emerges as a promising solution. The incorporation of machine learning, driven by powerful computing platforms, is posited to introduce innovative problem-solving approaches in dynamic and heterogeneous communication environments. This integration envisions making significant contributions to intelligent system management and optimization through predictive capabilities and data-driven decision making.

This Topic endeavours to explore the intersection of machine learning and communication research, presenting a collection of state-of-the-art contributions which underscore the potential of machine learning as a catalyst for adaptive and intelligent communication. The manuscripts presented in this Topic have undergone a rigorous peer-review process and have been selected for publication in the Topic "Machine Learning in Communication Systems and Networks" by various MDPI journals, including *Applied Sciences, Sensors, Electronics, Photonics, Journal of Sensor and Actuator Networks*, and *Telecom*. Comprising twenty-one articles, including an editorial and twenty research papers, this Topic offers insights into current challenges and innovative solution approaches involving machine learning adaptation in mobile communication and networks.

Yichuang Sun, Haeyoung Lee, and Oluyomi Simpson
Editors

sensors

MDPI

Editorial

Machine Learning in Communication Systems and Networks

Yichuang Sun, Haeyoung Lee * and Oluyomi Simpson

School of Physics, Engineering and Computer Science, University of Hertfordshire, Hatfield AL10 9AB, UK; y.sun@herts.ac.uk (Y.S.); o.simpson@herts.ac.uk (O.S.)
* Correspondence: h.lee@herts.ac.uk

1. Introduction

The landscape of communication environments is undergoing a revolutionary transformation, driven by the relentless evolution of technology and the growing demands of an interconnected world. The proliferation of mobile devices, the rise of IoT applications, and the deployment of 5G networks have ushered in an era where communication environments are not only increasingly complex but also highly dynamic. With the capability of 5G networks to support various forms of vertical integration, the landscape is poised for diverse applications and enhanced connectivity across industries [1]. Furthermore, even for 6G networks, the provision of ubiquitous and 3D coverage in the form of an integrated space–air–ground–sea network is envisioned [2]. In this rapidly evolving technological ecosystem, the need for intelligent solutions to adaptively manage the intricacies of communication systems is more pressing than ever [3]. As we stand on the cusp of these transformative changes, the integration of machine learning techniques emerges as a pivotal catalyst poised to revolutionize the way we address challenges and harness opportunities in communication systems and networks [4].

Traditionally, communication systems heavily relied on model-based approaches, wherein various components were meticulously modeled based on data analysis or measurement data. While these model-based approaches have been successful, they face challenges in accurately modeling dynamic and complex communication environments [5]. Machine learning (ML), capable of extracting characteristics and identifying hidden relationships, becomes a powerful tool in scenarios where traditional designs may falter due to model mismatches [6]. Moreover, the data-driven essence of ML enables inference about network traffic, service requirements, user behavior, and dynamic channels, leading to improved resource provisioning and network operation [3]. ML, with its real-time adaptability and ability to extract insights from vast datasets, promises to reshape communication. The increasing volume and diversity of data in dynamic communication systems demand innovative approaches for efficient operation and optimal performance. From predicting environmental or system status changes to optimizing resource allocation and addressing security threats [7], ML spans applications like intelligent traffic management [8] and automatic reconfiguration in communication infrastructure [9,10].

In this editorial, we explore the intersection of ML and communication, unraveling how these technologies synergize to meet current challenges and leverage opportunities in our highly connected world. In the subsequent section, we provide concise summaries of key points covered in the twenty articles collected in this Special Issue.

Citation: Sun, Y.; Lee, H.; Simpson, O.; Machine Learning in Communication Systems and Networks. *Sensors* **2024**, *24*, 1925. https://doi.org/10.3390/s24061925

Received: 1 March 2024
Accepted: 14 March 2024
Published: 17 March 2024

2. An Overview of Published Articles

In the dynamic realm of communication systems, achieving precise prediction and estimation of communication channels is paramount for optimizing overall system performance. The following five articles concentrate on leveraging ML techniques to effectively address the challenges of channel estimation. In the research conducted by Gaballa et al. (Contribution 1), the primary focus lies in predicting channel coefficients for users in the Non-Orthogonal Multiple Access (NOMA) system. Within the NOMA system, these coefficients assume a critical role in optimizing power distribution at the base station (BS) and streamlining the retrieval of desired data at the user end. The authors employ a deep Q-network (DQN) approach for the BS, enabling it to learn an optimal channel prediction policy. This policy is designed to maximize the sum rates for all users in the NOMA network, leveraging pertinent information such as user states, user distance, channel path loss, and power distribution. Similarly, the study by Gaballa et al. (Contribution 2) delves into channel estimation in power domain NOMA systems. In this investigation, the prediction of channel status information (CSI) is coupled with the determination of power factors for each user, achieved through a Q-learning-based reinforcement learning (RL) approach. In the study by Camana et al. (Contribution 3), the dynamic update of a radio environment map (REM) is explored through the prediction of received signal strength indicator (RSSI) values. The REM proves invaluable in detecting shadow areas with potential for improved network planning and accurate indoor localization. In the study, devices exhibiting similar signal strengths are grouped into clusters using the K-means algorithm, and the dynamic REM update is then orchestrated through a random forest (RF)-based ML algorithm. This model predicts RSSI values for each location, incorporating historical measurement data, including user location and RSSI values. The ML model is designed for real-time updates, facilitated by data collected from a mobile robot, ensuring a seamless and continuous adaptation of the REM, effectively responding to alternations in the wireless environment. The study by Phaiboon et al. (Contribution 4) focuses on path loss prediction within smart agriculture sensor networks, aiming to provide effective coverage areas and system capacity. For challenging environments like plantations, where signal paths are obstructed by trees and vegetation, the authors introduce an adaptive neuro-fuzzy inference system (ANFIS) that combines fuzzy logic and neural networks to learn path loss. Utilizing path loss measurement data and incorporating information such as sensor node distances and antenna heights, the ANFIS model provides an efficient means of estimating path loss. In the article by Ribouh et al. (Contribution 5), the focus is on identifying the distinctive characteristics of the CSI of received signals in vehicular communication by employing convolutional neural network (CNN)-based learning. The study aims to develop a model capable of discerning a vehicles' surroundings among five categories: rural line-of-sight (LoS), urban LoS, urban nLoS (non-LoS), highway LoS, and highway nLoS. The ultimate goal of this environment detection model is to empower autonomous vehicles to make informed speed limit decisions based on their surroundings.

ML is expected to play an important role in the demodulation process of communication systems since it can adaptively learn and extract complex patterns from received signals, particularly in dynamic and challenging environments. The following four articles are dedicated to the integration of ML into the demodulation process. In the investigation by Harper et al. (Contribution 6), automatic modulation classification (AMC) is used to estimate the modulation scheme employed by the transmitter. AMC proves invaluable in predicting the module schemes of a transmit signal when they are unknown. The authors examine the impact of a variety of architecture changes and propose the design of neural network (NN)-based AMC models. The scenario considered in the study by Zhang et al. (Contribution 7) involves decoding low-density parity check (LDPC) codes. LDPC codes, prevalent in modern communication systems on account of their extended code lengths and versatile combinations, present challenges in decoding and coding blind recognition. To address these challenges, the authors propose an architecture for coding the blind recognition of LDPC codes using deep learning (DL), incorporating a cascade network structure with denoising and blind recognition networks. This innovative approach enhances encoding performance even under poor signal-to-noise ratio (SNR) conditions. In Lamilla et al.'s study (Contribution 8), the attention shifts to a coherent optical encoding system. The authors introduce a robust coding algorithm based on laser intensity profile recognition, utilizing support vector machine (SVM)-based ML for data symbol classification and recognition. This strategy proves effective in mitigating the signal noise added to communication channels. While the above three articles focus on the decoding accuracy performance, the paper by Cho et al. (Contribution 9) considers how to improve the decoding speed for short-length Reed–Muller (RM) codes. Acknowledging the simplistic structure of RM codes and their potential use as control channels in wireless communication, the authors employ a revised auto-encoder scheme, a supervised ML technique, to design an ML-based decoding scheme for faster decoding.

Intelligent resource allocation, empowered by ML, is capable of taking on complex challenges related to the efficiency and adaptability of communication systems and networks. The integration of ML not only ensures the effective utilization of individual resource domains but also facilitates the joint optimization of multiple resource allocations, elevating decision-making processes and overall system performance. The following articles explore ML applications for intelligent resource allocation. The investigation by Pu et al. (Contribution 10) focuses on optimal transmission channel selection in jamming environments. Employing wideband spectrum sensing and Q-learning, the authors design transmitters to dynamically adapt to jamming issues by learning effective channel selection strategies, resulting in high success rates. In Ding et al.'s study (Contribution 11), they employ ML in the routing optimization of low-Earth-orbit (LEO) constellation networks. Satellite nodes, functioning as learning agents, dynamically adapt to changes in topology and channel conditions. Through a collaborative multi-agent reinforcement learning (MARL) framework, satellites share their learning experiences using Q-tables. The proposed three-step routing approach involves neighbor node discovery, followed by offline and online training to ensure that satellites swiftly acquire network link status and adjust their routing strategies accordingly. In the article by Zhang et al. (Contribution 12), the authors delve into the joint optimization of bandwidth and power allocation using multi-agent learning. The proposed approach targets to maximize the system throughput by addressing co-channel interference and ensuring adherence to quality of service (QoS) constraints. Within a large-scale uplink system, individual users act as learning agents, each striving for an optimal strategy in bandwidth and power allocation for their uplink transmission. The collaborative learning process involves sharing users' past training experiences, leading to the centralized training of all agents aligned with a common objective, maximizing the system's throughput. In Liu et al.'s work (Contribution 13), aerial edge computing networks, comprising low-altitude aerial base stations (AeBSs) and a high-altitude node, are considered. The study focuses on minimizing task processing delay and energy consumption through the control of AeBSs' deployment and computation offloading in this two-level aerial network. Utilizing deep RL (DRL), the optimization of low-altitude AeBSs and offloading strategies is carried out by considering factors such as their computational capacity, the number of associated users, the number of computational tasks required by users, and the channel gain with users. Sharing learning model parameters with a high-altitude node, the proposed RL mechanism enables collaborative control among AeBSs, while the high-altitude node serves as a global aggregator, improving training efficiency within the federated DRL framework. In the study by Camana et al. (Contribution 14), a DNN is applied to jointly optimize the beamforming vectors and power-splitting ratios in a multi-input, single-output (MISO) simultaneous wireless information and power transfer (SWIPT) system. The optimization objective is to minimize overall transmission power while ensuring compliance with predefined requirements for energy harvest and minimum data rate within the multi-user system.

ML could also offer benefits for communication network management, including dynamic network configuration, network traffic analysis, and efficient resource allocation. In the article by Hamdan et al. (Contribution 15), Open RAN (O-RAN), recognized for its potential in interoperability, scalability, and cost efficiency, is studied. Despite its advantage, the intricate management of the O-RAN system poses challenges, and the article conducts a thorough survey of current research endeavors while outlining research opportunities about how to uses ML for network automation in O-RAN. In the paper by Baek et al. (Contribution 16), ML is employed to monitor and analyze network traffic, providing benefits in various domains including traffic control, network security, and resource planning. The focus of this paper lies in web services, which are a combination of multiple applications where various application traffic sflows can be intertwined within service traffic. For web services, classifying traffic solely based on service units may lead to high errors in misclassification. To tackle this challenge, a DL-based algorithm performing multitask classification is proposed. This algorithm aims to classify application traffic by considering the relationships between browser, protocol, service, and application tasks within web services.

By leveraging data-driven insights, ML can be useful for service-specific decision making. The following two papers consider distinct service contents, focusing on e-Health and vehicular communication, respectively. In the contribution by AlZailaa et al. (Contribution 17), the emphasis is on addressing the real-time urgency inherent to critical tasks within e-Health applications. Operating within hierarchical fog–cloud networks, the paper employs a support vector machine (SVM)-based ML approach to classify and schedule tasks efficiently. A SVM-based task classification method is introduced, tailored for handling of latency-sensitive critical tasks. Building upon task classification outcomes, the study devises a task priority assignment and resource mapping algorithm. The overarching objective is to minimize latency and enhance the overall resource utilization in fog–cloud networks. In the work by Huang et al. (Contribution 18), the focus shifts to vehicular networks. ML is harnessed for precise vehicle arrival time estimation. Employing support vector regression (SVR)-based learning, the ML model incorporates factors like average vehicle speed, weather conditions, time, and the real-time road traffic information from roadside units (RSUs). Vehicles utilizing this learning algorithm predict their arrival times at specific road sections, transmitting this information to the RSUs. The significance of these data lies in their utilization by RSUs to efficiently manage bandwidth, particularly for supporting reliable real-time video applications. When vehicle users compete for bandwidth, RSUs leverage arrival information to prioritize services, optimizing overall user experiences by offloading traffic to vehicle-to-vehicle (V2V) links.

The traditional approach to analyzing extensive datasets using ML involves centralized ML models. However, the surge in data generation from diverse end devices and concerns over privacy issues have sparked significant interest in federated and distributed learning. Federated learning (FL) allows clients to cooperate to generate a global model without sharing sensitive client data with a server. In the work by Seol et al. (Contribution 19), the impact of statistical heterogeneity indicating non-independent and identical distribution (non-IID) of the training datasets (generated by clients) is highlighted, which clients will use for local training in an FL framework. A novel approach is proposed to reduce statistical heterogeneity and dynamically control batch size and learning rate, aiming to enhance FL performance. In the investigation by Bemani et al. (Contribution 20), the emphasis is on understanding the impact of communication-induced noise during FL training on the convergence and accuracy performance of the ML mode. The paper proposes the use of analog over-the-air aggregation to effectively manage noise in communication channels, ultimately contributing to improved convergence in ML algorithms.

3. Conclusions

This compilation of articles sheds light on the transformative impact of machine learning on communication systems and networks. As evident from the diverse range of contributions, ML not only enhances traditional aspects of communication networks but also paves the way for novel applications and optimizations. The showcased articles emphasize the role of ML in addressing intricate challenges, from intelligent resource allocation and dynamic network management to efficient channel estimation and service-specific decision making. The application domains span across e-Health, transportation, agriculture, and more, highlighting the versatility of ML in shaping the future of communication technologies.

Despite significant strides, challenges in applying ML persist. The heterogeneity of communication environments and the ever-evolving nature of network dynamics present ongoing hurdles. Issues related to the privacy, security, and interoperability of ML models in communication contexts also call for further research. Additionally, the scalability and adaptability of ML algorithms to handle the burgeoning volume of data generated in real-time pose continuous challenges.

Looking ahead, collaborative efforts between the ML and communication technology communities will be essential to address these challenges. Interdisciplinary research, harmonization of data formats, standardization of ML methodologies in communication protocols, and the development of scalable, privacy-preserving algorithms will be crucial for the sustainable advancement of ML applications in communication environments.

Author Contributions: All authors contributed equally to this editorial. All authors have read and agreed to the published version of the manuscript.

Conflicts of Interest: The authors declare no conflicts of interest.

List of Contributions:

1. Gaballa, M.; Abbod, M. Simplified Deep Reinforcement Learning Approach for Channel Prediction in Power Domain NOMA System. *Sensors* **2023**, *23*, 9010. https://doi.org/10.3390/s23219010
2. Gaballa, M.; Abbod, M.; Aldallal, A. A Study on the Impact of Integrating Reinforcement Learning for Channel Prediction and Power Allocation Scheme in MISO-NOMA System. *Sensors* **2023**, *23*, 1383. https://doi.org/10.3390/s23031383
3. Camana, M.R.; Garcia, C.E.; Hwang, T.; Koo, I. A REM Update Methodology Based on Clustering and Random Forest. *Appl. Sci.* **2023**, *13*, 5362. https://doi.org/10.3390/app13095362
4. Phaiboon, S.; Phokharatkul, P. Applying an Adaptive Neuro-Fuzzy Inference System to Path Loss Prediction in a Ruby Mango Plantation. *J. Sens. Actuator Netw.* **2023**, *12*, 71. https://doi.org/10.3390/jsan12050071
5. Ribouh, S.; Sadli, R.; Elhillali, Y.; Rivenq, A.; Hadid, A. Vehicular Environment Identification Based on Channel State Information and Deep Learning. *Sensors* **2022**, *22*, 9018. https://doi.org/10.3390/s22229018
6. Harper, C.A.; Thornton, M.A.; Larson, E.C. Automatic Modulation Classification with Deep Neural Networks. *Electronics* **2023**, *12*, 3962. https://doi.org/10.3390/electronics12183962
7. Zhang, X.; Zhang, W. A Cascade Network for Blind Recognition of LDPC Codes. *Electronics* **2023**, *12*, 1979. https://doi.org/10.3390/electronics12091979
8. Lamilla, E.; Sacarelo, C.; Alvarez-Alvarado, M.S.; Pazmino, A.; Iza, P. Optical Encoding Model Based on Orbital Angular Momentum Powered by Machine Learning. *Sensors* **2023**, *23*, 2755. https://doi.org/10.3390/s23052755
9. Cho, H.W.; Song, Y.J. High Speed Decoding for High-Rate and Short-Length Reed–Muller Code Using Auto-Decoder. *Appl. Sci.* **2022**, *12*, 9225. https://doi.org/10.3390/app12189225
10. Pu, Z.; Niu, Y.; Xiang, P.; Zhang, G. Sightless but Not Blind: A Non-Ideal Spectrum Sensing Algorithm Countering Intelligent Jamming for Wireless Communication. *Electronics* **2022**, *11*, 3402. https://doi.org/10.3390/electronics11203402
11. Ding, Z.; Liu, H.; Tian, F.; Yang, Z.; Wang, N. Fast-Convergence Reinforcement Learning for Routing in LEO Satellite Networks. *Sensors* **2023**, *23*, 5180. https://doi.org/10.3390/s23115180
12. Zhang, C.; Lv, T.; Huang, P.; Lin, Z.; Zeng, J.; Ren, Y. Joint Optimization of Bandwidth and Power Allocation in Uplink Systems with Deep Reinforcement Learning. *Sensors* **2023**, *23*, 6822. https://doi.org/10.3390/s23156822
13. Liu, L.; Zhao, Y.; Qi, F.; Zhou, F.; Xie, W.; He, H.; Zheng, H. Federated Deep Reinforcement Learning for Joint AeBSs Deployment and Computation Offloading in Aerial Edge Computing Network. *Electronics* **2022**, *11*, 3641. https://doi.org/10.3390/electronics11213641
14. Camana, M.R.; Garcia, C.E.; Koo, I. Beamforming Optimization with the Assistance of Deep Learning in a Rate-Splitting Multiple-Access Simultaneous Wireless Information and Power Transfer System with a Power Beacon. *Electronics* **2024**, *13*, 872. https://doi.org/10.3390/electronics13050872.
15. Hamdan, M.Q.; Lee, H.; Triantafyllopoulou, D.; Borralho, R.; Kose, A.; Amiri, E.; Mulvey, D.; Yu, W.; Zitouni, R.; Pozza, R.; et al. Recent Advances in Machine Learning for Network Automation in the O-RAN. *Sensors* **2023**, *23*, 8792. https://doi.org/10.3390/s23218792

16. Baek, U.-J.; Kim, B.; Park, J.-T.; Choi, J.-W.; Kim, M.-S. A Multi-Task Classification Method for Application Traffic Classification Using Task Relationships. *Electronics* **2023**, *12*, 3597. https://doi.org/10.3390/electronics12173597
17. AlZailaa, A.; Chi, H.R.; Radwan, A.; Aguiar, R.L. Service-Aware Hierarchical Fog–Cloud Resource Mapping for e-Health with Enhanced-Kernel SVM. *J. Sens. Actuator Netw.* **2024**, *13*, 10. https://doi.org/10.3390/jsan13010010
18. Huang, C.-J.; Hu, K.-W.; Cheng, H.-W. An Adaptive Bandwidth Management Algorithm for Next-Generation Vehicular Networks. *Sensors* **2023**, *23*, 7767. https://doi.org/10.3390/s23187767
19. Seol, M.; Kim, T. Performance Enhancement in Federated Learning by Reducing Class Imbalance of Non-IID Data. *Sensors* **2023**, *23*, 1152. https://doi.org/10.3390/s23031152
20. Bemani, A.; Björsell, N. Low-Latency Collaborative Predictive Maintenance: Over-the-Air Federated Learning in Noisy Industrial Environments. *Sensors* **2023**, *23*, 7840. https://doi.org/10.3390/s23187840

References

1. 5G PPP Technology Board and 5G IA Verticals Task Force. Empowering Vertical Industries through 5G Networks—Current Status and Future Trends. *White Paper* **2020**, 1–108. [CrossRef]
2. Wang, C.X.; You, X.; Gao, X.; Zhu, X.; Li, Z.; Zhang, C.; Wang, H.; Huang, Y.; Chen, Y.; Haas, H.; et al. On the Road to 6G: Visions, Requirements, Key Technologies, and Testbeds. *IEEE Commun. Surv. Tutor.* **2023**, *25*, 905–974. [CrossRef]
3. Noor-A-Rahim, M.; Liu, Z.; Lee, H.; Khyam, M.O.; He, J.; Pesch, D.; Moessner, K.; Saad, W.; Poor, H.V. 6G for Vehicle-to-Everything (V2X) Communications: Enabling Technologies, Challenges, and Opportunities. *Proc. IEEE* **2022**, *110*, 712–734. [CrossRef]
4. Morocho-Cayamcela, M.E.; Lee, H.; Lim, W. Machine Learning for 5G/B5G Mobile and Wireless Communications: Potential, Limitations, and Future Directions. *IEEE Access* **2019**, *7*, 137184–137206. [CrossRef]
5. Gündüz, D.; de Kerret, P.; Sidiropoulos, N.D.; Gesbert, D.; Murthy, C.R.; van der Schaar, M. Machine Learning in the Air. *IEEE J. Sel. Areas Commun.* **2019**, *37*, 2184–2199. [CrossRef]
6. Dai, L.; Jiao, R.; Adachi, F.; Poor, H.V.; Hanzo, L. Deep Learning for Wireless Communications: An Emerging Interdisciplinary Paradigm. *IEEE Wirel. Commun.* **2020**, *27*, 133–139. [CrossRef]
7. Harahsheh, K.; Chen, C.H. A Survey of Using Mchine Learning in IoT Security and the Challenges faced by Researchers. *Informatica* **2023**, *47*, 1–54. [CrossRef]
8. Boutaba, R.; Salahuddin, M.A.; Limam, N.; Ayoubi, S.; Shahriar, N.; Estrada-Solano, F.; Caicedo, O.M. A Comprehensive Survey on Machine Learning for Networking: Evolution, Applications and Research Opportunities. *J. Internet Serv. Appl.* **2018**, *9*, 16. [CrossRef]
9. Luong, N.C.; Hoang, D.T.; Gong, S.; Niyato, D.; Wang, P.; Liang, Y.C.; Kim, D.I. Applications of Deep Reinforcement Learning in Communications and Networking: A Survey. *IEEE Commun. Surv. Tutor.* **2019**, *21*, 3133–3174. [CrossRef]
10. Han, S.; Chih-Lin, I.; Li, G.; Wang, S.; Sun, Q. Big Data Enabled Mobile Network Design for 5G and Beyond. *IEEE Commun. Mag.* **2017**, *55*, 150–157. [CrossRef]

sensors

MDPI

Article

Simplified Deep Reinforcement Learning Approach for Channel Prediction in Power Domain NOMA System

Mohamed Gaballa * and Maysam Abbod

Department of Electronic and Electrical Engineering, Brunel University London, Uxbridge UB8 3PH, UK; maysam.abbod@brunel.ac.uk
* Correspondence: mohamedgaballa.gaballa@brunel.ac.uk

Abstract: In this work, the impact of implementing Deep Reinforcement Learning (DRL) in predicting the channel parameters for user devices in a Power Domain Non-Orthogonal Multiple Access system (PD-NOMA) is investigated. In the channel prediction process, DRL based on deep Q networks (DQN) algorithm will be developed and incorporated into the NOMA system so that this developed DQN model can be employed to estimate the channel coefficients for each user device in NOMA system. The developed DQN scheme will be structured as a simplified approach to efficiently predict the channel parameters for each user in order to maximize the downlink sum rates for all users in the system. In order to approximate the channel parameters for each user device, this proposed DQN approach is first initialized using random channel statistics, and then the proposed DQN model will be dynamically updated based on the interaction with the environment. The predicted channel parameters will be utilized at the receiver side to recover the desired data. Furthermore, this work inspects how the channel estimation process based on the simplified DQN algorithm and the power allocation policy, can both be integrated for the purpose of multiuser detection in the examined NOMA system. Simulation results, based on several performance metrics, have demonstrated that the proposed simplified DQN algorithm can be a competitive algorithm for channel parameters estimation when compared to different benchmark schemes for channel estimation processes such as deep neural network (DNN) based long-short term memory (LSTM), RL based Q algorithm, and channel estimation scheme based on minimum mean square error (MMSE) procedure.

Keywords: DRL; DQN; Q-learning; LSTM; NOMA

Citation: Gaballa, M.; Abbod, M. Simplified Deep Reinforcement Learning Approach for Channel Prediction in Power Domain NOMA System. *Sensors* **2023**, *23*, 9010. https://doi.org/10.3390/s23219010

Academic Editors: Yichuang Sun, Haeyoung Lee and Oluyomi Simpson

Received: 19 August 2023
Revised: 19 October 2023
Accepted: 27 October 2023
Published: 6 November 2023

1. Introduction

It can be noticed that the high energy consumption by the connected terminals in the current wireless networks can create an essential challenge in designing the upcoming 6G wireless systems [1]. Therefore, it is important to consider this energy consumption issue in future wireless communication networks, and at the same time, we need to maintain the required quality of service (QoS) for devices or services in that networks. Basically, NOMA system utilizes a superposition coding (SC) procedure that involves multiplexing different signals related to different users before transmission, which can contribute to the energy efficient transmission scheme. Moreover, NOMA system can also be designated to ensure the desired quality of service (QoS) levels for all superimposed user devices. Numerous research efforts have been dedicated to NOMA system in order to find an efficient strategy for different challenging tasks such as power allocation, beamforming, and channel assignment [2].

Recently, many authors have investigated different machine learning algorithms and artificial intelligence tools to optimize the resource allocation problems in NOMA system [3]. Furthermore, reinforcement learning (RL) based Q-learning algorithm and deep reinforcement learning based Q network (DQN) have gained a remarkable interest among authors in various fields. The Q-learning algorithm is a subclass of reinforcement learning that depends on Q-tables to store the optimal Q-values for each state-action pair in order to maximize the future reward in the system. Alternatively, deep reinforcement learning-based Q network (DQN) algorithm is mainly dependent on adopting hidden layers that can effectively enhance network convergence and system performance.

1.1. Related Works

In the context of optimizing communication systems, several works have employed the Q-learning algorithm to enhance the performance of wireless networks based on different perspectives. The work in [4] applied the Q-learning algorithm to introduce a framework for enabling mobile edge computing in NOMA system. In [5], authors suggested a dynamic reinforcement learning scheme for power allocation in order to jointly maximize the sum rate and the spectral efficiency in MIMO-NOMA system when smart jamming is considered. The authors applied the Q-learning algorithm to allocate a certain power level to each user terminal, to mitigate the jamming effects.

Basically, by incorporating deep learning into RL, deep reinforcement learning (DRL) can address the challenges associated with Q-learning in terms of Q-table storage. Based on that, the work in [6] introduced a deep Q-network (DQN) to model a multiuser NOMA offloading problem, while the work in [7], proposed a power allocation technique based on deep reinforcement learning in cache-assisted NOMA system. Furthermore, authors in [8] introduced a DRL based actor-critic algorithm to handle the dynamic power allocation policy. Likewise, DRL based actor-critic algorithm was also applied in [9] to attain the optimal policy for user scheduling and resource allocation in HetNets. In [9], the authors designed the actor network in order to decide the policy that can select a stochastic action based on Gaussian distribution, while the critic network role is to evaluate the value function and guides the actor network to discover or learn the optimal policy.

Deep reinforcement learning was also introduced in [10] to arrive at a sub-optimal power allocation scheme for an uplink multicarrier NOMA cell. The work in [11], considered a joint channel assignment and power distribution procedure in NOMA system. Authors in [11], derived a near-optimal power allocation scheme by considering two users per channel, and the channel assignment was performed using deep reinforcement learning algorithm to boost the overall sum rate while the minimum rate for each user device is considered.

1.2. Research Gap and Significance

Several machine learning (ML) algorithms have been suggested to clearly address diverse issues in wireless networks such as channel assignment, beamforming, and power allocation. Also, several RL algorithms have been proposed to handle the channel estimation task in wireless communication systems. However, most of the current research that covers the channel prediction task in the NOMA system is mainly dependent on deep neural networks (DNN) which include some sort of complexity in the network structure. Hence, in this work, we aim to introduce a deep reinforcement learning scheme based on a simplified DQN approach to reduce the complexity structure and at the same time enhance the channel estimation process. Furthermore, to the best of the authors' knowledge, there is no study that explores the utilization of deep reinforcement learning (DRL) based deep Q network (DQN) algorithm for estimating the channel parameters for user devices in the NOMA system. In addition, and to the best of the authors' knowledge, there is no study that has investigated the performance of NOMA system when both the DQN algorithm that used as channel estimator and the optimized power scheme are jointly implemented for user detection in NOMA system.

It is worth mentioning that unlike classical deep learning algorithms, which mainly depend on learning from a training data set, the proposed DQN algorithm is developed based on the LSTM network to adapt to the variations in the channel and to dynamically enhance the system performance based on the interaction with the environment.

1.3. Contributions to Knowledge

In this work, the contributions can be summed up as shown:

- A simplified DQN structure is proposed to demonstrate how RL based DQN algorithm is developed to predict the channel parameters for each user in the NOMA cell in Rayleigh fading channels.
- Investigate the combination between the RL algorithm and the LSTM model, to compose the simplified DQN structure in order to be utilized as a channel estimator.
- Validate the efficiency of the proposed DQN scheme, by establishing different benchmark schemes for comparison. Three different simulation environments are established as follows: (1) Channel prediction scheme based on standard minimum mean square error (MMSE) procedure [12]; (2) Standard DNN based on LSTM network for channel prediction applied in [13], (3) The RL based Q-algorithm for channel prediction applied in [14]. The simulation outcomes of these benchmark schemes were compared with the results of our proposed DQN model, and the results emphasized that reliability can be guaranteed by our developed DQN algorithm for predicting channel parameters even when the number of users in NOMA cell is increased.
- Simulate the impact of integrating the simplified DQN structure for channel prediction and the optimized power scheme derived in [13] for the purpose of multiuser detection in the power domain NOMA system.

The remainder of this paper is structured as follows. Section 2 describes the system model. The Deep Reinforcement Learning Framework is presented in Section 3. The Channel Estimation Based DQN Algorithm is discussed in Section 4. DQN Operation and framework are discussed in Section 5. DQN Dataset Generation is introduced in Section 6. Section 7 discusses the DQN Policy and Algorithm. DQN state space, action space, and reward are introduced in Section 8. Detailed DQN Procedure and workflow are listed in Section 9. Complexity analysis is also discussed in Section 10. The simulation environment is described in Section 11, and simulation results are presented in Section 12. Finally, conclusions are given in Section 13.

2. System Model

In a NOMA cell, numerous user devices can be served via the same resource block (RB) by employing the power domain (PD) in both uplink and downlink transmissions. In this paper, we are considering a downlink NOMA cell, where the BS can serve distinct types of users or devices at the same time via different fading channels. At the transmitter side, the BS can assign a specific channel or subcarrier to every set of user devices, and the signals of these devices can be multiplexed using unique power levels. At the receiver side, each user device will receive the desired signal beside the undesirable signals related to other devices in the same channel that will be considered either as interference or noise. The undesirable received signals will be considered as noise if the power level of the desired signal is high, otherwise, these additional signals will be regarded as interference. To decode the desired signal, each user device will use the successive interference cancelation (SIC) procedure. The SIC technique will first decode the signal with the highest power level and then subtract that signal from the principal signal, and this process will continue until the desired signal is decoded.

Typically, before applying the SIC procedure at the receiver side, the channel parameters for each user need to be available or estimated to perform the equalization process. Also, to calculate the data rate or channel capacity for each user, we need to calculate the signal to interference plus noise ratio (SINR), and SINR itself includes the channel gain $|h_i|^2$, where h_i represents the fading channel between the BS and user device i. In the NOMA scenario, the data rate R_i for user device i can be expressed as follows:

$$R_i = \log_2\left(1 + \frac{P_T\alpha_i\eta_i}{\sum_{j=1}^{i-1} P_T\alpha_j\eta_i + 1}\right) \tag{1}$$

where α_i is the power allocation factor for user device i, and η_i is the channel to noise ratio (CNR) for user i and P_T is the total power assigned by the BS. The channel to noise ratio η_i for user i, can be expressed as follows:

$$\eta_i = \frac{|h_i|^2}{\sigma_n^2} \tag{2}$$

where $|h_i|^2$ is the channel gain for user device i, and σ_n^2 is the noise power. In this work, we are considering a downlink NOMA system, and the total number of devices in the cell is N. In the NOMA cell, all signals related to the N devices are combined, and the BS will transmit this composed signal to all users in the cell. The composed signal X can be represented as follows [15]:

$$X = \sum_{i=1}^{N} \sqrt{P_T \alpha_i} x_i \; i = 1, 2, \ldots., N \tag{3}$$

where x_i is the desired signal for user device i. The composed transmitted signal X can be received at the receiver side of each user terminal, with path loss and Additive White Gaussian noise (AWGN), hence the received signal Y can be represented as

$$Y = \sum_{i=1}^{N} \sqrt{P_T \alpha_i} h_i x_i + n \; i = 1, 2, \ldots., N \tag{4}$$

where h_i is the fading channel between BS and user device i and n denotes the AWGN component. After receiving the composed signal and estimating the channel parameters, the receiver at each user device will activate the SIC procedure to decode the desired signal. In PD-NOMA, distinct power levels will be given to user terminals in the cell, and the highest power level will be given to the user device with the lowest CNR, while the lowest power level will be given to the user device with the highest CNR. Therefore, if user devices have the following CNRs:

$$\eta_1 > \eta_2 > \ldots. > \eta_N \tag{5}$$

Then, these user devices will be assigned power levels as follows:

$$P_1 < P_2 < \ldots. < P_N \tag{6}$$

The SINR for user device i can be represented as shown:

$$SINR_i = \frac{P_T \alpha_i \eta_i}{\sum_{j=1}^{i-1} P_T \alpha_j \eta_i + 1} \; i = 1, 2, \ldots., N \tag{7}$$

The BS can allocate power P_i to any user terminal as shown in the following expression [15]:

$$P_i = \left(P_T - \left(\sum_{j=1}^{i-1} P_T \alpha_j \right) \right) \geq P_{th} \tag{8}$$

The expression in (8), can be interpreted as follows: for proper achievement for the SIC process, the user device with low CNR must have a higher power level than the sum of power levels for other devices that have high CNR.

Based on the aforementioned analysis, in what follows we will consider the scenario for three users downlink PD-NOMA system, and we will provide some sort of mathematical analysis for the achievable capacity for each user when both perfect SIC and imperfect SIC are applied [16]. As indicated before, BS can send the superposition coded signal X which can be expressed as

$$X = \sqrt{P_t} \left(\sqrt{\alpha_n} x_n + \sqrt{\alpha_m} x_m + \sqrt{\alpha_f} x_f \right) \tag{9}$$

where α_n, α_m, and α_f are the power factors allocated to the near user, middle user, and far user, respectively. Likewise, x_n, x_m, and x_f denote the desired symbols related to the near, middle, and far users respectively. Hence, the signal received at far user can be represented as follows:

$$y_f = Xh_f + n_f \tag{10}$$

where h_f represent the fading channel among BS and the far user, while n_f represents the AWGN noise component at far user with zero mean and σ^2 variance. The received signal at far user can be expressed in details as follows:

$$y_f = \sqrt{P_t \alpha_f} x_f h_f + \sqrt{P_t}(\sqrt{\alpha_m} x_m + \sqrt{\alpha_n} x_n) h_f + n_f \tag{11}$$

The 1st term in (11) represents the desired signal for far user, but the 2nd term denotes the interference term from the middle and near users. Far user is usually described by poor channel condition and his particular signal x_f can be assigned additional power by BS compared to other users. Thus, according to the SIC scheme, far user can directly decode his own signal x_f from received signal y_f. The possible rate for far user R_f could be expressed as follows:

$$R_f = \log_2\left(1 + \frac{\eta_f\ P_t \alpha_f}{\eta_f\ P_t(\alpha_n + \alpha_m) + 1}\right) \tag{12}$$

Likewise, the attainable bit rate for the middle user R_m in the case of perfect SIC, can be expressed as follows:

$$R_m = \log_2\left(1 + \frac{\eta_m\ P_t \alpha_m}{\eta_m\ P_t(\alpha_n) + 1}\right) \tag{13}$$

Typically, the user near the BS has a good channel condition; therefore, his signal x_n is usually assigned low power level. Therefore, at near user side when perfect SIC is applied, firstly immediate decoding for far user signal x_f is accomplished, then it is removed from the composite signal. Next, the middle user signal x_m is decoded and removed from the remaining signal. Finally, the near user achieved rate R_n can be expressed as follows:

$$R_n = \log_2(1 + \eta_n\ P_t \alpha_n) \tag{14}$$

In the case of imperfect SIC, the attainable bit rate for the middle user can be expressed as:

$$R_m = \log_2\left(1 + \frac{\eta_m\ P_t \alpha_m}{\epsilon\ \eta_m\ P_t\left(\alpha_f\right) + \eta_m\ P_t(\alpha_n) + 1}\right) \tag{15}$$

where $\epsilon\ \eta_m\ P_t\left(\alpha_f\right)$ represents the error residual term from far user signal decoding. Likewise, the attainable bit rate for the near user in case of imperfect SIC can be expressed as:

$$R_n = \log_2\left(1 + \frac{\eta_n\ P_t \alpha_n}{\epsilon\ \eta_n\ P_t\left(\alpha_f\right) + \epsilon\ \eta_n\ P_t(\alpha_m) + 1}\right) \tag{16}$$

where $\epsilon\ \eta_n\ P_t\left(\alpha_f\right)$ is the error residual term from far user signal decoding and $\epsilon\ \eta_n\ P_t(\alpha_m)$ is the error residual term from middle user signal decoding.

3. Deep Reinforcement Learning Framework

In this section, we will introduce the concept of deep reinforcement learning (DRL), which is a special case of reinforcement learning procedure [17,18]. Reinforcement learning is a fork of machine learning, where an agent interacts with the environment to carry out the best sequences of actions that can maximize the expected future reward in an interactive environment. Generally, reinforcement learning can be classified as single-agent or multi-agent based on the quantity of agents in the environment. In the scenario of a single agent RL, the agent needs to recognize the entire states in the environment and the decision-making task can be modeled as a Markov decision process (MDP) framework. In this work, our proposed DQN structure assumes a single agent, and the best sequence of actions that can be chosen by the agent will be generated based on the adopted deep neural network (DNN).

The fundamental elements in the deep reinforcement learning (DRL) algorithm can be listed as follows [14,18]:

1. **Observations**: the continuous measurements of the properties of the environment, and all of the observed properties in the environment can be included in the state space $\boldsymbol{S}$.
2. **States**: the discret observation at time step t can be denoted as state $s_t \in \boldsymbol{S}$.
3. **Actions**: an action a_t is one of the valid decisions that the agent can select at time step t from the action space $\boldsymbol{A}$.
4. **Policy**: a policy denoted by $\pi(.)$, is the criteria that control how to select a certain action at any given state while interacting with the environment.
5. **Rewards**: the immediate reward r_t, is obtained after an agent carries out a specific action a_t in a given state s_t, which leads to moving to a new state s_{t+1}.
6. **State-action value function**: denoted by $Q_\pi(s,a)$, and represents the expected discounted reward when the agent starts at a certain state s_t and selects a specific action a_t based on the policy π.

In the DQN framework, when an agent selects an action a_t at a given time step t, the agent's state will change from the current state s_t to the subsequent state s_{t+1} and as a result of this transition, the agent will receive an immediate reward r_t from the environment. Based on that scenario, the network can generate an experience tuple $e = (s_t,\ a_t,\ r_t,\ s_{t+1})$ that can be stored in the experience replay buffer $\boldsymbol{\mathcal{D}}$. The primary target of the agent in RL scheme is to maximize the long-term cumulative discounted reward R_t^γ, which can be defined as follows [14,18]:

$$R_t^\gamma = \sum_{i=0}^{\infty} \gamma^i r_{t+i} \tag{17}$$

where γ is the discount factor. To enhance the R_t^γ, an optimal policy π^* is essential to map the best actions to states. In other words, the optimal policy π^* can significantly assist the agent in deciding which action should be selected at any given state, to satisfy the optimal long-term cumulative reward. Typically, the state action Q-value function is defined as the expectation of the cumulative discounted reward R_t^γ. Overall, we can notice that based on the current state s_t, the considered policy π, and the selected action a_t, the state-action Q value function can be further expressed as follows [14,19]:

$$\begin{aligned} Q_\pi(s_t,a_t) &= \mathbb{E}\left[R_t^\gamma | s_t, a_t\right] = \mathbb{E}\left[\sum_{i=0}^{\infty} \gamma^i r_{t+i} | s_t, a_t\right] \\ &= \mathbb{E}\left[r_t + \gamma Q_\pi(s_{t+1}, a_{t+1}) | s_t, a_t\right] \end{aligned} \tag{18}$$

where $\mathbb{E}[\,.]$ denotes the expectation parameter. When the optimal policy π^* is applied for maximizing all states and action pairs, then the optimal Q-value function $Q_{\pi^*}(s_t,a_t)$ that follows the optimal policy π^* can be expressed as follows:

$$Q_{\pi^*}(s_t,a_t) = \mathbb{E}\left[r_t + \gamma Q_{\pi^*}(s_{t+1}, a_{t+1}) | s_t, a_t\right] \tag{19}$$

The expression in (19) is known as the Bellman equation. The benefit of the Bellman equation is to represent the state-action Q-value function into two components: the instantaneous reward r_t and the long-term discounted reward. However, the Bellman equation is nonlinear, and hence, there is no closed form solution to it. As a result, an iterative procedure such as the Q-learning algorithm has emerged to converge the Bellman equation to obtain the optimal Q-value function [18,19]. On the other hand, the computation of the Q-learning algorithm may become more complex in multi-user environments that have huge state and action spaces, and as a result, the size of the Q-table will be extremely large. Hence, the regular solution to this limitation is to estimate the Q-values using a function approximator, by adopting hidden layers, which is the core component in our developed deep Q network.

The basic DQN architecture is shown in Figure 1, and it consists of three main phases: The first phase represents the input layer that includes the current states of the environment. The second stage includes the hidden layers that act as a function approximator. Mainly in the hidden layers, the Rectified Linear Unit (ReLU) activation function is applied to compute the hidden layer values. The primary gain of utilizing ReLU as an activation function is the computational efficiency [20], which may lead to faster convergence. At the end phase, the output layer is responsible for predicting the optimal state-action value function, $Q_{\pi^*}(s, a, W_t)$, where W_t is the updated weights of the hidden layers at time instant t.

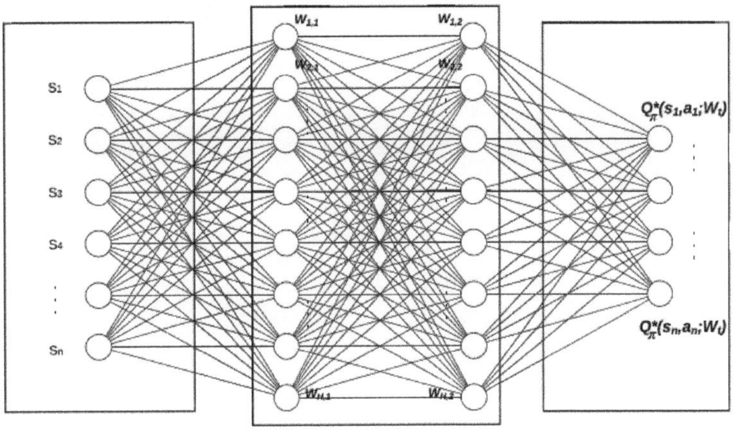

Figure 1. DQN basic structure with two hidden layers.

4. Channel Estimation Based DQN Algorithm

In this section, the simplified DRL structure will be introduced, and Figure 2 illustrates the architecture of the simplified DRL scheme that mainly relies on the DQN algorithm and LSTM network to achieve the most appropriate performance. The DQN network will be trained, and the weights of the hidden layers will be updated to approximate the state-action value function $Q_\pi(s, a)$. As indicated in the aforementioned discussion, each experience tuple is described as $e_t = (s_t, a_t, r_t, s_{t+1})$, and all experience tuples will be stored in an experience replay buffer $\mathcal{D} = \{e_1\, e_2\, e_3\, \ldots\, e_t\}$, and these experience tuples can be utilized to train the DQN using the gradient descent algorithm [21]. It is optimum for the DQN algorithm to exploit all available experience tuples in each training iteration, but this will be costly when the training set is huge. A more effective procedure is to update the DQN weights in each iteration using an arbitrary subset from the replay buffer $\mathcal{D}$, and this subset is described as a mini batch. Based on the architecture of the proposed DQN structure shown in Figure 2, it can be noticed that the loss function can be computed based on the difference between the output of the target DNN and the output of the policy DNN. Hence, the loss function can be defined as follows [18,19]:

$$\mathcal{L}(W) = \sum_{e \in \mathcal{D}} \left(r_t + \gamma\, maxQ_{\pi^*}(s_{t+1}, a_{t+1}, \hat{W}) - Q_{\pi^*}(s_t, a_t, W)\right)^2 \tag{20}$$

where $\mathcal{L}(W)$ denotes the DQN loss function for a random mini batch sampled from the replay buffer $\boldsymbol{\mathcal{D}}$ at time slot t and $\hat{W}$ represents the nearly static weights in the target DNN and these weights are mainly updated every T time steps. To minimize the loss function $\mathcal{L}(W)$, the weights W of the policy DNN will be updated every t time step using a stochastic gradient descent (SGD) algorithm applied on a batch of random samples selected from the replay buffer $\boldsymbol{\mathcal{D}}$. Typically, the SGD algorithm can update the weights of the policy DNN W in an iterative process with a learning rate of $\mu > 0$ as follows [21]:

$$W_{t+1} = W_t - \mu \nabla \mathcal{L}_t(W_t) \tag{21}$$

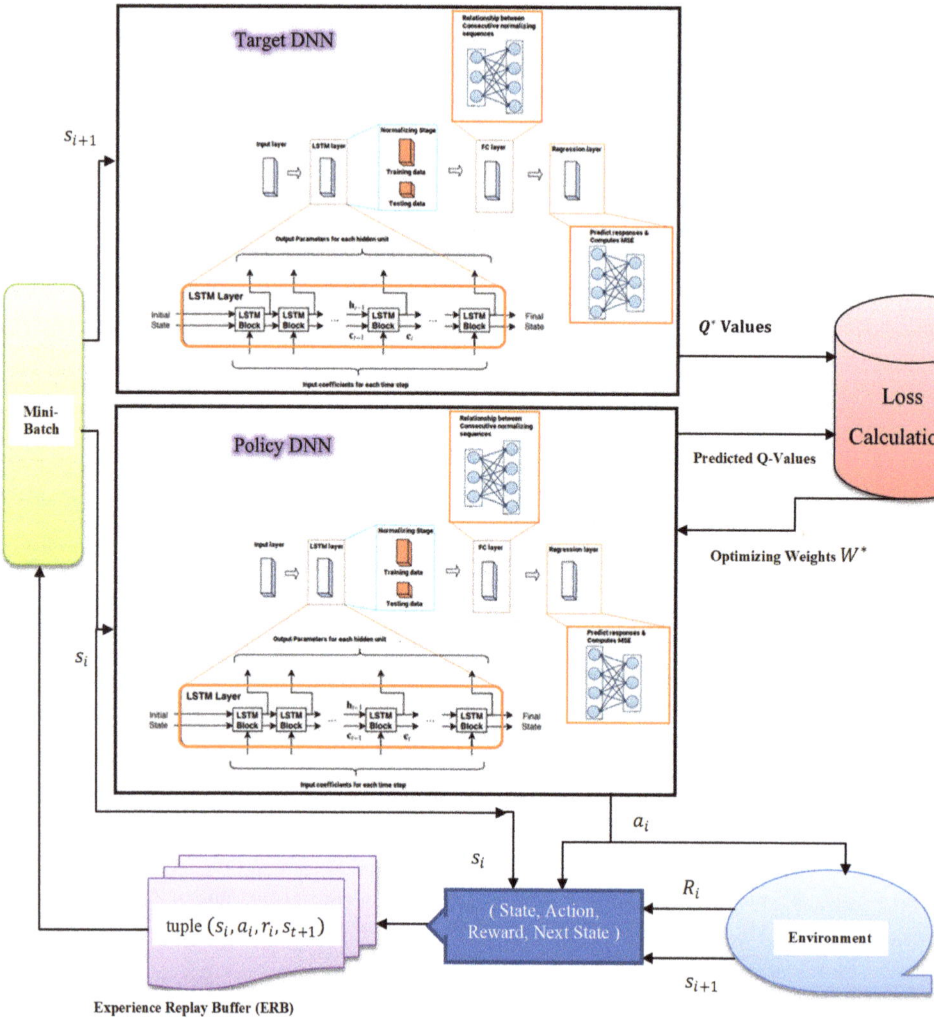

Figure 2. Proposed DQN Architecture.

5. Proposed DQN Operation and Phases

Phase 1: Initialization and generation of training data

1. Perform a few random actions with the environment to initialize the experience replay data.

2. Initialize the weights for the policy DNN and copy these weights to the Target DNN.
3. Starting with the first time step,
    - Based on the initial interaction with the environment, random states can be generated to be used as input for the policy DNN.
    - The policy DNN will predict the Q-values for all actions that can be decided in the current state, and then those Q-values will be inspected to select or identify a certain Q-value based on the most suitable action.
    - Based on the selected and executed action, the experience replay will receive the reward and move to the next state.
    - The experience replay will store the results in the replay buffer.
    - Each result will be considered as a sample training data, that can be later used as training data.

Phase 2: Select a random batch for training

1. Select a batch of random samples from the replay buffer and use these samples as an inputs for both the policy DNN and the target DNN.
2. From the random sample, use the current state as input to the policy DNN.
3. The policy DNN can predict the Q-values for all actions that can be performed in the current state.
4. Based on the decided or selected action, the policy DNN will identify the predicted Q-value.
5. The next state from the selected random sample will be used as input to the Target DNN.
6. The Target network will predict the Q-values for all actions that can be performed in the next state, then the Target DNN will select the maximum of those Q-values.

Phase 3: Get the Target Q-value

1. The Target Q-value can be decided based on two components
    - The immediate reward from the environment
    - The max Q value that has been predicted by the target DNN in the next state

Phase 4: Compute the Loss function

1. Compute the loss function between the Target Q value and the predicted Q Value in terms of mean squared error (MSE).

Phase 5: Back-propagate the Loss function

1. Back-propagate the loss in order to update the weights of the policy DNN using SGD.
2. At this stage, the weights of the Target DNN are not updated and remain fixed, and this completes the processing for this time step.

Phase 6: Repeat for next time step

1. The process will be repeated for the next time step.
    - The policy DNN weights have been updated but not the Target DNN.
    - This allows the policy DNN to learn to predict more accurate Q-values, while the weights for the target DNN remain fixed for a while.
2. After T time steps, copy the policy DNN weights to the Target DNN. This step will enable the Target DNN get the updated weights so that it can also predict more accurate target Q-values.

Long-short term memory (LSTM) network is a developed design from the recurrent neural network (RNN), which can inspect long-term dependencies and has the ability to remember previous information for future usage. The LSTM network has a chain structure consisting of multiple LSTM cells and the proposed DQN structure shown in Figure 2 is clearly adopting the LSTM network as the DNN hidden layers. The DNN based LSTM in Figure 2 is mainly consists of four layers, and each layer contains several neurons, and the weighted sum of each neuron will be the input to an activation function. In our proposed DQN approach, the length of each training sequence is specified as L, which is the dimension of the input layer. In our scenario, we choose the input layer of the DNN to include 128 neurons, and the input states to the input layer will be shifted to the subsequent layer after updating the weight parameters [13,22].

As shown in Figure 2, we have applied one LSTM layer as the second layer in both the policy DNN and the target DNN, and the LSTM layer itself includes 300 hidden cells. For each hidden cell, the learnable weights are specified as follows: the input weights W, the recurrent weights R, and the bias b.

The third layer in both the policy DNN and the target DNN is a fully connected layer that processes the outputs of the LSTM layer, and it assembles all of the characteristics and internal information gathered by the prior layers. The fully connected layer behaves separately at each time step, and all neurons in a fully connected layer are connected to all the neurons in the previous layer.

The last adopted layer in both the policy DNN and the target DNN is the regression layer, which is responsible for computing the mean square error (MSE), improving the cell status, and updating the cell weights. A regression layer can also predict the response of the trained network. It is worth mentioning that normalizing the training data in the LSTM network enables the stabilization and acceleration of the training process for neural networks. It is shown in Figure 2, that in the simplified DQN structure, the input states are established according to the size of the input layer, then these states will be passed into both the policy DNN and the target DNN and the state action value functions will be predicted at the output.

The design of a single LSTM cell is basically shown in Figure 3 [13,22]. Each LSTM cell has three inputs and two output parameters. The hidden state h_{t-1} and the cell state c_{t-1} are the shared parameters between inputs and outputs and the other parameter is the current input. The LSTM cell also includes three sigmoid functions and two tanh functions to regulate the flow of information. In the initialization stage, random hidden states will be generated along with the input for the first LSTM cell. Then the current outputs that include the current hidden state h_t and current cell state c_t and the new input x_t will comprise the three inputs to the next cell.

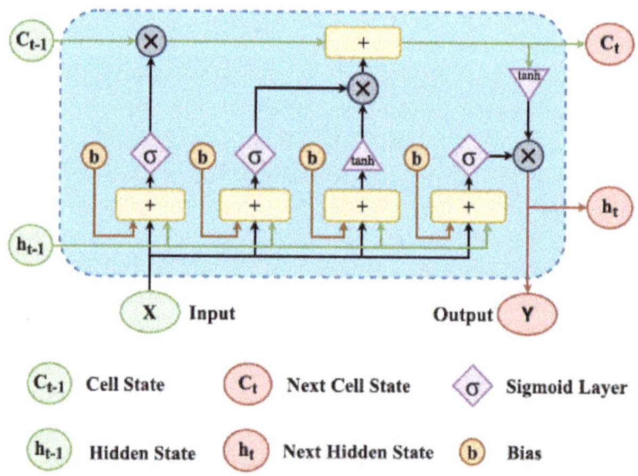

Figure 3. LSTM Cell Structure.

6. DQN Dataset Generation

Typically, the DQN framework involves an agent, a deep neural network (DNN), and the environment. The agent will interact with the environment via the DNN and decide which action to take. In our proposed DQN framework, the BS will be considered as an agent, and it will interact with the environment, which includes the user devices and fading channels. At the start, the agent (BS) will start exploring the environment to collect the information or the states for each user device in the cell, such as power distribution, user distance, channel model, and path loss [23,24].

Typically, at each time step t, and based on the current state s_t for each user device, the agent can decide on a certain action a_t using the DNN to maximize the sum rates for all users in the NOMA network. Accordingly, the agent (BS) will receive an instant reward r_t and move to the next state s_{t+1} in the environment. By taking decisions on actions, the agent (BS) can learn more about the environment to achieve an optimal channel prediction policy π_c. In our scenario, we aim that this optimal policy π_c for predicting or estimating the channel parameters for each user device can be learned and updated at each time step t via the simplified DQN structure illustrated in Figure 2. Furthermore, the agent (BS) can further enhance the policy π_c by repeating the channel estimation process for multiple episodes. Based on the proposed DQN architecture shown in Figure 2, it is clearly noticed that the DNN based LSTM replaces the Q-table to estimate the Q-values for each state–action pair in the environment, and this designed DNN can be considered as the policy controller for the channel estimation procedure.

7. DQN Policy

The period of time in which the agent interacts with the environment via the proposed DQN scheme is termed an episode, and every episode has a total duration of T time steps. At each episode, the main aim is to estimate the channel parameters for each user in order to maximize the sum rates for all users in the NOMA cell. In our simplified DQN approach, the dimension of the input layer for the DNN based LSTM is set equal to the available states in state space S for each user, and correspondingly the dimension of the output layer is equal to the number of possible actions in the action space A for each user. As indicated in Figure 2, The LSTM layer, and the fully connected layer are both comprising the hidden layers part of the proposed DQN model, and this may provide a reasonable balance between the network performance and computational complexity. Typically, the Q learning procedure is considered an off-policy algorithm, which means that without applying any greedy policy, the Q algorithm can iteratively estimate the best action for maximizing the future reward. In our developed DQN algorithm, we decide to apply a near-greedy action selection policy, that has two approaches as shown in Figure 4 [25]:

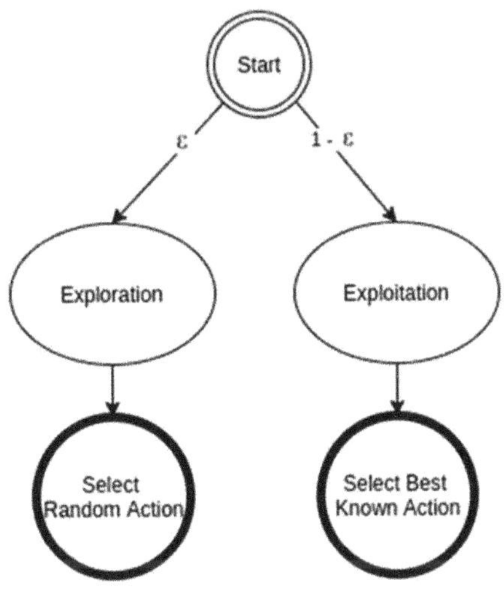

Figure 4. Near-greedy action selection scheme.

The first approach is the exploration, where the agent discovers and carries out random actions at a time step t. The second approach is the exploitation, where the agent can decide on an action to maximize the state-action value function $Q_\pi(s_t, a_t, W_t)$ based on the previous experience and the current network weights.

In our proposed near-greedy action selection policy, the agent has an exploration rate of ϵ and an exploitation rate of $(1-\epsilon)$ where $0 < \epsilon < 1$, and ϵ is considered a hyperparameter that can control the trade-off between exploitation and exploration during the training process. Hence, based on that designated action selection policy, the agent (BS) can select an explicit action a_t at a given state s_t at every time step t and correspondingly, the agent can receive a positive or negative reward and move to a new state s_{t+1}.

8. DQN State Space, Action Space, and Reward

Initially, the distance between each user device and the BS and channel path loss needs to be specified in the dataset to facilitate the random generation of the channel coefficients for every user in the examined NOMA system [13,14,22]. In addition, pilot symbols will be created, transmitted, and identified at both the BS and at the receiver side of each device to also assist in the initial channel parameters estimation process. As well, the power factor for each device in the NOMA cell needs to be initially assigned in the dataset. To set up the Q values, the channel parameters for every user device in the cell can be initialized either using the random generation of the channel parameters based on the path loss and the distance or using the pilot symbols. In our simplified DQN algorithm, we initialize the channel parameters based on both schemes, the random generation and pilot symbols. Throughout the DQN algorithm iterations, the Q-values will be predicted according to the DQN algorithm procedure.

As previously mentioned, in our channel estimation procedure, we need to efficiently predict the channel parameters for each user device in the examined NOMA cell to facilitate the maximization of the sum rates for all users in the considered NOMA system at each time step t. Hence, the state space S can be created to include the following states:

(a) The current power factor α_i for each user in the NOMA cell,
(b) The current user distance d_i that represents the distance between BS (agent) and the user device i.
(c) The present channel path loss φ.

Accordingly, the resultant state space S for N users NOMA system can be represented as [13,14,25]

$$S = \{\alpha_1\ \alpha_2\ \alpha_3\ \ldots \alpha_N\ d_1\ d_2\ d_3 \ldots d_N\ ,\ \varphi\ \} \tag{22}$$

For each user, all the actions that can be chosen by the agent (BS) can be selected from action space A. In our scanario, the possible actions in the action space A can be introduced as follows:

(a) Change the distance of the user device within a limited range of 5 m.
(b) Increase or decrease the power distribution factor α_i by a certain step size of 0.05.

The reward function also plays an principal role in the DQN algorithm, and there are many ways to assign the rewards based on the selected action. In our developed DQN scenario, we decided to calculate the rate for each user in the NOMA system using (1), to reflect the immediate reward r returned from the environment to the agent (BS) after choosing a certain action a_t at state s_t. Hence, based on the selected action, if the calculated data rate is higher than a specified threshold R_{th}, this will reflect a positive reward for the agent, while a lower data rate will reflect a negative reward. Based on the aforementioned discussion, Algorithm 1 can summarize the algorithm steps for estimating the channel parameters for each user in the NOMA cell, based on our simplified DQN structure.

Algorithm 1 Proposed DQN Algorithm for channel parameters estimation

1. Initialize policy DNN and target DNN networks with random weights $(W,\ \hat{W})$.
2. Initialize experience replay memory (ERM).
3. Randomly generate the exploration rate ϵ.
4. **for** each episode do
5. **for** each step do
6. **for** each user device do based on ϵ, and based on the current state s_i, Select the channel parameters and add to action space a_i
7. **end** for
8. Observe the immediate rewards r_i and move to the next state s_{t+1}.
9. Insert (s_i, a_i, r_i, s_{t+1}) in experience replay memory (ERM).
10. Create a mini batch with random sample of tuple (s_i, a_i, r_i, s_{t+1}) from ERM.
11. **for** each tuple in mini batch do
12. Predict the Q-values using policy DNN.
13. Approximate Q^* values using target DNN.
14. Calculate the loss between Q^* values generated from Target DNN and Q values generated from Policy DNN.
15. Update the weights W of the policy DNN using SGD.
16. **end** for
17. **end** for
18. $\hat{W} \leftarrow W$ after a certain number of T steps.
19. **end** for

9. Detailed DQN Procedure and Workflow

In this section, we can list the detailed workflow for the developed DQN algorithm that is responsible for estimating the channel parameters for each user in the examined NOMA system:

- Initialize the weights for both the policy DNN and the target DNN.
- Initialize the ERM with a typical size of 10,000 (it can be 10^6).
- Initialize the ϵ parameter for near-greedy action selection policy with a large value of $\epsilon = 0.999$ (start by exploration then decay).
- Initialize data records (tuples).
    - (a) Generate a random channel coefficients based on the fading model parameters with size = 120).
    - (b) Based on the pilot symbols, approximate the channel coefficients with size = 8).
    - (c) For each user, both the randomly generated channel parameters and the coefficients generated based on the pilot symbols will be combined and used as initial channel parameters.
- Assign initial distance, initial power factor, and path loss, and prepare the state space S for each user.
- Select a random state s_t from the sate space and used it as an input for policy DNN.
- The policy DNN will select a random action and correpondigly select a random Q-value, and based on this step, the policy DNN can predict the channel coefficients.
- Calculate the rate, and based on the calculated rate the reward can be assigned.
- Go to the next state s_{t+1}
- Compose a tuple $e_1 = (s_t,\ a_t,\ r,\ s_{t+1})$
- Store a tuple e_1 in ERM.
- Generate experience tuples = 1000, and store these tuples in ERM.
- Select a random batch of the tuples from ERM with batch size 32 tuples.
- Number of episodes = 20, and number of steps = 10^4
- For each tuple in the random batch do the following:
    - (a) From the policy DNN, select the Q-values (channel coefficients) randomly.
    - (b) From the Target DNN select the Q-values based on the greedy policy

(c) Assign the Reward.
(d) Calculate the Loss function as follows: (Target Q-value (Reward + Q_{max} value) − policy Q-value).
(e) Update the weights of the policy DNN based on the SGD.

- Every $T = 10^2$ steps, copy the weights of the policy DNN to the Target DNN.
- Activation functions used in LSTM layers are (sigmoid and tanh), while activation functions used at the output layer are (linear or Relu).
- SGD optimizer is utilized for weight updates.

10. Complexity Analysis

It is important to quantify the computational complexity of the proposed algorithm. Overall, deep learning algorithms are mainly dependent on hyperparameters, hence, applying analytical methods to guarantee the convergence of the proposed DQN algorithm usually has some sort of difficulty. Hence, it is a common challenge in literature to prove the optimality and convergence of the algorithm in an analytical way [26–28]. Alternatively, in this section we can focus on showing the amount of work per iteration in the developed DQN algorithm. For the NOMA system with N users and K base stations, the computational complexity of the proposed DQN algorithm can be introduced as follows: it is known that the size of the state space is denoted by S and the size of the action space is denoted by A and both have a significant role in the complexity of the deep Q-learning algorithm. Following [14,29], the computational complexity of the Q-learning algorithm with the greedy policy is estimated to be $\mathcal{O}(S \times A \times M)$ for each iteration, where S is the number of states, A is the number of actions, and M is the number of steps per episode. In our proposed DQN scenario, it can be shown that the size of the state space is $K + N$, and the size of the action space is $2(K + N)$. Therefore, the amount of the work per iteration can be described as follows $\mathcal{O}\left(\left(2K^2 + \times 4NK + 2N^2\right) \times M\right)$. According to [12], the corresponding computational complexity for the traditional channel estimation method based on MMSE procedure can achieve a relatively low complexity $\mathcal{O}\left(N^{2.37}\right)$ [12,30] but at the cost of performance degradation. Based on the aforementioned analysis, it can be shown that the developed DQN algorithm has some sort of complexity but at the cost of performance improvement as will be verified in the simulations results.

11. Simulation Parameters and Environment

Discussion for the simulation parameters and settings is described in this section. The simulated downlink NOMA system includes three distinct user devices and one BS. The BS is equipped with a single antenna and each user device in the cell is also equipped with a single antenna. In the simulated NOMA environment, the modulated signal related to each user in the downlink transmission will be superimposed and transmitted by the BS to each user device via independent Rayleigh fading channels, and the path loss is set to 3.5. At the receiver side, we assume that a perfect SIC procedure is applied and AWGN is considered and the noise power density is set to $N_0 = -174$ dBm/Hz.

MATLAB simulation tool is employed to realize the following: (1) inspect, characterize, and evaluate the performance of the proposed deep reinforcement learning based DQN algorithm which developed to be utilized as a channel estimator in the examined NOMA system, (2) Diverse performance metrics will be measured to evaluate the efficiency of the proposed DQN algorithm when being utilized in the channel estimation process. Simulations are accomplished with 10^4 iterations, and limited pilot symbols are generated and recognized at the BS and each user device to assist in the estimation process. The main simulation parameters can be summarized as shown in Table 1.

Table 1. Summary of Simulation Parameters.

Parameter	Value
Simulation Tool	MATLAB
Modulation type	QPSK
Number of Users	3, [2–20]
System Bandwidth B	1000 kHz
Fading distribution	Rayleigh
Path loss φ	3.5
Number of Iterations	10^4
Noise PSD N_0	−174 dBm/Hz
Learning Rate α	0.01
Discount factor γ	0.9
Batch size	32
Initial exploration rate ϵ	0.999
Optimizer	SGD
R_{th}	2 b/s

The simulation figures are created based on the assumption that the channel parameters for each user will be estimated based on the simplified DQN algorithm. Therefore, in order to examine the impact of utilizing the proposed DQN approach, the channel estimation technique based on standard minimum mean square error (MMSE) procedure [12] is also simulated for the sake of comparison. As indicated in Section 9, initially both the randomly generated channel parameters and the channel coefficients generated based on the pilot symbols will be combined and used in the simulation environment, to model the Rayleigh fading channel. In our developed DQN algorithm, at the end of each training episode, the predicted $Q(s, a)$ values generated from the policy DNN will be employed as an approximated channel coefficients for each user device to recover the desired signal. Different power factors are initially assigned for every user device according to the current distance from the BS and the present channel condition. Power factors α_n, α_m, and α_f are assigned for near, middle, and far users, respectively. In a fixed power allocation setup, we initially assign $\alpha_f = 0.65$, $\alpha_m = 0.3$, and $\alpha_n = 0.05$. In the simulation files, the transmission distance for every user device with respect to BS is initially defined as follows: $d_f = 1000$ m, $d_m = 500$ m, and $d_n = 100$ m. User data and pilot symbols are modulated using the Quadrature phase shift keying (QPSK) modulation format and the applied transmitted power range is set to vary from 0 to 30 dBm for many reasons, firstly, to match with the benchmark environments that simulated from the literature, secondly, most of the simulation environments are applying this classical range, and thirdly, on average, the performance metric behavior can be certainly predictable after 30 dBm power level.

12. Simulation Results and Analysis

Simulation results that describe the comparison between the proposed DRL based DQN algorithm and the MMSE procedure when both being utilized to estimate the channel parameters for each device are shown in Figure 5 in terms of BER versus power. The estimated channel parameters using both procedures will be employed for the signal recovery for each user and the simulated results are generated where a fixed power allocation scheme is considered. It is clearly noticed that when the developed DQN algorithm is applied for predicting the channel parameters, each user device in the examined NOMA cell shows the ability to provide a visible enhancement in lowering the BER compared to the MMSE technique. As an example, at a particular transmitted power of 20 dBm, the realized BER value for far user device using the MMSE procedure is 10^{-1}, while the achieved BER in the case of DQN is 10^{-2}. Similarly, the improvment in the BER for middle and near user devices is obviously observed when the simplified DQN algorithm is applied compared to the MMSE procedure.

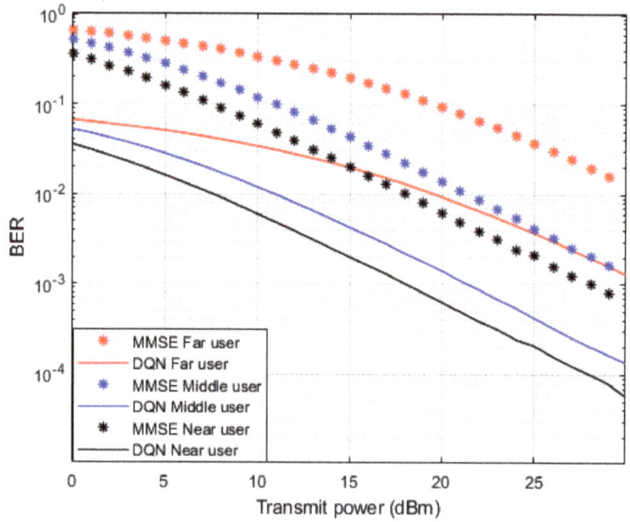

Figure 5. BER vs. power (DQN—MMSE).

In terms of the outage probability against applied power, Figure 6 illustrates the simulation results for the inspected user devices in NOMA cell when both the simplified DQN algorithm and the standard MMSE technique are implemented separately as a channel estimators. Similar to BER results, all user devices simulation outcomes indicate about 10 dBm enhancement in the power saving when the proposed DQN algorithm is applied compared to the MMSE technique. The reduction in the power transmitted also supports the improvement achived in minimizing the outage probability when the DQN algorithm is adopted. These visible improvements verify the advantage of usage the simplified DQN scheme as a channel estimator compared to the traditional MMSE procedure.

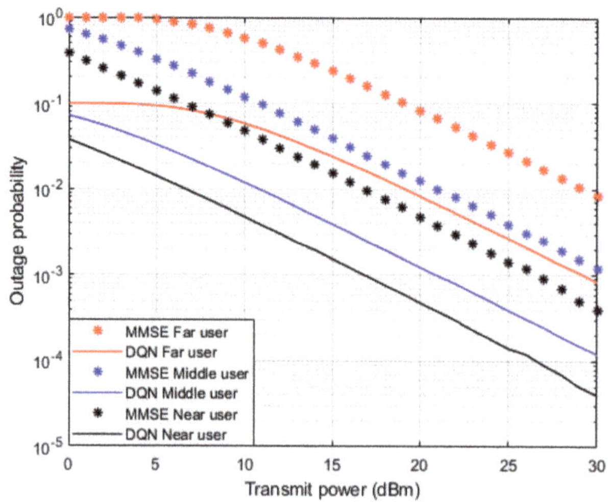

Figure 6. Outage Probability vs. power (DQN—MMSE).

Figure 7 presents the simulation results for the attainable capacity for each user in the examined NOMA system when both the simplified DQN algorithm and the standard MMSE channel estimation procedures are applied separately. The achieved rate for the near device shows significant enhancement by about 20 bit/s/Hz compared to far and middle users' rates. The dominance of the near user in terms of the possible rate may be justified by the stable channel condition for the near user compared to other users in NOMA system. Moreover, the results indicate that the proposed DQN algorithm still can deliver a stable bit rate compared to the MMSE technique for far and middle users' scenarios, and this slight improvment can be justified by the interference factor and inadequate link conditions for far and middle users.

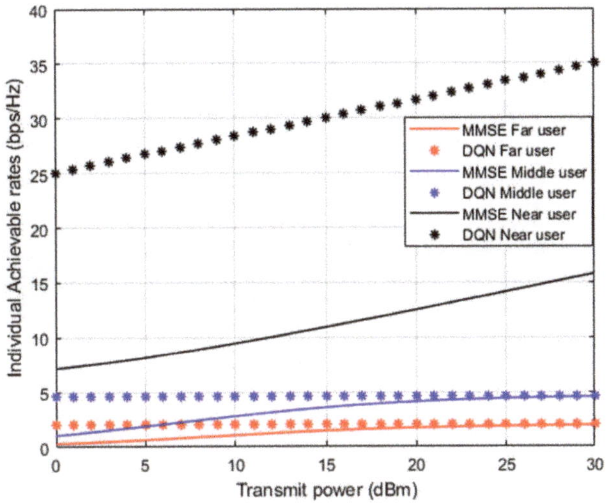

Figure 7. Achievable rates vs. power (DQN—MMSE).

In Figure 8, three distinct channel prediction schemes are investigated here as a benchmark comparison: (1) standard minimum mean square error (MMSE) procedure for channel estimation [12]; (2) DL based LSTM network for channel prediction applied in [13]; and RL based Q algorithm for channel estimation applied in [14]. Figure 8 displays the simulation outcomes for the sum rate for all user devices in the examined NOMA cell versus the applied power. It is apparent that the developed DRL based DQN algorithm shows superiority over the standard MMSE procedure approximately by more than 20 bit/s/Hz. Furthermore, the simplified DQN algorithm shows an improvement over the DL based LSTM procedure presented in [13] by nearly 10 bit/s/Hz. For the third benchmark applied in [14], the simplified DQN procedure shows a performance enhancement by 8 bit/s/Hz, approximately compared to the RL based Q algorithm. These findings support that this simplified DQN algorithm can be a strong candidate technique compared to other procedures when it is being utilized as a channel estimator.

Simulation results for the sum rate performance metric against different numbers of users in the examined NOMA cell are also illustrated in Figure 9, where the reference power is assigned to be 1 dBm. Similar to the simulation environment in Figure 8, three distinct channel prediction schemes are also investigated here as a benchmark comparison: (1) channel estimation based on standard minimum mean square error (MMSE) procedure [12]; (2) DL based LSTM structure for channel prediction applied in [13]; and RL based Q algorithm for channel estimation applied in [14]. As revealed from the results in Figure 9, it is clearly noticed that our simplified DQN algorithm can realize a substantially greater sum rate with respect to the MMSE procedure, by at least 4 bit/s/Hz when the cell capacity is initialized with 2 users. It can also be noticed that as the number of user devices in the cell keeps increasing, the developed DQN algorithm still shows dominance in accomplishing higher sum rates compared to the DL based LSTM scheme by 2 bit/s/Hz approximately. Similarly, the hidden layers feature in the simplified DQN scheme play a sufficient role in providing a noticeable enhancement in the sum rates compared to the Q-learning algorithm while the number of user devices in the NOMA cell is increasing. Generally, these findings reveal that dependability can be ensured by our simplified DQN algorithm even when the user devices in the cell increase. Furthermore, it is worth saying that while increasing the user devices in the cell, the interference will also grow up, thus the performance and the sum rate could be affected.

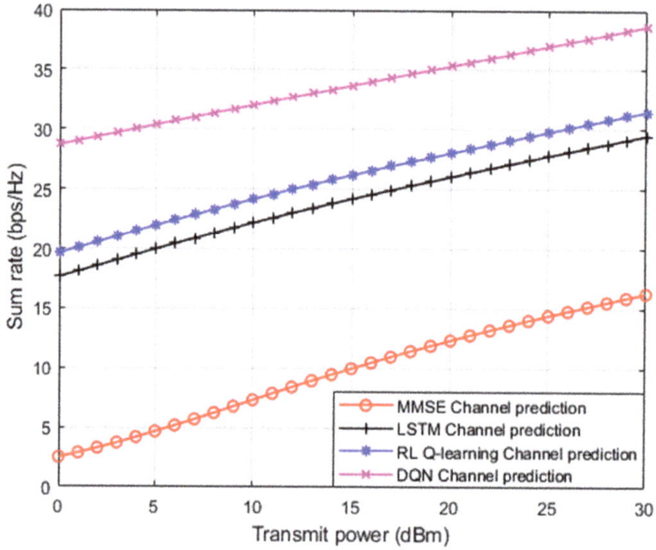

Figure 8. Sum rate vs. power (MMSE, LSTM, RL Q-learning, DQN).

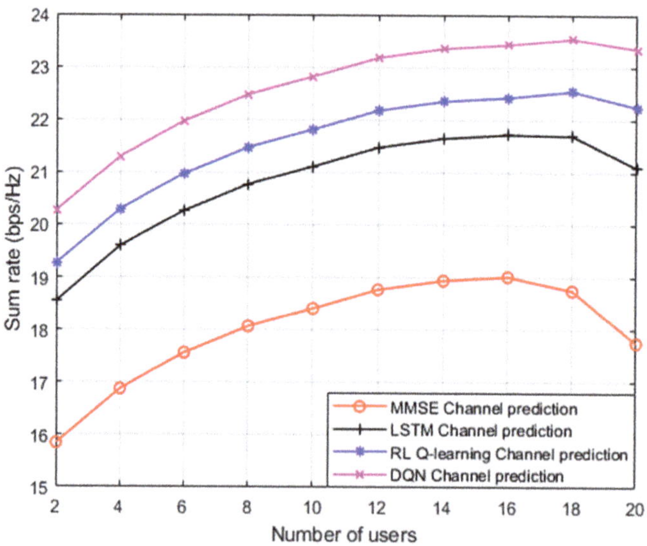

Figure 9. Sum rate v. number of users (MMSE, LSTM, RL Q-learning, DQN).

Simulation results in terms of BER against the applied power are also shown in Figure 10, where both the proposed DQN approach and the RL based Q algorithm [14] are utilized as different approaches for channel parameters estimation. Moreover, the optimized power coefficients derived in [13] for the examined NOMA cell are also applied in this simulation environment. Simulation outcomes indicate that all user devices in the cell can provide a perceivable enhancement in the performance when the simplified DQN algorithm is applied as a channel estimator compared to the case when the Q learning algorithm is implemented when the optimized power scheme is considered. Based on the simulation results, it can be clearly noticed that the developed DQN algorithm for channel estimation and the optimized power scheme can both provide an imprvment in the power saving by more than 5 dBm compared to the case when Rl based Q algorithm and optimized power scheme are both applied.

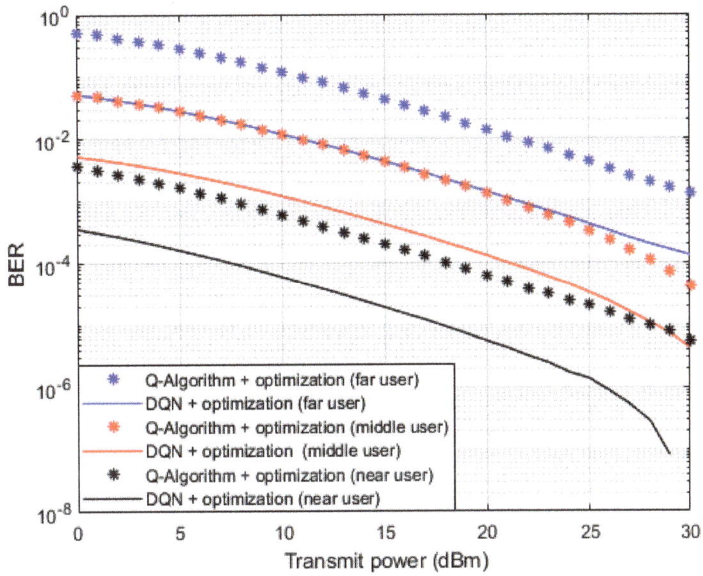

Figure 10. BER vs. power (DQN—Q learning—Optimization).

13. Conclusions

In this paper, the impact of utilizing a simplified deep reinforcement learning based DQN algorithm to specifically estimate the channel parameters for each user device in the NOMA system is discussed. In the proposed algorithm, the DQN model is initialized based on generating a random channel parameters then the weights of the simplified DQN model are updated based on the interaction between the agent and the environment in order to maximize the received downlink sum rates and at the same time minimize the loss function. The reliability of the developed DQN structure to estimate the channel parameters is examined by comparing the performance of the proposed DQN algorithm with a diverse benchmark schemes. A selective benchmark schemes were simulated, such as MMSE procedure for channel estimation, DNN based LSTM for channel estimation, and RL based Q algorithm for channel estimation. Simulation outcomes have proven that the simplified DQN algorithm can provide a noticeable enhancement in terms of the system performance compared to the simulated benchmark schemes. Furthermore, various performance metrics have been examined, and the simulation results also verified the superiority of the simplified DQN structure even when the cell capacity is increased.

Author Contributions: Conceptualization, M.G.; Methodology, M.A.; Validation, M.A.; Formal analysis, M.G. and M.A.; Supervision, M.A. All authors have read and agreed to the published version of the manuscript.

Funding: This research received no external funding.

Institutional Review Board Statement: Not applicable.

Informed Consent Statement: Not applicable.

Data Availability Statement: Not applicable.

Conflicts of Interest: The authors declare no conflict of interest.

Abbreviations

The following abbreviations are used in this manuscript:

AWGN	Additive White Gaussian Noise
BER	bit error rate
BS	Base Station
CSI	Channel state information
DL	Deep Learning
DNN	Deep Neural Network
FPA	Fixed Power Allocation
OPS	Optimized Power structure
LSTM	Long Short-Term Memory
DQN	Deep Q networks
ML	Machine Learning
MSE	Mean Square Error
MMSE	Minimum Mean Square Error
MUD	Multiuser detection
PD-NOMA	Power Domain Non-Orthogonal Multiple Access
QoS	Quality of Service
SIC	Successive interference cancellation
RL	Reinforcement Learning
DRL	Deep Reinforcement Learning

References

1. Alsabah, M.; Naser, M.A.; Mahmmod, B.M.; Abdulhussain, S.H.; Eissa, M.R.; Al-Baidhani, A.; Noordin, N.K.; Sait, S.M.; Al-Utaibi, K.A.; Hashim, F. 6G wireless communications networks: A comprehensive survey. *IEEE Access* **2021**, *9*, 148191–148243. [CrossRef]
2. Almekhlafi, M.; Arfaoui, M.A.; Assi, C.; Ghrayeb, A. Joint Resource and Power Allocation for URLLC-eMBB Traffics Multiplexing in 6G Wireless Networks. In Proceedings of the ICC 2021—IEEE International Conference on Communications, Montreal, QC, Canada, 14–23 June 2021; pp. 1–6. [CrossRef]
3. Du, J.; Jiang, C.; Wang, J.; Ren, Y.; Debbah, M. Machine Learning for 6G Wireless Networks: Carrying Forward Enhanced Bandwidth, Massive Access, and Ultrareliable/Low-Latency Service. *IEEE Veh. Technol. Mag.* **2020**, *15*, 122–134. [CrossRef]
4. Yang, Z.; Liu, Y.; Chen, Y.; Al-Dhahir, N. Cache-aided NOMA mobile edge computing: A reinforcement learning approach. *IEEE Trans. Wirel. Commun.* **2020**, *19*, 6899–6915. [CrossRef]
5. Xiao, L.; Li, Y.; Dai, C.; Dai, H.; Poor, H.V. Reinforcement learning-based NOMA power allocation in the presence of smart jamming. *IEEE Trans. Veh. Technol.* **2018**, *67*, 3377–3389. [CrossRef]
6. Yang, P.; Li, L.; Liang, W.; Zhang, H.; Ding, Z. Latency optimization for multi-user NOMA-MEC offloading using reinforcement learning. In Proceedings of the 28th Wireless and Optical Communications Conference (WOCC), Beijing, China, 9–10 May 2019; pp. 1–5. [CrossRef]
7. Doan, K.N.; Vaezi, M.; Shin, W.; Poor, H.V.; Shin, H.; Quek, T.Q.S. Power allocation in cache-aided NOMA systems: Optimization and deep reinforcement learning approaches. *IEEE Trans. Commun.* **2020**, *68*, 630–644. [CrossRef]
8. Zhang, S.; Li, L.; Yin, J.; Liang, W.; Li, X.; Chen, W.; Han, Z. A dynamic power allocation scheme in power-domain NOMA using actor-critic reinforcement learning. In Proceedings of the IEEE/CIC International Conference on Communications in China (ICCC), Beijing, China, 16–18 August 2018; pp. 719–723.
9. Gaballa, M.; Abbod, M.; Alnasur, S. Hybrid Deep Learning for Channel Estimation and Power Allocation for MISO-NOMA System. In Proceedings of the 2022 IEEE Future Networks World Forum (FNWF), Montreal, QC, Canada, 10–14 October 2022; pp. 361–366. [CrossRef]

10. Giang, H.T.H.; Hoan, T.N.K.; Thanh, P.D.; Koo, I. Hybrid NOMA/OMA-based dynamic power allocation scheme using deep reinforcement learning in 5G networks. *Appl. Sci.* **2020**, *10*, 4236. [CrossRef]
11. He, C.; Hu, Y.; Chen, Y.; Zeng, B. Joint power allocation and channel assignment for NOMA with deep reinforcement learning. *IEEE J. Sel. Areas Commun.* **2019**, *37*, 2200–2210. [CrossRef]
12. Neumann, D.; Wiese, T.; Utschick, W. Learning the MMSE Channel Estimator. *IEEE Trans. Signal Process.* **2018**, *66*, 2905–2917. [CrossRef]
13. Gaballa, M.; Abbod, M.; Aldallal, A. Investigating the Combination of Deep Learning for Channel Estimation and Power Optimization in a Non-Orthogonal Multiple Access System. *Sensors* **2022**, *22*, 3666. [CrossRef] [PubMed]
14. Gaballa, M.; Abbod, M.; Aldallal, A. A Study on the Impact of Integrating Reinforcement Learning for Channel Prediction and Power Allocation Scheme in MISO-NOMA System. *Sensors* **2023**, *23*, 1383. [CrossRef]
15. Rezvani, S.; Jorswieck, E.A.; Joda, R.; Yanikomeroglu, H. Optimal Power Allocation in Downlink Multicarrier NOMA Systems: Theory and Fast Algorithms. *IEEE J. Sel. Areas Commun.* **2022**, *40*, 1162–1189. [CrossRef]
16. Ding, Z.; Schober, R.; Poor, H.V. Unveiling the Importance of SIC in NOMA Systems—Part 1: State of the Art and Recent Findings. *IEEE Commun. Lett.* **2020**, *24*, 2373–2377. [CrossRef]
17. Luong, N.C.; Hoang, D.T.; Gong, S.; Niyato, D.; Wang, P.; Liang, Y.-C.; Kim, D.I. Applications of Deep Reinforcement Learning in Communications and Networking: A Survey. *IEEE Commun. Surv. Tutor.* **2019**, *21*, 3133–3174. [CrossRef]
18. Cao, Y.; Zhang, G.; Li, G.; Zhang, J. A Deep Q-Network Based-Resource Allocation Scheme for Massive MIMO-NOMA. *IEEE Commun. Lett.* **2021**, *25*, 1544–1548. [CrossRef]
19. Chu, M.; Liu, A.; Jiang, C.; Lau, V.K.N.; Yang, T. Wireless Channel Prediction for Multi-user Physical Layer with Deep Reinforcement Learning. In Proceedings of the 2022 IEEE 95th Vehicular Technology Conference: (VTC2022-Spring), Helsinki, Finland, 19–22 June 2022.
20. Parhi, R.; Nowak, R.D. The Role of Neural Network Activation Functions. *IEEE Signal Process. Lett.* **2020**, *27*, 1779–1783. [CrossRef]
21. Tian, Y.; Zhang, Y.; Zhang, H. Recent Advances in Stochastic Gradient Descent in Deep Learning. *Mathematics* **2023**, *11*, 682. [CrossRef]
22. Gaballa, M.; Abbod, M.; Aldallal, A. Deep Learning and Power Allocation Analysis in NOMA System. In Proceedings of the 2022 Thirteenth International Conference on Ubiquitous and Future Networks (ICUFN), Barcelona, Spain, 5–8 July 2022; pp. 196–201. [CrossRef]
23. Ling, J.; Xia, J.; Zhu, F.; Gao, C.; Lai, S.; Balasubramanian, V. DQN-based resource allocation for NOMA-MEC-aided multi-source data stream. *EURASIP J. Adv. Signal Process.* **2023**, *2023*, 44. [CrossRef]
24. Dai, L.; Wang, B.; Ding, Z.; Wang, Z.; Chen, S.; Hanzo, L. A survey of non-orthogonal multiple access for 5G. *IEEE Commun. Surv. Tuts.* **2018**, *20*, 2294–2323. [CrossRef]
25. Zhai, Q.; Bolić, M.; Li, Y.; Cheng, W.; Liu, C. A Q-Learning-Based Resource Allocation for Downlink Non-Orthogonal Multiple Access Systems Considering QoS. *IEEE Access* **2021**, *9*, 72702–72711. [CrossRef]
26. Challita, U.; Dong, L.; Saad, W. Proactive resource management for LTE in unlicensed spectrum: A deep learning perspective. *IEEE Trans. Wirel. Commun.* **2018**, *17*, 4674–4689. [CrossRef]
27. Wei, Y.; Yu, F.R.; Song, M.; Han, Z. User scheduling and resource allocation in HetNets with hybrid energy supply: An actor-critic reinforcement learning approach. *IEEE Trans. Wirel. Commun.* **2018**, *17*, 680–692. [CrossRef]
28. Sun, Y.; Peng, M.; Mao, S. Deep reinforcement learning-based mode selection and resource management for green fog radio access networks. *IEEE Internet Things J.* **2019**, *6*, 1960–1971. [CrossRef]
29. Jin, C.; Allen-Zhu, Z.; Bubeck, S.; Jordan, M.I. Is Q-learning provably efficient? *Proc. Int. Conf. Adv. Neural Inf. Process. Syst.* **2018**, *31*, 4868–4878.
30. Liao, Y.; Hua, Y.; Cai, Y. Deep learning based channel estimation algorithm for fast time-varying MIMO-OFDM systems. *IEEE Commun. Lett.* **2019**, *24*, 572–576. [CrossRef]

sensors

MDPI

Article

A Study on the Impact of Integrating Reinforcement Learning for Channel Prediction and Power Allocation Scheme in MISO-NOMA System

Mohamed Gaballa [1,*], Maysam Abbod [1] and Ammar Aldallal [2]

[1] Department of Electronic & Electrical Engineering, Brunel University London, Uxbridge UB8 3PH, UK
[2] Department of Telecommunication Engineering, Ahlia University, Manama 10878, Bahrain
* Correspondence: mohamedgaballa.gaballa@brunel.ac.uk

Abstract: In this study, the influence of adopting Reinforcement Learning (RL) to predict the channel parameters for user devices in a Power Domain Multi-Input Single-Output Non-Orthogonal Multiple Access (MISO-NOMA) system is inspected. In the channel prediction-based RL approach, the Q-learning algorithm is developed and incorporated into the NOMA system so that the developed Q-model can be employed to predict the channel coefficients for every user device. The purpose of adopting the developed Q-learning procedure is to maximize the received downlink sum-rate and decrease the estimation loss. To satisfy this aim, the developed Q-algorithm is initialized using different channel statistics and then the algorithm is updated based on the interaction with the environment in order to approximate the channel coefficients for each device. The predicted parameters are utilized at the receiver side to recover the desired data. Furthermore, based on maximizing the sum-rate of the examined user devices, the power factors for each user can be deduced analytically to allocate the optimal power factor for every user device in the system. In addition, this work inspects how the channel prediction based on the developed Q-learning model, and the power allocation policy, can both be incorporated for the purpose of multiuser recognition in the examined MISO-NOMA system. Simulation results, based on several performance metrics, have demonstrated that the developed Q-learning algorithm can be a competitive algorithm for channel estimation when compared to different benchmark schemes such as deep learning-based long short-term memory (LSTM), RL based actor-critic algorithm, RL based state-action-reward-state-action (SARSA) algorithm, and standard channel estimation scheme based on minimum mean square error procedure.

Keywords: RL; Q-learning; MISO-NOMA; KKT conditions

Citation: Gaballa, M.; Abbod, M.; Aldallal, A. A Study on the Impact of Integrating Reinforcement Learning for Channel Prediction and Power Allocation Scheme in MISO-NOMA System. *Sensors* **2023**, *23*, 1383. https://doi.org/10.3390/s23031383

Academic Editors: Yichuang Sun, Haeyoung Lee and Oluyomi Simpson

Received: 1 December 2022
Revised: 13 January 2023
Accepted: 19 January 2023
Published: 26 January 2023

1. Introduction

The Non-Orthogonal Multiple Access (NOMA) system has been characterized as an inspiring multiple access form for upcoming wireless approaches to enhance the spectral efficiency and throughput [1]. NOMA system can develop the available resources more realistically by efficiently, taking into consideration the users' channel environments and also giving support to several users with distinctive Quality of Service (QoS) needs [2]. The integration of NOMA and multiple antenna techniques can be exploited to improve and reinforce system performance [3], therefore, inspecting Multiple Input-Single Output (MISO) NOMA system can be a good example in the direction of characterizing the expected upgrade in achievable data rates [4]. In downlink NOMA structure, the receiver device can receive a multiplexing of signals transmitted to several user terminals in the NOMA cell, thus eliminating the interference generated by other user devices come to be essential for coordinated detection. Frequently in power domain NOMA (PD-NOMA), multiuser detection can be handled via successive interference cancellation (SIC) [5]. In the SIC procedure, symbols from numerous users are decoded successively on the basis of the

Channel State Information (CSI) and power percentage designated for each user. A broad investigation of CSI for various users is demanding because pilot data that can be exploited in channel prediction, might interfere with symbols from other user terminals, therefore affecting the performance of a conventional prediction scheme, such as the Minimum Mean Square Error (MMSE) estimator [6]. Furthermore, power allocation policy is considered an essential issue for user devices when PD-NOMA is considered [7].

Deep Learning (DL) or Reinforcement Learning (RL) techniques, have the ability to track the differences in the channels among users and BS, thus, they are recently considered a powerful tool for upcoming radio systems [8,9]. Hence, allocating the power factors or estimating the CSI for user devices with the assistance of Machine Learning (ML) algorithms, triggered the authors for more deep investigations into this field in order to enhance the performance and detection process.

1.1. Related Works

Different techniques were introduced by authors in [10] to realize the optimal MMSE channel estimator in the Reconfigurable Intelligent Surfaces (RIS)-based MISO system. In the first technique, the authors suggest an analytical linear estimator to adjust the phase shift matrix of the RIS during the training phase, and the estimator based on that technique is shown to produce sensible accuracy compared to the least-squares method when the statistical properties of the applied channel and noise are considered. In the other approach, authors have expressed the channel prediction problem as an image denoising problem, then they introduce a Convolutional Neural Network (CNN) to achieve the denoising and predict the optimal MMSE channel parameters. Numerical outcomes have clarified that the proposed estimator based CNN algorithm can offer improved performance compared to the linear estimation method and low computational intricacy is preserved.

Toward enhancing the link reliability, a neural network model for a wireless channel estimator is proposed in [11] to be used with uncoded space-time diversity procedure in Multi Input Multi Output (MIMO) system. Based on the neural network ML structure, a channel estimator is suggested, and a mathematical scheme is presented to derive an optimum power transmission factors that can assist in lessening the channel prediction bandwidth utilization. Simulation results revealed that the channel estimator based on the proposed neural network structure can deliver an improvement in Bit Error Rate (BER) and Mean Square Error (MSE) compared to the standard MMSE channel estimation technique.

In a massive MIMO system and on the basis of a deep autoencoder scheme, authors in [12] performed experimental verifications on two tasks, one task for channel estimation modelling for wireless links, and the other task is belonging to a power allocation policy. The proposed deep learning autoencoder is also used to manage the issue raised from inadequate training datasets that may cause critical overfitting problems and consequently affect the model's reliability. Results based on the autoencoder procedure clarified that the suggested scheme could successfully enhance performance when the extent of the training dataset is mainly within a specified threshold selection.

To get over limitations raised when standard iterative power control techniques are utilized, such as high complexity and unnecessary latency, the work in [13] introduced a deep learning framework to manage these issues. In the presented structure, the outdated and partial CSI is exploited, and a Deep Neural Network (DNN) framework is created to construct an optimization problem to boost the spectral efficiency in device-to-device communication systems. User fairness and energy efficiency constraints were examined, and simulation outcomes showed that the proposed DNN model can attain better spectral and energy efficiency compared to the MMSE procedure when numerous channel correlation factors are considered.

Based on CSI, the position of each user device with respect to BS, and the path loss, a deep learning framework labelled PowerNet is introduced in [14]. The authors attempt to prove that it is possible to avoid the time consumption involved with intricate channel estimation procedures, and at the same time, power control can be managed. Different from traditional DNNs that employ a fully connected structure, the presented PowerNet method utilizes a CNN layers to recognize the interference model through several links in wireless networks. Simulation outcomes revealed that the suggested PowerNet scheme can realize a stable performance without explicit channel estimation.

Recently, approximating the channel parameters or predicting the power factors with the assistance of Reinforcement learning (RL), is investigated by many researchers. The authors of [15] proposed an end-to-end channel estimation framework for a downlink multiuser multiple antenna system. The authors presented an RL-based actor-critic scheme for channel estimation without the assumption of ideal CSI. The authors mainly depend on the agent to bring and utilize the pilot symbols into the estimation process and then employ the estimated channel parameters to create downlink beamforming matrices. To satisfy the purpose of maximizing the sum rate reward, network parameters are adjusted based on the deep policy gradient method. The results proved that the suggested channel estimation algorithm can provide convergence and stable performance under various channel statistics and can perform better than the typical MMSE procedure when the sum rate metric is examined.

In [16], the authors developed a Deep Reinforcement Learning (DRL) method for device-to-device pairing to understand the correlation patterns between wireless networks. The introduced RL algorithm is adopted to explore the joint channel selection and power control problem for device-to-device pairing and to boost the weighted sum rate. Based on the suggested DRL learning procedure, each device-to-device pair can make use of the outdated and local information to understand the network parameters and perform decisions independently. Results showed that without a global CSI, the suggested DRL scheme is capable to attain a stable performance close to that achieved using standard analytical approaches.

The combination between a DNN as a tool for channel prediction and an optimized power scheme is explored in [17] for the purpose of multiuser detection in the NOMA system. The DNN based Long Short-Term Memory (LSTM) network is developed for channel prediction based on complex data processing. The DNN network is trained on the basis of both the correlation between successive training sequences and the normalised channel statistics. The efficiency of the suggested DNN based LSTM for channel prediction is inspected using different fading models and simulation outcomes, in terms of different performance metrics, have proved that the presented DNN scheme for channel estimation can provide a consistent performance compared to the MMSE procedure even when cell capacity is expanded.

1.2. Research Gap and Significance

Based on the preceding works, many of the proposed schemes that consider predicting the channel parameters task are mainly focused on implementing several deep neural networks (DNN) while applying RL approaches, which in turn leads to an increase in the number of hidden layers with a massive number of neurons in each layer. The significance of this study is to illuminate that we can eliminate the need for such DNN approaches, and instead, we can adopt the RL based developed Q-learning algorithm to predict the channel coefficients for each user device in MISO-NOMA cell, and at the same time, a notable improvement in system performance and network convergence is realized. The most prominent gain of the developed channel estimator scheme is that it can enhance the system performance without the need for hidden layers or an external training set.

In addition, several RL algorithms have been proposed to explicitly address the issues associated with channel state information (CSI), beamforming, and power allocation. To the best of the authors' knowledge, there is no study that explores the incorporation between Q-learning algorithm for channel prediction and the power allocation policy as an integrated scheme for multiuser detection in downlink MISO-NOMA system in fading channels.

Furthermore, it is worth mentioning that unlike deep learning algorithms, that mainly depend on learning from a training data set, the proposed Q-learning algorithm in our study is developed to dynamically enhance the system performance and adjust to the variations in the channel based on the feedback from the environment.

1.3. Contributions to Knowledge

The channel prediction problem in downlink NOMA systems was considered in numerous works. In addition, there have been several works that apply machine learning (ML) to handle the channel estimation task in wireless communication systems. However, most of the current research on channel prediction in the NOMA systems based on ML is introduced via deep neural networks. To the best of the author's knowledge, currently, there is no research that manages the channel approximation task in a multiuser multi-input single-output NOMA system through an RL based Q-learning algorithm. The RL based Q-learning algorithm is developed based on maximizing the sum rates for all users in the network such that it can be used efficiently to predict the channel parameters for each user in the MISO-NOMA cell.

In addition, in this work, a structured mathematical analysis is introduced to formulate a non-complex analytical form for the power allocation for user devices in the examined MISO-NOMA system based on boosting the sum rate of the system while considering the constraints of the total power budget in the system, and the QoS for each user. Furthermore, the performance of the MISO-NOMA system is investigated when both the developed Q-learning algorithm for channel estimation and the derived power allocation scheme are jointly implemented. In this work, the contributions can be summed up as shown:

- In this study, a framework is proposed to illuminate how RL based Q-learning algorithm is developed based on maximizing the sum rates for all users in a MISO-NOMA system in order that it can be used dynamically to predict the channel parameters for each user in the MISO-NOMA cell.
- As a reference comparison, four further simulation environments are established. (1) the standard minimum mean square error (MMSE) based channel prediction scheme (Neumann et al.); (2) the DNN algorithm based on LSTM network for channel prediction applied in [17], (3) the RL based actor-critic procedure for channel prediction applied in [15], (4) the fourth simulation environment is dependent on applying RL based State-Action-Reward-State-Action (SARSA) procedure (Ahsan et al. and Mu et al.). The simulation outcomes of these environments are compared with the results of our proposed RL based Q-learning scheme, and the results emphasized that dependability can be assured by our developed Q-model for predicting channel parameters even when the number of devices in the cell is increased.
- To validate the efficacy of the developed Q-learning algorithm for channel prediction, the developed Q-model is investigated using Rayleigh and Rician fading channels.
- Evaluate the beneficial impact of cooperatively integrating the RL based Q-learning algorithm for channel prediction and the derived power allocation scheme for the purpose of multiuser recognition in the power domain MISO-NOMA system.
- The optimized power allocation scheme and the fixed power allocation scheme are both compared when the developed Q-learning scheme is implemented as a channel estimator.

The remainder of this paper is structured as follows. Section 2 describes the system model. Analysis of the optimization problem is presented in Section 3. The optimization framework and procedure are discussed in Section 4. The RL structure is introduced in Section 5. Section 6 discusses the Q-learning algorithm-based channel prediction. The RL-based Q-model architecture and channel estimation algorithm are summarized in Section 7. The simulation environment is described in Section 8, and simulation results are presented in Section 9. Lastly, conclusions are shown in Section 10.

Notation: bold lower-case letters denote vectors, bold upper-case letters denote matrices, and lower-case letters denote scalars. The subscript on a lower-case letter x_i represent ith element of vector $\boldsymbol{x}$. $E(\cdot)$ refers to the expectation and $(\cdot)^{\mathrm{T}}$ refers to the transpose of the vector. For two real numbers a ≤ b, [a, b] is the set for all real numbers in the range from a to b.

2. System Model

2.1. Multiuser Environment

In this work, a multiuser environment with a single Base Station (BS) and multiple user devices (UDs) is considered. The BS is supplied with N antennas and all the UDs are supplied with a single antenna. The network is assumed to work with equal length time intervals and each time interval includes one transmission, which contains either uplink or downlink transmissions. The pilot-assisted channel prediction is considered in this work, where pilot symbols can be identified by BS and UDs [15,17]. Each user device initially transmits its pilot symbols to BS via an uplink channel. Then, prior to data transmission, the BS can inspect the pilot symbols and the available network information to facilitate estimating the downlink CSI. The main aim of this work is to model the channel prediction task and to manage the power allocation scheme. We can refer to the matrix of downlink channel coefficients from BS with N antennas to UD i as:

$$\boldsymbol{H}_i = [\boldsymbol{h}_{1i};\ \boldsymbol{h}_{2i};\ldots;\boldsymbol{h}_{Ni}] \tag{1}$$

where $\boldsymbol{h}_{ji}$ represents the vector channel parameters from jth antenna at BS to the ith UD, with $j \in [1,\ 2,\ \ldots, N]$ and $i \in [1,\ 2,\ \ldots, M]$, where N is the number of antennas at BS and M is the number of users in MISO-NOMA cell. Furthermore, we can denote the data signal transmitted to UD i as

$$\boldsymbol{s}_i = [s_{i1},\ s_{i2},\ \ldots, s_{iK}] \tag{2}$$

where K is the length of the signal. Then, the matrix of all the UD's sequences can be expressed as

$$\boldsymbol{S} = [\boldsymbol{s}_1;\boldsymbol{s}_2;\ldots;\boldsymbol{s}_M] \tag{3}$$

The received kth signal at jth UD can be denoted as:

$$y_{kj} = \sum_{i=1}^{N} \boldsymbol{h}_{ij} s_{ki} + z_{kj} \tag{4}$$

where z_{kj} denotes the AWGN with zero mean and variance σ^2 at jth UD through kth signal duration. The received kth symbol at all UDs is:

$$\boldsymbol{Y}_k = \sum_{i=1}^{N} \boldsymbol{h}_i s_{ki} + \boldsymbol{z}_k \tag{5}$$

where

$$\boldsymbol{Y}_k = [y_{k1}; y_{k2}\ ;\ \ldots\ ; y_{kM}] \tag{6}$$

$$\boldsymbol{z}_k = [z_{k1};\ z_{k2};\ \ldots; z_{kM}] \tag{7}$$

Many of the current works depend on pilot symbols to approximate the uplink channel parameters and then utilize channel reciprocity to realize the prediction of downlink channel weights [15,18]. These schemes for CSI prediction may not be reliable, especially in cases of inadequate channel reciprocity owing to hardware constraints. Furthermore, this kind of estimator may introduce estimation errors in case the uplink and downlink channel parameters are not stationary within a certain transmission.

In the developed Q-learning procedure, we plan to get assistance from the pilot symbols, and network information to explicitly predict the downlink channel parameters. The set of estimated channel coefficients among BS and M UDs can be indicated as

$$\hat{\mathbf{H}} = [\hat{\mathbf{H}}_1;\ \hat{\mathbf{H}}_2;\ldots;\hat{\mathbf{H}}_M] \tag{8}$$

where $\hat{\mathbf{H}}_i$ is the predicted matrix channel coefficients between BS that contains N antennas and ith UD, and can be expressed as follows:

$$\hat{\mathbf{H}}_i = \left[\hat{h}_{1i};\ \hat{h}_{2i}; \ldots; \hat{h}_{Ni}\right] \tag{9}$$

where $\hat{h}_{ji}$ represents the predicted channel parameters between jth antenna at BS and the ith UD.

2.2. MISO-NOMA Environment

The fundamental idea of NOMA is to achieve non-orthogonal resource allocation between users while increasing the processing at the receiver side [19]. With non-orthogonal resource allocation, NOMA can attain massive connectivity and accomplish higher spectral efficiency. Existing research on the NOMA system mainly focuses on the code domain and power domain. In the code domain NOMA, distinct spread-spectrum codes are designated to different users and then multiplexed over the same time-frequency resource block. In the power domain NOMA (PD-NOMA) [19], the transmitter superimposes signals with different power levels to be sent to several users on the shared spectrum. At the receiver, each user can decode his own desired signal by means of successive interference cancellation (SIC).

In this subsection, the downlink MISO-NOMA system is explored where user devices and BS are linked by different fading channels. NOMA cell is assumed where one BS with two antennas is implemented to assist user devices (UDs), and each device terminal has one antenna. In PD-NOMA [19], user devices receive the superimposed signal sent from BS which involves target and interfering signals sent through the same resources. Thus, combining different signals supported by unique power portions is critical to distinguish signals and strengthen the successive interference cancellation (SIC) technique. The system structure for the basic components implemented in the examined MISO-NOMA system is shown in Figure 1.

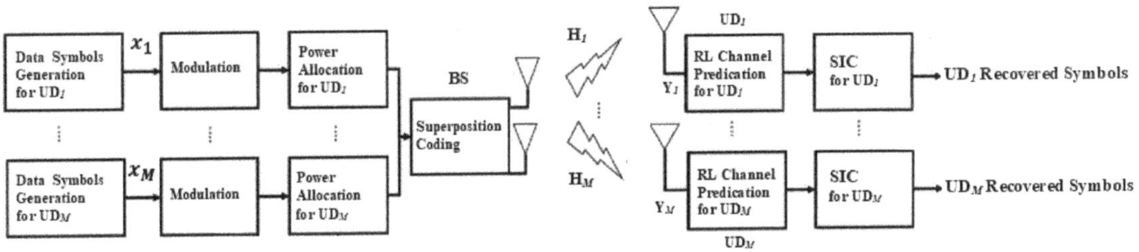

Figure 1. MISO-NOMA system basic Structure based RL channel prediction.

In our observed MISO-NOMA cell, three user devices are considered in the cell, and the examined user devices are identified corresponding to their fading channels and the distances from BS. Fading channels with Rayleigh distribution are adopted to characterize the channel model for every user. The user terminal at the boundary of the cell is realized as a far user, while the nearest user equipment is designated as a near user terminal. The examined cell contains three user devices and the fading path can be distinguished for every user as follows [3]: $h_n \sim \left(0, d_n^{-k}\right)$ for near users, $h_m \sim \left(0, d_m^{-k}\right)$ for the middle user, and $h_f \sim \left(0, d_f^{-k}\right)$ for the user at the edge of the cell, where h_i implies a vector represents the fading path coefficients among BS and user i. Path loss exponent is represented by k, and AWGN is considered with noise power indicated as σ^2. In terms of channel gains, the relation between user devices can be indicated as $|h_n|^2 > |h_m|^2 > \left|h_f\right|^2$ [20] and overall power transmitted from BS to all users in the cell is labelled as P_t. Every user device contains a receiving element that can activate the SIC process to get rid of signals related to other devices with bad channel environments. In contrast, signals related to user devices with good link conditions may not be separated and regarded as interference. According to the aforementioned assumptions, the superposition-coded signal $\boldsymbol{x}$ sent from BS can be stated as follows [3,17]:

$$\boldsymbol{x} = \sqrt{P_t}\left(\sqrt{\eta_n}x_n + \sqrt{\eta_m}x_m + \sqrt{\eta_f}x_f\right) \tag{10}$$

where η_f, η_m and η_n represent power factors given for a far device, middle device, and near device separately. Furthermore, $\boldsymbol{x}_f$, $\boldsymbol{x}_m$ and $\boldsymbol{x}_n$ refer to the signal vectors related to far, middle, and near users respectively. The received downlink signal at a far device in the MISO-NOMA cell can be shown as:

$$\boldsymbol{y}_f = \boldsymbol{x}\boldsymbol{h}_{f1} + \boldsymbol{x}\boldsymbol{h}_{f2} + z_f \tag{11}$$

where $\boldsymbol{h}_{f1}$ represents the channel coefficients among a far device and the 1st antenna at BS, $\boldsymbol{h}_{f2}$ represents the channel coefficients among the far device and 2nd antenna at BS and z_f is AWGN noise component at the far device with mean zero and variance σ^2. The far user is signified by weak link condition, and signal $\boldsymbol{x}_f$ is usually given further power percentage by BS where $\eta_f > \eta_m > \eta_n$. The obtained signal at a far device can be formulated as:

$$\boldsymbol{y}_f = \sqrt{P_t\eta_f}\boldsymbol{x}_f\left(\boldsymbol{h}_{f1} + \boldsymbol{h}_{f2}\right) + \left(\sqrt{P_t\eta_m}\boldsymbol{x}_m + \sqrt{P_t\eta_n}\boldsymbol{x}_n\right)\left(\boldsymbol{h}_{f1} + \boldsymbol{h}_{f2}\right) + z_f \tag{12}$$

The 1st term in (12) implies the target signal for far device and the 2nd term indicates the interference term from other user devices. The possible bit rate for a far device could be shown as [3,21]:

$$R_f = \log_2\left(1 + \frac{\left|\boldsymbol{h}_{f1} + \boldsymbol{h}_{f2}\right|^2 P_t\eta_f}{\left|\boldsymbol{h}_{f1} + \boldsymbol{h}_{f2}\right|^2 P_t(\eta_n + \eta_m) + \sigma^2}\right) \tag{13}$$

Typically, the near user device has a good link status alongside BS, therefore, a low power factor can be assigned to x_n, and the near user received signal can be stated as

$$\boldsymbol{y}_n = \sqrt{P_t\eta_n}\boldsymbol{x}_n(\boldsymbol{h}_{n1} + \boldsymbol{h}_{n2}) + \left(\sqrt{P_t\eta_m}\boldsymbol{x}_m + \sqrt{P_t\eta_f}\boldsymbol{x}_f\right)(\boldsymbol{h}_{n1} + \boldsymbol{h}_{n2}) + z_n \tag{14}$$

In Equation (14), the 1st term represents the anticipated signal, and the 2nd term implies interference from other devices. It can be noted from Equation (14), that the interference can be principal since the far user may be assigned a further power percentage. Thus, at a near device, SIC is accomplished, where direct decoding for the far user signal $\boldsymbol{x}_f$ is implemented first, then eliminated from the aggregate signal. After that, the middle device signal $\boldsymbol{x}_m$ is decoded and gets rid of it from the resultant signal and the possible rate for a near user R_n can be shown as:

$$R_n = \log_2\left(1 + \frac{|\boldsymbol{h}_{n1} + \boldsymbol{h}_{n2}|^2 P_t\eta_n}{\sigma^2}\right) \tag{15}$$

3. Optimization Problem Characterization

The key objective here is to maximize the sum rates for user devices in the MISO-NOMA cell. Sum rate maximization is considered based on optimizing the power coefficients for each user terminal in compliance with the status of the channel between each user and the BS. In downlink MISO-NOMA, the objective function or the sum rates for M user devices can be formulated as [3,22]:

$$R_{sum} = \sum_{i=1}^{M}\log_2\left(1 + \frac{|\boldsymbol{h}_{i1} + \boldsymbol{h}_{i2}|^2 P_t\eta_i}{|\boldsymbol{h}_{i1} + \boldsymbol{h}_{i2}|^2 P_t\sum_{j=1}^{i-1}\eta_j + \sigma^2}\right) \tag{16}$$

In the optimization problem, the constraints can be presented as follows:

3.1. Power Constraint

The power designated for every user device in the cell is a fraction of the whole power P_t sent from BS, therefore the power percentage for each device must conform with [22]:

$$\sum_{i=1}^{M} \eta_i \leq 1 \tag{17}$$

where η_i is the power percentage allocated for the ith user.

3.2. QoS Constraints

In our analysis, we consider that all the user devices in the examined MISO-NOMA cell need to satisfy a QoS requirement where the minimum rate R_{min} is required to be realised in the system [22,23], this constraint can be expressed as follows:

$$Log_2(1 + SINR_i) \geq R_{min} \tag{18}$$

where $SINR_n$ is the signal-to-interference plus noise ratio for ith user and R_{min} is the minimum required transmission rate in the examined MISO-NOMA cell. The expression in (18) can be redeveloped as follows [24]:

$$|\boldsymbol{h}_{i1} + \boldsymbol{h}_{i2}|^2 \rho \left(\eta_i - (2^{R_{min}} - 1) \sum_{j=1}^{i-1} \eta_j \right) > (2^{R_{min}} - 1) \tag{19}$$

where ρ represents the SNR and η_j is the power percentage given for jth user device.

4. Optimization Framework

The main aims in this part include the following: (1) present the objective function and the constraints in a standard form, (2) find a general expression for the 1st and 2nd derivative of the objective function, (3) based on the mathematical analysis and the derived formulas, we can inspect that $\frac{\partial^2 R_{Sum}}{\partial \eta_i^2}$ is a negative function, which validates that the objective function is a concave with distinctive global maximum, and (4) finally, we deduce the optimal power factors for each user based on applying the Lagrange function and the KKT necessary conditions.

On the basis of the objective function in (16) and the constraints in (17) & (19) and the fact that there are two antennas at the BS and one antenna at each user terminal, the standard optimization problem can be generally reformulated as follows [24,25]:

$$\max_{\eta} R_{sum} = \sum_{i=1}^{M} \log_2 \left(\frac{|\boldsymbol{h}_{i1} + \boldsymbol{h}_{i2}|^2 P_t \sum_{j=1}^{i-1} \eta_j + \sigma^2 + |\boldsymbol{h}_{i1} + \boldsymbol{h}_{i2}|^2 P_t \eta_i}{|\boldsymbol{h}_{i1} + \boldsymbol{h}_{i2}|^2 P_t \sum_{j=1}^{i-1} \eta_j + \sigma^2} \right) \tag{20}$$

such that

$$\sum_{j=1}^{M} \eta_j \leq 1$$

$$(2^{R_{min}} - 1) - \rho |\boldsymbol{h}_{i1} + \boldsymbol{h}_{i2}|^2 \left(\eta_i - (2^{R_{min}} - 1) \sum_{j=1}^{i-1} \eta_j \right) \leq 0$$

$$\eta_i \geq 0 \quad \forall i = 1, 2, \ldots, M$$

In this part, the power optimisation framework is accomplished with regards to three user devices in the MISO-NOMA cell, therefore, the examined constraints can be represented as shown [25,26]:

$$\psi_1(\eta) = \eta_n + \eta_m + \eta_f - 1 \tag{21}$$

$$\psi_2(\eta) = (2^{R_{min}} - 1) - \rho\left|\boldsymbol{h}_{f1} + \boldsymbol{h}_{f2}\right|^2\left(\eta_f - (2^{R_{min}} - 1)(\eta_m + \eta_n)\right) \tag{22}$$

$$\psi_3(\eta) = (2^{R_{min}} - 1) - \rho|\boldsymbol{h}_{m1} + \boldsymbol{h}_{m2}|^2\left(\eta_m - (2^{R_{min}} - 1)(\eta_n)\right) \tag{23}$$

Since the constraints $\psi_1(\eta)$, $\psi_2(\eta)$ & $\psi_3(\eta)$ are linear in terms of η, they are considered convex.

Typically, to prove that the objective function R_{Sum} is concave with a distinctive global maximum, we need to find the first derivative $\frac{\partial R_{Sum}}{\partial \eta_i}$ and the second derivative $\frac{\partial^2 R_{Sum}}{\partial {\eta_i}^2}$ of the objective function [3,24]. The first derivative of the objective function can be deuced in general form as follows [23]:

$$\begin{aligned}\frac{\partial R_{Sum}}{\partial \eta_i} = \frac{1}{ln2} & \left(\frac{|\boldsymbol{h}_{i1}+\boldsymbol{h}_{i2}|^2 P_t}{\left|\boldsymbol{h}_{i1}+\boldsymbol{h}_{i2}\right|^2 P_t \sum_{j=1}^{i} \eta_j + \sigma^2}\right) \\ & - \frac{1}{ln2}\sum_{k=1}^{M-i}\left\{\left(\frac{(|\boldsymbol{h}_{(i+k)1}+\boldsymbol{h}_{(i+k)2}|^2 P_t)^2 \eta_{i+k}}{(\left|\boldsymbol{h}_{(i+k)1}+\boldsymbol{h}_{(i+k)2}\right|^2 P_t \sum_{j=1}^{i+k} \eta_j + \sigma^2)}\right) \right. \\ & \left. \times \left(\frac{1}{(\left|\boldsymbol{h}_{(i+k)1}+\boldsymbol{h}_{(i+k)2}\right|^2 P_t \sum_{j=1}^{i+k-1} \eta_j + \sigma^2)}\right)\right\}\end{aligned} \tag{24}$$

Similarly, the second derivative of the objective function can be derived in general form as follows [23,24]:

$$\begin{aligned}\frac{\partial^2 R_{Sum}}{\partial {\eta_i}^2} = -\frac{1}{ln2} & \left\{\left(\frac{(|\boldsymbol{h}_{i1}+\boldsymbol{h}_{i2}|^2 P_t)^2}{(\left|\boldsymbol{h}_{i1}+\boldsymbol{h}_{i2}\right|^2 P_t \sum_{j=1}^{i} \eta_j + \sigma^2)^2}\right)\right. \\ & - \sum_{k=1}^{M-i}\left\{\left(\left|\boldsymbol{h}_{(i+k)1} + \boldsymbol{h}_{(i+k)2}\right|^2 P_t\right)^3 \eta_{i+k}\right. \\ & \times \left(\frac{[2(\left|\boldsymbol{h}_{(i+k)1}+\boldsymbol{h}_{(i+k)2}\right|^2 P_t \sum_{j=1}^{k+i-1} \eta_j + \sigma^2) + \left|\boldsymbol{h}_{(i+k)1}+\boldsymbol{h}_{(i+k)2}\right|^2 P_t \eta_{i+k}]}{(\left|\boldsymbol{h}_{(i+k)1}+\boldsymbol{h}_{(i+k)2}\right|^2 P_t \sum_{j=1}^{i+k} \eta_j + \sigma^2)^2}\right) \\ & \left.\left. \times \left(\frac{1}{(\left|\boldsymbol{h}_{(i+k)1}+\boldsymbol{h}_{(i+k)2}\right|^2 P_t \sum_{j=1}^{i+k-1} \eta_j + \sigma^2)^2}\right)\right\}\right\}\end{aligned} \tag{25}$$

Based on the above mathematical analysis and the derived formulas, we can inspect that $\frac{\partial^2 R_{Sum}}{\partial {\eta_i}^2}$ is a negative function, which verifies that the objective function is a concave with a distinctive global maximum [3,24,27]. To derive the optimal power factors, the Lagrange function and the KKT necessary conditions can be applied [28].

$$\mathcal{L}\left(\eta_n, \eta_m, \eta_f, \mu_1, \mu_{2,}, \mu_3\right) = R_{Sum} - \mu_1\psi_1(\eta) - \mu_2\psi_2(\eta) - \mu_3\psi_3(\eta) \tag{26}$$

where μ_1 , μ_2, and μ_3 represent Lagrange multipliers for the 3 users' scenario.

- Optimality conditions can be written as follows [3,24,27]:

$$\frac{\partial R_{Sum}}{\partial \eta_n} - \mu_1\frac{\partial \psi_1(\eta)}{\partial \eta_n} - \mu_2\frac{\partial \psi_2(\eta)}{\partial \eta_n} - \mu_3\frac{\partial \psi_3(\eta)}{\partial \eta_n} = 0 \tag{27}$$

$$\frac{\partial R_{Sum}}{\partial \eta_m} - \mu_1\frac{\partial \psi_1(\eta)}{\partial \eta_m} - \mu_2\frac{\partial \psi_2(\eta)}{\partial \eta_m} - \mu_3\frac{\partial \psi_3(\eta)}{\partial \eta_m} = 0 \tag{28}$$

$$\frac{\partial R_{Sum}}{\partial \eta_f} - \mu_1\frac{\partial \psi_1(\eta)}{\partial \eta_f} - \mu_2\frac{\partial \psi_2(\eta)}{\partial \eta_f} - \mu_3\frac{\partial \psi_3(\eta)}{\partial \eta_f} = 0 \tag{29}$$

Given the fact that $|h_n|^2 > |h_m|^2 > \left|h_f\right|^2$, we can demonstrate that the analyzed constraints are feasible [3] and after a few mathematical manipulations the closed form for the power factors η_f, η_m, and η_n can be deduced as follows [27]:

$$\eta_f = \left(\frac{(2^{R_{min}} - 1)}{2^{R_{min}}}\right)\left(1 + \frac{1}{\rho\left|h_{f1} + h_{f2}\right|^2}\right) \tag{30}$$

$$\eta_m = \left(\left(\frac{(2^{R_{min}}-1)}{2^{R_{min}}}\right)\left(1 + \frac{1}{\rho|h_{m1}+h_{m2}|^2}\right) - \left(\frac{2^{R_{min}}-1}{2^{R_{min}}}\right)^2\left(1 + \frac{1}{\rho\left|h_{f1}+h_{f2}\right|^2}\right)\right) \tag{31}$$

$$\eta_n = 1 - \left(\eta_m + \eta_f\right)$$
$$\eta_n = \frac{1}{(2^{R_{min}})}\left(\left(\frac{1+\rho\left|h_{f1}+h_{f2}\right|^2}{(2^{R_{min}})\rho\left|h_{f1}+h_{f2}\right|^2}\right) + \left(\frac{(2^{R_{min}}-1)}{\rho|h_{m1}+h_{m2}|^2} - \frac{1}{\rho\left|h_{f1}+h_{f2}\right|^2}\right)\right) \tag{32}$$

5. Reinforcement Learning Framework

Typically, RL is developed on the basis of a Markov Decision Process (MDP) design, that contains basic elements [29,30]: a state space '$\boldsymbol{S}$', which is the set of states or observations in the environment and these states can be observed by the agent. An action space '$\boldsymbol{A}$', which is the set of actions that can be selected by the agent at each state. An instantaneous reward '$\boldsymbol{R}$', which is the direct reward that is given to the agent after selecting an action a $\in \boldsymbol{A}$ to transfer to a state $s \in \boldsymbol{S}$. Policy '$\boldsymbol{P}$' represents the mapping criteria to move from the current observed state to a new state based on the action that will be taken by an agent. Another important element in the RL process is the State-action value function $Q(s,a)$, which is formally described as the expectation or the average of cumulative discounted rewards when an action $\mathbf{a} \in \boldsymbol{A}$ is selected by an agent in the state $\mathbf{s} \in \boldsymbol{S}$ when a certain policy is considered. Furthermore, RL can be considered a method of understanding the agent's interaction in a stochastic environment by successively selecting actions during a sequence of time periods. Therefore, the main aim of reinforcement learning is to train an agent to carry out a certain task within an uncertain environment [30].

The interaction between the agent and the environment can be described as follows: at each time period, the agent can recognize the observations or states in the environment, and based on the current observation, the agent can identify and carry out a specific action. Then, an immediate reward will be sent from the environment to the agent. The reward is a measure of how effective the action is, when the agent performs a certain action to achieve a specific goal [31]. Basically, at each learning time interval, the RL agent interacts with the environment by following a particular policy that controls the transition between state space to action space.

Based on the aforementioned discussion and as shown in Figure 2, the RL agent can be essentially represented by two elements: a policy and a learning algorithm [32]. The policy is the mapping criterion that chooses actions on the basis of the observations or status observed in the environment. Usually, the policy can be represented as a function with tunable parameters, such as DNN, while the learning algorithm constantly improves the parameters of the policy based on observations, actions, and rewards [33]. In general, the objective of the learning algorithm is to realize the best possible policy that can maximize the expected cumulative long-term reward received during the task.

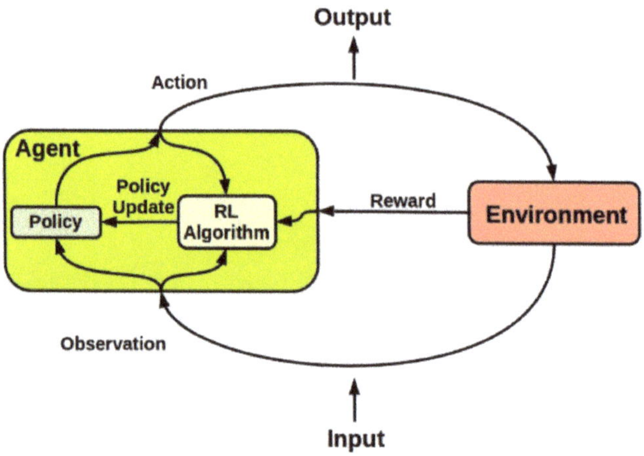

Figure 2. Reinforcement Learning Framework.

6. Channel Estimation Based Q-Learning Algorithm

In the considered channel prediction scheme, it is assumed that the action spaces are discrete, therefore, we manage to use an RL-based Q-learning procedure as one of the candidates of RL schemes for parameters update in our examined cell [34,35]. The Q-learning algorithm is categorized as a model-free, and off-policy reinforcement learning procedure, also a Q-learning agent is characterized as a value-based RL agent that has the role of updating a specific critic value function to enhance the future rewards. At a certain state, the agent can inspect and select the action for which the expected reward is maximized. In this section, RL based Q-learning is employed for channel prediction tasks in the MISO-NOMA cells where pilot symbols are also adopted to assist in the channel estimation process [36]. Therefore, it is assumed that there is coordination between BS and user devices such that the pilot symbols can be recognized at the BS and user terminals. In our work, we have considered the BS as the Q-learning agent, and we assume that the BS will start estimating the channel parameters for each user after user devices complete sending the pilot signals [37]. Therefore, in our developed RL based Q-learning algorithm can be utilized to estimate the CSI after the BS receives the pilot signals.

The scenario for the channel prediction process based on the developed Q-learning model can be outlined in this way [38]. Firstly, at the start of each transmission time slot, user devices can send pilot symbols to BS across the uplink channel. Secondly, on the basis of the developed RL based Q-learning algorithm and availability of network information such as user's distance and path loss, BS (agent) can predict the downlink CSI for user devices. Thirdly, BS will generate the superposition coding signal and performs downlink data transmission. Finally, the receiver of each user terminal will receive the downlink transmitted data and the estimated channel parameters based on Q-learning algorithm will be utilized to decode the desired signal. In addition, each user device can feedback the signal-to-interference plus noise ratio (SINR) or the achieved rate to the BS to enhance the detection process.

In this study, the main objective of the developed RL based Q-learning algorithm is to maximize the downlink sum rate and reduce the estimation loss. Instead of estimating the received signal, we primarily concentrate on incorporating the developed Q-learning model in the NOMA system for the purpose of channel estimation [39]. The RL-based Q-agent is designed to estimate the channel parameters by interacting with the environment, hence strict orthogonal pilot symbols are not required as shown in the standard procedures. Throughout the learning iteration, the Q-learning agent decides on the action that can enhance the approximated state-action value function $Q(s,a)$ therefore, the expected long-term reward can be also maximized in the neural networks. It is worth mentioning that when increasing the number of learning iterations, updating Q-values becomes more sufficient, and an improved channel approximation and sum rate reward can also be achieved [34,36,40].

In the proposed Q-learning scheme, the sum rate is presented at the learning time interval t as R_t, hence, the instantaneous sum rate at time instant t can be shown as follows [15,34]

$$R_t = \sum_{i=1}^{M} \log(1 + SINR_{it}) \tag{33}$$

where $SINR_{it}$ is the signal-to-interference plus noise ratio of user i at time instant t and M is the number of users in the MISO-NOMA cell. In this work, the optimum goal of the developed Q-learning algorithm is to maximize the total discounted reward R^{γ} starting from time instant t, which can be denoted as

$$R_t^{\gamma} = \sum_{k=t}^{\infty} \gamma^{k-t} R_{k+1} \tag{34}$$

where R_t^{γ} is the discounted reward at time slot t, and γ is the discount factor. Substituting the sum rate from (33) into (34), the discounted sum rate reward, can be expressed as [41]:

$$R_t^{\gamma} = \sum_{l=t}^{\infty} \gamma^{l-t} \sum_{i=1}^{M} \log\left(1 + SINR_{i(l+1)}\right) \tag{35}$$

As previously stated, the Q-learning agent is the BS, whose aim is to boost the accumulative transmission sum rate. Therefore, two value functions can be inspected while considering the RL maximization problem [34,36,42], the first one is the state value function $V(s)$

$$V(s) = E[R^{\gamma}/(S_t = s)] \tag{36}$$

and the other one is the state-action value function $Q(s,a)$

$$Q(s,a) = E[R^{\gamma}/(S_t = s, A_t = a)] \tag{37}$$

where E denotes the expected value given that the agent follows a certain policy within the applied procedure. Due to unspecified transition probabilities and limited observed states, an optimal policy is difficult to achieve. Therefore, the Q-learning procedure is developed to approximately achieve the best possible policy. In the developed Q-learning procedure, the state-action value function $Q(s,a)$ values are learned via trial and error and are updated according to the following formula [15,34,36,42]:

$$Q(s,a) \leftarrow (1-\alpha)Q(s,a) + \alpha\left[R(s,a) + \gamma \max_{a' \in A} Q(s',a')\right] \tag{38}$$

where α is the learning rate, s' denotes the new state, and a' is the new action that will be considered by the agent from the action space A to maximize the new state-action value function $Q(s',a')$.

7. Q-Learning Network Architecture

Basically, in data transmission, the frame transmitted includes data and pilot symbols. It is supposed that the implemented channel model is stationary throughout one frame transmission of data and pilot signals and the channel parameters are varying from one frame to another. The basic architecture of the channel prediction scenario based on the developed Q-learning procedure employed in our examined network is illustrated in Figure 3, which primarily consists of several stages [17,43].

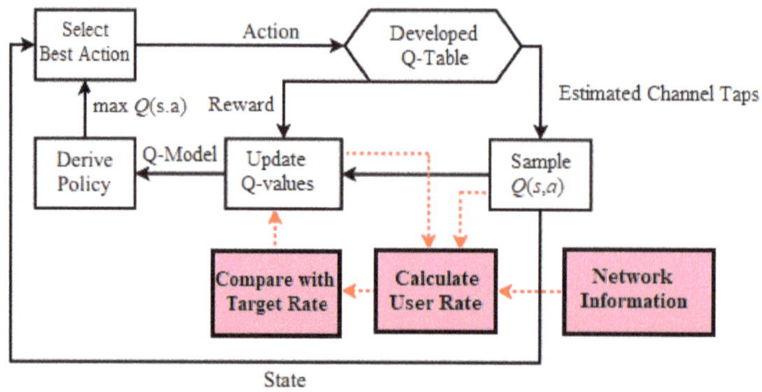

Figure 3. The architecture of the proposed Channel prediction scheme based developed Q-Learning algorithm.

In the first stage, initial channel parameters will be created based on a distinct Rayleigh channel models. In the second stage, we initialize the $\boldsymbol{Q}$ table and initialize the reward matrix $\boldsymbol{R}$ with zero values. The signal-to-interference plus noise ratio (SINR), and the minimum required rate R_t, can be calculated for every user device in the MISO-NOMA cell with the aid of the availability of the network information such as the initial assigned power percentage for each user terminal, and the entire power transmitted from BS P_T. Primarily, the Q-values can be adjusted based on the difference between the assigned target rate $\boldsymbol{R_T}$ and the initial generated user rate for each device. In the third stage, the best action will be explored and implemented by the $\boldsymbol{Q}$ agent, and then updating the values for the Q-table that represent the observation action pair $Q(s,a)$. Furthermore, the values for the reward matrix $\boldsymbol{R}$ will be dynamically assigned according to the actions executed by the Q-agent.

In the fourth stage, the state action value function $\boldsymbol{Q(s,a)}$ that represent the values for the Q-table will be modified according to a Q-learning procedure with the aid of the following parameters, the discount factor γ, the assigned immediate reward matrix $\boldsymbol{R}$, and the learning rate $\boldsymbol{\alpha}$. Throughout the learning phase, the generated state action values $Q(s,a)$ will be sampled to calculate the new channel rate and at the same time update the Q-table until the optimum rate or the terminal state is achieved.

Dataset Preparation

Essentially, path loss and the distance between every user terminal and the BS need to be specified in the dataset to facilitate the random generation of the channel weights for every user device in the examined MISO-NOMA network [43]. In the beginning, pilot symbols are created, transmitted, and identified at the BS and at the receiver of every device. Additionally, power factors for every device in the cell need to be initially assigned. The channel weights for every device in the cell are initialized to set up the Q-table values, and during the algorithm iterations, the Q-values are modified according to a Q-learning procedure [34–36].

Throughout the learning process and for the sake of updating the Q-table, the discount factor γ, learning rate $\boldsymbol{\alpha}$, the target rate $\boldsymbol{R_T}$, current state, and the terminal state should be identified. In our developed Q-learning algorithm, the Q-agent will choose the next state at random and set it as the next state, then the Q-learning agent will inspect all possible actions available to move to the next state. Next, the Q-learning agent will carefully identify the best action $\boldsymbol{a}$, that satisfies the maximum value for $Q(s,a)$ to move to the new state. After moving to the new state, a reward value will be assigned to the agent as a measure of how successful this transition was in order to move to the new state [44]. During the update phase, we compute ΔQ, which represents the difference between the new generated value function and the preceding value function of $Q(s,a)$. Then, update the resultant $Q(s,a)$ value in the Q-table according to the following formula.

$$Q(s,a) = Q(s,a) + \alpha \cdot \Delta Q \tag{39}$$

Based on the updated Q-values in Q-table and the updated channel gain, a new achieved rate can be calculated and compared to the target rate for each user device in the cell. In the developed Q-learning algorithm, once the optimum rate or the terminal state is reached, the developed Q-matrix will be employed to compose the channel taps for each user device. The developed Q-learning procedure for channel approximation can be summarized as presented in Algorithm 1.

Algorithm 1: Developed Q-learning Channel Prediction Structure.

1. Initialize the $\boldsymbol{Q}$ table values and initialize the reward matrix $\boldsymbol{R}$ with zeroes.

Inputs

2. Number of Iterations and the size for the channel parameters for every user device.
3. Initial distance "$\boldsymbol{d}_i$" of every user device from the BS.
4. Path loss parameter "$\boldsymbol{\vartheta}$".
5. Design random pilot symbols.
6. Initialize the random channel parameters for each user "$\boldsymbol{h}_{ij}$" based on fading model, $j \in [1, 2, \ldots, N]$ and $i \in [1, 2, \ldots, M]$. N is the number of antennas at BS and M is the number of devices in the cell.
7. Designate the power percentage "$\boldsymbol{\eta}_i$" for each user.
8. Determine system bandwidth "$\boldsymbol{B}$", Total transmit power "$\boldsymbol{P}_T$", and noise spectral density "$\boldsymbol{N}_o$"
9. Assign the desired channel parameters "$\boldsymbol{h}_{id}$" and the target rate "$\boldsymbol{R}_T$"

Procedure

10. Based on the channel gain $\left|\boldsymbol{h}_{ij}\right|^2$, total transmit power "$\boldsymbol{P}_T$", and initial power factor for each user "$\boldsymbol{\eta}_i$", signal to interference noise ratio "$\boldsymbol{SINR}_i$", minimum required rate "$\boldsymbol{R}_i$" can be calculated for each device.
11. At each iteration, compare the initial generated rate "$\boldsymbol{R}_i$" with the target rate "$\boldsymbol{R}_T$".
12. Update the values for the $\boldsymbol{Q}$-table that represent the current state and action pair $\boldsymbol{Q}(\boldsymbol{s},\boldsymbol{a})$.

Q-algorithm

13. identify discount factor "γ", learning rate "$\boldsymbol{\alpha}$", the current state, and the terminal state.
14. Choose the next state at random and set it as the next new state.
15. Inspect all possible actions "$\boldsymbol{a}_i$" to move to the new state.
16. Select the best action $\boldsymbol{a}_i \in \boldsymbol{A}$, which satisfies the maximum value for the $\boldsymbol{Q}$-value function argmax $\boldsymbol{Q}(\boldsymbol{s},\boldsymbol{a})$ to move to the new state.
17 Identify the immediate Reward "$\boldsymbol{R}$", based on the action implemented to move to the new state.
18. Based on the following: (1) maximum Q-value $\boldsymbol{Q}(\boldsymbol{s},\boldsymbol{a})$ obtained in (16), (2) the corresponding reward "$\boldsymbol{R}$", (3) the discount factor "γ", then $\boldsymbol{Q}(\boldsymbol{s},\boldsymbol{a})$ can be updated based on bellman's equation

$\boldsymbol{Q}(\boldsymbol{s},\boldsymbol{a}) \leftarrow \boldsymbol{R} + \gamma$ argmax $\boldsymbol{Q}(\boldsymbol{s},\boldsymbol{a})$

Outputs

19. Based on the updated $\boldsymbol{Q}(\boldsymbol{s},\boldsymbol{a})$ values in $\boldsymbol{Q}$-table, the channel coefficients "$\boldsymbol{h}_{ij}$" and channel gain $\left|\boldsymbol{h}_{ij}\right|^2$ can be updated and a new user rate can be calculated and compared to the target rate "$\boldsymbol{R}_T$".
20. Compute the difference "$\boldsymbol{\Delta Q}$" between the updated value function $\boldsymbol{Q}_{new}(\boldsymbol{s},\boldsymbol{a})$ and the previous $\boldsymbol{Q}(\boldsymbol{s},\boldsymbol{a})$.
21. Based on (20), $\boldsymbol{Q}(\boldsymbol{s},\boldsymbol{a})$ value in the $\boldsymbol{Q}$-table can be further updated according to $\boldsymbol{Q}(\boldsymbol{s},\boldsymbol{a}) \leftarrow \boldsymbol{Q}(\boldsymbol{s},\boldsymbol{a}) + \alpha \cdot \boldsymbol{\Delta Q}$
22. Check whether the terminal state has been reached or the episode has been completed.
23. Compose predicted channel taps $\hat{\boldsymbol{h}}_i$

8. Simulation Environment

Characterization of the simulation parameters and settings is discussed in this section. The examined downlink MISO-NOMA system contains three distinct user devices and one BS in which the BS is supplied with two antennas and every user device in the cell is provided with a single antenna. In the examined NOMA structure, the modulated signals in downlink transmission are superimposed and transferred by BS to user devices via independent Rayleigh or Rician fading channels that are influenced by AWGN with noise power density assigned as $N_0 = -174$ dBm/Hz and the path loss is set to 3.5. MATLAB software is utilized as a simulation tool to satisfy the following aims, (1) inspect, characterize, and evaluate the performance of the developed RL based Q-learning algorithm when implemented as a channel estimator in the considered MISO-NOMA system, (2) investigate the reliability of incorporating the developed Q-algorithm as channel estimator scheme with the optimized power scheme in the examined MISO-NOMA network, and performance metrics are considered to explore the impact of this integration. (3) optimized power allocation scheme and fixed power allocation scheme are both compared when the developed Q-learning scheme is utilized as a channel estimator in the cell. Monte-Carlo simulations are performed with $N = 10^5$ iterations, and at the outset of each iteration, pilot symbols are randomly generated and recognized at the BS and each device. The main simulation parameters are summarized in Table 1.

Table 1. Summary of Simulation Parameters.

Parameter	Value
Simulation Tool	MATLAB
Modulation type	QPSK
Number of Users	3, [2–10]
System Bandwidth B	1000 kHz
Fading channel	(Rayleigh, Rician)
Path loss exponent	3.5
Number of Iterations	10^5
Noise PSD N_0	-174 dBm/Hz
Learning Rate α	0.1
Discount factor γ	0.9

The presented Simulation figures are generated based on the assumption that the channel coefficients are not available at each user device. Thus, in order to examine the effectuality of the developed RL based Q-learning procedure, and for the sake of comparison, four further simulation environments are established, (1) standard minimum mean square error (MMSE) based channel prediction scheme [45]; (2) DL algorithm based on LSTM network for channel prediction applied in [17], (3) RL based actor-critic procedure for channel prediction applied in [15], (4) the fourth simulation environment is dependent on applying RL based State-Action-Reward-State-Action (SARSA) procedure (Ahsan et al., Mu et al. and Jiang et al.). Throughout the simulations, we point out to MMSE technique as conventional NOMA, to denote that user devices are applying the MMSE technique for predicting the channel state information (CSI) prior to reconstructing the desired signal.

In the simulation environment, NOMA parameters are generated on the basis of the LTE standard [46,47], and channel parameters are created to initially model the Rayleigh fading channels based on the ITU models. In our developed Q-learning algorithm, at the end of the training episode, or if the terminal state is reached, the updated $Q(s, a)$ values in the Q-table will be employed as a practical channel coefficients for the user devices. Different power percentages are initially assigned for every user device according to channel gain and based on the existing distance from the BS. Power factors η_n, η_m, and

η_f are specified for near, middle, and far users respectively. In a fixed power allocation setup, we designate $\eta_f = 0.65$, $\eta_m = 0.25$, and $\eta_n = 0.1$. In the optimized power structure (OPS), power factors are allocate d for user devices in proportion to the analytical formula concluded previously for every device in Section 4. In the simulation files, the transmission distance for each user device with respect to BS is assigned as follows: $d_f = 900$ m, $d_m = 400$ m, and $d_n = 100$ m. Data and pilot symbols are modulated using Quadrature phase shift keying (QPSK) as the modulation format and the applied transferred power is mostly varying from 0 to 30 dBm.

9. Simulation Results and Discussion

Simulation outcomes that clarify the comparison between the developed RL based Q-learning algorithm and the conventional NOMA scheme that applies MMSE method to predict the channel coefficients for each device are shown in Figure 4 in terms of BER versus power transmitted. The predicted channel parameters using both schemes are employed for the signal detection for each user device and the simulated results are shown where fixed power allocation (FPA) is considered. When the developed Q-algorithm is applied for channel estimation, each user device in the examined MISO-NOMA cell provides a noticeable improvement in lowering the BER compared to the MMSE procedure. At particular BER values such as 10^{-2}, the attained power saving by the Q-learning algorithm is within 2 dBm for far and middle user devices, while a power reduction within 1 dBm is recorded for the near user.

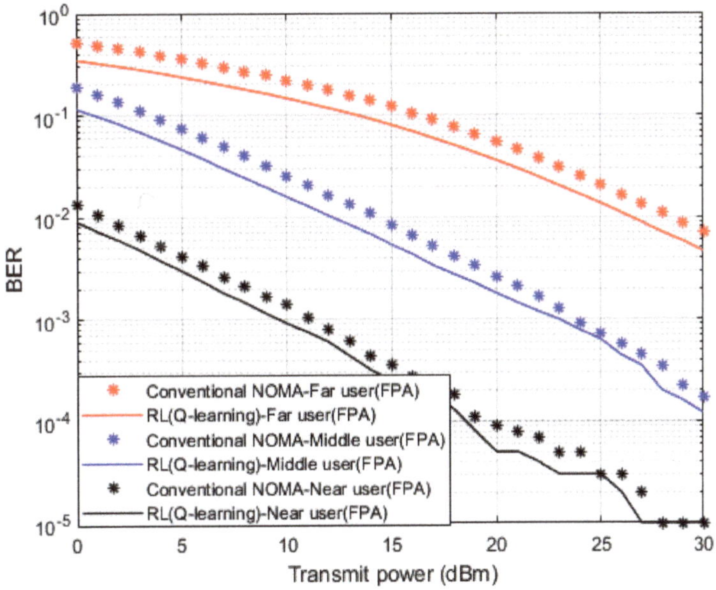

Figure 4. BER vs. power (Q-learning & Conventional NOMA (MMSE)).

In terms of the outage probability against applied power, Figure 5 illustrates the results for the inspected user devices in the MISO-NOMA cell when the developed Q-learning and standard MMSE are considered as a channel estimator schemes. Far, and middle devices simulation outcomes indicate about 2 dBm enhancement in saving power to realize 10^{-2} outage probability when the developed Q-learning algorithm scenario is applied compared to the MMSE procedure. Similarly, a near user with the developed Q-learning algorithm displays a 1 dBm improvement in power saving with respect to the MMSE scheme. This enhancement in power saving verifies the advantage of the developed Q-model as a channel estimator compared to the MMSE technique.

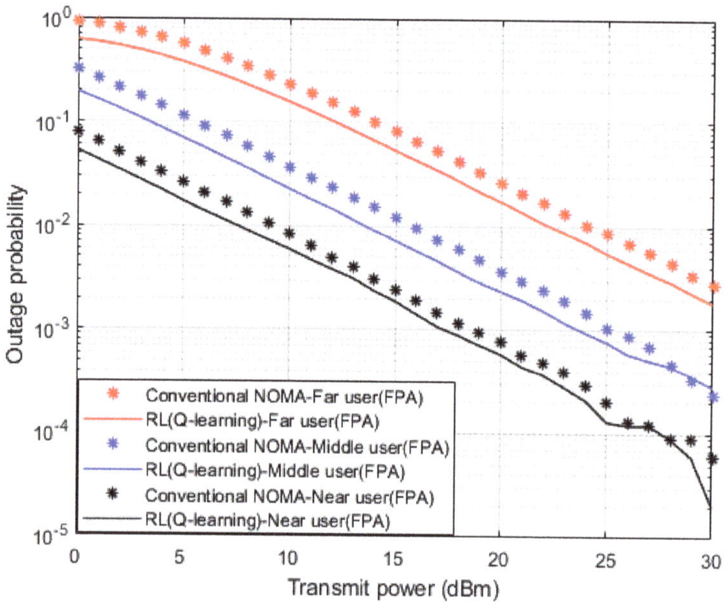

Figure 5. Outage Prob. vs. power (Q-learning & Conventional NOMA (MMSE)).

In Figure 6, we implement three baselines for comparisons: (1) standard minimum mean square error (MMSE) based channel prediction scheme [45]; (2) DL algorithm based on LSTM network for channel prediction applied in [17]; and RL based actor-critic procedure for channel prediction applied in [15]. This figure shows simulation results for the sum rate for all the user devices in the MISO-NOMA network versus applied power. Based on the simulation outcomes, it is evidently shown that the developed RL based Q-learning algorithm reveals superiority over standard MMSE procedure by 12 b/s/Hz approximately. Furthermore, the developed Q-learning scheme performs an enhancement over the DL based LSTM procedure presented in [17] by 2 b/s/Hz. For the third benchmark in [15], we generate the simulation environment according to the following: the actor and critic networks are both composed of two hidden layers with 400 and 300 nodes, respectively. The learning rate for actor and critic networks are 10^{-4} and 10^{-3} respectively. The discount factor γ is set to be 0.9 and has a buffer size of 10^5 [15]. Our developed RL based Q-learning procedure, shows superiority over the RL based actor-critic procedure at low power levels while starting from 23 dBm the actor-critic procedure starts showing some enhancement in terms of sum rates compared to the Q-learning process. These findings can validate that the developed Q-learning algorithm can be a competitive scheme compared to other algorithms that mainly depend on hidden layers to predict channel parameters.

Simulation outcomes for the sum rate against different number of users in the applied MISO-NOMA cell are illustrated in Figure 7, where the reference power is chosen to be 1 dBm. In addition to our proposed Q-learning algorithm, three distinct channel prediction methods are investigated as a benchmark comparison: (1) standard minimum mean square error (MMSE) based channel prediction scheme [45]; (2) DL algorithm based on LSTM network for channel prediction applied in [17]; and RL based actor-critic technique for channel estimation applied in [15]. As revealed from the results, our developed RL based Q-learning algorithm can achieve a substantial greater sum rate with respect to standard MMSE procedure, by at least 2 b/s/Hz. It can be observed that as the number of user devices in the cell is increasing, the suggested RL based Q-learning algorithm still shows dominance in accomplishing higher rates with respect to MMSE and DL based LSTM channel estimation methods. Similar to Figure 6, the RL actor-critic procedure applied in [15] is created in our MISO-NOMA environment with the following parameters: the actor and critic networks are both composed of two hidden layers with 400 and 300 nodes, respectively. The learning rate for actor and critic networks are 10^{-4} and 10^{-3} respectively. The discount factor γ is set to be 0.9 and has a buffer size of 10^5 [15]. As shown in the results, the developed Q-learning scheme is showing an advantage over the actor-critic scheme with up to 6 users in the cell. Then, the hidden layers feature in the actor-critic procedure starts producing some sort of improvement in the sum rates compared to the Q-learning algorithm while the number of user terminals in the cell is increasing. Overall, these outcomes reveal that dependability can be assured by the suggested Q-learning algorithm even when the user devices in the cell are increased. In addition, it is worth saying that while increasing the user devices in the system, the interference will also grow up, thus the sum rate could be degraded.

Figure 8 illustrates simulation outcomes for the achievable capacity for every device in the examined MISO-NOMA system when both the developed Q-learning algorithm and MMSE channel estimation procedures are implemented. The attained rate for near devices reveals substantial improvement by 10 b/s/Hz over far and middle users' rates. The superiority of the near user in terms of the achievable rate is anticipated, due to the stable channel situation for the near user compared to other devices in the system. Additionally, the suggested Q-learning algorithm still can deliver few visible improvements compared to the MMSE technique for far and middle users' environments, this slight improvement is associated with the interference and inadequate link conditions for far and middle devices.

In addition to the three baselines comparisons implemented in Figures 6 and 7, we also create and implement RL based State-Action-Reward-State-Action (SARSA) algorithm [48–50] in Figures 9–11 for the purpose of more investigations and benchmark comparisons. The features and parameters of the SARSA algorithm are adapted in order that the SARSA procedure can be used as a channel estimator and compare the results of SARSA algorithm with the results obtained based on our developed Q-learning algorithm.

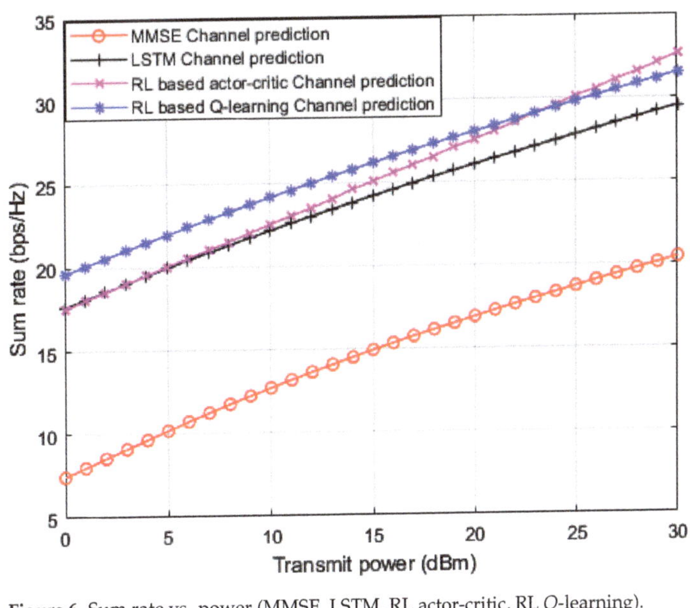

Figure 6. Sum rate vs. power (MMSE, LSTM, RL actor-critic, RL Q-learning).

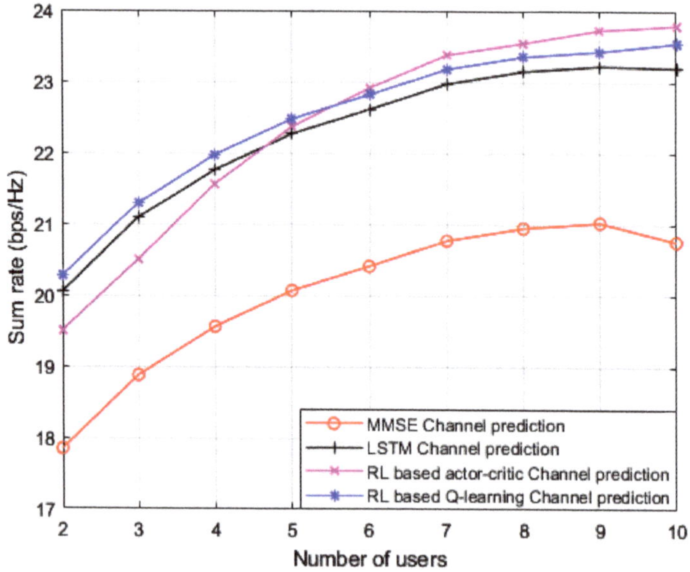

Figure 7. Sum rate vs. number of users (MMSE, LSTM, RL actor-critic, RL Q-learning).

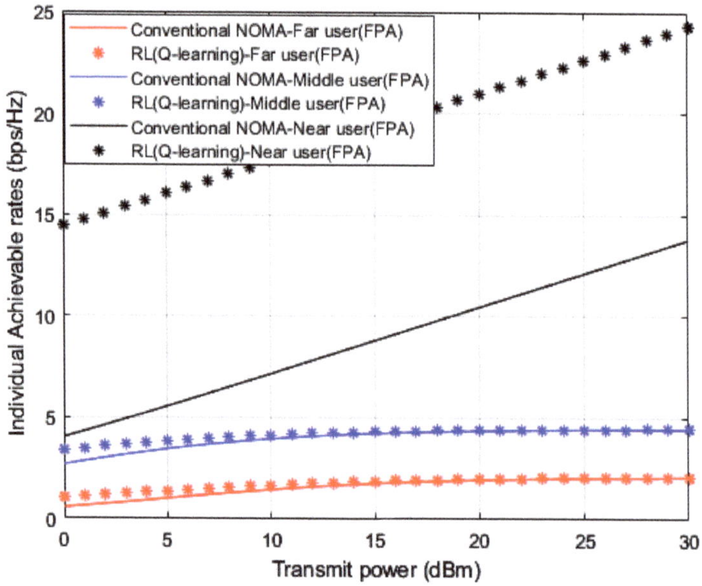

Figure 8. Individual rate vs. power (Q-learning, Conventional NOMA (MMSE)).

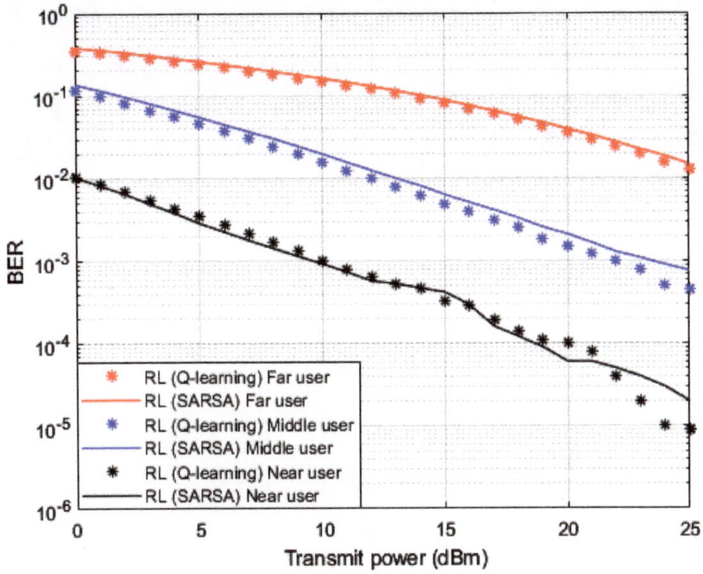

Figure 9. BER vs. power (*Q*-learning, SARSA).

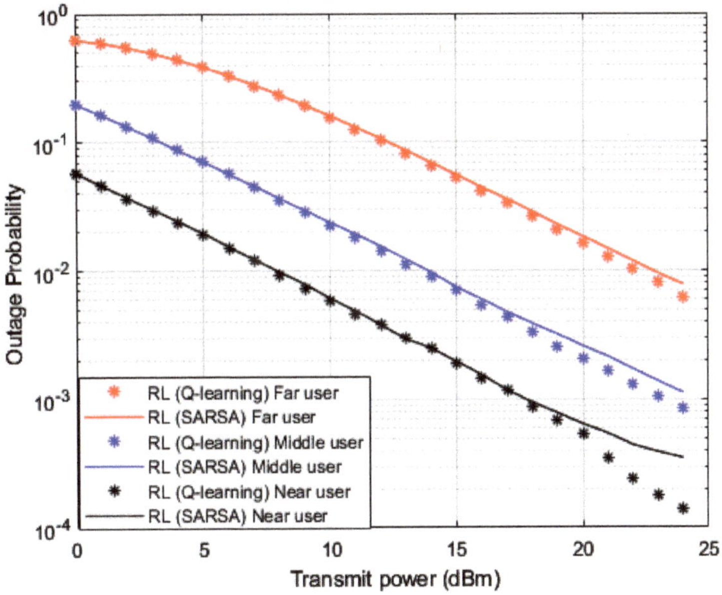

Figure 10. Outage Prob. vs. power (*Q*-learning, SARSA).

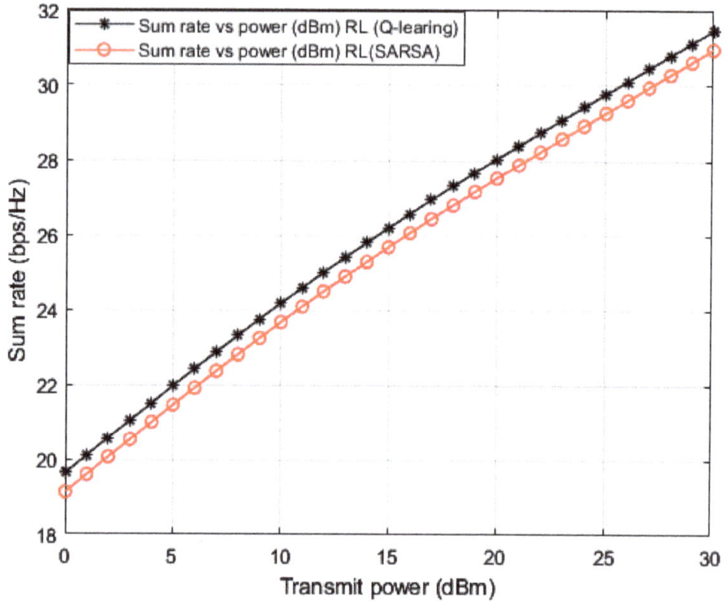

Figure 11. Sum rate vs. power (Q-learning, SARSA).

The Q-learning algorithm and SARSA algorithm are two efficient RL algorithms, they are both table-based procedures with a Q-table to record equivalent Q-values of each state-action pair. However, when the size of state space increases, it will need a considerable amount of memory. Similar to the Q-learning algorithm, the SARSA algorithm also has exploration and exploitation processes, and it also needs a Q-table to record $Q(s_t, a_t)$ value corresponding to state s_t and action a_t. Differently, the running steps of the SARSA algorithm are as follows. First, according to the action selection scheme, the gent at the current state s_t, will select the action a_t. Then, the agent gets an immediate reward $\boldsymbol{R}$ based on the corresponding $Q(s_t, a_t)$ value. Finally, s_t will transfer to s_{t+1} and the agent will choose the next action a_{t+1}. Hence, the SARSA algorithm is a bit different from the Q-learning procedure, where the Q-value in the SARSA method is updated based on the action a_t implemented by the agent at the state s_t. While in the Q-learning algorithm, the action with the greatest Q-value in the next state s_{t+1} is employed to update Q-table.

In Figures 9 and 10, where BER and outage probability metrics are simulated against transmitted power, both our developed Q-learning and SARSA algorithms show comparable performance. However, at high power levels, the suggested Q-learning algorithm shows little improvement compared to the SARSA algorithm, which may be justified that the Q agent deciding the greedy action, which is the action that provides the maximum Q-value for the state. More investigations for the comparison between SARSA and the developed Q-learning algorithms are shown in Figure 11. Sum rates versus applied power are simulated in Figure 11, and it is noticed that the suggested Q-learning scheme provides an advantage over the SARSA algorithm, and a power saving is recorded by 1–2 dB approximately.

The proposed Q-learning method and traditional MMSE technique will be further examined when the Rician channel is applied for the path between BS and each user device. Rician channel is a stochastic model for wireless transmission where the signal reaches the receiver device via various scattered paths. Figure 12, illustrate simulation outcomes for BER against power transmitted when the Rician fading channel is applied. In the Rician simulation environment, we assign parameter $K = 10$, where K is described as the fraction of the signal power of the line-of-sight path to the signal power of the remaining scattered components. In addition, maximum doppler shift = 100 and sample rate = 9600 Hz are used. Results for the Rician channel indicate that the Q-learning algorithm still can provide some sort of enhancement in decreasing the BER compared to the MMSE procedure. This slight improvement can be explained by the existence of a line of site component among BS and user terminal which can enhance the work of the MMSE procedure.

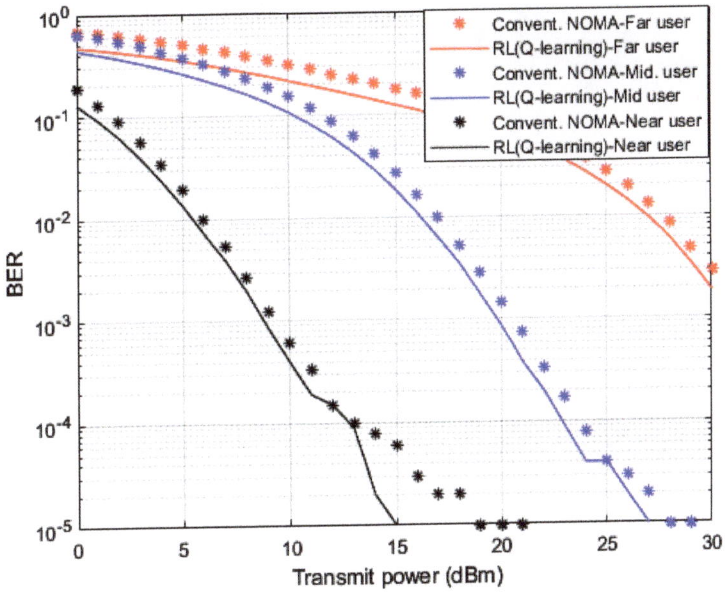

Figure 12. BER vs. Power (*Q*-learning, Conventional NOMA (MMSE)—Rician channel).

In Figure 13, two separate simulation setups are accomplished here to produce these results. In the first setup, the Fixed Power Allocation (FPA) structure is assigned for every user terminal in the MISO-NOMA cell. The second setup depends on the Optimized Power Structure (OPS) applied in accordance with the analytical power scheme that previously concluded in Section 4. FPA or OPS will be applied in conjunction with the suggested Q-learning algorithm as a channel estimator. Simulation outcomes in terms of BER indicate that far and middle users show the dominance of the OPS over the FPA. It can be noted that at specific BER values such as 10^{-2}, the achieved power saving by OPS policy is about 5 dBm for the far user, and 1–2 dBm approximately for the middle user. For near user results, the developed Q-learning algorithm jointly with the FPA scheme provide evident improvement in terms of BER over OPS, this could be clarified that for near device scenario, the stable channel condition provides more advantageous for the performance than the assigned power.

Outage probability results versus power are shown in Figure 14, where OPS and FPA schemes are also implemented. Both arrangements of OPS and FPA are implemented in conjunction with the proposed Q-learning algorithm as a channel estimator in the MISO-NOMA cell. Both far user and middle user results reveal an improvement in outage probability where a power reduction can be observed within 1–2 dBm when OPS is applied compared to the FPA scheme. On the other hand, near user with a Q-learning algorithm and FPA scenario shows a considerable outage improvement compared to the OPS case. A power reduction within 5 dBm is achieved when the FPA scheme is applied. These findings verify the results obtained for BER in Figure 13, which indicate that the FPA scheme is more adequate for user devices with high channel gains.

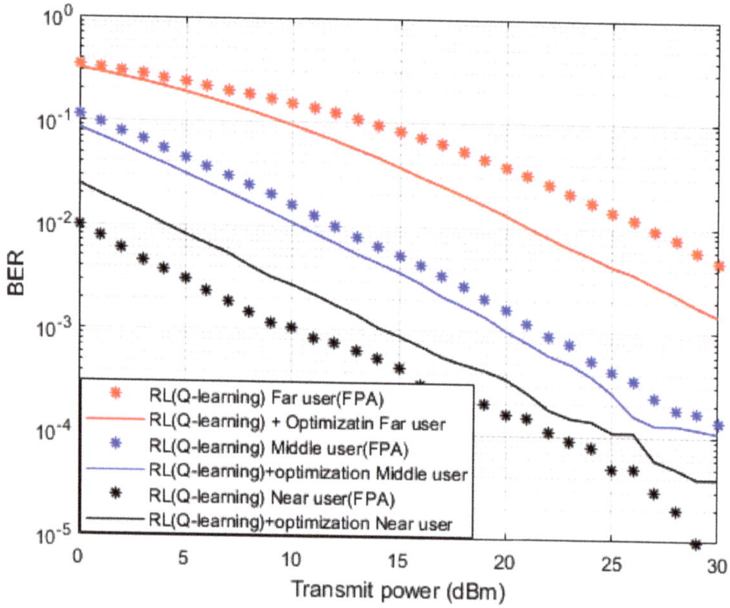

Figure 13. BER vs. Power (*Q*-learning, Optimization, FPA).

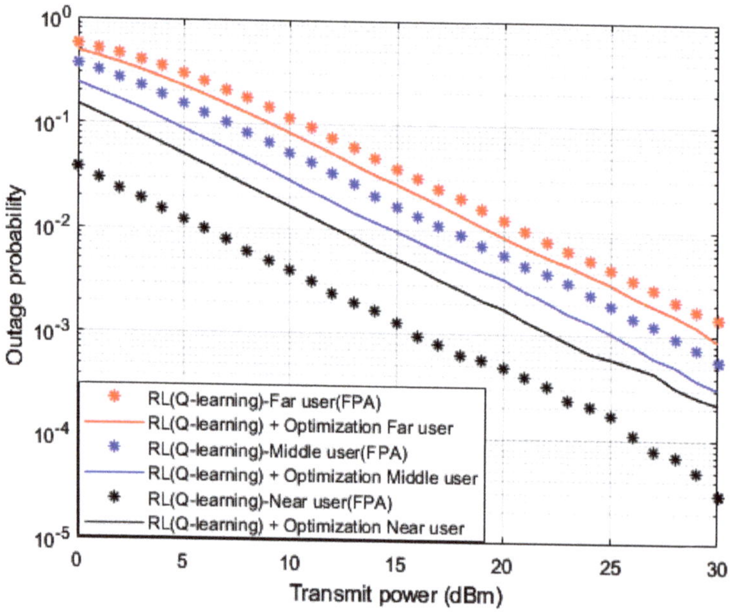

Figure 14. Outage Prob. vs. Power (*Q*-learning, Optimization, FPA).

In Figure 15, attainable rates for user devices are simulated against power transmitted when OPS and FPA schemes are applied in conjunction with the proposed Q-learning algorithm that is applied as a channel estimator. Results for far and middle devices point out that OPS provides 1 b/s/Hz improvement compared to the FPA scheme. This limited improvement might be clarified where the management of the power allocation for devices is not necessarily sufficient enough to alleviate the influence of interference particularly for far and middle devices that mainly experience unstable links environments. As expected, results for near user device reveal superiority in achieved rate with respect to middle and far devices with at least 10 b/s/Hz. Furthermore, the results for the near user with FPA indicate a noticeable improvement compared to OPS, which validates the results obtained in Figures 13 and 14.

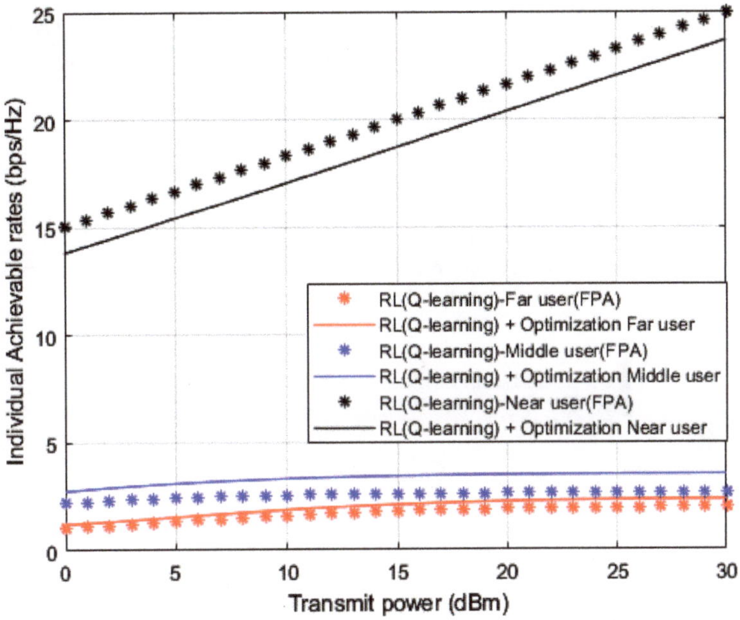

Figure 15. Individual rate vs. Power (*Q*-learning, Optimization, FPA).

In the end, we can further provide the analysis of the computational complexity as follows: The complexity of the reinforcement learning algorithm mainly depends on the size of the state space and the size of the action space [51]. According to [51], we can approximate the computational complexity of the Q-learning algorithm as $\mathcal{O}(SAH)$ per iteration, where S is the number of states, A is the number of actions, and H is the number of steps per episode. According to the state space and action space defined in our simulation environment, the amount of work per iteration can be approximated as $\mathcal{O}(NMK)$ [51,52], where N represents a number of antennas in BS, M represent a number of user devices in the cell, and K represents the size of channel coefficients. On the other hand, the computational complexity for the benchmark scheme implemented in [15], is described as follows: the sizes of the input layer, the first hidden layer, the second hidden layer, and the output layer for each network implemented in [15] is denoted as I, h_1, h_2, and U respectively. Thus, the total number of parameters in each network can be denoted as $\theta = I + h_1 + h_2 + U$, therefore, the complexity of this scheme regarding the channel estimation task can be approximated as $\mathcal{O}(MN_A(I + h_1 + h_2 + U)$ [15], where M represents the number of user terminals and N_A represent the number of antennas at BS. According to [53], the corresponding computational complexity for the traditional channel estimation method based MMSE can achieve a relatively low complexity, $\mathcal{O}(M^{2.37})$ [45,53], but, at the cost of performance degradation. Based on the aforementioned analysis, it can be shown that the complexity of the developed RL based Q-learning algorithm is competitive compared to other procedures.

10. Conclusions

In this study, the influence of adopting a developed RL based Q-learning algorithm to distinctly predict the channel parameters for every user device in the MISO-NOMA system is analyzed. In the developed Q-learning algorithm, the Q-model is created on the basis of the initialized channel statistics then updated based on the interaction between the Q-agent and the environment to maximize the received downlink sum rate and minimize the estimation loss. The efficacy of the developed Q-learning procedure is investigated by inspecting the performance of the proposed algorithm against different benchmark channel estimation schemes. The first benchmark scheme is based on standard MMSE procedure, the second approach is applying DL based LSTM network, the third scheme is implementing RL based actor-critic algorithm, and the fourth benchmark scheme is using RL based SARSA algorithm. In addition, the reliability of the proposed Q-learning procedure is validated by analyzing the behavior of the developed Q-learning algorithm in different fading channels. Furthermore, we provided a scenario that explores how the proposed channel prediction method based on Q-learning algorithm and the derived power allocation structure are both cooperatively employed for multiuser recognition in the MISO-NOMA network. Simulation results emphasized that dependability can be ensured by the developed Q-model even when the number of users in the cell is increased. Furthermore, the simulation outcomes in terms of BER, Outage probability, and individual user rate have demonstrated that the developed Q-learning algorithm for channel estimation jointly with an optimized power scheme can both realize consistent performance.

Author Contributions: Conceptualization, M.G.; Methodology, M.A.; Validation, A.A.; Formal analysis, M.G. and M.A.; Supervision, M.A. All authors have read and agreed to the published version of the manuscript.

Funding: This research received no external funding.

Institutional Review Board Statement: Not applicable.

Informed Consent Statement: Not applicable.

Data Availability Statement: Not applicable.

Conflicts of Interest: The authors declare no conflict of interest.

Abbreviations

The following abbreviations are used in this manuscript:

AWGN	Additive White Gaussian Noise
BER	bit error rate
BS	Base Station
CSI	Channel state information
DL	Deep Learning
DNN	Deep Neural Network
FPA	Fixed Power Allocation
OPS	Optimized Power structure
KKT	Karush-Kuhn–Tucker
LSTM	Long Short-Term Memory
LTE	Long Term Evolution
ML	Machine Learning
MSE	Mean Square Error
MMSE	Minimum Mean Square Error
MUD	Multiuser detection
PD-NOMA	Power Domain Non-Orthogonal Multiple Access
QoS	Quality of Service
SIC	Successive interference cancellation

MISO	Multi-input single-output
SARSA	State-Action-Reward-State-Action
RL	Reinforcement Learning

References

1. Dai, L.; Wang, B.; Ding, Z.; Wang, Z.; Chen, S.; Hanzo, L. A survey of non-orthogonal multiple access for 5G. *IEEE Commun. Surveys Tuts.* **2018**, *20*, 2294–2323.
2. Ding, Z.; Liu, Y.; Choi, J.; Sun, Q.; Elkashlan, M.; Chih-Lin, I.; Poor, H.V. Application of non-orthogonal multiple access in LTE and 5G networks. *IEEE Commun. Mag.* **2017**, *55*, 185–191. [CrossRef]
3. Gaballa, M.; Abbod, M.; Jameel, A. Power Optimization Analysis using Throughput Maximization in MISO Non-Orthogonal Multiple Access System. In Proceedings of the 2021 IEEE Globecom Workshops, Madrid, Spain, 7–11 December 2021.
4. Gaballa, M.; Abbod, M.; Albasman, M. Power Allocation & MRC Analysis for Single Input Multi Output Non-Orthogonal Multiple Access System. In Proceedings of the 2021 IEEE International Conferences on Internet of Things (iThings) and IEEE Green Computing & Communications (GreenCom) and IEEE Cyber, Physical & Social Computing (CPSCom) and IEEE Smart Data (SmartData) and IEEE Congress on Cybermatics (Cybermatics), Melbourne, Australia, 6–8 December 2021.
5. Wang, X.; Zhu, P.; Li, D.; Xu, Y.; You, X. Pilot-Assisted SIMO-NOMA Signal Detection with Learnable Successive Interference Cancellation. *IEEE Commun. Lett.* **2021**, *25*, 2385–2389. [CrossRef]
6. AbdelMoniem, M.; Gasser, S.M.; El-Mahallawy, M.S.; Fakhr, M.W.; Soliman, A. Enhanced NOMA system using adaptive coding and modulation based on LSTM neural network channel estimation. *Appl. Sci.* **2019**, *9*, 3022. [CrossRef]
7. Fu, Y.; Salaün, L.; Sung, C.W.; Chen, C.S. Distributed Power Allocation for the Downlink of a Two-Cell MISO-NOMA System. In Proceedings of the IEEE 87th Vehicular Technology Conference, Porto, Portugal, 3–6 June 2018; pp. 1–6.
8. Emir, A.; Kara, F.; Kaya, H.; Li, X. Deep learning-based flexible joint channel estimation and signal detection of multi-user OFDM-NOMA. *Phys. Commun.* **2021**, *48*, 101443. [CrossRef]
9. Jeon, Y.-S.; Li, J.; Tavangaran, N.; Poor, H.V. Data-aided channel estimator for MIMO systems via reinforcement learning. In Proceedings of the ICC 2020-2020 IEEE International Conference on Communications (ICC), Dublin, Ireland, 7–11 June 2020.
10. Kundu, N.K.; McKay, M.R. Channel Estimation for Reconfigurable Intelligent Surface Aided MISO Communications: From LMMSE to Deep Learning Solutions. *IEEE Open J. Commun. Soc.* **2021**, *2*, 471–487. [CrossRef]
11. Mthethwa, B.; Xu, H. Deep Learning-Based Wireless Channel Estimation for MIMO Uncoded Space-Time Labeling Diversity. *IEEE Access* **2020**, *8*, 224608–224620. [CrossRef]
12. Li, L.; Zhang, Z.; Yang, L. Influence of Autoencoder-Based Data Augmentation on Deep Learning-Based Wireless Communication. *IEEE Wirel. Commun. Lett.* **2021**, *10*, 2090–2093. [CrossRef]
13. Kim, D.; Jung, H.; Lee, I.H. Deep Learning-Based Power Control Scheme with Partial Channel Information in Overlay Device-to-Device Communication Systems. *IEEE Access* **2021**, *9*, 122125–122137. [CrossRef]
14. Zhang, T.; Mao, S. Energy-Efficient Power Control in Wireless Networks with Spatial Deep Neural Networks. *IEEE Trans. Cogn. Commun. Netw.* **2020**, *6*, 111–124. [CrossRef]
15. Chu, M.; Liu, A.; Jiang, C.; Lau, V.K.N.; Yang, T. Wireless Channel Prediction for Multi-user Physical Layer with Deep Reinforcement Learning. In Proceedings of the 2022 IEEE 95th Vehicular Technology Conference: (VTC2022-Spring), Helsinki, Finland, 19–22 June 2022.
16. Tan, J.; Liang, Y.C.; Zhang, L.; Feng, G. Deep Reinforcement Learning for Joint Channel Selection and Power Control in D2D Networks. *IEEE Trans. Wirel. Commun.* **2021**, *20*, 1363–1378. [CrossRef]
17. Gaballa, M.; Abbod, M.; Aldallal, A. Investigating the Combination of Deep Learning for Channel Estimation and Power Optimization in a Non-Orthogonal Multiple Access System. *Sensors* **2022**, *22*, 3666. [CrossRef] [PubMed]
18. Ma, W.; Qi, C.; Zhang, Z.; Cheng, J. Sparse Channel Estimation and Hybrid Precoding Using Deep Learning for Millimeter Wave Massive MIMO. *IEEE Trans. Commun.* **2020**, *68*, 2838–2849. [CrossRef]
19. Ghous, M.; Hassan, A.K.; Abbas, Z.H.; Abbas, G.; Hussien, A.; Baker, T. Cooperative Power-Domain NOMA Systems: An Overview. *Sensors* **2022**, *22*, 9652. [CrossRef]
20. Dai, L.; Wang, B.; Yuan, Y.; Han, S.; Chih-Lin, I.; Wang, Z. Non-orthogonal multiple access for 5G: Solutions, challenges, opportunities, and future research trends. *IEEE Commun. Mag.* **2015**, *53*, 74–81. [CrossRef]
21. Tang, Z.; Wang, J.; Wang, J.; Song, J. On the achievable rate region of NOMA under outage probability constraints. *IEEE Commun. Lett.* **2019**, *23*, 370–373. [CrossRef]
22. Yang, Z.; Xu, W.; Pan, C.; Pan, Y.; Chen, M. On the Optimality of Power Allocation for NOMA Downlinks With Individual QoS Constraints. *IEEE Commun. Lett.* **2017**, *21*, 1649–1652. [CrossRef]
23. Ding, Z.; Yang, Z.; Fan, P.; Poor, H.V. On the Performance of Non-Orthogonal Multiple Access in 5G Systems with Randomly Deployed Users. *IEEE Signal Process. Lett.* **2014**, *21*, 1501–1505. [CrossRef]
24. Gaballa, M.; Abbod, M.; Jameel, A.; Khaled, N. Throughput Maximization & Power Optimization Analysis in Non-Orthogonal Multiple Access System. In Proceedings of the 2021 IEEE 4th 5G World Forum, Montreal, QC, Canada, 13–15 October 2021.

25. Li, S.; Derakhshani, M.; Lambotharan, S. Outage-constrained robust power allocation for downlink MC-NOMA with imperfect SIC. In Proceedings of the IEEE International Conference on Communications (ICC), Kansas City, MO, USA, 20–24 May 2018; pp. 1–7.
26. Zhu, J.; Wang, J.; Huang, Y.; He, S.; You, X.; Yang, L. On Optimal Power Allocation for Downlink Non-Orthogonal Multiple Access Systems. *IEEE J. Sel. Areas Commun.* **2017**, *35*, 2744–2757. [CrossRef]
27. Boyd, S.; Vandenberghe, L. *Convex Optimization*; Cambridge University Press: Cambridge, UK, 2004.
28. Ben-Tal, A.; Nemirovski, A. *Lecture on Modern Convex Optimization: Analysis, Algorithms, and Engineering Applications*; MPS-SIAM: Philadelphia, PA, USA, 2018.
29. Sutton, R.S.; Barto, A.G. *Reinforcement Learning: An Introduction*; MIT Press: Cambridge, MA, USA, 2018.
30. Rothmann, M.; Porrmann, M. A Survey of Domain-Specific Architectures for Reinforcement Learning. *IEEE Access* **2022**, *10*, 13753–13767. [CrossRef]
31. Naeem, M.; De Pietro, G.; Coronato, A. Application of Reinforcement Learning and Deep Learning in Multiple-Input and Multiple-Output (MIMO) Systems. *Sensors* **2022**, *22*, 309. [CrossRef] [PubMed]
32. Amiri, R.; Almasi, M.A.; Andrews, J.G.; Mehrpouyan, H. Reinforcement learning for self-organization and power control of two-tier heterogeneous networks. *IEEE Trans. Wireless Commun.* **2019**, *18*, 3933–3947. [CrossRef]
33. Jeon, Y.S.; Lee, N.; Poor, H.V. Robust Data Detection for MIMO Systems with One-Bit ADCs: A Reinforcement Learning Approach. *IEEE Trans. Wirel. Commun.* **2019**, *19*, 1663–1676. [CrossRef]
34. Nie, J.; Haykin, S. A Q-learning-based dynamic channel assignment technique for mobile communication systems. *IEEE Trans. Veh. Technol.* **1999**, *48*, 1676–1687.
35. Ghavamzadeh, M.; Kappen, H.; Azar, M.; Munos, R. Speedy Q-Learning. In *Advances in Neural Information Processing Systems*; Shawe-Taylor, J., Zemel, R., Bartlett, P., Pereira, F., Weinberger, K.Q., Eds.; Curran Associates: Red Hook, NY, USA, 2011; Volume 24.
36. Zhai, Q.; Bolić, M.; Li, Y.; Cheng, W.; Liu, C. A Q-Learning-Based Resource Allocation for Downlink Non-Orthogonal Multiple Access Systems Considering QoS. *IEEE Access* **2021**, *9*, 72702–72711. [CrossRef]
37. Dong, M.; Tong, L.; Sadler, B.M. Optimal Insertion of Pilot Symbols for Transmissions over Time-Varying Flat Fading Channels. *IEEE Trans. Signal Process.* **2004**, *52*, 1403–1418. [CrossRef]
38. da Silva, M.V.; Montejo-Sánchez, S.; Souza, R.D.; Alves, H.; Abrão, T. D2D Assisted Q-Learning Random Access for NOMA-Based MTC Networks. *IEEE Access* **2022**, *10*, 30694–30706. [CrossRef]
39. da Silva, J.V.; Souza, R.D.; Alves, H.; Abrao, T. A NOMA-based Q-learning random access method for machine type communications. *IEEE Wireless Commun. Lett.* **2020**, *9*, 1720–1724. [CrossRef]
40. Mete, E.; Girici, T. Q-learning based scheduling with successive interference cancellation. *IEEE Access* **2020**, *8*, 172034–172042. [CrossRef]
41. Zhou, Y.; Zhou, F.; Wu, Y.; Hu, R.Q.; Wang, Y. Subcarrier Assignment Schemes Based on Q-Learning in Wideband Cognitive Radio Networks. *IEEE Trans. Veh. Technol.* **2019**, *69*, 1168–1172. [CrossRef]
42. Xi, B.; Lei, D. Q-Learning-Based Teaching-Learning Optimization for Distributed Two-Stage Hybrid Flow Shop Scheduling with Fuzzy Processing Time. *Complex Syst. Model. Simul.* **2022**, *2*, 113–129. [CrossRef]
43. Gaballa, M.; Abbod, M.; Aldallal, A. Deep Learning and Power Allocation Analysis in NOMA System. In Proceedings of the 2022 Thirteenth International Conference on Ubiquitous and Future Networks (ICUFN), Barcelona, Spain, 5–8 July 2022; pp. 196–201.
44. Youssef, M.-J.; Nour, C.A.; Lagrange, X.; Douillard, C. A Deep Q-Learning Bisection Approach for Power Allocation in Downlink NOMA Systems. *IEEE Commun. Lett.* **2022**, *26*, 316–320. [CrossRef]
45. Neumann, D.; Wiese, T.; Utschick, W. Learning the MMSE Channel Estimator. *IEEE Trans. Signal Process.* **2018**, *66*, 2905–2917. [CrossRef]
46. Cirik, A.C.; Balasubramanya, N.M.; Lampe, L.; Vos, G.; Bennett, S. Toward the Standardization of Grant-Free Operation and the Associated NOMA Strategies in 3GPP. *IEEE Commun. Stand. Mag.* **2019**, *3*, 60–66. [CrossRef]
47. *Recommendation ITU-R M.1225*; Guidelines for Evaluation of Radio Transmission Technologies for IMT-2000. International Telecommunication Union (ITU): Geneva, Switzerland, 1997. Available online: https://www.itu.int/dms_pubrec/itu-r/rec/m/R-REC-M.1225-0-199702-I!!PDF-E.pdf (accessed on 15 November 2022).
48. Ahsan, W.; Yi, W.; Qin, Z.; Liu, Y.; Nallanathan, A. Resource Allocation in Uplink NOMA-IoT Networks: A Reinforcement-Learning Approach. *IEEE Trans. Wirel. Commun.* **2021**, *20*, 5083–5098. [CrossRef]
49. Mu, X.; Zhao, X.; Liang, H. Power Allocation Based on Reinforcement Learning for MIMO System with Energy Harvesting. *IEEE Trans. Veh. Technol.* **2020**, *69*, 7622–7633. [CrossRef]
50. Jiang, H.; Gui, R.; Chen, Z.; Wu, L.; Dang, J.; Zhou, J. An Improved SARSA (λ) Reinforcement Learning Algorithm for Wire-less Communication Systems. *IEEE Access* **2019**, *7*, 115418–115427. [CrossRef]
51. Jin, C.; Allen-Zhu, Z.; Bubeck, S.; Jordan, M. Is Q-learning provably efficient? In Proceedings of the 31st International Conference on Neural Information Processing Systems, Long Beach, CA, USA, 4–9 December 2017; Curran Associates: Red Hook, NY, USA, 2017; pp. 4863–4873.

52. Li, Z.; Pan, S.; Qin, Y. Multiuser Scheduling Algorithm for 5 G IoT Systems Based on Reinforcement Learning. *IEEE Trans. Veh. Technol.* **2022**. [CrossRef]
53. Liao, Y.; Hua, Y.; Cai, Y. Deep learning based channel estimation algorithm for fast time-varying MIMO-OFDM systems. *IEEE Commun. Lett.* **2019**, *24*, 572–576. [CrossRef]

applied sciences

MDPI

Article

A REM Update Methodology Based on Clustering and Random Forest

Mario R. Camana , Carla E. Garcia, Taewoong Hwang and Insoo Koo *

Department of Electrical, Electronic and Computer Engineering, University of Ulsan, Ulsan 44610, Republic of Korea; mario_camana@hotmail.com (M.R.C.); carli.garcia27@hotmail.com (C.E.G.); yuio124@naver.com (T.H.)
* Correspondence: iskoo@ulsan.ac.kr

Abstract: In this paper, we propose a radio environment map (REM) update methodology based on clustering and machine learning for indoor coverage. We use real measurements collected by the TurtleBot3 mobile robot using the received signal strength indicator (RSSI) as a measure of link quality between transmitter and receiver. We propose a practical framework for timely updates to the REM for dynamic wireless communication environments where we need to deal with variations in physical element distributions, environmental factors, movements of people and devices, and so on. In the proposed approach, we first rely on a historical dataset from the area of interest, which is used to determine the number of clusters via the *K*-means algorithm. Next, we divide the samples from the historical dataset into clusters, and we train one random forest (RF) model with the corresponding historical data from each cluster. Then, when new data measurements are collected, these new samples are assigned to one cluster for a timely update of the RF model. Simulation results validate the superior performance of the proposed scheme, compared with several well-known ML algorithms and a baseline scheme without clustering.

Keywords: radio environment map (REM); random forest (RF); machine learning; clustering

Citation: Camana, M.R.; Garcia, C.E.; Hwang, T.; Koo, I. A REM Update Methodology Based on Clustering and Random Forest. *Appl. Sci.* **2023**, *13*, 5362. https://doi.org/10.3390/app13095362

Academic Editors: Yichuang Sun, Haeyoung Lee and Oluyomi Simpson

Received: 4 March 2023
Revised: 17 April 2023
Accepted: 23 April 2023
Published: 25 April 2023

1. Introduction

Recently, the dramatic increase in the number of wireless devices and the required data rates to satisfy QoS for users' applications have made it essential to guarantee high-quality communication links. In this regard, the multipath fading effect is one of the main considerations for wireless signals in indoor scenarios. The presence of obstacles such as walls, roofs, furniture, and so on leads to attenuation of the received signal and variations in the received power on the user side while impacting the coverage area of the wireless transmitter.

To solve this issue, a radio environment map (REM) is proposed to detect shadow areas with a poor quality signal. Construction of the REM is based on real measurements from the area of interest in order to characterize the behavior of the wireless environment by representing it as a temperature map [1]. The detection of shadow areas contributes to successful network planning and leads to an increase in the quality of communication for users. A REM can also provide useful information about the positions of wireless devices, interference levels, and other related information that can be used to improve the performance of the services by using resource allocation schemes [2]. In [3], the authors proposed construction of a REM based on machine learning (ML) methods, which showed better performance compared with conventional statistical models. A REM framework for IoT networks was presented in [1], where the authors considered ML algorithms to construct the REM. Their results proved the superior performance of ML learning models compared with conventional interpolation methods such as Kriging and nearest neighbor. However, the aforementioned work did not consider a methodology to update the REM in a timely manner.

One important aspect of REM management is the need for a timely update mechanism responding to wireless environment changes due to variations in physical element distributions, environmental factors, movements of people and devices, and so on. Updating radio maps can help network planners adjust the network configuration to compensate for these changes. Moreover, in indoor localization systems, updating radio maps can significantly improve the accuracy of localization, which is crucial for applications such as asset tracking and indoor navigation. On the other hand, clustering is considered a meaningful, energy-efficient technique because it contributes to reducing power consumption by organizing user nodes into groups denominated as clusters [4]. Clustering can be used in radio maps to group similar signal strength measurements into clusters based on their proximity to each other in physical space. This allows for a more efficient representation of the radio map since similar signal strength measurements can be processed together to create a more accurate estimate of signal strength in a particular area. Our objective is to update the REM in real time by using a combination of the *K*-means algorithm and a machine learning model. Specifically, the *K*-means algorithm constructs clusters and, for each cluster, we develop a ML model that is trained using actual data measurements specific to that cluster. As new data are collected, they are assigned to their respective cluster to update the corresponding ML model.

1.1. Related Work

For indoor localization systems, mobile crowd sensing has become a popular method for updating radio maps [5–12]. In [5–7], the authors propose an approach where mobile users generate reports at their current locations to update the radio maps. The radio map is based on a database, and the predicted location for a query point is obtained by using the nearest neighbors. The method to update the radio map is based on simply appending the new fingerprints into the database and removing those that are older than a certain period. In [8], the authors use an integrity check algorithm to determine whether to update the radio map used for indoor localization. In this method, several new fingerprints are accumulated before updating the received signal strength indicator (RSSI) value at a particular location. The update is performed by using the average value of the accumulated points to update the database.

In [9,10], the authors propose a method for adapting radio maps used for indoor localization to changes in the environment. The proposed method utilizes data gathered from typical wireless users' devices that were stationary at certain locations. RSSI values received from several reference locations are used to update the mapping of the RSSI value to a particular location. In [11], the decision of whether to update the radio map for each reference point is made by a periodic adaptive estimate algorithm. The authors represent the radio map using a matrix expression, and the update process involves updating the fingerprints in the radio map with the average of valid RSSI measurements collected from users. However, the success of the previous approaches heavily depends on the location and behavior of mobile users to gather new measurements, and their proposed algorithms require collecting new data for each location to update the predictive relationship between the RSS value and corresponding positions. In [12], the authors propose an updating method for signal maps based on Bayesian compressive sensing (BCS). Several crowdsourced samples are first mapped to the nearest reference point, and the BCS-based approach computes the signal change for each reference point based on the correlation between the new samples and reference points. In the aforementioned works, the radio map construction is based on collecting data and organizing it into a database, and the methodology to update the radio is focused on indoor localization approaches, where new measurements in each of the reference locations are required to correctly update the radio map. Unlike the previously mentioned methods, our proposed method only requires newly collected data in specific sectors to update a large area of interest. Additionally, the prediction of RSSI values is performed using powerful ML algorithms.

The authors in [13] propose a scheme for REM updates based on hypothesis testing, which is used to decide whether to update the REM each time. In [14], the authors propose a REM update mechanism based on Siamese neural networks to determine the level of similarity between an already-constructed REM and a new REM. However, previous work only considered a simple average of collected measurements to construct the REM. In [15], the authors propose a REM update scheme based on clustering and Gaussian process regression (GPR). They manually collected RSSI data with a smartphone in particular areas of the experiment room, and they use the collected dataset to predict the RSSI at specific reference points by using GPR with clustering. However, the objective of the GPR-based scheme is the RSSI prediction of only reference points to later be used for indoor localization, where a complete prediction of the whole area of interest is not obtained. Moreover, the authors only consider scenarios by varying the number of training samples and do not analyze the errors under changes in the wireless conditions such as in the presence of new obstacles or under the relocation of the AP. This is contrary to our proposed method, which is able to obtain the RSSI prediction of any point of the area of interest and represent it as a temperature map. Moreover, we extensively evaluate the proposed scheme by considering several comparative ML methods and several different scenarios, including the presence of obstacles and relocation of the AP.

1.2. Main Contributions

In this paper, we propose a REM update methodology based on clustering and ML algorithms. In particular, we consider the K-means algorithm to create K clusters in the area of interest, and the random forest (RF) algorithm is applied to predict the quality of the wireless signal for each cluster. The proposed REM update can be applied in different scenarios, such as technology industries, where several sensor nodes interact to carry out different tasks, such as maintenance or scheduled programming. Many times, it is difficult to collect measurements in the whole area to evaluate coverage prediction and build the REM because sensor nodes need to move frequently. Similarly, in hospitals, users storing measurements on every floor for coverage prediction may disturb other users, and measurement collecting tasks cannot be done as many people walk around the area. Therefore, the application of REM updating via clustering can reduce time-consuming tasks and optimize network resources. The main contributions of this paper are summarized as follows:

- We propose an efficient methodology to update a REM based on clustering and RF in a timely manner. In the proposed scheme, the K-means algorithm is applied to divide the area of interest in K clusters, where one RF model is deployed per cluster. The REM is constructed to cover every point within the area of interest, where the prediction of the RSSI values for each location is obtained by the corresponding RF model in each cluster.
- The RSSI measurements were collected by a mobile robot, which can reduce the risk of human error because the robot can be programmed to move in a controlled manner. This can help ensure that RSSI measurements are taken at consistent intervals and under consistent conditions, while improving the accuracy and reliability of the measurements. Moreover, mobile robots can operate autonomously, which can save time and resources compared to manual data collection methods.
- In the REM construction, to avoid abrupt changes in the border areas between the clusters, we propose a methodology that utilizes the weighted average of the RF model predictions from the two nearest centroids to determine the RSSI value of the points within the border areas. Moreover, when new measurements are available, only the RF models for clusters that have enough measurement samples are updated.
- We extensively evaluate the proposed scheme for different scenarios, including the presence of obstacles and relocating the AP, and we consider several comparative ML methods, including the case without clusters. Moreover, the computational complexity of the proposed scheme is analyzed along with the comparative schemes.

The simulation results demonstrated the superior performance of the proposed scheme compared to the baseline methods in effectively adapting to changes in the wireless environment. Moreover, the proposed approach requires only newly collected data in specific sectors to update a large area of interest.

The rest of this paper is organized as follows. Section 2 describes the measurement methodology. Section 3 presents the proposed REM updating framework. In Section 4, we present comparative models and an evaluation of simulation results. Finally, Section 5 concludes the paper.

2. Measurement Methodology

The scenario considered for indoor REM analysis was Room 302 in the Engineering Building at the University of Ulsan. The equipment used for the analysis included the TurtleBot3 [16] mobile robot, which is equipped with sensors, navigation systems, and decision-making algorithms that enable it to operate and complete tasks without human intervention. In particular, TurtleBot3 is composed of several parts, including an embedded controller, a light detection and ranging (LiDAR) sensor (laser distance sensor LDS-02), an inertial measurement unit (IMU) sensor, an encoder, a single board computer (SBC), and the robot operating system (ROS). A Raspberry Pi powers the SBC, which configures algorithms in a Linux environment and relies on ROS to enable communication between different processes. Meanwhile, the embedded controller, which utilizes OpenCR, is responsible for controlling the movement of the mobile robot through various sensors. The LiDAR sensor utilizes laser beams to calculate the distance to objects in its surroundings. By scanning the laser beams in a 360-degree horizontal field of view, LiDAR can generate a 2D point cloud map of the robot's surroundings. This map can be utilized for obstacle detection, mapping, and localization. Moreover, TurtleBot3 utilizes the odometry technique that employs sensor data from the encoders and the IMU to estimate the position and orientation of the robot in relation to its starting position. The odometry data are then combined with the LiDAR data to provide a more accurate and robust localization system, where the coordinates of the location points are presented in a 2D Cartesian coordinate system [17].

RSSI was utilized to estimate the link quality between the transmitter and receiver. RSSI is a measurement of the power present in a received radio signal, and it is commonly used in wireless communication systems like Wi-Fi, Bluetooth, and cellular networks to determine the signal strength received from a transmitter. The RSSI value is typically expressed in dBm (decibel-milliwatts) and reflects the power of the signal received by the receiver's antenna. A higher RSSI value indicates a stronger signal, while a lower RSSI value indicates a weaker signal. In our experiments, we used the built-in Wi-Fi module of the Raspberry Pi to collect RSSI data, along with the SSID, signal frequency, and link quality. The RSSI and location data were assembled in Ubuntu 22.04LTS through a shell script and synchronized based on timestamps.

The access point (AP) for the experiments was the IPTime N704M at 2.4 GHz. Figure 1 is a floor plan of the room used for the experiments, indicating the location of the AP. The floor plan presented in Figure 1 was obtained with the Hovermap HF1 [18], which is a mobile LiDAR 3D scanner, to map GPS-denied environments. Hovermap uses innovative simultaneous localization and mapping (SLAM) algorithms along with LiDAR data to produce 3D point clouds of the scanned area.

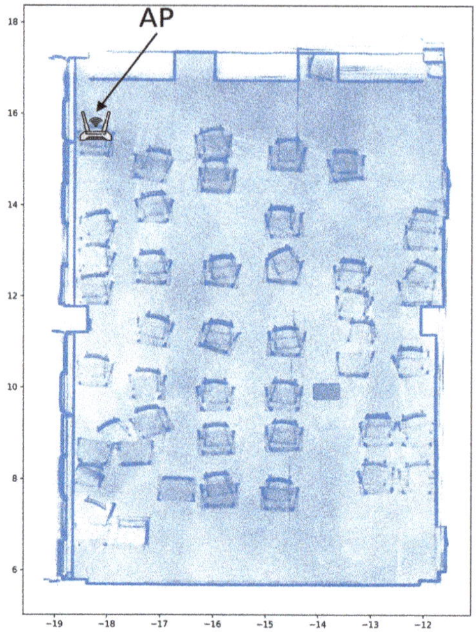

Figure 1. Floor plan of the room used for the experimental evaluations.

3. Proposed Approach for REM Updates

3.1. Overview

Figure 2 illustrates an overview of the proposed approach to REM updates based on clustering and the RF algorithm. First, assume the initial dataset is $D_H = \{\mathbf{z}_1, ..., \mathbf{z}_n, ..., \mathbf{z}_N\}$, where N is the total number of collected historical measurements and $\mathbf{z}_n = \{x_n, y_n, R_n\}$, in which x_n is the location in the x-axis, y_n is the location in the y-axis, and R_n is the RSSI value at position (x_n, y_n). The initial module of the proposed approach conducts clustering. In particular, we consider the K-means algorithm where N samples are separated into K groups. A detailed description of the K-means algorithm can be found in Section II of [19]. The K-means algorithm is applied to initial dataset D_H to obtain K cluster centers, which are used to assign each n-th sample to one of the K clusters based on the nearest centroid.

The second module, based on the RF algorithm, is where we train one RF model per cluster. Once each n-th measurement of dataset D_H has been assigned to a cluster, we have $D_k = \left\{\mathbf{z}_{1_k}, ..., \mathbf{z}_{m_k}, ..., \mathbf{z}_{M_k}\right\}$ at the k-th cluster with $k = 1, ..., K$ and $\mathbf{z}_{m_k} = \left\{x_{m_k}, y_{m_k}, R_{m_k}\right\}$. Then, K RF models are trained based on the corresponding dataset D_k. A detailed description of the RF algorithm is presented in Section 3.2.

Once we have the K-trained RF models, we construct a grid, $G = \{\mathbf{f}_1, ..., \mathbf{f}_h, ..., \mathbf{f}_H\}$, where H is the total number of samples in the grid, and $\mathbf{f}_h = \left\{x_h, y_h\right\}$. The grid is created to cover the whole area of interest and corresponds to the positions to be estimated by the RF algorithm. Next, each $\mathbf{f}_h$ is assigned to one cluster, and the corresponding RSSI value in that cluster is predicted by the RF model. Then, we construct the REM by using the location coordinates and by mapping the RSSI values to a specific color in a color map. A detailed description of REM construction is presented in Section 3.3.

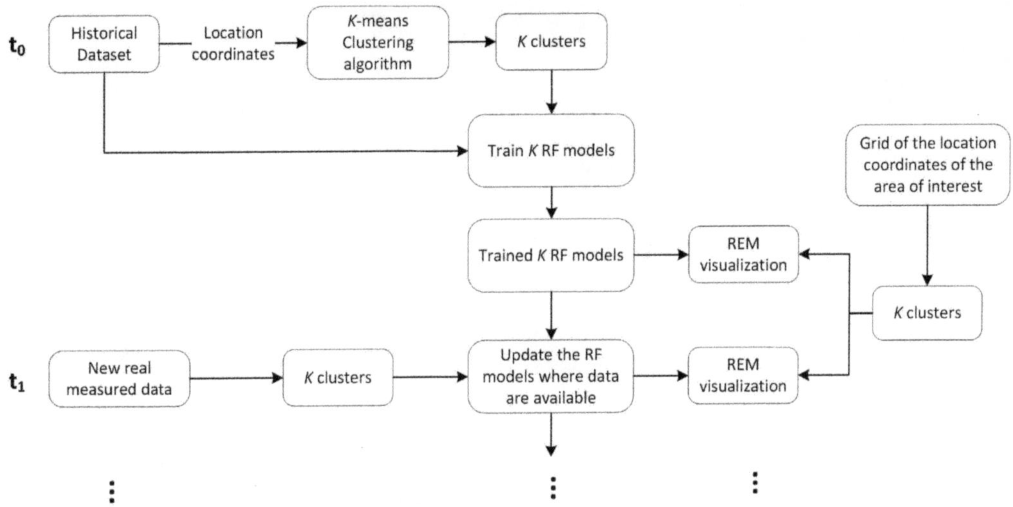

Figure 2. Proposed approach for REM update.

Next, we consider the REM update, denoted as $\mathbf{t}_1$ in Figure 2, based on newly collected measurements. Once we have trained the RF models based on the historical dataset, we apply the procedure to update the REM because the wireless environment constantly changes over time. For instance, in a smart warehouse, the positions of the products constantly change during the day. In general, the proposed scheme for the REM updates is based on training the RF models with newly collected data only in clusters with enough samples. We denote the newly collected dataset as $D_{\mathbf{t}_1} = \{\mathbf{z}_{1_{\mathrm{t}1}}, \ldots, \mathbf{z}_{n_{\mathrm{t}1}}, \ldots, \mathbf{z}_{N_{\mathrm{t}1}}\}$, where $N_{\mathrm{t}1}$ is the number of newly collected measurements at the time $\mathbf{t}_1$, and $\mathbf{z}_{n_{\mathrm{t}1}} = \{x_{n_{\mathrm{t}1}}, y_{n_{\mathrm{t}1}}, R_{n_{\mathrm{t}1}}\}$. First, we assign each $n_{\mathrm{t}1}$-th sample to one of the K clusters, which creates K possible datasets, each of them denoted $D_{k_{\mathrm{t}1}} = \left\{\mathbf{z}_1, \ldots, \mathbf{z}_{m_{k,\mathrm{t}1}}, \ldots, \mathbf{z}_{M_{k,\mathrm{t}1}}\right\}$. Since the newly collected measurements are not guaranteed to cover the whole area of interest, some datasets may contain no (or a very small number of) samples, which can lead to degradation when training the new RF model. Therefore, at the k-th cluster, to replace the k-th RF model with a new RF model trained with dataset $D_{k_{\mathrm{t}1}}$, we establish a condition calling for a minimum number of samples needed in the dataset to train a new k-th RF model: $|D_{k_{\mathrm{t}1}}| \geq N_{\mathrm{min}}$, where N_{min} is the minimum number of samples. If $|D_{k_{\mathrm{t}1}}| \geq N_{\mathrm{min}}$ is satisfied, the k-th RF model is trained based on the corresponding new dataset, $D_{k_{\mathrm{t}1}}$, replacing the old RF model in the k-th cluster. On the other hand, if dataset $D_{k_{\mathrm{t}1}}$ does not contain the minimum number of samples, N_{min}, the previous k-th RF model is used to predict the points in the k-th cluster. Finally, we update the REM by using the current K RF models to predict the RSSI values of grid G.

3.2. Random Forest

An RF regressor is a type of ensemble learning method composed of W decision tree regressors, where the predicted RF value corresponds to the average value of the predictions for each independent decision tree regressor. Figure 3 shows the structure of a decision tree composed of a root node as the starting point, split nodes where a split rule is applied, and leaf nodes.

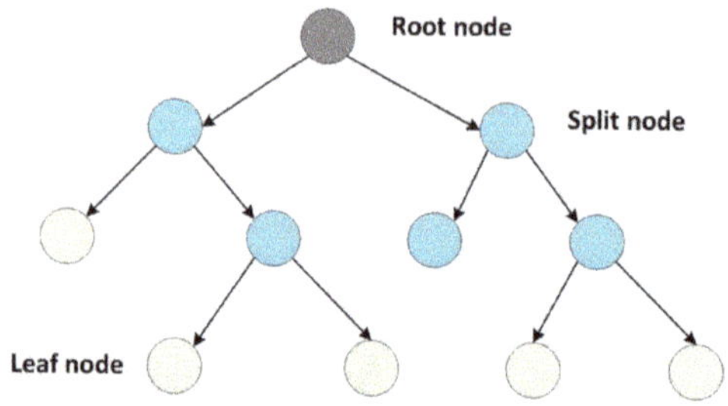

Figure 3. Example of a decision tree regressor.

We denote the training dataset of the RF algorithm as $D_{RF} = \{\mathbf{z}_1, \ldots, \mathbf{z}_m, \ldots, \mathbf{z}_M\}$, where M is the total number of training samples, and $\mathbf{z}_m = \{x_m, y_m, R_m\}$. The features correspond to position (x_m, y_m), and R_m is the target RSSI value at position (x_m, y_m). First, the RF algorithm creates W datasets, $\{D_{RF,w}\}$, from the original dataset, D_{RF}, with the bootstrap aggregation technique. Then, each w-th decision tree regressor is trained with corresponding dataset $D_{RF,w}$.

In each decision tree regressor, dataset $D_{RF,w}$ is used to select a split rule in each split node until reaching a leaf node. Let us define $D^b_{RF,w}$ as the subset of training samples at the b-th split node of the w-th decision tree regressor. Then, the best-split rule to be used at the b-th split node is determined with dataset $D^b_{RF,w}$. A candidate split rule is defined as follows:

$$s^{b,w}_{f,\alpha}(z_i) = \begin{cases} 1, & \text{if } f_{z_i} > \alpha \\ 0, & otherwise, \end{cases} \tag{1}$$

where f_{z_i} is the value of feature $f \in \{x_i, y_i\}$ in sample z_i, with $z_i \in D^b_{RF,w}$ during the training procedure and α representing a threshold. In the training process, at the b-th split node, a small pool of random features is selected, and a set of possible thresholds for each feature is evaluated to select the best-split rule based on the lowest mean squared error (MSE). Split rule (1) divides the samples in dataset $D^b_{RF,w}$ into two groups: $DR^b_{RF,w}$ contains the training samples satisfying the b-th split rule, and $DL^b_{RF,w}$ contains the rest. The MSE of each candidate split rule is evaluated as follows [3,20]:

$$\text{MSE}\left(s^{b,w}_{f,\alpha}\right) = \frac{1}{\left|DR^b_{RF,w}\right|} \sum_{i \in DR^b_{RF,w}} \left(R_i - \hat{R}^{DR}_b\right)^2 + \frac{1}{\left|DL^b_{RF,w}\right|} \sum_{i \in DL^b_{RF,w}} \left(R_i - \hat{R}^{DL}_b\right)^2, \tag{2}$$

where $\left|DR^b_{RF,w}\right|$ represents the number of samples in dataset $DR^b_{RF,w}$, $\left|DL^b_{RF,w}\right|$ represents the number of samples in dataset $DL^b_{RF,w}$, R_i is the true target value of the i-th sample, while $\hat{R}^{DR}_b$ and $\hat{R}^{DL}_b$ are the predicted values based on the average RSSI of the training samples in datasets $DR^b_{RF,w}$ and $DL^b_{RF,w}$, respectively, when the candidate split test, $s^{b,w}_{f,\alpha}$, is applied. Then, the candidate split test with the lowest MSE is selected as the best-split rule for the b-th split node. The aforementioned procedure is repeated in the next split node until reaching a leaf node, which is determined by the minimum number of training samples required to split a node, $m_{\min}$. Finally, the RSSI value associated with the leaf node corresponds to the average of the RSSI values of the training samples in that leaf node.

Once the RF model is successfully trained, the test sample goes to each w-th decision tree regressor and evaluates the split rule at each split node to continue to the next split node. The process finishes when reaching the leaf node, where the associated value of that leaf node determines the predicted RSSI value for the sample in the w-th decision tree regressor. The final prediction of the RF model is the average of the RSSI values predicted by all the W decision tree regressors.

3.3. REM Construction

Given K cluster centers, the K-trained RF models, and grid G, we construct the REM for the area of interest. First, each $\mathbf{f}_h$ sample from grid G is assigned to one cluster among the K clusters available. Next, the border samples in the area of the intersection of the clusters are determined to avoid a hard change in the REM between clusters. Then, P neighbor samples are obtained for each sample in the border area based on Euclidean distance. Next, we group all the P neighbor samples of the border areas, and we remove duplicate points to create grid G_{border}. Finally, we remove the samples of G_{border} from original grid G to obtain grid G_{in}. Therefore, we have divided G into grids of the border areas, G_{border}, and a grid with the internal points in each cluster, G_{in}.

To obtain the RSSI value for the points of grid G_{in}, we assign each sample from G_{in} to one of the K clusters, and we use the corresponding RF model to predict the RSSI value. In the samples in grid G_{border}, we determine the two nearest cluster centers for each sample, and the predicted RSSI value is the weighted average of the corresponding two RF models. In detail, let us consider sample $\mathbf{f}_{h,border}$ from grid G_{border}. First, we evaluate the distance from $\mathbf{f}_{h,border}$ to all K cluster centers, selecting two clusters with the nearest Euclidean distance, denoted as $d_{\mathbf{f}_{h,boder},C_a}$ and $d_{\mathbf{f}_{h,boder},C_b}$, where $d_{\mathbf{f}_{h,boder},C_a} < d_{\mathbf{f}_{h,boder},C_b}$ and where C_a represents one of the K available clusters. Then, we use the RF model of cluster C_a to predict the first RSSI value of $\mathbf{f}_{h,border}$, denoted as $R_{h,border,C_a}$, and the RF model of cluster C_b predicts the second RSSI value of $\mathbf{f}_{h,border}$, denoted as $R_{h,border,C_b}$. Finally, the RSSI value for $\mathbf{f}_{h,border}$ is determined with the following weighted average:

$$R_{h,border} = \frac{\left(\frac{d_{\mathbf{f}_{h,boder},C_b}}{d_{\mathbf{f}_{h,boder},C_a}}\right)^5 R_{h,border,C_a} + \left(\frac{d_{\mathbf{f}_{h,boder},C_a}}{d_{\mathbf{f}_{h,boder},C_b}}\right)^5 R_{h,border,C_b}}{\left(\frac{d_{\mathbf{f}_{h,boder},C_b}}{d_{\mathbf{f}_{h,boder},C_a}}\right)^5 + \left(\frac{d_{\mathbf{f}_{h,boder},C_a}}{d_{\mathbf{f}_{h,boder},C_b}}\right)^5}. \quad (3)$$

Once all RSSI values are obtained for all samples in grids G_{in} and G_{border}, we use Matplotlib in Python to create the REM based on the color map. Moreover, we superpose the SLAM map of the room by carefully adjusting the opacity of the REM with the alpha parameter from Matplotlib.

4. Evaluation

4.1. Historical Dataset and Comparative Models

Initial dataset D_H is composed of measurements collected inside Room 302, illustrated in Figure 1, with a total of $N = 1160$ samples covering the whole room. We considered three different error metrics to evaluate the performance of the proposed scheme: mean absolute percentage error (MAPE), root mean square error (RMSE), and R2 score. As comparative schemes, we considered the support vector regression (SVR) algorithm [21], multilayer perceptron (MLP) [22], and the AdaBoost regressor [23]. Moreover, we obtained error results by considering different numbers of clusters. The RF model and the AdaBoost regressor were trained with 200 decision tree regressors; the comparative SVR algorithm considered the amount of regularization, $C = 1000$, and the radial basis function (RBF) kernel; and MLP had three hidden layers with 100 hidden units per layer from using a rectified linear unit (ReLU) activation function. The computer used for the simulations had an AMD Ryzen 9 5900X 12-Core processor and 48 GB of RAM.

Table 1 presents the aforementioned error metrics of the proposed scheme and several comparative approaches by using 5-fold cross-validation with historical dataset D_H, where the presented results are averaged over several independent simulations. We can see that the RF-based scheme achieved the fewest errors among the comparative methods, and the AdaBoost algorithm was the second-best scheme. In addition, the impact from the number of clusters in the RF algorithm was very small since the same error rate can be obtained by using one cluster or four clusters. Note that Table 1 analyzes the error results on dataset D_H, where measurements for the whole area of interest are available. However, in Section 4.2, we analyze the performance from different numbers of clusters in a realistic case when only partial data from the area of interest are available.

Table 1. Error evaluation of the dataset D_H.

		MAPE		
Model for Prediction	**4 Clusters**	**3 Clusters**	**2 Clusters**	**1 Cluster**
RF	1.511%	1.513%	1.507%	1.515%
SVR	3.565%	3.718%	3.828%	4.504%
MLP	4.325%	4.174%	4.254%	4.835%
AdaBoost	3.001%	3.178%	3.603%	4.064%
		RMSE		
Model for Prediction	**4 Clusters**	**3 Clusters**	**2 Clusters**	**1 Cluster**
RF	1.295	1.290	1.290	1.285
SVR	2.418	2.486	2.537	2.793
MLP	2.648	2.556	2.602	2.977
AdaBoost	1.810	1.912	2.150	2.409
		R2 Score		
Model for Prediction	**4 Clusters**	**3 Clusters**	**2 Clusters**	**1 Cluster**
RF	0.948	0.949	0.949	0.949
SVR	0.819	0.809	0.801	0.760
MLP	0.783	0.799	0.791	0.727
AdaBoost	0.899	0.887	0.857	0.821

Figure 4 illustrates the REMs obtained for the historical dataset by using 4 clusters with the RF model, SVR, and AdaBoost. We can see that the REMs obtained with the RF algorithm and AdaBoost have similar patterns since both algorithms use the decision tree regressor as the base estimator. However, the REM obtained with RF is considered the most realistic because it has the lowest error, as we can observe in Table 1.

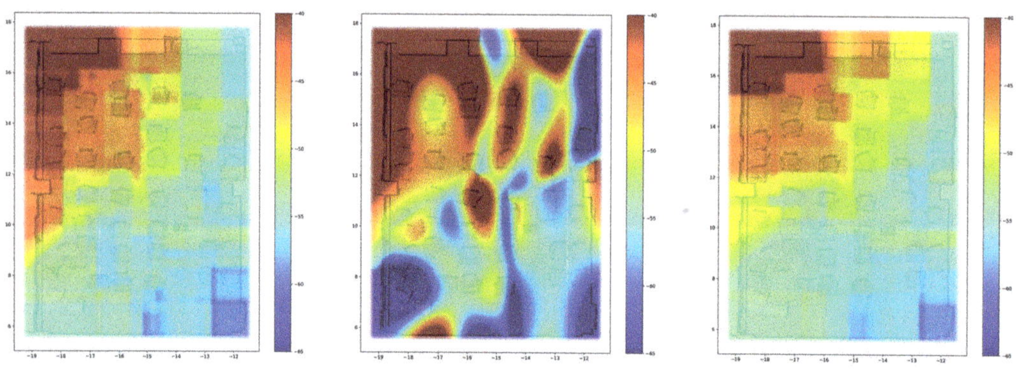

(a) (b) (c)

Figure 4. REM for the initial measurements by considering four clusters. (**a**) RF; (**b**) SVR; (**c**) AdaBoost.

4.2. REM Update Evaluation

In this subsection, we present the performance of the proposed scheme for REM updates. In particular, we considered a practical scenario where obstacles are added around the AP, which degrades coverage of the area of interest. Figure 5 shows the REM obtained by using RF in the ideal case where we can collect data measurements of the whole room after adding the obstacles. The data collected for this ideal case were used as testing data to analyze the error in the proposed scheme compared with the baseline method. By comparing Figures 4a and 5, we can see that adding obstacles around the AP degraded the coverage, particularly in the quality of the received signal in the bottom area of the room, compared with no obstacles.

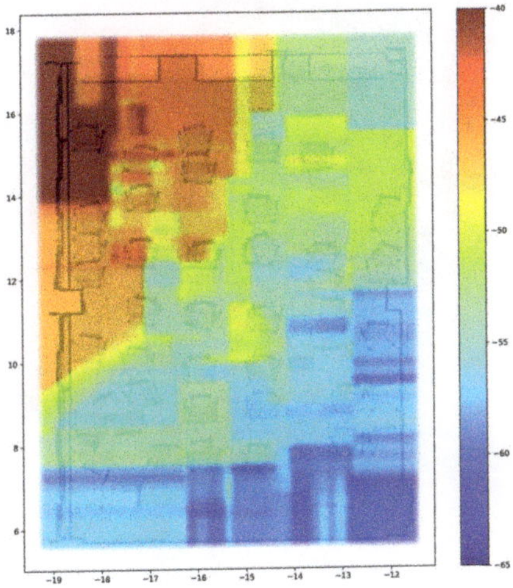

Figure 5. Target REM after changing the physical environment of the room by using RF.

Next, we analyzed a scenario where newly collected data were obtained only from a partial area of the room, which is a realistic assumption since it is not always possible to measure the whole area of interest each time. We considered three sets of newly collected data (at times $\mathbf{t}_1$, $\mathbf{t}_2$, and $\mathbf{t}_3$). It was assumed that during $\mathbf{t}_1$, $\mathbf{t}_2$, and $\mathbf{t}_3$, coverage by the Wi-Fi signal was stable in the whole room. Figure 6 shows the newly collected measurements at the three different times.

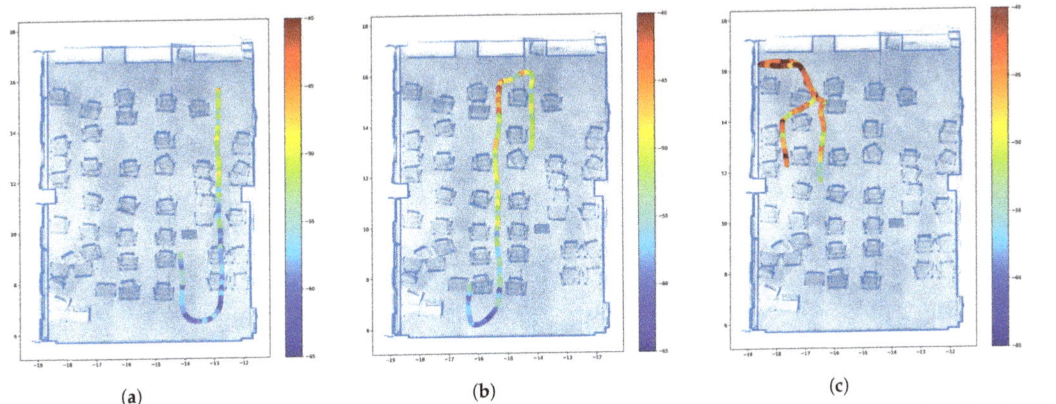

Figure 6. The three areas considered for newly collected measurements. (**a**) Data collected at time $\mathbf{t}_1$; (**b**) Data collected at time $\mathbf{t}_2$; (**c**) Data collected at time $\mathbf{t}_3$.

Figure 7 illustrates the proposed REM update mechanism for the three sets of newly collected data that followed the scheme in Figure 2. In detail, data collected at time $\mathbf{t}_1$ were used to train the RF model only for the corresponding clusters, while the remaining clusters used the previous RF model. We observe in Figure 7a that only the right area of the room was updated to the new scenario, which can be confirmed from Figure 5. On the other hand, the left part of the room was still predicted as having the historical measurements.

The reason is that, at time $\mathbf{t}_1$, there were no data available to update the left part of the REM. Next, at time $\mathbf{t}_2$, we could update the bottom area of the room while the upper left area was still not updated due to the lack of measurements in that area. Finally, at time $\mathbf{t}_3$, we updated the upper left area, leading to a complete update of the whole room, compared with Figure 5.

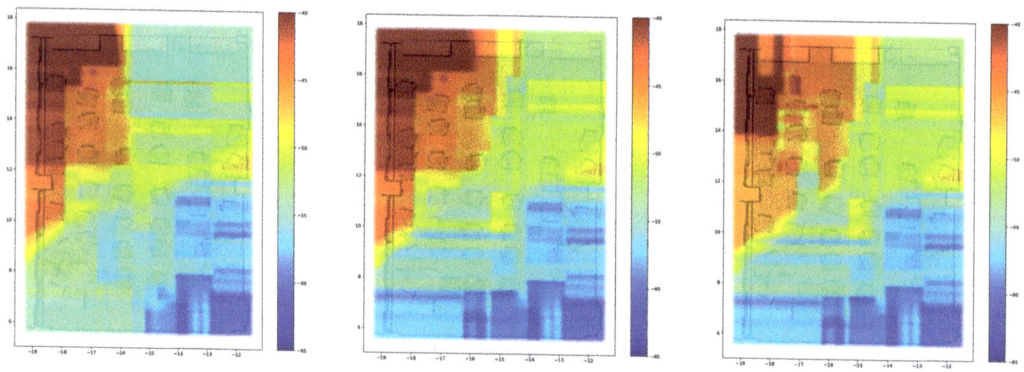

(**a**) (**b**) (**c**)

Figure 7. The proposed REM update mechanism considering four clusters. (**a**) The REM at time $\mathbf{t}_1$; (**b**) The REM at time $\mathbf{t}_2$; (**c**) The REM at time $\mathbf{t}_3$.

Figure 8 shows a baseline REM update mechanism without clustering for the 3 newly collected datasets. In this baseline scheme, the newly collected data were used to train an RF model for the whole room without using clustering. First, at time $\mathbf{t}_1$, the newly collected measurements were not enough to show the position of the AP; i.e., the upper left area was poorly predicted, leading to a high error. Next, at time $\mathbf{t}_2$, the collected measurements at the center of the room provided some small insight into coverage of the room, but the obtained REM is not ideal when we compare it with Figure 5. Finally, at time $\mathbf{t}_3$, the measurements collected in the area close to the AP make the RF model trained with those data overestimate the coverage of the room, leading to high error and an absence of shadow areas.

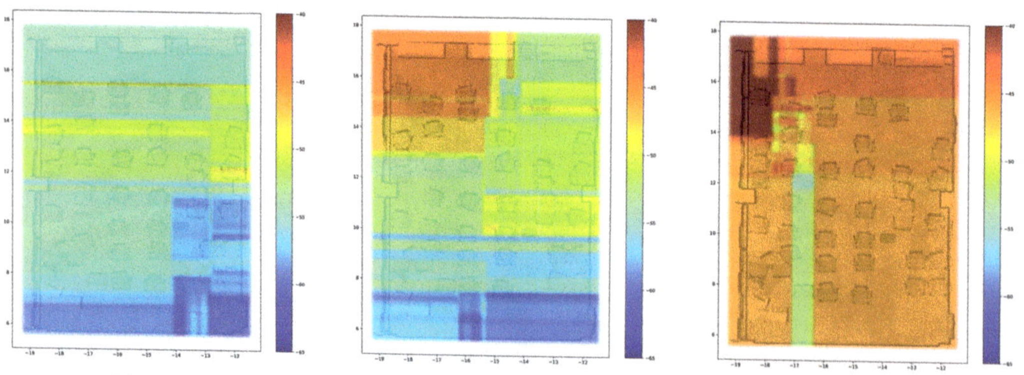

(**a**) (**b**) (**c**)

Figure 8. The baseline REM update mechanism without clustering. (**a**) The REM at time $\mathbf{t}_1$; (**b**) The REM at time $\mathbf{t}_2$; (**c**) The REM at time $\mathbf{t}_3$.

Table 2 shows error evaluations for the 3 times when data were newly collected. In the four clusters, we can see that the error lessened as time passed because new clusters could be updated as new measurement data were collected. As presented in Figure 7, at time $\mathbf{t}_3$, we could update the whole room, leading to a significant reduction in the error metrics. In the case of the baseline method without clustering, the error was not reduced as time passed, and the errors obtained at each time significantly depended on the area where the new data were collected. For instance, we observe that the baseline method without clustering obtained the lowest error at time $\mathbf{t}_2$ for most of the compared methods because the data were collected in the middle of the room, which can provide a small insight into the coverage of the area of interest. The above-mentioned observation is consistent with the results shown in Figure 8 where, among the 3 time cases, the REM at time $\mathbf{t}_2$ is the closest representation to the target REM illustrated in Figure 5.

Table 2. Error evaluations from newly collected measurements.

MAPE		**4 Clusters**			**Without Clustering**	
Model for Prediction	t_1	t_2	t_3	t_1	t_2	t_3
RF	6.10%	5.03%	1.92%	10.52%	6.60%	12.40%
SVR	7.90%	6.01%	4.90%	15.89%	8.74%	15.54%
MLP	12.98%	5.76%	4.55%	27.27%	18.56%	7.97%
AdaBoost	6.40%	5.49%	3.82%	13.37%	6.78%	9.03%
RMSE		**4 Clusters**			**Without Clustering**	
Model for Prediction	t_1	t_2	t_3	t_1	t_2	t_3
RF	3.731	3.401	2.069	6.100	4.019	8.167
SVR	4.952	3.823	3.439	8.996	5.174	9.514
MLP	9.959	3.538	3.018	15.811	11.235	5.193
AdaBoost	3.952	3.422	2.527	7.513	3.960	6.064
R2 Score		**4 Clusters**			**Without Clustering**	
Model for prediction	t_1	t_2	t_3	t_1	t_2	t_3
RF	0.646	0.735	0.842	0.055	0.630	−1.463
SVR	0.563	0.598	0.440	−1.056	0.387	−2.342
MLP	−1.520	0.714	0.664	−5.351	-1.888	0.004
AdaBoost	0.603	0.732	0.764	−0.434	0.641	−0.358

Table 3 presents the computational time for the training and prediction phases of the historical dataset and the newly collected data using the proposed approach and the baseline methods. The computational time of the K-means algorithm to select the best centroids based on the historical dataset D_H (consisting of 1160 samples) is 0.097 s, and the computational time to predict the nearest cluster for the total number of samples is 0.013 s. The number of samples in the newly collected datasets at times $\mathbf{t}_1$, $\mathbf{t}_2$, and $\mathbf{t}_3$ are 1800, 1800, and 2400, respectively. In Table 3, we observe that the computational time of the historical dataset is higher than the time when updating the model at time $\mathbf{t}_1$ because the historical dataset is used to train all the ML models for each cluster, while the data at time $\mathbf{t}_1$ are only used to update the ML model of the corresponding cluster. Moreover, we see that the computational time in the training phase increased as the number of samples increased because, in general, more samples mean more computations and larger memory requirements, leading to longer processing times. The higher computational time of the RF with four clusters compared to the RF without clustering is due to the need to split the samples into their respective clusters and the requirement of training four different instances of RF.

In the grid prediction phase, the computational time corresponds to the time to split the samples of the grid into their respective clusters and the time to predict the total number of points in the grid with their respective ML algorithm. For instance, in the case of RF with 4 clusters and a 100×100 grid, a total of 10,000 samples were assigned to their respective clusters, and the corresponding RF model predicted the RSSI value for each sample, resulting in a total time of 0.112 s. Moreover, we see that the proposed RF algorithm achieved the second lowest computational time among the compared ML methods with clustering. Note that MLP has the lowest computational time for the grid prediction with clustering but also has the highest error in the prediction as presented in Table 2.

Table 3. Computational time for the training and prediction phases.

	Training Time (s)			
Model for Prediction	**Historical Data D_H**	**Data at t_1**	**Data at t_2**	**Data at t_3**
RF 4 clusters	0.539	0.366	0.375	0.418
RF without clusters	0.213	0.215	0.234	0.341
SVR 4 clusters	0.111	0.205	0.231	0.374
MLP 4 clusters	1.14	0.669	1.48	1.13
AdaBoost 4 clusters	0.502	0.234	0.369	0.301
	Grid Prediction Time (s)			
Model for Prediction	**100 × 100 grid**		**300 × 300 grid**	
RF 4 clusters	0.112		0.285	
RF without clusters	0.04		0.251	
SVR 4 clusters	0.195		1.25	
MLP 4 clusters	0.038		0.218	
AdaBoost 4 clusters	0.134		0.625	

Next, we consider a second scenario by moving the location of the AP. In particular, the AP is relocated to the other corner of the room, which significantly changes the coverage condition in the area of interest. Figure 9 illustrates the REM that was obtained with RF under ideal conditions when data measurements for the entire room were collected after the relocation of the AP. The colormap limits for the current axes have been adjusted for better data visualization. By comparing Figures 4a and 9, we can observe the significant changes in the wireless conditions of the area of interest and the importance of timely REM updates.

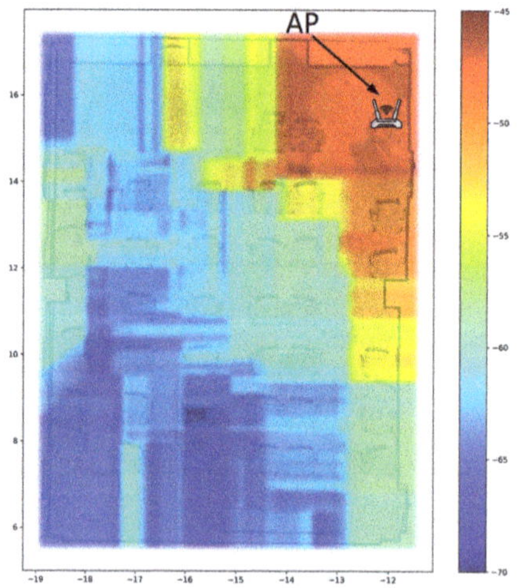

Figure 9. Target REM after the relocation of the AP by using RF.

Similar to the previously considered scenario, new data were acquired at different times as illustrated in Figure 10. Then, the proposed update methodology was performed based on the three datasets collected at times t_1, t_2 and t_3. Figure 11 shows the results of the proposed update methodology for the three sets of newly collected data. In Figure 11a, we can see that only the left area of the room was updated to the actual wireless coverage condition because the newly collected data at time $\mathbf{t}_1$ do not have information about the remaining areas of the room. Next, by using the newly collected data at time $\mathbf{t}_2$, the upper area of the room was updated, while the bottom right area remained unaltered because of insufficient measurements. Finally, the entire room was updated with the newly collected data at time $\mathbf{t}_3$, as evidenced by comparing it to Figure 9.

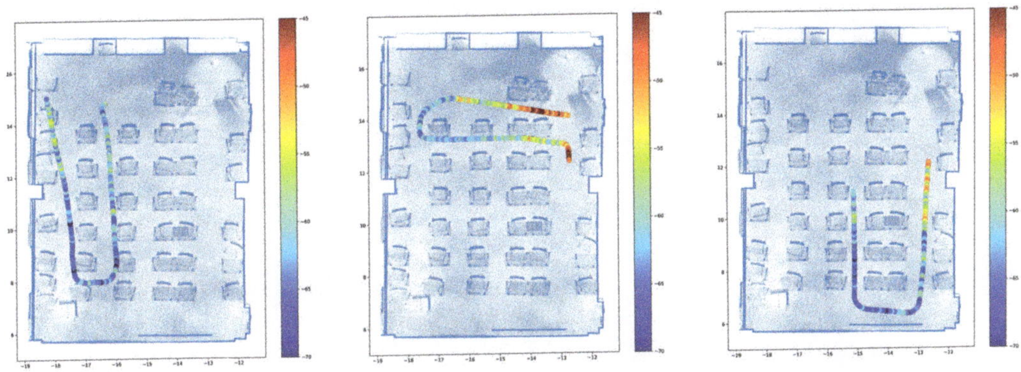

(**a**) (**b**) (**c**)

Figure 10. The three areas considered for newly collected measurements after the relocation of the AP. (**a**) Data collected at time $\mathbf{t}_1$; (**b**) Data collected at time $\mathbf{t}_2$; (**c**) Data collected at time $\mathbf{t}_3$.

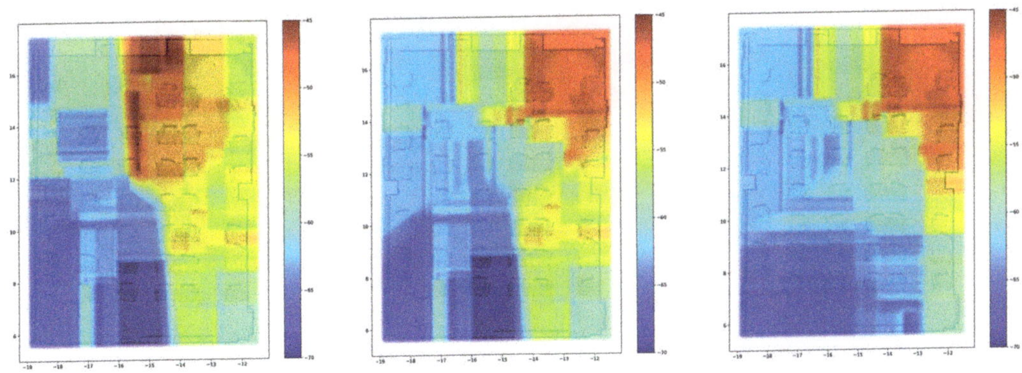

(**a**) (**b**) (**c**)

Figure 11. REMs obtained with the proposed update mechanism after the relocation of the AP. (**a**) The REM at time $\mathbf{t}_1$; (**b**) The REM at time $\mathbf{t}_2$; (**c**) The REM at time $\mathbf{t}_3$.

Table 4 presents the error evaluation for the newly collected data after the relocation of the AP at three different times by using the RF algorithm. Similar to the results obtained in Table 2, the error in the prediction decreased as time passed because new clusters could be updated as new measurement data were collected. For instance, the low error observed at time $\mathbf{t}_3$ is in concordance with the REM presented in Figure 11c, which is very similar to the ideal REM of Figure 9. In the case of the baseline scheme without clustering, the error remains consistent over time, and the errors at each time are greatly influenced by the location of the newly collected data.

Table 4. Error evaluations from newly collected measurements after the relocation of the AP.

	4 Clusters			Without Clustering		
Error metric	t_1	t_2	t_3	t_1	t_2	t_3
MAPE	7.11%	4.24%	2.19%	8.61%	7.45%	5.96%
RMSE	5.806	4.018	2.346	6.317	6.422	4.749
R2 score	0.044	0.255	0.826	−0.132	−0.902	0.288

Recently, reconfigurable intelligent surfaces (RIS) have been explored as a potential solution to enable a smart radio environment that can be dynamically configured using software. RIS is a type of metasurface that can be used to manipulate radio waves in order to control wireless communication [24]. An RIS is essentially a two-dimensional array of small, controllable elements that can adjust the phase and amplitude of incident electromagnetic waves to create a desired wavefront. Using RIS can enable the creation of smart environments that can adapt to changing electromagnetic conditions in real time. Field-programmable gate array can be used to control the operation of the RIS, or it can be manually controlled, as proposed in [25], by using touch controls. Therefore, analyzing the information in the REM can enable the optimization of the placement and reflection properties of an RIS to maximize the performance of wireless communication systems. Moreover, the REM can facilitate the dynamic adjustment of the reflection coefficients of RIS elements to adapt to changes in the wireless environment. For example, a REM can be used to identify areas with poor wireless coverage, such as areas with high levels of interference or signal attenuation. Then, an RIS can be strategically deployed in these areas to improve wireless coverage by reflecting and redirecting signals in the desired direction. Moreover, by using the information contained in a REM, we can optimize the design of the reflecting elements in the RIS to avoid interference with other wireless signals in the environment.

5. Conclusions

In this paper, we proposed a REM update methodology based on clustering and the RF algorithm. The proposed approach divides the area of interest into several clusters by using the K-means algorithm, and it trains one RF model per cluster based on real data measurements assigned to the corresponding clusters. A mobile robot was used to collect the RSSI measurements, which can reduce the risk of human error while improving the accuracy and reliability of the measurements. Next, only the RF models for clusters with enough measurement samples were updated when newly collected measurements became available. As time passed, the proposed scheme could update the whole REM by sector while reducing the error each time. Simulation results proved the superior performance of the proposed scheme compared to several well-known ML models, as well as the conventional case without clustering, in various scenarios, including the presence of obstacles and AP relocation. Subsequently, the proposed framework will be very useful for REM management in wireless scenarios where the physical element distribution constantly changes. For future research directions, an exciting topic is the utilization of information from the REM to optimize the allocation of resources in wireless networks, including spectrum, power, and antennas. This can improve overall system performance by leveraging the knowledge of the wireless propagation environment provided by the REM. For instance, the development of schemes to optimize the placement and the reflection properties of an RIS based on the information contained in a REM can help achieve desired wireless communication performance.

Author Contributions: Conceptualization, M.R.C., C.E.G. and I.K.; methodology, M.R.C.; software, M.R.C., C.E.G. and T.H.; validation, C.E.G., T.H. and I.K.; formal analysis, M.R.C.; investigation, M.R.C. and C.E.G.; resources, T.H. and I.K.; data curation, M.R.C. and T.H.; writing—original draft preparation, M.R.C.; writing—review and editing, C.E.G., T.H. and I.K.; visualization, M.R.C. and T.H.; supervision, I.K.; project administration, I.K.; funding acquisition, I.K. All authors have read and agreed to the published version of the manuscript.

Funding: This work was supported in part by the National Research Foundation of Korea (NRF) through the Korean Government's Ministry of Science and ICT (MSIT) under Grant NRF-2021R1A2B5 B01001721, and in part by the Regional Innovation Strategy (RIS) through the NRF funded by the Ministry of Education (MOE) under Grant 2021RIS-003.

Institutional Review Board Statement: Not applicable.

Informed Consent Statement: Not applicable.

Data Availability Statement: The data presented in this study are available on request from the corresponding author.

Conflicts of Interest: The authors declare no conflict of interest.

References

1. Chou, S.-F.; Yen, H.-W.; Pang, A.-C. A REM-Enabled Diagnostic Framework in Cellular-Based IoT Networks. *IEEE Internet Things J.* **2019**, *6*, 5273–5284. [CrossRef]
2. Bi, S.; Lyu, J.; Ding, Z.; Zhang, R. Engineering Radio Maps for Wireless Resource Management. *IEEE Wirel. Commun.* **2019**, *26*, 133–141. [CrossRef]
3. Garcia, C.E.; Camana, M.R.; Koo, I. Prediction of Digital Terrestrial Television Coverage Using Machine Learning Regression. *IEEE Trans. Broadcast.* **2019**, *65*, 702–712.
4. Han, T.; Bozorgi, S.; Orang, A.; Hosseinabadi, A.; Sangaiah, A.; Chen, M.-Y. A Hybrid Unequal Clustering Based on Density with Energy Conservation in Wireless Nodes. *Sustainability* **2019**, *11*, 746. [CrossRef]
5. Gallagher, T.; Li, B.; Dempster, A.G.; Rizos, C. Database updating through user feedback in fingerprint-based Wi-Fi location systems. In Proceedings of the 2010 Ubiquitous Positioning Indoor Navigation and Location Based Service, Kirkkonummi, Finland, 14–15 October 2010.
6. Lim, J.-S.; Jang, W.-H.; Yoon, G.-W.; Han, D.-S. Radio Map Update Automation for WiFi Positioning Systems. *IEEE Commun. Lett.* **2013**, *17*, 693–696. [CrossRef]
7. Luo, C.; Hong, H.; Chan, M.C.; Li, J.; Zhang, X.; Ming, Z. MPiLoc: Self-Calibrating Multi-Floor Indoor Localization Exploiting Participatory Sensing. *IEEE Trans. Mob. Comput.* **2018**, *17*, 141–154. [CrossRef]
8. Liu, X.; Cen, J.; Zhan, Y.; Tang, C. An Adaptive Fingerprint Database Updating Method for Room Localization. *IEEE Access* **2019**, *7*, 42626–42638. [CrossRef]
9. Wu, C.; Yang, Z.; Xiao, C. Automatic Radio Map Adaptation for Indoor Localization Using Smartphones. *IEEE Trans. Mob. Comput.* **2018**, *17*, 517–528. [CrossRef]
10. Wu, C.; Yang, Z.; Xiao, C.; Yang, C.; Liu, Y.; Liu, M. Static power of mobile devices: Self-updating radio maps for wireless indoor localization. In Proceedings of the 2015 IEEE Conference on Computer Communications (INFOCOM), Hong Kong, China, 26 April–1 May 2015.
11. Liu, X.; Cen, J.; Hu, H.; Yu, Z.; Huang, Y.; A radio map self-updating algorithm based on mobile crowd sensing. *J. Netw. Comput. Appl.* **2021**, *194*, 103225. [CrossRef]
12. Yang, B.; He, S.; Chan, S.-H.G. Updating Wireless Signal Map with Bayesian Compressive Sensing. In Proceedings of the 19th ACM International Conference on Modeling, Analysis and Simulation of Wireless and Mobile Systems, Malta, 13–17 November 2016.
13. Katagiri, K.; Fujii, T. Radio Environment Map Updating Procedure Considering Change of Surrounding Environment. In Proceedings of the 2020 IEEE Wireless Communications and Networking Conference Workshops (WCNCW), Seoul, Republic of Korea, 6–9 April 2020.
14. Zhen, P.; Zhang B.; Xie, C.; Guo, D. A Radio Environment Map Updating Mechanism Based on an Attention Mechanism and Siamese Neural Networks. *Sensors* **2022**, *22*, 6797. [CrossRef] [PubMed]
15. Zhao, J.; Gao, X.; Wang, X.; Li, C.; Song, M.; Sun, Q. An Efficient Radio Map Updating Algorithm based on K-Means and Gaussian Process Regression. *J. Navig.* **2018**, *71*, 1055–1068. [CrossRef]
16. TurtleBot3 Manual, Open-Source Robotics Foundation, Mountain View, CA, USA. Available online: https://emanual.robotis.com/docs/en/platform/turtlebot3/overview/ (accessed on 20 February 2023).
17. Khan, M.U.; Zaidi, S.A.A.; Ishtiaq, A.; Bukhari, S.U.R.; Samer, S.; Farman, A. A Comparative Survey of LiDAR-SLAM and LiDAR based Sensor Technologies. In Proceedings of the 2021 Mohammad Ali Jinnah University International Conference on Computing (MAJICC), Karachi, Pakistan, 15–17 July 2021.
18. *Hovermap Mapping and Autonomy Payload User Manual*; Emesent (PTY) LTD: Milton, QLD , Australia, April 2021.

19. Na, S.; Xumin, L.; Yong, G. Research on k-means Clustering Algorithm: An Improved k-means Clustering Algorithm. In Proceedings of the 2010 Third International Symposium on Intelligent Information Technology and Security Informatics, Jian, China, 2–4 April 2010.
20. Phan, H.; Maaß, M.; Mazu, R.; Mertins, A. Random regression forests for acoustic event detection and classification. *IEEE/ACM Trans. Audio Speech Lang. Process.* **2015**, *23*, 20–31. [CrossRef]
21. Smola, A.J.; Schölkopf, B. A tutorial on support vector regression. *Stat. Comput.* **2004**, *14*, 199–222. [CrossRef]
22. Aggarwal, C.C. *Neural Networks and Deep Learning: A Textbook*, 1st ed.; Springer: Cham, Switzerland, 2018.
23. Drucker, H. Improving regressors using boosting techniques. In Proceedings of the Fourteenth International Conference on Machine Learning, San Francisco, CA, USA, 8–12 July 1997.
24. Dajer, M.; Ma, Z.; Piazzi, L.; Prasad, N.; Qi, X.-F.; Sheen, B.; Yang, J.; Yue, G. Reconfigurable intelligent surface: Design the channel—A new opportunity for future wireless networks. *Digit. Commun. Netw.* **2022**, *8*, 87–104. [CrossRef]
25. Chen, L.; Ma, Q.; Luo, S.S.; Ye, F.J.; Cui, H.Y.; Cui, T.J. Touch-Programmable Metasurface for Various Electromagnetic Manipulations and Encryptions. *Small* **2022**, *18*, 2203871. [CrossRef] [PubMed]

Journal of
*Sensor and
Actuator Networks*

MDPI

Article

Applying an Adaptive Neuro-Fuzzy Inference System to Path Loss Prediction in a Ruby Mango Plantation

Supachai Phaiboon [1,*] and Pisit Phokharatkul [2]

[1] Department of Electrical Engineering, Faculty of Engineering, Mahidol University, Nakhon Pathom 73170, Thailand
[2] Department of Electrical Engineering and Energy Management, Faculty of Engineering, Kasem Bundit University, Bangkok 10250, Thailand
* Correspondence: supachai.pai@mahidol.ac.th

Abstract: The application of wireless sensor networks (WSNs) in smart agriculture requires accurate path loss prediction to determine the coverage area and system capacity. However, fast fading from environment changes, such as leaf movement, unsymmetrical tree structures and near-ground effects, makes the path loss prediction inaccurate. Artificial intelligence (AI) technologies can be used to facilitate this task for training the real environments. In this study, we performed path loss measurements in a Ruby mango plantation at a frequency of 433 MHz. Then, an adaptive neuro-fuzzy inference system (ANFIS) was applied to path loss prediction. The ANFIS required two inputs for the path loss prediction: the distance and antenna height corresponding to the tree level (i.e., trunk and bottom, middle, and top canopies). We evaluated the performance of the ANFIS by comparing it with empirical path loss models widely used in the literature. The ANFIS demonstrated a superior prediction accuracy with high sensitivity compared to the empirical models, although the performance was affected by the tree level.

Keywords: adaptive neuro-fuzzy inference system; path loss prediction; Ruby mango; 433 MHz wireless sensor network

Citation: Phaiboon, S.; Phokharatkul, P. Applying an Adaptive Neuro-Fuzzy Inference System to Path Loss Prediction in a Ruby Mango Plantation. *J. Sens. Actuator Netw.* **2023**, *12*, 71. https://doi.org/10.3390/jsan12050071

Academic Editors: Yichuang Sun, Haeyoung Lee and Oluyomi Simpson

Received: 25 July 2023
Revised: 23 September 2023
Accepted: 25 September 2023
Published: 7 October 2023

1. Introduction

A wireless sensor network (WSN) can be deployed in plantations to control the crop quality and quantity. However, the presence of trees and vegetation in the signal path can substantially degrade the performance of radio communication systems by causing signal attenuation, diffraction, scattering, and polarization [1]. Plantations comprise rows of densely foliated trees that can cause a significant path loss. Moreover, tree leaves tend to absorb water, which can cause further scattering of the signal. Low frequencies such as 240 MHz are less likely to be affected by weather conditions, such as rain and strong winds. Understanding path loss in the presence of vegetation is critical to the application of WSNs for smart agriculture.

Recently, empirical models, such as the log-distance and exponential decay models, have been developed for predicting the path loss around different types of vegetation in different frequency bands [2–13]. Raheemah et al. [2] proposed an empirical path loss model for a mango greenhouse at a frequency of 2.425 GHz with seven different antenna heights of 0.5, 1.0, 1.5, 2, 2.5, 3, and 3.5 m. The greenhouse contained 13 mango trees in three rows with a separation distance of 3.2 m between trees in the same row and 2.2 m between each row. The trees were 5 years old with a mean maximum height of 2 m, main trunk height of 1 m, and mean trunk diameter of 0.16 m. They demonstrated that their model had a better prediction accuracy than previous models in the literature. Anzum et al. [3] proposed a log-distance model with multiwall attenuation based on long-range (LoRa) measurement data at 433 MHz for oil palm trees planted in a symmetric pattern. Anderson et al. [4] characterized the path loss at a low antenna height (1.3 m) for ultrawideband propagation

(830–4200 MHz) in four forest environments: light brush, light forest, medium forest, and dense forest. Azevedo et al. [5] proposed an empirical path loss model at frequencies of 870 and 2.414 MHz that considers the tree trunks of different trees. They multiplied the tree density and average trunk diameter to obtain a coefficient for the path loss exponent. Barrios-Ulloa et al. [6] reviewed log-distance and exponential decay models for the path loss at 200 MHz–95 GHz and compared their performances in terms of the root mean squared error (RMSE). Meng et al. [7,8] proposed a plane-earth model for near-ground wireless communication in the very high frequency (VHF) and ultrahigh-frequency (UHF) bands. They limited their interest to specific phenomena, such as the impact of near-ground or surface components on the signal propagation in different environments. Tang et al. [9] analyzed path loss models with the breakpoint distance on, near, and above the ground (heights of 5 cm, 50 cm, and 1 m, respectively) at a frequency of 470 MHz. Jong et al. [10] proposed a scattering model for a single oak tree at a frequency of 1.9 GHz. Pinto et al. [11] proposed a semi-deterministic model depending on the distance between transceivers of WSN nodes and the vegetation height for tomato greenhouses and demonstrated a superior prediction accuracy. Leonor et al. [12,13] proposed a raytracing scattering model for Ficus benjamina and Thuja pelicata trees at frequencies of 20 and 62.4 GHz.

Some researchers have used artificial intelligence (AI) technologies such as machine learning (ML) to improve the accuracy of empirical models in specific areas [14–22]. Chiroma et al. [14] reviewed the performances of AI models, such as support vector machines, neural networks (NNs), genetic algorithms, and the adaptive neuro-fuzzy interference system (ANFIS) in several types of communication environments: urban, suburban, and rural. Hakim et al. [15] developed ANFIS path loss models for forest, jungle, and open dirt road environments at frequencies of 433, 868, and 920 MHz and obtained a superior prediction accuracy compared to conventional empirical models. Faruk et al. [16] applied ML models including ANFIS to VHF and UHF bands in a typical urban environment and obtained a high prediction accuracy. Nunez et al. [17] showed that an artificial neural network (ANN) improved the prediction accuracy for indoor communication at 26.5–40 GHz [21]. Famoriji et al. [18] applied a backpropagation NN to path loss prediction in a tropical region and achieved a better prediction accuracy than a conventional log-distance model. Cruz et al. [19] applied an ANN and neuro-fuzzy system to the path loss prediction of a long-term evolution signal transmitted at a frequency of 1.8 GHz and compared the RMSE with those of commonly used empirical models in the literature. Wu et al. [20] applied a multilayer perceptron neural network (MLPNN) to predicting the path loss of three base stations at 2.5 GHz and achieved a better prediction accuracy than empirical log-distance models. Ostlin et al. [21] applied an ANN to predicting the path loss of a code-division multiple-access mobile network in a rural area and obtained good results. Egi et al. [22] proposed an ANN to predict the received signal level according to the detected tree canopy and location. Their model achieved an error of 4.26%, while the empirical model had an error of 6.29–16.9%.

Non-uniform vegetation is a common source of error for both empirical and intelligent path loss models. Some empirical models have been developed, providing a good prediction performance in uniform environments, such as an oil palm plantation and mango greenhouse. However, they still provided an error because of fast fading. The AI models are generally applied to specific non-uniform environments, such as urban, suburban, and rural areas, to improve the large errors at some measurement points. The models were trained with the measured data to provide a good prediction. However, in the case of a specific environment, such as the Ruby mango plantation, the AI model needs an expert system to create a rule base for training. Therefore, we applied an ANFIS in this study to predict the path loss of a signal in a Ruby mango plantation. Ruby mango trees are trimmed to limit their height and are planted in symmetric patterns for a uniform environment.

The ANFIS only requires two inputs: the distance and antenna height in relation to the tree level (i.e., trunk and canopy). The regular pattern of trees should allow the ANFIS to predict the path loss with high accuracy. To evaluate the performance of the proposed model, we compared it against existing empirical models. The contributions of this article are summarized as follows:

- An accurate semi-deterministic path loss prediction for a uniform Ruby mango plantation with an ANFIS engine, which consists of two inputs, namely, the distance between the transceivers of WSN nodes and vegetation height together, and an output of path loss prediction.
- The validation of the model using RMSE, MAE, and MAPE against benchmark models.

The rest of this paper is organized as follows. Section 2 presents related path loss models. Section 3 presents the proposed model. Section 4 presents the experimental procedure. Section 5 presents the results. Section 6 concludes the paper.

2. Related Path Loss Models

Four empirical models are widely used to predict the path loss considering vegetation and are presented here.

2.1. ITU-R Model

The International Telecommunications Union Recommendations (ITU-R) model [23] was developed from measurements mainly at the UHF band and was proposed for cases where either the transmit or receive antenna is near a small grove of trees through which the signal propagates. This model is commonly used for frequencies between 200 MHz and 95 GHz, and it is expressed as

$$\mathrm{ITU-R(dB)} = 0.2\mathrm{f}^{0.3}\mathrm{d}^{0.6} \tag{1}$$

where f is the frequency (MHz) and d is the tree depth.

2.2. COST 235 Model

The COST 235 model [24] is based on measurements at the millimeter-wave frequency band (9.6–57.6 GHz) through a small grove of trees performed over two seasons: when the trees were in-leaf and out-of-leaf. This model is also applicable to frequencies between 200 MHz and 95 GHz, and it is expressed by

$$\mathrm{COST\ 235(dB)} = \begin{cases} 26.6\mathrm{f}^{-0.2}\mathrm{d}^{0.5} \text{ out of leaf} \\ 15.6\mathrm{f}^{-0.009}\mathrm{d}^{0.26} \text{ in leaf} \end{cases} \tag{2}$$

2.3. FITU-R Model

The fitted ITU-R (FITU-R) model is based on datasets collected during the in-leaf and out-of-leaf states at 11.2 and 20 GHz [7]:

$$\mathrm{FITU-R(dB)} = \begin{cases} 0.37\mathrm{f}^{0.18}\mathrm{d}^{0.59} \text{ out of leaf} \\ 0.39\mathrm{f}^{0.39}\mathrm{d}^{0.25} \text{ in leaf} \end{cases} \tag{3}$$

This model was developed because the lateral wave becomes dominant in both the VHF and UHF bands at relatively large forest depths, especially when both the transmit and receive antennas are inside the forest. Based on measurement data from an oil palm tree plantation, the FITU-R model becomes the following:

$$\mathrm{LITU(dB)} = 0.48\mathrm{f}^{0.43}\mathrm{d}^{0.13} + 40\log(\mathrm{d}) - 20\log(\mathrm{h_t}) - 20\log(\mathrm{h_r}) \tag{4}$$

where f is the carrier frequency (MHz), $\mathrm{h_t}$ is the height of the transmit antenna (m), $\mathrm{h_r}$ is the height of the receive antenna (m), and d is the distance between the transmit and receive antennas (m). The model in (4) is for a theoretically free space, which can be used to obtain a reference model for estimating the path loss in different environments:

$$PL_{free}(dB) = 32.4 + 20log_{10}(f) + 20log_{10}(d) \tag{5}$$

where f is the frequency (MHz) and d is the distance between the transmit and receive antennas (km). In a forest environment, Equation (5) can be modified to

$$PL_{forest}(dB) = Af^B d^C + 32.4 + 20log_{10}(f) + 20log_{10}(d) \tag{6}$$

For near-ground path loss, a plane-earth model is often used to consider both line-of-sight (LOS) and ground-reflected rays received by the receive antenna [7]:

$$PL_{PlaneEarth}(dB) = 40log(d) - 20log(h_t) - 20log(h_r) \tag{7}$$

where h_t is the height of the transmit antenna (m), h_r is the height of the receive antenna (m), and d is the distance between the transmit and receive antennas (m). If the excess loss is considered, then (6) becomes

$$PL_{forest}(dB) = Af^B d^C + 40log_{10}(d) - 20log_{10}(h_t) - 20log_{10}(h_r) \tag{8}$$

Because of lateral wave propagation from diffraction over treetops and beside trees, especially in VHF and UHF bands, the effect induced by the perfect plane-earth model is reduced. Thus, the fitted ground reflection model becomes applicable:

$$L_{FGR}(dB) = 10nlog_{10}(d) - 20log_{10}(h_t) - 20log_{10}(h_r) \tag{9}$$

where n is an empirical path loss exponent for LOS ground reflection. Then, the path loss model becomes

$$PL_{forest}(dB) = Af^B d^c + L_{FGR}(dB) \tag{10}$$

In this study, we used the path loss model in (6) because the excess loss (i.e., first term) includes the near-ground effect of different antenna heights. Table 1 summarizes the A, B, and C parameters.

Table 1. Path loss exponent parameters at 433 MHz (SF 7, BW 125 kHz).

Antenna Height (m)	$PL(d_0)$ (dB)	PLE (NLOS)	A	B	C
0.3	26.57	3.79	0.98	0.39	0.34
1.2	23.2	3.84	0.8	0.39	0.35
2.2	17.54	4.33	0.98	0.39	0.33
2.7	22.1	3.71	1.0	0.39	0.3

2.4. Log-Distance Model

The log-distance model specific to forest environments is given by [25,26]

$$PL_{forest}(dB) = PL(d_0) + 10n_{NLOS}log_{10}(d) \tag{11}$$

where $PL(d_0)$ is the path loss at a distance of 1 m (floating intercept) and n_{NLOS} is the non-LOS (NLOS) path loss exponent (PLE).

3. Proposed ANFIS Model

From the path loss model involved in Section 2, various empirical models have been developed. These empirical models are created using mathematical/statistical methods. In some cases, the environment is complex and finding a mathematical model is not easy. However, if a researcher has expertise in analyzing the nature of electromagnetic propagation, researchers can use that expertise to create fuzzy rules to predict the propagation of electromagnetic waves. This research proposes to use an ANFIS to generate fuzzy rules for path loss model prediction.

The ANFIS is a hybrid of an NN and fuzzy system, which is an inferential linguistic processing that incorporates fuzzy rules into a knowledge base [27]. The knowledge base contains fuzzy rules obtained from experts and can be adapted to produce appropriate results. However, the input–output pair must be learned to optimize the output. The NN is used to learn input–output relationships to adjust the fuzzy rules until a suitable output is found. Figure 1 shows the general architecture of an ANFIS, which has five main layers [28]. The rectangular boxes are adaptive nodes, while round boxes are fixed nodes.

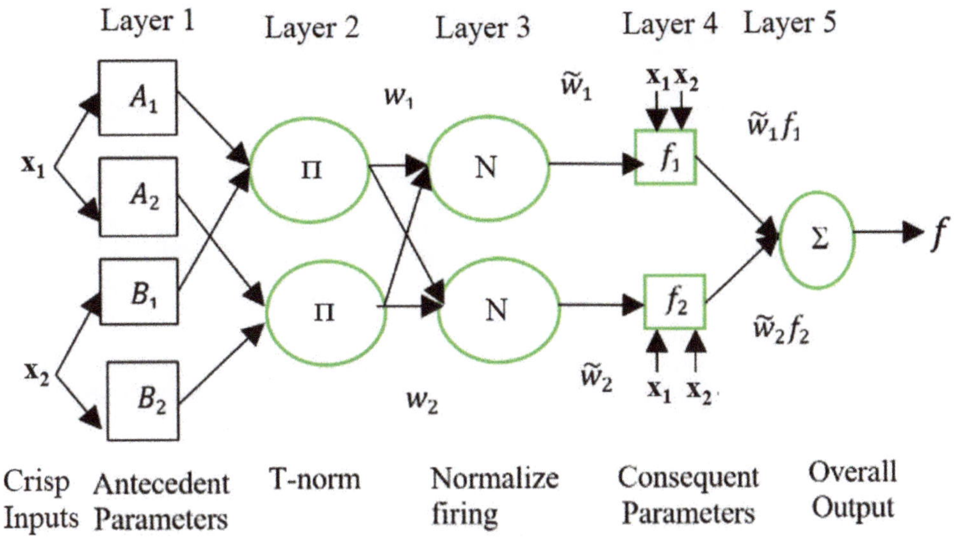

Figure 1. General architecture of the ANFIS.

Our ANFIS has two inputs and one output. Each input is divided into two fuzzy sets: A_1, A_2, and B_1, B_2. The output parameters are p_j, q_j, and r_j, with n rules:

Rule 1: IF x_1 is A_1^1 and x_2 is B_1^1 THEN $f_1 = p_1x_1 + q_1x_2 + r_1$

Rule 2: IF x_1 is A_2^2 and x_2 is B_2^2 THEN $f_2 = p_2x_1 + q_2x_2 + r_2$

...

Rule n: IF x_1 is A_i^n and x_2 is B_i^n THEN $f_n = p_nx_1 + q_nx_2 + r_n$

where, A_i^j and B_i^j are the fuzzy description of the input sets, and fj are the crisp description of the outputs.

Layer 1: This layer comprises antecedent parameters obtained via fuzzy determination from the Crisp input x to membership value μ_{A_i} or μ_{B_i} using the following membership function:

$$O_j^1 = \mu_{A_i}(x) \tag{12}$$

where O_j^1 is the membership of A_i derived from the input x. The membership function may be a triangular, inverted bell, or other shape.

Layer 2: This layer comprises the T-norm operator or fuzzy rule base, which associates fuzzy values from each dimension and sends the product as an output signal:

$$w_j = \mu_{j1}(x_1)\mu_{j2}(x_2) \tag{13}$$

where w_j is the firing strength from each rule and $\mu_{ji}(x_i)$ is the fuzzy value from the ith dimension of rule j.

Layer 3: This layer involves normalizing the firing strength or weighted layers so that all conditions from all rules can be combined into a single value:

$$\tilde{w}_j = \frac{w_j}{w_1 + w_2 + \ldots + w_n}, j = 1, 2, \ldots, n \tag{14}$$

Layer 4: The layer comprises consequent parameters obtained as follows:

$$\tilde{w}_j f_j = \tilde{w}_j (p_j x_1 + q_j x_2 + r_j) \tag{15}$$

Layer 5: This layer comprises the overall output, which includes all incoming signals and their defuzzification:

$$\tilde{w}^T f = \sum_{j=1}^{n} \tilde{w}_j f_j = \frac{\sum_{j=1}^{n} w_j f_j}{\sum_{j=1}^{n} w_j} = overall\ output \tag{16}$$

where $\tilde{w}^T = [\tilde{w}_1 \tilde{w}_2 \ldots \tilde{w}_n]$ is the fuzzy value normalized from $1 - n$ rules and $f^T = [f_1 f_2 \ldots f_n]$ is the output of $1 - n$ rules.

From the proposed model, there are two major issues, as mentioned in Sections 2 and 3: (1) the mathematical/statistical models, as in Section 2, and (2) the fact that the ANFIS is a black box method, and the fuzzy rules that are tuned from artificial neural networks (ANNs) are not easily understandable. Based on ANFIS concepts, many models have been proposed and applied to the subject of time series. The relationship between CO_2 emissions from the energy sector and global temperature increases was investigated using ANFIS, ANN, and fuzzy time series models. This research aimed to avoid strict assumptions and study the complex relationships between variables [29]. A time series forecasting model using a hybrid method of an autoregressive adaptive network fuzzy inference system (AR-ANFIS) was studied. The AR-ANFIS was trained by using particle swarm optimization, and fuzzification was performed using the fuzzy C-Means method [30]. Multivariate time series prediction using a neuro-fuzzy model was proposed. Gaussian membership functions and a learning algorithm were used in the consequent layer [31]. The reviewed literature included training and optimization methods using complex functions and learning algorithms. This technique increases the complexity of the analysis. However, in this study path loss is determined based on a trajectory with a linear relationship between the variables. Therefore, this research uses linear relationships in layer 4 as consequent parameters. This is enough to create an accurate model. This section uses numerical data, which consist of distances, the height of the antenna, and the signal strength of electromagnetic waves. A collection of related datasets consists of a training dataset and a testing dataset. The training dataset is used to generate fuzzy rules and adjust the fuzzy set using a neural network. In general, the principle of fuzzy rules is to optimize the input set of rules in a given operating environment. Traditional methods use experts' expertise to modify fuzzy rules. The ability to predict the results depend on the expert's expertise. Furthermore, fuzzy rules can be created by simulating real situations to learn to create rules, which is inconvenient in the case of electromagnetic wave propagation. In this research, the fuzzy tuning setup has been developed using a neural network to achieve more accurate predictions. Adapting fuzzy rules will require training from the training data to optimize fuzzy sets and fuzzy rules, as detailed in the section above.

The antenna height and distance between the communication nodes influences the propagation path loss. Therefore, two inputs of the ANFIS are the antenna height in meters and logarithm of distance in meters. The output is the propagation path loss in dBm. A flowchart of the ANFIS path loss modeling is shown in Figure 2. The next section describes the site survey and measurement setup in detail.

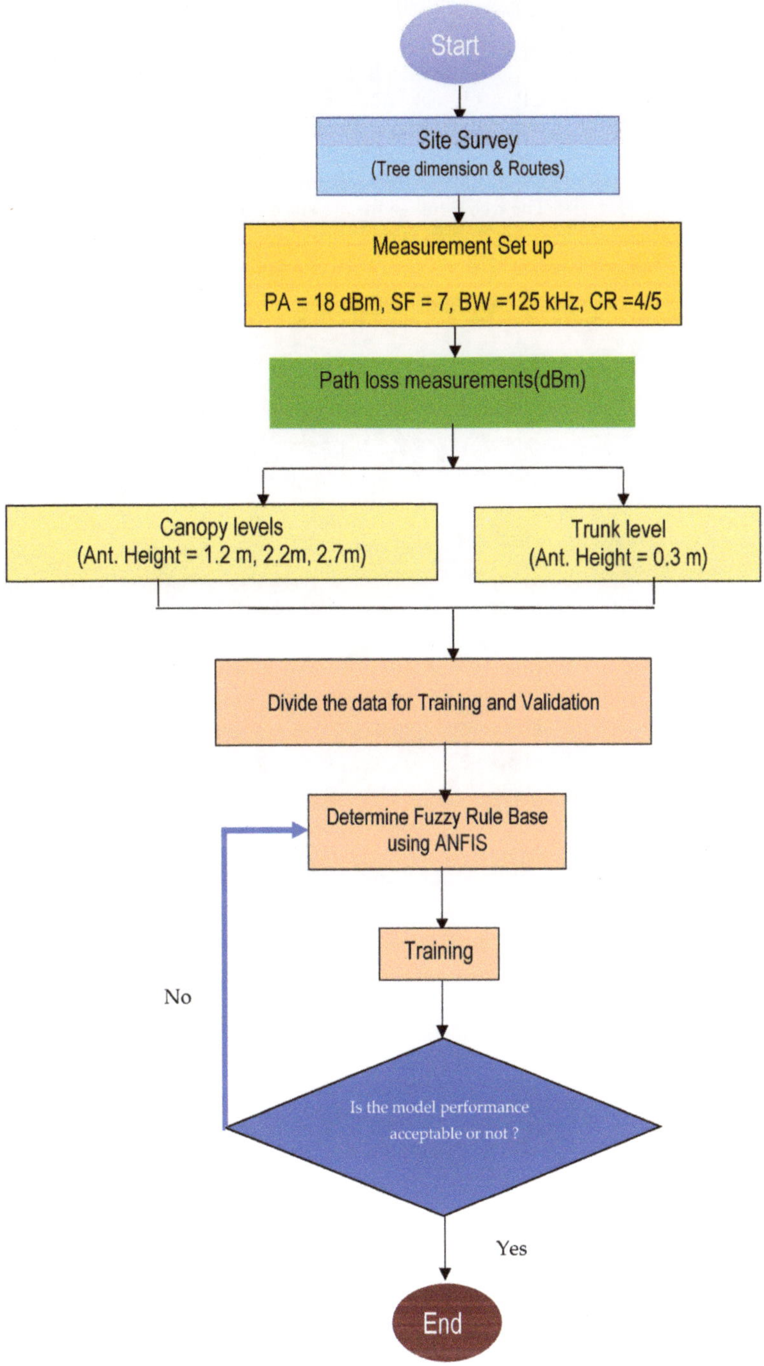

Figure 2. Flowchart of the ANFIS path loss model.

4. Experimental

4.1. Study Site

The study site was a Ruby mango plantation in Sakaeo Province, Thailand, with the GPS coordinates 13.4166954 N, 102.1368925 E. To ensure a good harvest, Ruby mango trees must be planted at a certain density. Thus, the mango tree plantation follows a specific pattern. The trees are planted in straight rows with 6 m between rows and 5 m between their trunks in the same row, as shown in Figure 3. The plantation has 320 trees per hectare. The tree dimensions are summarized in Table 2. The average total tree height was about 4.5 m, which comprised a trunk height of 0.55 m, a trunk diameter of 0.51 m, a canopy depth of 3.96 m, and an average canopy diameter of 5.69 m.

Figure 3. Plan of Ruby mango plantation.

Table 2. Measured dimensions of Ruby mango trees.

No.	Total Height	Trunk Height	Trunk Diameter	Canopy Depth	Canopy Diameter
Tree 1	3.82	0.56	0.4	3.4	5.5
Tree 2	4.66	0.66	0.56	4.0	6.0
Tree 3	4.79	0.49	0.45	4.3	5.6
Tree 4	5.15	0.65	0.64	4.5	6.5
Tree 5	4.77	0.47	0.63	4.3	6.2
Tree 6	3.96	0.46	0.46	3.5	4.7
Tree 7	4.85	0.65	0.54	4.2	6.0
Tree 8	3.97	0.47	0.43	3.5	5.0
Average	4.50	0.55	0.51	3.96.	5.69

All numerical values are in meters.

4.2. Measurement Setup

The measurement equipment comprised a fixed 433 MHz LoRa module (transceiver and omnidirectional antenna) as the receiving station and a portable LoRa module as the transmitter. These modules were connected to a microcontroller (Arduino board) that programmed the transmitter to send a data packet containing the word "hello" with the received signal strength indicator (RSSI) wirelessly to the receive antenna every 1.5 s. Three physical layer parameters were considered that influence the effective bit rate during modulation to cause noise and signal interference in a communication channel: the spreading factor (SF), bandwidth (BW), and coding rate (CR) [32].

Spreading factor: The SF is the ratio between the symbol rate and chip rate. A higher SF increases the sensitivity and transmission range with a lower packet error rate (PER) and RSSI, but it increases the airtime of the transmitted packet. Therefore, a lower SF should result in a higher PER and minimum RSSI.

Bandwidth: A higher BW increases the transmission range and data rate and thus decreases the airtime, but it also decreases the sensitivity by integrating additional noise. A lower BW increases the sensitivity but decreases the data rate. A typical LoRa network operates at a BW of 125, 250, or 500 kHz.

Coding rate: A LoRa network sets a CR to protect against bursts of interference. The CR is usually set to 4/5, 4/6, 4/7, or 4/8. A higher CR better protects the system against decoding errors by transmitting more redundant data bits but increases the airtime.

In this experiment, we only focused on the wave propagation characteristics of Ruby mango trees. Therefore, we used an SF of 7, BW of 125 kHz, and CR of 4/5 to increase the minimum RSSI, PER, sensitivity, and protection against decoding errors. The RSSI can be converted to the path loss by [33]

$$PL(dB) = P_t + G_t + G_r - (RSSI + K) \tag{17}$$

where K is an offset that depends on the characteristics of the transceiver chips used, the frequency, and the chosen technology and its features. However, K may be obtained via calibration. Table 3 summarizes the equipment parameters. To model the path loss, the RSSI data were captured via a notebook computer at the receiving station while the portable transmitting node was moved in 5 m intervals to a maximum distance of 40 m in both the forward and reverse directions. The heights of the transmit and receive antennas were set equal but varied at 0.3, 1.2, 2.2, and 2.7 m above the ground, as shown in Figure 4. Table 4 shows the path loss parameters of Equation (11), which was used for comparison with the ANFIS model.

Table 3. Parameter setup.

No.	*Parameters*	*Value*	*Unit*
1	Power amplifier (PA)	18	dBm
2	Antenna gain	2.2	dBi
3	Frequency	433	MHz
4	Bandwidth (BW)	125	kHz
5	Spreading factor	7	-
6	Code rate (CR)	4/5	-
7	Offset factor (K)	28	dBm

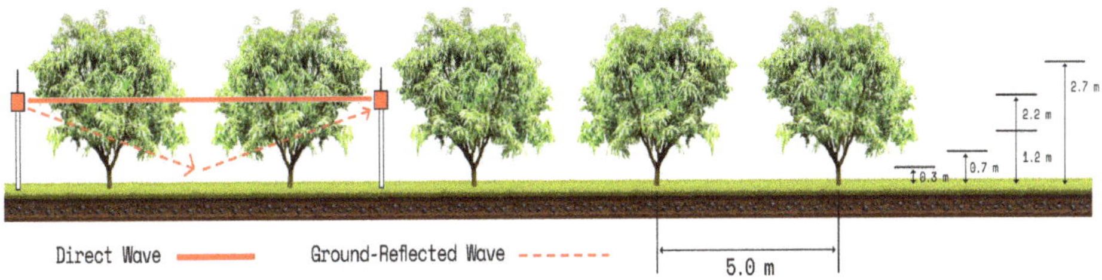

Figure 4. Propagation measurement.

Table 4. RMSE of ANFIS model and validation.

Antenna Height (m)	ANFIS Validation	
0.3 (Trunk)	3.17	3.31
1.2 (Canopy_bottom)	1.34	1.58
2.2 (Canopy_middle)	1.65	1.57
2.7 (Canopy_top)	2.61	2.60

The following metrics were used to evaluate the model performances based on their deviation from the measurement data: the absolute mean error (*AME*), mean absolute error (*MAE*), mean absolute percentage error (*MAPE*), and root mean square error (*RMSE*). These metrics were calculated as follows:

$$AME = \frac{1}{N}\left|\sum_{i=1}^{N} M_i - PL_i\right| \tag{18}$$

$$MAE = \frac{1}{N}\sum_{i=1}^{N}|M_i - PL_i| \tag{19}$$

$$MAPE = \frac{100}{N}\sum_{i=1}^{N}\left|\frac{M_i - PL_i}{M_i}\right| \tag{20}$$

$$RMSE = \sqrt{\frac{\sum_{i=1}^{N}(M_i - PL_i)^2}{N}} \tag{21}$$

where M_i is the measured path loss, PL_i is the predicted path loss, N is the total number of data, and the subscript i is the number of a given data.

5. Results and Discussion

5.1. ANFIS Model and Validation

The measurement data described in Section 4 were used for the training and validation of the proposed ANFIS model. The training dataset comprised input–output data pairs, such as the antenna height (m), distance (m), and measured path loss (dB). A first-order Sugeno fuzzy model was used for the ANFIS structure, where the inputs were the antenna height and distance and the output was the path loss. The membership functions used the psigmf model with a mixed learning process (Hybrid), [5, 5] mfs, and 100 epochs for calculation. The results are shown in Figures 5 and 6. The crisp input was divided into five fuzzy sets to obtain the minimum error. Each set contained {in1mf1, in1mf2, in1mf3, in1mf4, and in1mf5} for the antenna height input and {in2mf1, in2mf2, in2mf3, in2mf4, and in2mf5} for the log-distance input, and the output {out1mf1-out1mf25} was obtained according to the following 25 rules:

Rule 1: IF x_1 is mf1 and x_2 is mf1, THEN y is mf1
Rule 2: IF x_1 is mf1 and x_2 is mf2, THEN y is mf2
Rule 3: IF x_1 is mf1 and x_2 is mf3, THEN y is mf3
Rule 4: IF x_1 is mf1 and x_2 is mf4, THEN y is mf4
Rule 5: IF x_1 is mf1 and x_2 is mf5, THEN y is mf5
Rule 6: IF x_1 is mf2 and x_2 is mf1, THEN y is mf6
Rule 7: IF x_1 is mf2 and x_2 is mf2, THEN y is mf7
Rule 8: IF x_1 is mf2 and x_2 is mf3, THEN y is mf8
Rule 9: IF x_1 is mf2 and x_2 is mf4, THEN y is mf9
Rule 10: IF x_1 is mf2 and x_2 is mf5, THEN y is mf10
Rule 11: IF x_1 is mf3 and x_2 is mf1, THEN y is mf11
Rule 12: IF x_1 is mf3 and x_2 is mf2, THEN y is mf12
Rule 13: IF x_1 is mf3 and x_2 is mf3, THEN y is mf13
Rule 14: IF x_1 is mf3 and x_2 is mf4, THEN y is mf14
Rule 15: IF x_1 is mf3 and x_2 is mf5, THEN y is mf15
Rule 16: IF x_1 is mf4 and x_2 is mf1, THEN y is mf16

Rule 17: IF x_1 is mf4 and x_2 is mf2, THEN y is mf17
Rule 18: IF x_1 is mf4 and x_2 is mf3, THEN y is mf18
Rule 19: IF x_1 is mf4 and x_2 is mf4, THEN y is mf19
Rule 20: IF x_1 is mf4 and x_2 is mf5, THEN y is mf20
Rule 21: IF x_1 is mf5 and x_2 is mf1, THEN y is mf21
Rule 22: IF x_1 is mf5 and x_2 is mf2, THEN y is mf22
Rule 23: IF x_1 is mf5 and x_2 is mf3, THEN y is mf23
Rule 24: IF x_1 is mf5 and x_2 is mf4, THEN y is mf24
Rule 25: IF x_1 is mf5 and x_2 is mf5, THEN y is mf25

Figure 5a shows the ANFIS structure for training comprising five membership functions with the two adjusted inputs (Figure 5b,c) for 25 rules and 25 membership functions to obtain the output. Figure 6 shows the inference engine based on 25 rules with the first input as the antenna height of 1.2 m and the second input as the log-distance (d) of 0.698 (5 m), which provides a path loss output of 57.0 dB. Table 4 shows a good agreement of the ANFIS with validation.

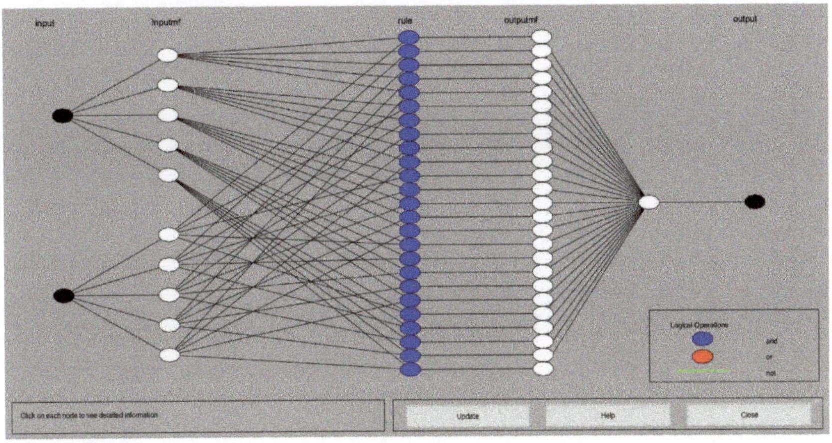

(**a**)

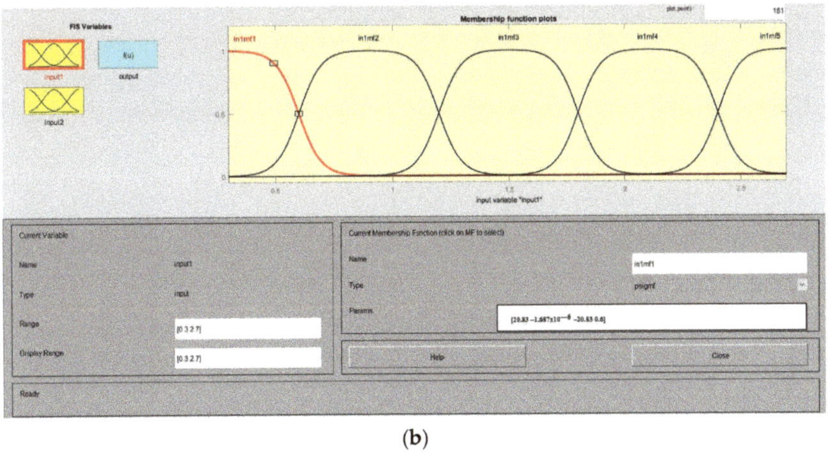

(**b**)

Figure 5. *Cont.*

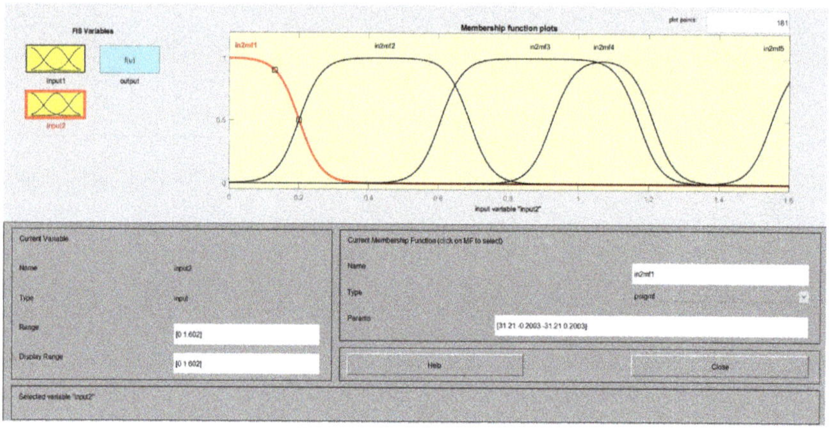

(c)

Figure 5. ANFIS with two inputs: (**a**) structure, (**b**) antenna height as input 1, and (**c**) log-distance as input 2.

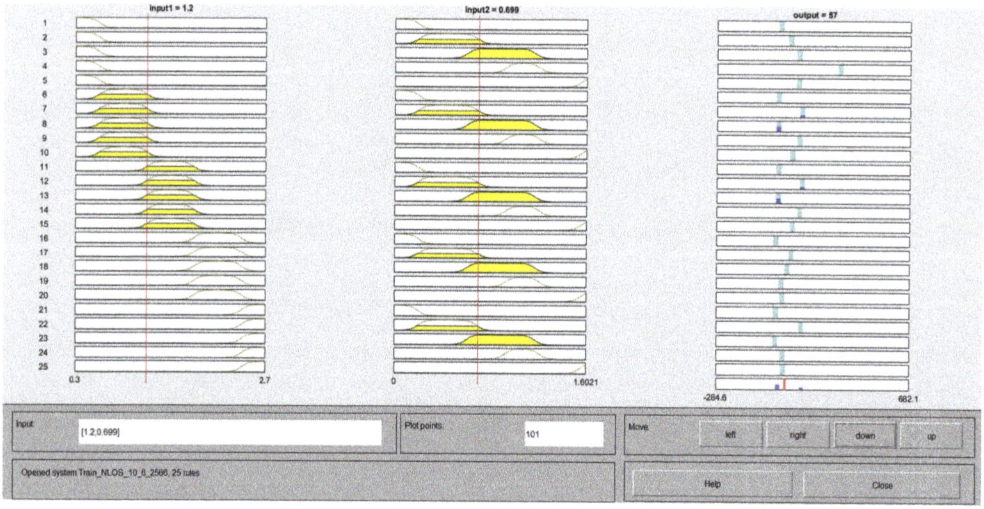

Figure 6. Inference engine.

5.2. Data Analysis of Proposed Model

The performance of the proposed ANFIS was evaluated in terms of the metrics presented in Equations (18)–(21). Tables 5–8 compare the ANFIS and measured data described in Section 4 according to the AME, MAE, MAPE, and RMSE, respectively. From the graphs in Figures 7–10, the proposed ANFIS model can predict output values for the accurate estimation of measured path loss because mathematical models are obtained by averaging data to create a model. The real environment of the mango trees consists of different trunks and canopies. They are completely asymmetrical. Moreover, the wave attenuation value is not constant at each distance between the mango trees. It is difficult to find mathematic equations to explain these differences. However, the ANFIS model is a machine learning model that uses fuzzy rules to predict outcomes. The nature of the fuzzy rule depends on the linguistic variables used to describe the mango tree's environment, which affects wave propagation. The fuzzy set is adjusted with a neural network to obtain the fuzzy sets with suitable values for prediction. Therefore, the ANFIS model makes predictions more accurate. Overall, the proposed ANFIS provided better prediction accuracy than the empirical models, especially at the bottom canopy, where it achieved AME, MAE, MAPE, and RMSE values of 0.01, 1.03, 1.64, and 1.34, respectively. However, it obtained a relatively large error at the trunk level with AME, MAE, MAPE, and RMSE values of 0.32, 2.43, 3.81, and 3.17, respectively. The MAE was smaller than the RMSE because of deviations in the measured path loss. The AME had the smallest values because the upper measurement data refuted the lower measurement data, while the MAPE had similar values to the RMSE.

Table 5. Model comparison using AME.

Antenna Height (m)	AME					
	Exponential Decay Equation (6)	Log-Distance Equation (11)	ITU-R	COST235	FITU-R	ANFIS
0.3 (Trunk)	0.11	5.73	19.33	5.41	20.26	0.32
1.2 (Canopy_bottom)	0.28	3.14	15.91	7.91	15.69	0.01
2.2 (Canopy_middle)	1.71	5.36	17.64	5.76	17.24	0.06
2.7 (Canopy_top)	0.77	7.3	16.82	6.54	16.47	0.02

Table 6. Model comparison using MAE.

Antenna Height (m)	MAE					
	Exponential Decay Equation (6)	Log-Distance Equation (11)	ITU-R	COST235	FITU-R	ANFIS
0.3 (Trunk)	6.17	6.32	19.63	10.19	20.55	2.43
1.2 (Canopy_bottom)	2.66	3.36	16.39	7.91	16.49	1.03
2.2 (Canopy_middle)	4.71	5.52	19.08	6.86	19.09	1.27
2.7 (Canopy_top)	0.77	7.61	17.69	7.45	17.84	2.08

Table 7. Model Comparison Using MAPE.

Antenna Height (m)	MAPE					
	Exponential Decay Equation (6)	Log-Distance Equation (11)	ITU-R	COST235	FITU-R	ANFIS
0.3 (Trunk)	11.91	8.89	25.09	15.49	26.20	3.81
1.2 (Canopy_bottom)	5.8	4.9	22.3	14.04	22.73	1.64
2.2 (Canopy_middle)	12.48	7.51	27.57	17.11	28.22	1.76
2.7 (Canopy_top)	11.08	10.51	24.42	15.5	25.28	3.16

Table 8. Model comparison using RMSE.

Antenna Height (m)	RMSE					
	Exponential Decay Equation (6)	Log-Distance Equation (11)	ITU-R	COST235	FITU-R	ANFIS
0.3 (Trunk)	7.74	8.59	21.65	11.77	22.59	3.17
1.2 (Canopy_bottom)	3.69	4.08	16.96	8.61	17.1	1.34
2.2 (Canopy_middle)	6.7	7.05	19.84	8.63	19.87	1.65
2.7 (Canopy_top)	6.52	9.1	18.62	9.09	18.53	2.61

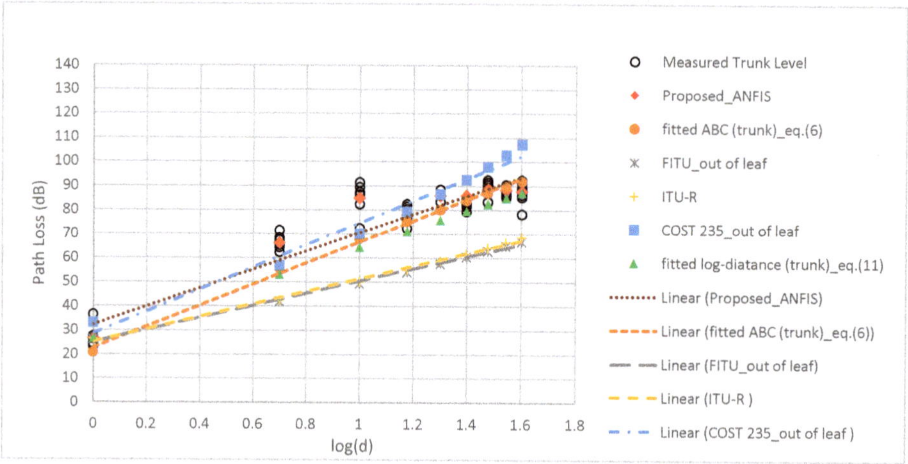

Figure 7. Predicted and observed path losses at an antenna height of 0.3 m (trunk).

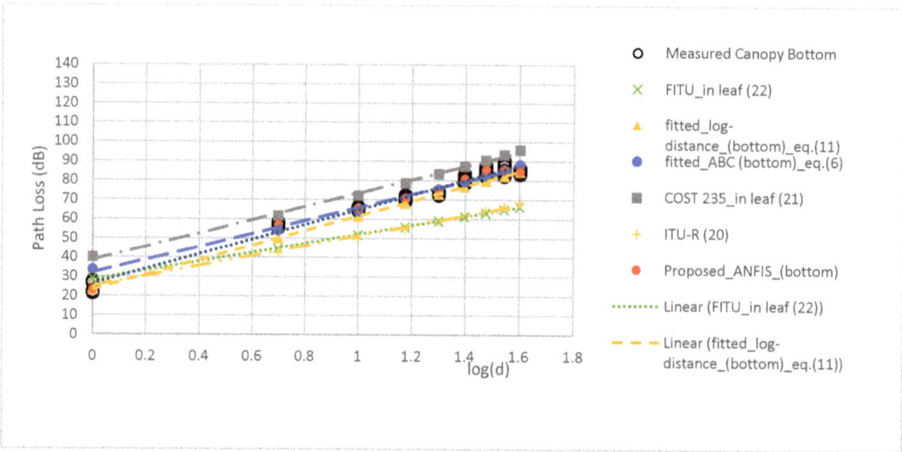

Figure 8. Predicted and observed path losses at an antenna height of 1.2 m (bottom canopy).

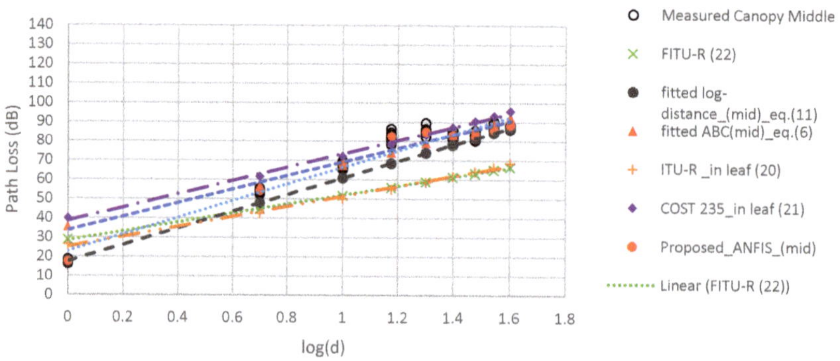

Figure 9. Predicted and observed path losses at an antenna height of 2.2 m (middle canopy).

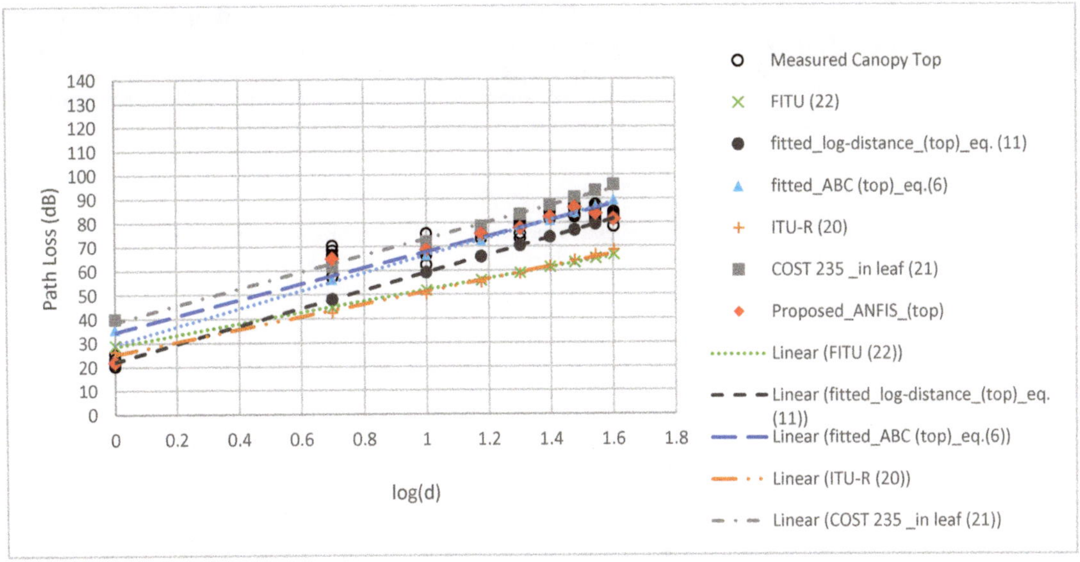

Figure 10. Predicted and observed path losses at an antenna height of 2.7 m (top canopy).

5.3. Comparison with Empirical Path Loss Models

The ANFIS demonstrated good agreement with the measurement data compared with the empirical models in Section 2. The empirical model in Equation (6) provides the best prediction with an AME, MAE, MAPE, and RMSE values of 0.11, 0.77, 5.8, and 3.69, respectively (please see Tables 5–8). In the case of the three conventional models, ITU-R (1), COST 235 (2), and FITU-R (3) provide large error prediction where either the transmit or receive antenna is near a small grove of trees. However, the COST 235 provides better prediction in the UHF band, as shown in Tables 5–8, which the signal propagates at different antenna heights with the selected AME, MAE, MAPE, and RMSE values of 5.41, 6.86, 14.04, and 8.61, respectively. Additionally, the proposed ANFIS model provides a very high sensitivity compared with the empirical models, as indicated by the red dots in Figures 7–10 for the trunk level and bottom, middle, and top canopy levels, respectively. These confirm the advantage of the proposed ANFIS model as well.

6. Conclusions

In this study, we applied an ANFIS to predict the path loss of a WSN in a Ruby mango plantation at 433 MHz. It is a combination of a neuro-fuzzy system and a learning algorithm. The ANFIS is able to learn from data and make predictions based on those data. The learning algorithm is able to adjust the weights of the connections between the neurons in the network and the parameters of fuzzy sets. This allows the ANFIS to learn and adapt to new data. We performed path loss measurements with two transceiver nodes between the trees at difference antenna heights. The ANFIS requires two inputs to predict the path loss: the antenna height corresponding to the tree level and the distance between the transmitter and receiver nodes. Each input was classified into five adjected fuzzy sets. The influence engine with 25 rule bases predicted the path loss with minimum error. We compared the performance of the proposed ANFIS with empirical models. The results showed that the proposed ANFIS demonstrated a superior prediction accuracy and high sensitivity with the best AME, MAE, MAPE, and RSME values of 0.01, 1.03, 1.64, and 1.34, respectively, at the antenna height of 1.2 m above ground (canopy bottom), although the performance fluctuated with the tree level. In future work, the capacity of the ANFIS will be improved to predict the path loss of inflorescence and fruit on tree, especially at a frequency of 2.4 GHz with a wave length of approximately 0.125 cm, which is smaller than the dimension of the inflorescence and fruit. This will significantly improve the signal attenuation, enabling the detection of inflorescence and fruit using the ANFIS for monitoring as well. The advantage of the ANFIS model is that is combines both numerical and linguistic knowledge. The ANN ability of the ANFIS is used to classify data and identify patterns. Compared to the ANN, the ANFIS model is more transparent to the user and causes less recognition errors.

Future research could extend the intelligence model by using a larger dataset to examine if better predictions can be obtained. The number of inputs in terms of attendance determinants, type of tree, difference frequency of wireless nodes, and amount of data can be expanded to develop more accurate and intelligent models and several applications.

Author Contributions: Conceptualization, S.P. and P.P.; methodology, S.P.; software, P.P.; validation, P.P.; formal analysis, S.P. and P.P.; investigation, S.P. and P.P.; resources, S.P. and P.P.; data curation, S.P.; writing—original draft preparation, S.P.; writing—review and editing, S.P.; visualization, S.P. and P.P.; supervision, P.P.; project administration, S.P.; funding acquisition, S.P. All authors have read and agreed to the published version of the manuscript.

Funding: Mahidol University for APC.

Data Availability Statement: Not applicable.

Acknowledgments: The authors thank Orapin Pitaksakorn for allowing them to conduct experiments in the Ruby mango plantation.

Conflicts of Interest: The authors declare no conflict of interest.

References

1. Attenuation in Vegetation, Radiocommunication Assembly, document ITU-R P.833-8, ITU-R, 2013.
2. Raheemah, A.; Sabri, N.; Salim, M.S.; Ehkan, P.; Ahmad, R.B. New empirical path loss model for wireless sensor networks in mango greenhouses. *Comput. Electron. Agric.* **2016**, *127*, 553–560. [CrossRef]
3. Anzum, R.; Habaebi, M.H.; Islam, R.; Hakim, G.P.N.; Khandaker, M.U.; Osman, H.; Alamri, S.; Elrahim, E.A. A multiwall path-loss prediction model using 433 MHz LoRa-WAN frequency to characterize foliage's in-fluence in a Malaysian palm oil plantation environment. *Sensors* **2022**, *22*, 5397. [CrossRef] [PubMed]
4. Anderson, C.R.; Volos, H.I.; Buehrer, R.M. Characterization of low-antenna ultrawideband propagation in a forest environment. *IEEE Trans. Veh. Technol.* **2013**, *62*, 2878–2895. [CrossRef]
5. Azevedo, J.; Santos, F.E.S. An empirical propagation model for forest environments at tree trnk level. *IEEE Trans. Antennas Propag.* **2011**, *59*, 2357–2367. [CrossRef]
6. Barrios-Ulloa, A.; Ariza-Colpas, P.P.; Sánchez-Moreno, H.; Quintero-Linero, A.P.; De la Hoz-Franco, E. Modeling radio wave propagation for wireless sensor networks in vegetated environments: A systematic literature review. *Sensors* **2022**, *22*, 5285. [CrossRef]
7. Meng, Y.S.; Lee, Y.H.; Ng, B.C. Empirical near ground path loss modeling in a forest at VHF and UHF bands. *IEEE Trans. Antennas Propag.* **2009**, *57*, 1461–1468. [CrossRef]
8. Meng, Y.S.; Lee, Y.H. Investigations of foliage effect on modern wireless communication systems: A review. *Prog. Electromagn. Res.* **2010**, *105*, 313–332. [CrossRef]
9. Tang, W.; Ma, X.; Wei, J.; Wang, Z. Measurement and analysis of near-ground propagation models under different terrains for wireless sensor networks. *Sensors* **2019**, *19*, 1901. [CrossRef]
10. de Jong, Y.L.C.; Herben, M.H.A.J. A tree-scattering model for improved propagation prediction in urban microcells. *IEEE Trans. Veh. Technol.* **2004**, *53*, 503–513. [CrossRef]
11. Pinto, D.C.; Damas, M.; Holgado-Terriza, J.A.; Arrabal-Campos, F.M.; Gómez-Mula, F.; Martínez-Lao, J.A.M.; Cama-Pinto, A. Empirical Model of Radio Wave Propagation in the Presence of Vegetation inside Greenhouses Using Regularized Regressions. *Sensors* **2020**, *20*, 6621. [CrossRef]
12. Leonor, N.; Caldeirinha, R.; Fernandes, T.; Ferreira, D.; Sánchez, M.G. A 2D ray-tracing based model for micro-and millimeter-wave propagation through vegetation. *IEEE Trans. Antennas Propag.* **2014**, *62*, 6443–6453. [CrossRef]
13. Leonor, N.; Sánchez, R.M.G.; Fernandes, T.; Ferreira, D. A 2D ray-tracing based model for wave propagation through forests at micro and millimeter wave frequencis. *IEEE Access* **2018**, *6*, 32097–32108. [CrossRef]
14. Chiroma, H.; Nickolas, P.; Faruk, N.; Alozief, E.; Olayinkaf, I.F.Y.; Adewole, K.S.; Abdulkarimh, A.; Oloyedef, A.A.; Sowandef, O.A.; Garbai, S.; et al. Large scale survey for radio propagation in developing machine learning model for path losses in communication systems. *Sci. Afr. J.* **2023**, *19*, e01550. [CrossRef]

15. Hakim, G.P.N.; Habaebi, M.H.; Toha, S.F.; Islam, M.R.; Yusoff, S.H.B.; Adesta, E.Y.T.; Anzum, R. Near Ground Pathloss Propagation Model Using Adaptive Neuro Fuzzy Inference System for Wireless Sensor Network Communication in Forest, Jungle and Open Dirt Road Environments. *Sensors* **2022**, *22*, 3267. [CrossRef]
16. Faruk, N.; Popoola, S.I.; Surajudeen-Bakinde, N.T.; Oloyede, A.A.; Abulkarim, A.; Olawoyin, L.A.; Ali, M.; Calafate, C.T.; Atayero, A.A. Path Loss predictions in the VHF and UHF bands within urban environments: Experimental investigation of empirical, heuristics and geospatial models. *IEEE Access* **2019**, *7*, 77293–77307. [CrossRef]
17. Nunez, Y.; Lovisolo, L.; Mello, L.S.; Orihuela, C. Path-Loss Prediction of Millimeter-wave using Machine Learning Techniques. In Proceedings of the 2022 IEEE Latin-American Conference on Communications (LATINCOM), Rio de Janeiro, Brazil, 30 November–2 December 2022; pp. 1–4.
18. Famoriji, O.J.; Shongwe, T. Path loss prediction in Tropical regions using machine learning techniques: A case study. *Electronics* **2022**, *11*, 2711. [CrossRef]
19. Cruz, H.A.O.; Nascimento, R.N.A.; Araujo, J.P.L.; Pelaes, E.G.; Cavalcante, G.P.S. Methodologies for path loss prediction in LTE-1.8 GHz networks using neuro-fuzzy and ANN. In Proceedings of the 2017 SBMO/IEEE MTT-S International Microwave and Optoelectronics Conference (IMOC), Aguas de Lindoia, Brazil, 27–30 August 2017.
20. Wu, L.; He, D.; Ai, B.; Wang, J.; Qi, H.; Guan, K.; Zhong, Z. Artificial Neural Network Based Path Loss Prediction for Wireless Communication Network. *IEEE Access* **2020**, *8*, 199523–199538. [CrossRef]
21. Ostlin, E.; Zepernick, H.J.; Suzuki, H. Macro cell Path-Loss Prediction Using Artificial Neural Networks. *IEEE Trans. Veh. Technol.* **2010**, *59*, 2735–2747. [CrossRef]
22. Egi, Y.; Otero, C.E. Machine-Learning and 3D Point-Cloud Based Signal Power Path Loss Model for the Deployment of Wireless Communication Systems. *IEEE Access* **2019**, *7*, 42507–42517. [CrossRef]
23. CCIR. *Influences of Terrain Irregularities and Vegetation on Troposphere Propagation*; CCIR: Geneva, Switzerland, 1986; pp. 235–236, CCIR Rep.
24. European Commission. *COST 235: Radio Propagation Effects on Next-Generation Fixed-Service Terrestrial Telecommunication Systems*; European Union: Luxembourg, 1996; Final Rep.
25. Parsons, J.D. *The Mobile Radio Propagation Channel*, 2nd ed.; Wiley: New York, NY, USA, 2000.
26. Rappaport, T.S. *Wireless Communication*; Prentic Hall Publishers: Upper Saddle River, NJ, USA, 1996.
27. Zadeh, L.A. The Concept of a Linguistic Variable and its Application to Approximate Reasoning. *Inf. Sci.* **1975**, *8*, 199–249. [CrossRef]
28. Jang, J.S.R. ANFIS: Adaptive-Network-Based Fuzzy Inference System. *IEEE Trans. Syst. Man Cybern.* **1993**, *23*, 665–685. [CrossRef]
29. Khan, M.Z.; Khan, M.F. Application of ANFIS, ANN, and fuzzy time series models to CO_2 emission from the energy sector and global temperature increase. *Int. J. Clim. Chang. Strateg. Manag.* **2019**, *11*, 622–642. [CrossRef]
30. Sarıca, B.; Eğrioğlu, E.; Aşıkgil, B. A new hybrid method for time series forecasting: AR–ANFIS. *Neural Comput. Appl.* **2018**, *29*, 749–760. [CrossRef]
31. Vlasenko, A.; Vlasenko, N.; Vynokurova, O.; Peleshko, D. A Novel Neuro-Fuzzy Model for Multivariate Time-Series Prediction. *Data* **2018**, *3*, 62. [CrossRef]
32. Citoni, B.; Fioranelli, F.; Imran, M.A.; Abbasi, Q.H. Internet of Things and LoRaWAN-Enabled Future Smart Farming. *IEEE Internet Things Mag.* **2019**, *2*, 14–19. [CrossRef]
33. Onykiienko, Y.; Popovych, P.; Yaroshenko, R.; Mitsukova, A.; Beldyagina, A.; Makarenko, Y. Using RSSI data for LoRa network path loss modeling. In Proceedings of the 2022 IEEE 41st International Conference on Electronics and Nanotechnology (ELNANO), Kyiv, Ukraine, 10–14 October 2022; pp. 576–580.

sensors

MDPI

Article

Vehicular Environment Identification Based on Channel State Information and Deep Learning

Soheyb Ribouh [1], Rahmad Sadli [2], Yassin Elhillali [2], Atika Rivenq [2] and Abdenour Hadid [3,*]

1 Normandie Université Rouen, LITIS (Laboratoire d'Informatique, de Traitement de l'Information et des Systèmes), Av. de l'Université le Madrillet, 76801 Saint Etienne du Rouvray, France
2 Département d'Opto-Acousto-Électronique, DOAE, Institut d'Électronique de Microélectronique et de Nanotechnologie, IEMN, Université Polytechnique Hauts-de-France, UMR 8520, 59300 Valenciennes, France
3 Sorbonne Center for Artificial Intelligence, Sorbonne University Abu Dhabi, Abu Dhabi P.O. Box 38044, United Arab Emirates
* Correspondence: abdenour.hadid@ieee.org

Abstract: This paper presents a novel vehicular environment identification approach based on deep learning. It consists of exploiting the vehicular wireless channel characteristics in the form of Channel State Information (CSI) in the receiver side of a connected vehicle in order to identify the environment type in which the vehicle is driving, without any need to implement specific sensors such as cameras or radars. We consider environment identification as a classification problem, and propose a new convolutional neural network (CNN) architecture to deal with it. The estimated CSI is used as the input feature to train the model. To perform the identification process, the model is targeted for implementation in an autonomous vehicle connected to a vehicular network (VN). The proposed model is extensively evaluated, showing that it can reliably recognize the surrounding environment with high accuracy (96.48%). Our model is compared to related approaches and state-of-the-art classification architectures. The experiments show that our proposed model yields favorable performance compared to all other considered methods.

Keywords: Vehicle-To-Everything (V2X) communications; channel state information; deep learning; vehicular network; autonomous vehicle; intelligent transportation systems

Citation: Ribouh, S.; Sadli, R.; Elhillali, Y.; Rivenq, A.; Hadid, A. Vehicular Environment Identification Based on Channel State Information and Deep Learning. *Sensors* **2022**, *22*, 9018. https://doi.org/10.3390/s22229018

Academic Editor: Yichuang Sun

Received: 21 October 2022
Accepted: 19 November 2022
Published: 21 November 2022

1. Introduction

Autonomous connected vehicles have been the focus of recent research works on intelligent transportation systems (ITS), in which autonomous vehicles are anticipated to be widely used as part of the smart road vision and the next generation of transportation systems. The development of autonomous driving system aims to achieve the highest level of autonomy, at which no driver is required. When this goal is met, Vehicle-To-Everything (V2X) communications will emerge as a paramount enabler for leveraging the full potential of these vehicles. Furthermore, V2X communication is mandatory to ensure the transition from self-autonomy to full collaborative autonomy [1–3]. Thus, to allow connectivity between vehicles, vehicular networks (VN) should be set up in ad hoc fashion by forming Vehicle Ad Hoc Networks (VANETs) and Mobile Ad Hoc Networks (MANETs) [4].

Because V2X communication is quite important, the automotive industry is declaring its intent to deploy V2X communication technology in their future cars. Moreover, it is further supported by transportation system governments, such as the proposed mandate from the National Highway Traffic and Safety Administration (NHTSA) that suggests all vehicles have V2X capability [5]. On the other hand, the goal of deploying autonomous vehicles is to improve road safety through cooperative driving that uses the available roadway efficiently and reduces road congestion.

According to the NHTSA, most crash accidents are caused by vehicles traveling over the speed limit. Consequently, in order to provide road safety, autonomous vehicles should be aware of the speed limit and the environment. Thus identifying the type of environment in which the vehicle is driving allows the vehicle to make good a self-decision as to the correct driving speed.

In this context, Artificial Intelligence (AI) has been established as a leading actor towards the developments of intelligent systems, enabling autonomous vehicles to make correct decisions [6,7]. Thus, in this paper we introduce a new approach towards vehicular environment identification without the need for specific sensors. The proposed method consists of using exchanged Cooperative Awareness Messages (CAM) between vehicles as well as between vehicles and infrastructure to explore channel characteristics, which are then used to recognize the vehicular environment, as shown in Figure 1.

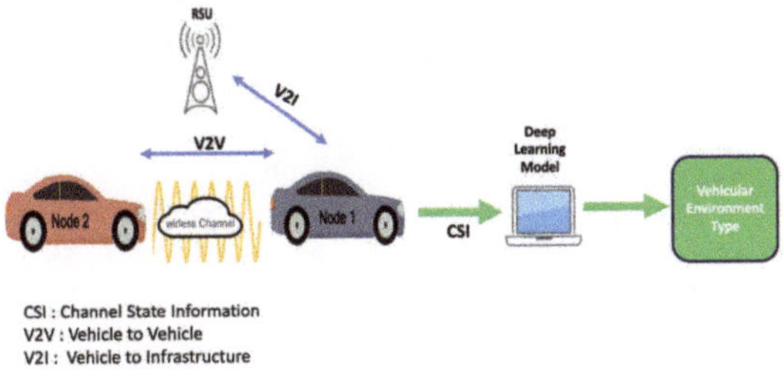

Figure 1. Vehicular environment identification process.

2. Related Work

In the literature, many research works have focused on deep learning-based environment perception in order to make critical decisions such as vehicle speed in the context of correct decision-making for autonomous cars. In [8], the authors proposed a new method called the integrated perception approach to construct the environment. They used road information data such as the distances to surrounding lane markings provided from video images. These data were used as the input features of a neural network model in order to reach the correct driving decisions.

A highway environment identification approach has been presented in [9]. The authors used video data of a highway area recorded under various weather conditions in order to develop a vision system for recognizing the bounds of highway areas and updating the vehicle with respect to the highway driving conditions. In [10], the authors presented a new perception method for urban environments. Their approach was based on the use of video images provided from an embedded camera in a vehicle, which were then used to train a neural network in order to develop a conditional navigation model that allows for prior reception of high-level directional commands.

A vehicular urban environment perception method for autonomous vehicles was presented in [11]. The approach consists of a Global Positioning System (GPS), Radar, and Light Detection And Ranging (LiDAR)-based data fusion algorithm for reaching safe driving decisions. An environment perception approach for a self-driving vehicle in an urban area was established in [6]. In this method, the authors used a 64-beam rotating LiDAR with a specific unsupervised algorithm, then generated high-resolution maps of the surrounding environment, allowing the vehicle to enable the suitable driving parameters for its environment. The authors of [12] established an approach based on the use of data fusion in order to obtain a presentation of the environment that includes a camera, 360-degree LiDAR, and GPS/Inertial Measurement Unit (IMU) sensors deployed in a vehicle. Thus, the vehicle can make correct self-driving decisions depending on the environment in which it drives. In [13], the authors proposed an environment perception framework to enhance the environmental awareness of autonomous vehicles. This framework incorporates Voxel Region-based Convolution Neural Network (PVRCNN)-based vision features and leverages Vehicle-to-Infrastructure (V2I) communication technology. The Normal

Distributions Transform (NDT) point cloud registration algorithm is used both onboard and at the roadside to obtain the position of autonomous vehicles and objects detected by the multi-sensor system at the roadside are sent back to the autonomous vehicles to improve their perception. An end-to-end machine learning model that combines control algorithms, convolutional neural networks (CNNs), and multitask (MT) learning for autonomous driving was introduced in [14]. The proposed model is able to simultaneously perform regression and classification tasks for estimating perception indicators and driving decisions, and can be used to evaluate inference efficiency and driving stability. In [15], a new approach of enhancement perception for Autonomous Driving Using Semantic and Geometric Data Fusion was presented based on low-level fusion of semantic scene information and geometry from LiDAR-based 3D point clouds. This method provides better range coverage and enables improved perception through 3D object classification and detection. In [16], the authors introduced real-time object identification, distance estimation, and instantaneous position tracking in all environmental conditions using a deep learning algorithm with no additional sensors. The proposed framework was implemented on a Raspberry Pi 4 Model B using the Raspberry Pi NoIR Camera Module V2.

Almost all of the approaches described above are essentially based on the use of specific sensors such as cameras, radars, and LIDARs. Data collection based on these sensors requires a significant amount of computing resources and power [17].

To avoid this, we propose a novel environment identification approach based on deep learning dedicated to autonomous vehicles without the need for specific sensors. For this, we exploit the shared wireless channel characteristics between vehicles communicating in vehicular networks.

Because the CSI values are the most accurate representation of wireless channel characteristics [18], we use the CSI values estimated from the packets exchanged between vehicles through Vehicle-to-Vehicle (V2V) communications as input features for our proposed convolutional neural network model. This model is able to reliably identify the surrounding environment by learning the channel characteristics (CSI) for each environment. Thus, the vehicle can set up the right automotive driving parameters (such as speed limits) corresponding to the identified environment.

The remainder of this paper is organized as follows. Section 3 describes our wireless communication model. Section 4 provides an overview of the proposed vehicular environment identification process, while Section 5 describes our tests setups and the evaluation of the performance of our proposed method. Finally, we provide our conclusions in Section 6.

3. System Model

To begin, we establish a wireless communication vehicular network model in which each vehicle uses a half-duplex transmitter/receiver pair to communicate with other vehicles. These vehicles exploit the wireless channel effect (characterized by CSI) on the received messages as the input features for the CNN model used to identify the vehicular environment.

The proposed V2X network operates on the IEEE 802.11p standard. The main physical (PHY) layer of this protocol is based on the Orthogonal Frequency Division Multiplex (OFDM) waveform. The exchanged frames in the vehicular network are constituted as shown in the figure below (Figure 2).

Figure 2. IEEE 802.11p PHY layer frame structure.

The vehicular wireless channel is structured as a double selective fading propagation channel, which is characterized by the delay spread and the Doppler spread [18]. The base-band time-varying response of the multi-path channel is provided by

$$h(t,\tau) = \sum_{l=0}^{L-1} A_l(t,\tau)\delta(\tau - \tau_l(t)), Where$$
$$A_l(t,\tau) = |A_l(t,\tau)| \exp[j(2\pi f_0 \tau_l(t) + \phi_l(t,\tau)] \tag{1}$$

where L represents the number of non-zero paths, $A_l(t)$ represents the time-varying complex amplitudes, and $\tau_l(t)$ represents the time-varying path delays. Moreover, note that the phase of the complex amplitude $A_l(t)$ in this instance depends on the variation of Doppler shift. In addition to the time delay, the signal's transmission over this channel may cause a Doppler shift in each path. As a result, the various delayed and frequency-shifted versions of the transmitted signal are superimposed at the receiver side [19].

We assume that the channel characteristics are static over a constant time T_c (coherence time) [20], which is inversely proportional to the maximum Doppler shift f_d:

$$T_c \approx \frac{0.423}{f_d} \tag{2}$$

In vehicular communication, f_d can be expressed by the speed difference between the two communicating vehicles ΔV, as shown below:

$$f_d = \frac{\Delta V}{c} f_0$$
$$\Delta V = |V_1 - V_2| \tag{3}$$

where c and f_0 represent the celerity (speed of light) and the communication center frequency, respectively.

Depending on the coherence bandwidth ($1/T_c$), when the channel's coherence bandwidth exceeds the signal's bandwidth, the channel exhibits flat fading. When the coherence bandwidth of the channel is smaller than the bandwidth of the signal, it is known as a frequency-selective fading channel (inter-symbol interference in the time domain).

According to the European Telecommunications Standards Institute (ETSI), the V2X scenario has a major impact on wave propagation, and thus the channel model [21]. We can consider five major vehicular environments depending on the different channel modeling characteristics of power, delay, and doppler [19,22,23]. These vehicular environment characteristics are shown in Table 1.

Because the vehicular environment is highly mobile, the transmitted messages are affected by the wireless channel. The received signal over the vehicular wireless channel can be written as

$$Y(k) = X(k)H(k) + W(k) \tag{4}$$

where $X(k)$ denotes the transmitted data symbols, W is the noise in the receiver, and $H(k)$ denotes the wireless channel response. This channel response is characterized by the CSI. At the receiver side, a channel estimation task is mandatory; this aims to calculate the CSI, which is required in order to recover the transmitted data. Because channel estimation is quite important in V2X communications, a great deal of research work has been carried out in this field [24–26]. The channel estimation approaches in the literature are mainly based on observation and long training sequences (LTS). The most common channel estimation method used in industrial implementations on V2X communication boards is the LS (Least Square) estimator, thanks to its low complexity; it can be expressed as

$$\hat{H}_{LS} = \underset{H_{LS}}{min} ||Y_t - X_t.H||_2^2 \tag{5}$$

where $||.||_2$ is the L_2 norm, X_t is the long training sequence vector, and Y_t denotes the corresponding observation vector. A close optimization of the LS estimator was established in [27], as follows:

$$\hat{H}_{LS} = X_t^{-1} Y_t \tag{6}$$

Table 1. Vehicular environment characteristics.

	Taps	Power [dB]	Delay [ns]	Doppler [Hz]
U-LOS	**Tap 1**	0	0	0
	Tap 2	−8	117	236
	Tap 3	−10	183	−157
	Tap 4	−15	333	492
U-NLOS	**Tap 1**	0	0	0
	Tap 2	−3	267	295
	Tap 3	−4	400	−98
	Tap 4	−10	533	591
R-LOS	**Tap 1**	0	0	0
	Tap 2	−14	83	492
	Tap 3	−17	183	−295
H-LOS	**Tap 1**	0	0	0
	Tap 2	−10	100	689
	Tap 3	−15	167	−492
	Tap 4	−20	500	886
H-NLOS	**Tap 1**	0	0	0
	Tap 2	−2	200	689
	Tap 3	−5	433	−492
	Tap 4	−7	700	886

4. Vehicular Environment Identification Methodology

In this work, we consider the vehicular environment identification process as a multi-class classification problem. Thus, we propose two methods to tackle it, as shown in Figure 3. The first is based on the use of the long training sequences (LTSs) of the received frame, while the second approach is based on the calculated CSI values. Both the LTSs and the CSI values include 128 data samples, and these samples are used as the input features of the CNN model.

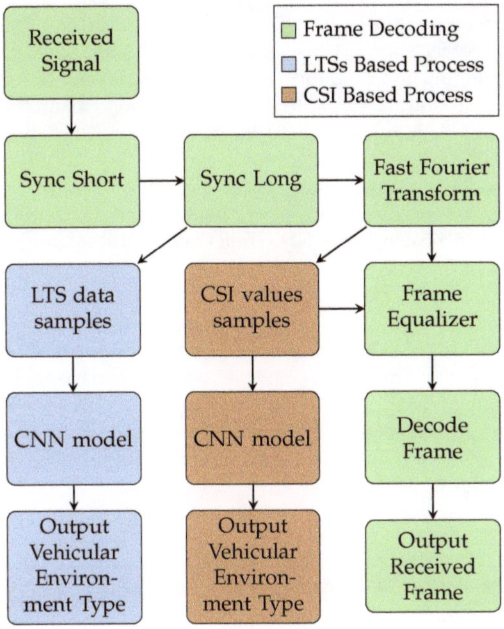

Figure 3. Flow chart describing vehicular environment identification process.

4.1. The Proposed Model

To tackle to problem of vehicular environment identification, we propose the Convolutional Neural Network (CNN) architecture shown in Figure 4.

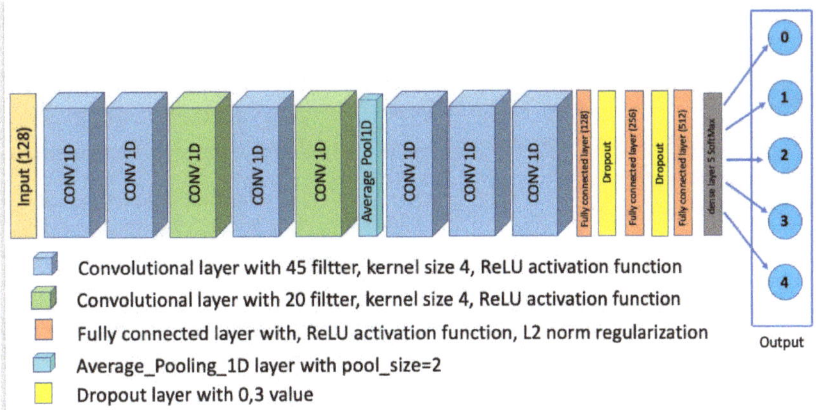

Figure 4. Proposed CNN Architecture.

The proposed CNN model is constructed as follows: First, it begins with two similar one-dimensional (1D) convolutional layers; these two layers include 45 filters. Then, we have a third 1D convolutional layer, including 20 filters. This is followed by two other 1D convolutional layers that include 45 and 20 filters, respectively. The size of the filters utilized in all the previous convolutional layers is (4×1). After that, we have an average pooling layer with a pool size of 2. This average pooling layer has three 1D convolutional layers and employs 45 fillers of (4×1) kernel size. The ReLu function is used as an activation layer for all the previous convolutional layers. These layers are followed by three fully connected layers, which include 128, 256, and 512 neurons respectively, with the ReLu activation function used for these three dense layers. To reduce the overfitting effect, we add two dropout layers (with $p = 0,3$) after the first and the second fully connected layer. Furthermore, second-norm regularization is used for all the fully connected layers [28]. Finally, the output layer is a fully connected layer in which the number of neurons is 5 (equal to the number of environment classes), with SoftMax used as the activation function.

The proposed architecture was built using the Tensorflow library [29], and we used 20 epochs and a batch size of 50 to train the model.

4.2. Data-Set Generation

For training, we considered 5 classes of vehicular environments: Rural LOS (Line-of-sight), Urban LOS, Urban NLOS (Non-Line-of-sight), Highway LOS, and Highway NLOS. Each environment is modeled by a wireless channel based on real-world vehicular environment measurement of the delay, gain, and Doppler frequency.

The vehicular channel characteristics of each environment can be found in Table 1. A label is assigned to each environment corresponding to the class outputs of the CNN model (Table 2).

Table 2. Vehicular environment labels and required speed limits.

Vehicular Environment	Label	Speed Limits
Highway NLOS	0	130 km/h
Highway LOS	1	130 km/h
Rural LOS	2	90 km/h
Urban LOS	3	50 km/h
Urban NLOS	4	50 km/h

To generate the dataset samples we employ a half-duplex V2V communication based on OFDM, which was developed using Matlab. To simulate the different vehicular environments (wireless channels models) we used the V2VChannel framework in Matlab, which is referenced in [19].

Several 802.11p packets are transmitted through the different channel models. For each environment, the packets are transmitted at a different value of the Signal-to-Noise Ratio (SNR), where the SNR range is from 15 dB to 40 dB with a step of 0.5 dB.

This process was repeated 400 times with different releases of the channel model for each environment. At each step, we computed the LTS in the received packet and the calculated CSI values, obtaining the 128 symbols (features) of both LTS and CSI associated with the specific label corresponding to each environment, as shown in Table 2.

The saved sequence features (F_i) for either CSI or LTS can be expressed as

$$F_i = [[A(1), A(2), ..., A(N)]] \tag{7}$$

where $A(i)$ is the CSI or LTS sample. At the end of the process, we had 100,000 dataset samples, of which we used 80% as the training set and 20% as the validation set.

5. Evaluation and Results

In order to assess the validity and accuracy of the proposed model, we evaluated the system on different datasets. We generated a test set by transmitting several 802.11p packets through the different channel models (V2VChannel framework of Matlab). For each environment, the SNR range was set from 15 dB to 40 dB with a step of 0.25 dB. This process was repeated 30 times with different releases of the channel model for each environment, resulting in 15,000 test sequences (LTS and CSI).

Before evaluating the proposed architecture, we trained our model using the categorical cross-entropy loss function and the Adam optimizer [30]. The training process was carried out on a machine including an NVIDIA Tesla P100 GPU.

Because both the CSI and LTS are complex numbers, we use training and test sets with three configurations depending on the input data format for each configuration. We use the magnitude for the first test-bed, the angle for the second, and a two-channel input for the third one, wherein the real part of the complex number is used for the first-channel input and the imaginary part is set for the second-channel input.

5.1. LTS Approach Performance Evaluation

As shown in Table 3, the two-channel configurations has high accuracy, achieving 93.42%, which is better than the magnitude and the angle configurations, which are 92.22% and 91.78% respectively.

Table 3. LTS approach accuracy for magnitude angle and two-channel configurations.

Configuration	Accuracy
Magnitude	92.22 %
Angle	91.78 %
2-Channel	93.42 %

Figure 5 represents the confusion matrix of the test samples for the proposed CNN architecture using the LTS as input features within a two-channel configuration. From this confusion matrix, it can be seen that our proposed CNN model is able to reliably recognize the different vehicular environment; it correctly identifies the H-NLOS and H-LOS environments with an individual accuracy of 98.3% and 86.7%, respectively, and the R-LOS, U-LOS, and U-NLOS environments with 94% accuracy.

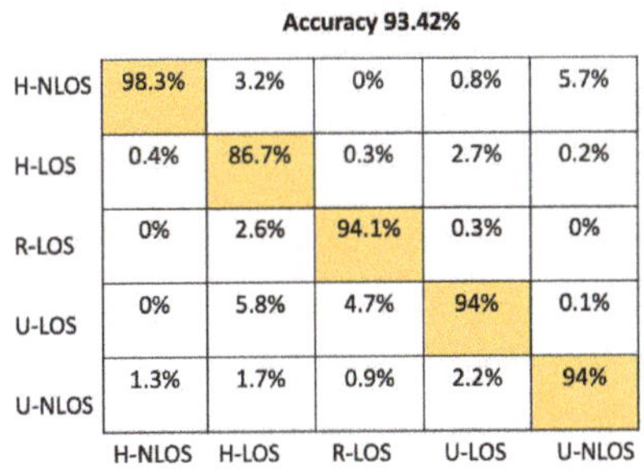

Figure 5. Confusion matrix based on LTS approach for the proposed CNN.

We compared the proposed CNN architecture to an ANN architecture containing four dense fully connected layers, including 64, 128, 256, and 512 neurons before the output layer, each of which have five neurons (equal to the number of environments to identify). Other machine learning classifier candidates used for comparison are the Random Forest classifier (RF, with 100 trees), K-Neighbors classifier (K-NN), where K was set to five neighbors, Gaussian Naive Bayes (GNB), and Support Vector Machine (SVM) with a linear kernel.

Table 4 shows the comparison between our proposed model and the approaches mentioned above based on test accuracy and environment identification prediction time. The prediction time has been determined using an NVIDIA Tesla P100 GPU. From Table 4, it is obvious that the prediction time of our proposed CNN Architecture has better performance than either SVM or K-NN, providing a prediction time of 51.33 μs. This time is comparable to the other approaches (ANN, RF, GNB) that have llowess prediction times; moreover, the accuracy of our model is significantly greater than these approaches; indeed, it has the best overall test accuracy at 93.42%.

Table 4. Classification accuracy and average prediction time comparison for LTS approach.

Approach	Accuracy (%)	Prediction Time (μs)
Proposed CNN	93.42	51.33
ANN	86.16	23.11
RF	68.34	25.71
K-NN	63.18	7180
GBN	20.62	4.11
SVM	31.38	10499

In order to provide more detail about the test accuracy classification, the confusion matrices of all the considered approaches are presented in Figures 5–10.

It can be seen that the K-NN and RF approaches can identify H-NLOS and U-LOS environments with acceptable individual accuracy (up to 80%) and provide less than 65% of individual test accuracy for H-LOS, R-LOS, and U-LOS environments. From Figures 9 and 10, it is clear that both the GNB and SVM classifiers fail to provide reliable environment identification.

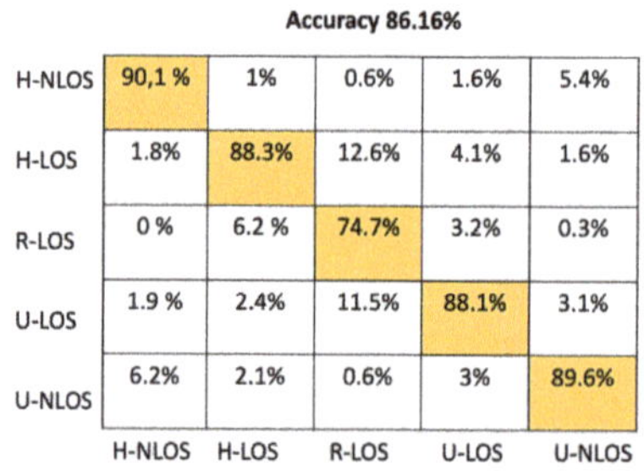

Figure 6. Confusion matrix for ANN based on LTS approach.

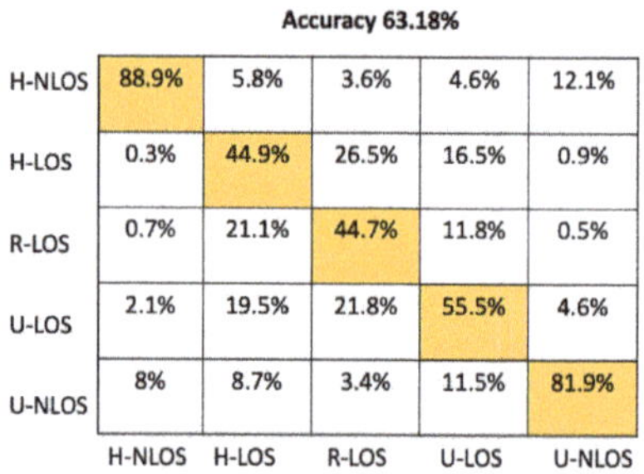

Figure 7. Confusion matrix for KNN based on LTS approach.

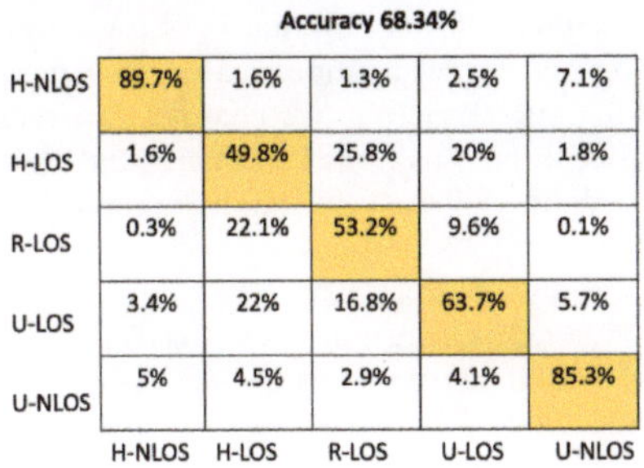

Figure 8. Confusion matrix for RF based on LTS approach.

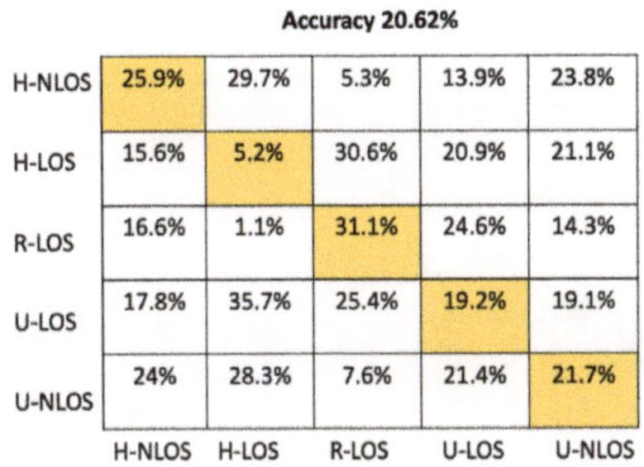

Figure 9. Confusion matrix for GNB based on LTS approach.

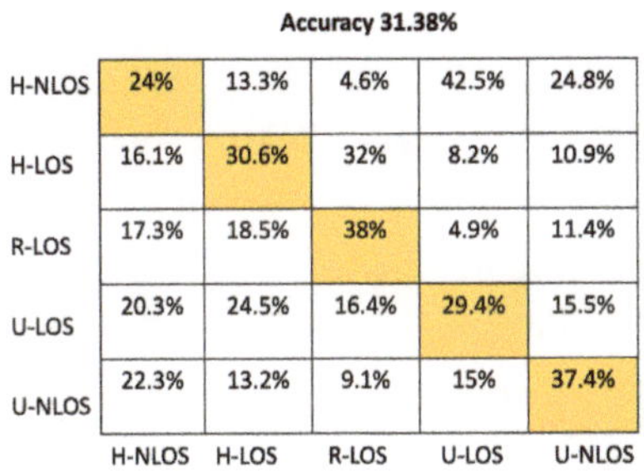

Figure 10. Confusion matrix for SVM based on LTS approach.

5.2. CSI Approach Performance Evaluation

Vehicular environment identification based on the CSI approach was performed using three input feature configuration: two-channel, magnitude, and angle.

From Table 5, it is clear that that the two-channel input feature configuration provides the best performances; it has 96.48% accuracy, which is greater than the accuracy provided by the LTS approach (93.42% in two-channel configuration).

Table 5. CSI approach accuracy for magnitude angle and two-channel configurations.

Configuration	Accuracy
Magnitude	90.63 %
Angle	91.50 %
2-Channel	96.48 %

Figure 11 presents the confusion matrix of the proposed CNN architecture on the test set when using CSI values as input features for the model. The presented results are calculated taking into account a two-channel input shape, as this provides the most accurate performance. From this confusion matrix, it is clearly apparent that our proposed model can reliably identify all the vehicular environments based on CSI values, with high individual accuracy up to 92%.

Our model achieves 99.9%, 95.2%, 92.7%, 97.4%, and 97.2% for H-NLOS, H-LOS, R-LOS, U-LOS, and U-LOS environments, respectively.

The proposed CNN model based on CSI was compared to the related machine learning classifiers RF, K-NN, GBN, and SVM, as well as to an ANN architecture, in terms of test accuracy and the average time required to identify the environment (prediction time, performed using an NVIDIA Tesla P100 GPU). The ANN architecture and parameter settings of these classifiers are the same as described in Section 5.1.

Accuracy 96.48%

	H-NLOS	H-LOS	R-LOS	U-LOS	U-NLOS
H-NLOS	99.9%	0.3%	0%	0%	2.6%
H-LOS	0.1%	95.2%	2.9%	1.8%	0%
R-LOS	0%	0%	92.7	0.2%	0%
U-LOS	0%	4.2%	4.1%	97.4%	0.2%
U-NLOS	0%	0.3%	0.3%	0.6%	97.2%

Figure 11. Confusion matrix for the proposed CNN based on CSI approach.

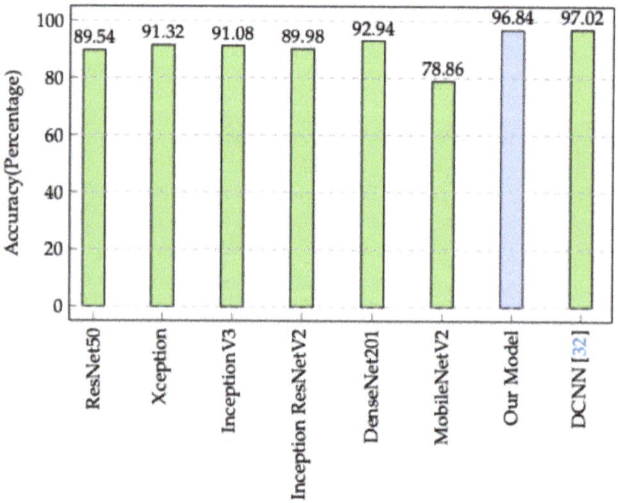

Figure 12. Comparison of our model's accuracy to state-of-the-art alternatives.

From Table 6, it can be seen that our proposed CNN model has better performance than either SVM or K-NN in terms of prediction time, as it yields in 39.56 μs. While this achieved prediction time is comparable to the other approaches (ANN, RF, GNB) that have low prediction time, in term of the test accuracy our CNN model highly outperforms all the other approaches, reaching 96.48%.

Table 6. Classification accuracy and average prediction time comparison for CSI approach.

Approach	Accuracy (%)	Prediction Time (μs)
Proposed CNN	96.48	39.56
ANN	85.64	21.11
RF	67.77	24.04
K-NN	59.26	8999
GNB	27.06	4.38
SVM	32.33	15756

According to these performance results, it is clear that the CSI approach is more accurate than the LTS approach in terms of both test accuracy and environment prediction time.

5.3. Comparison between Our Model and State-of-the-Art Architectures

Our proposed CNN architecture was compared against the popular state-of-the-art classification architectures ResNet50 [31], Xception [32], InceptionV3 [33], InceptionResNetV2 [34], DenseNet201 [35], and MobileNetV2 [36]. We trained these architectures on the same training datasets. Then, we evaluated their classification performances on the test set. Prior to this process, we updated the input shape of the input layer to fit our data inputs and updated the output layer size to five classes in order to equal the number of vehicular environments to identify.

Because the previously mentioned state-of-the-art architectures are designed to receive two-dimensional (2D) inputs for shape size, we considered a 2D channel matrix in the input features instead of a 1D channel vector. Thus, we rearranged our dataset from 1D to 2D, as follows:

$$\underset{[n*n]}{H_{2D}} = Diag(\underset{[1*n]}{H_{1D}}) \quad (8)$$

where $Diag()$ is the diagonal matrix, H_{2D} and H_{1D} are the channel matrix and corresponding channel vector, respectively, where their coefficients are the CSI values, and n is equal to 128, which is equivalent to the number of CSI values estimated per packet; thus, we have an input shape size of 128×128.

In Table 7, our proposed model is compared to the indicated state-of-the-art architectures in terms of average test accuracy (Acc), individual test accuracy for each environment, and the average time required to identify the environment (prediction time).

Table 7. Comparison between our model and state-of-the-art alternatives.

Architecture	H-NLOS Acc (%)	H-LOS Acc (%)	R-LOS Acc (%)	U-LOS Acc (%)	U-NLOS Acc (%)	Acc (%)	Prediction Time (μs)
Our Model	99.9	95.2	92.7	97.4	97.2	96.48	39.56
ResNet50	98.1	88.2	77.8	90.1	93.5	89.54	672
Xception	97.8	91.7	81.4	91.2	94.5	91.32	794
InceptionV3	99.1	79.8	86.9	96.1	93.9	91.08	683
Inception ResNetV2	98.5	89.1	80	86.5	95.8	89.98	1621
DenseNet201	98.5	92.7	85.7	91.2	96.6	92.94	1349
MobileNetV2	96.8	77.8	96	58.2	65.5	78.86	318
DCNN [37]	98.9	96.9	94.3	95.8	99.2	97.02	125

From Table 7, it is clear that our proposed model has the best performance in terms of prediction time compared to all the other architectures presented in the table, with a prediction time of 39.56 μs. The architectures ResNet50, Xception, and InceptionV3 have prediction times around of 700 μs, whereas the InceptionResNetV2, DenseNet201, MobileNetV2, and DCNN [37] architectures have 1621 μs, 1349 μs, 318 μs and 125 μs, respectively. Thus, the prediction time achieved by our model is significantly (at least three-fold) lower than the prediction time attained by the other architectures.

Regarding the overall test accuracy (Figure 12), our model reaches 96.48%. This achieved accuracy is greater than the test accuracy attained by ResNet50, Xception, InceptionV3, InceptionResNetV2, DenseNet201, and MobileNetV2, which have 89.54%, 91.32%, 91.08%, 89.98%, 92.94%, and 78.86%, respectively. Although the proposed DCNN architecture in [37] scores slightly higher than our model in term of test accuracy (by 0.54%), the prediction time achieved by our model is three times faster than DCNN [37].

To attain more insight into the classification performance of the considered architectures, the individual test accuracy of each vehicular environment is presented in Table 7. It can be seen that all the architectures are able to successfully discriminate the H-NLOS environment with accuracy of more than 96%; our model has the best accuracy, at 99.9%. Concerning the H-LOS environment, the individual accuracies are acceptable (close to 80%) for MobileNetV2 and InceptionV3, and the other architectures provide good individual accuracies of around 90%. The test accuracies for the R-LOS environment are less significant compared to other environments. This is due to the channel characteristics of this environment, which are close to the channel characteristics of the U-LOS and H-LOS environments. However, the accuracy for the R-LOS environment is considered good for our model, MobileNetV2, and DCNN (up to 90%), and is acceptable for the other models (around 80%). For the U-LOS environment, the test accuracy is generally around 90%, while our proposed model has the best accuracy at 97.4%. For the U-NLOS environment, all the architectures except MobileNetV2 are able to recognize this environment with high accuracy (up to 93%).

5.4. Minimum Performance Overhead and Reliability

Because our proposed vehicular environment identification approach based on CSI is intended to be implemented in autonomous connected cars for use in a time-critical setting, it is important that the environment identification prediction time consistently meet latency requirements in every scenario. According to [38], this hard time limit typically falls within a few milliseconds. As seen in the results, our proposed CSI-based vehicular environment identification model demonstrates a prediction time that is consistently within the **microsecond** range, which is well under the required range time.

In addition, our proposed CNN architecture using the CSI values as an inputs features proves that it can reliably identify different vehicular environments with high accuracy (96.48%) that meets the requirements for autonomous connected cars.

6. Conclusions

In this paper, we have presented a deep learning-based vehicular environment identification approach using vehicular wireless channel characteristics in the form of estimated CSI as input features for a CNN model. The results of our validation tests have demonstrate that our methodology can reliably recognize the surrounding environment with high accuracy (96.48%). These same results show that our approach has minimal performance overhead, measured in microseconds, which is well within the expected operational range across various autonomous driving scenarios. In addition, our CNN model has comparable performances to existing state-of-the-art architectures. In summary, our CSI-based vehicular environment identification approach is validated as a reliable solution to enable speed limit decisions for autonomous vehicles. However, because our vehicular environment identification method is based on the use of channel characteristics, it requires a continual exchange of messages. Moreover, we demonstrate promising results with our simulation scenarios. Hence, real-world CSI-based vehicular environment identification testbeds on a larger scale involving more vehicles in specific road conditions remains an open research problem worth investigating in the future.

Author Contributions: Conceptualization: S.R., R.S., Y.E., A.R. and A.H.; methodology: S.R., R.S. and A.H.; software: S.R.; validation: A.H.; formal analysis: S.R. and R.S.; data curation: S.R. and R.S.; writing—original draft preparation: S.R. and R.S.; writing—review and editing: S.R. and A.H.; supervision: A.H.; funding acquisition: A.R. All authors have read and agreed to the published version of the manuscript.

Funding: This research received no external funding.

Institutional Review Board Statement: Not applicable.

Informed Consent Statement: Not applicable.

Data Availability Statement: Not applicable.

Conflicts of Interest: The authors declare no conflict of interest.

References

1. Kockelman, K.; Boyles, S.; Stone, P.; Fagnant, D.; Patel, R.; Levin, M.W.; Sharon, G.; Simoni, M.; Albert, M.; Fritz, H.; et al. *An Assessment of Autonomous Vehicles: Traffic Impacts and Infrastructure Needs*; Technical Report; University of Texas at Austin, Center for Transportation Research: Austin, TX, USA, 2017.
2. Uhlemann, E. Time for autonomous vehicles to connect [connected vehicles]. *IEEE Veh. Technol. Mag.* **2018**, *13*, 10–13. [CrossRef]
3. Yue, L.; Abdel-Aty, M.; Wu, Y.; Wang, L. Assessment of the safety benefits of vehicles' advanced driver assistance, connectivity and low level automation systems. *Accid. Anal. Prev.* **2018**, *117*, 55–64. [CrossRef] [PubMed]
4. Nahar, A.; Das, D. MetaLearn: Optimizing routing heuristics with a hybrid meta-learning approach in vehicular ad-hoc networks. *Ad Hoc Netw.* **2023**, *138*, 102996. [CrossRef]
5. Abuelsamid, S. Toyota has big plans to get cars talking to each other and infrastructure in the US. *Forbes*, 16 April 2018.
6. Chen, C.; Seff, A.; Kornhauser, A.; Xiao, J. Deepdriving: Learning affordance for direct perception in autonomous driving. In Proceedings of the IEEE International Conference on Computer Vision, Santiago, Chile, 7–13 December 2015; pp. 2722–2730.
7. Okamoto, K.; Itti, L.; Tsiotras, P. Vision-based autonomous path following using a human driver control model with reliable input-feature value estimation. *IEEE Trans. Intell. Veh.* **2019**, *4*, 497–506. [CrossRef]
8. Zhang, Y.; Sun, P.; Yin, Y.; Lin, L.; Wang, X. Human-like autonomous vehicle speed control by deep reinforcement learning with double Q-learning. In Proceedings of the 2018 IEEE Intelligent Vehicles Symposium (IV), Changshu, China, 26–30 June 2018; pp. 1251–1256.
9. Seo, Y.W.; Lee, J.; Zhang, W.; Wettergreen, D. Recognition of highway workzones for reliable autonomous driving. *IEEE Trans. Intell. Transp. Syst.* **2014**, *16*, 708–718. [CrossRef]
10. Sauer, A.; Savinov, N.; Geiger, A. Conditional affordance learning for driving in urban environments. *arXiv* **2018**, arXiv:1806.06498.
11. Kim, B.; Kim, D.; Park, S.; Jung, Y.; Yi, K. Automated complex urban driving based on enhanced environment representation with GPS/map, radar, lidar and vision. *IFAC-PapersOnLine* **2016**, *49*, 190–195. [CrossRef]
12. Varga, R.; Costea, A.; Florea, H.; Giosan, I.; Nedevschi, S. Super-sensor for 360-degree environment perception: Point cloud segmentation using image features. In Proceedings of the 2017 IEEE 20th International Conference on Intelligent Transportation Systems (ITSC), Yokohama, Japan, 16–19 October 2017; pp. 1–8.
13. Duan, X.; Jiang, H.; Tian, D.; Zou, T.; Zhou, J.; Cao, Y. V2I based environment perception for autonomous vehicles at intersections. *China Commun.* **2021**, *18*, 1–12. [CrossRef]
14. Lee, D.H.; Chen, K.L.; Liou, K.H.; Liu, C.L.; Liu, J.L. Deep learning and control algorithms of direct perception for autonomous driving. *Appl. Intell.* **2021**, *51*, 237–247. [CrossRef]
15. Florea, H.; Petrovai, A.; Giosan, I.; Oniga, F.; Varga, R.; Nedevschi, S. Enhanced perception for autonomous driving using semantic and geometric data fusion. *Sensors* **2022**, *22*, 5061. [CrossRef]
16. Kabir, M.F.; Roy, S. Real-time vehicular accident prevention system using deep learning architecture. *Expert Syst. Appl.* **2022**, *206*, 117837. [CrossRef]
17. Zhu, H.; Yuen, K.V.; Mihaylova, L.; Leung, H. Overview of environment perception for intelligent vehicles. *IEEE Trans. Intell. Transp. Syst.* **2017**, *18*, 2584–2601. [CrossRef]
18. Ribouh, S.; Phan, K.; Malawade, A.V.; El Hillali, Y.; Rivenq, A.; Al Faruque, M.A. Channel State Information Based Cryptographic KeyGeneration for Intelligent Transportation Systems. *IEEE Trans. Intell. Transp. Syst.* **2020**. *to appear.*
19. Alexander, P.; Haley, D.; Grant, A. Cooperative intelligent transport systems: 5.9-GHz field trials. *Proc. IEEE* **2011**, *99*, 1213–1235. [CrossRef]
20. Wan, J.; Lopez, A.B.; Al Faruque, M.A. Exploiting wireless channel randomness to generate keys for automotive cyber-physical system security. In Proceedings of the 2016 ACM/IEEE 7th International Conference on Cyber-Physical Systems (ICCPS), Vienna, Austria, 11–14 April 2016; pp. 1–10.
21. ETSI, T. 103 257-1 V1. 1.1 (2019-05) Intelligent Transport Systems (ITS). Access Layer. Available online: https://www.etsi.org/deliver/etsi_tr/103200_103299/10325701/01.01.01_60/tr_10325701v010101p.pdf (accessed on 20 October 2022)

22. Bernado, L.; Zemen, T.; Tufvesson, F.; Molisch, A.F.; Mecklenbräuker, C.F. Delay and Doppler spreads of nonstationary vehicular channels for safety-relevant scenarios. *IEEE Trans. Veh. Technol.* **2013**, *63*, 82–93. [CrossRef]
23. Tan, I.; Tang, W.; Laberteaux, K.; Bahai, A. Measurement and analysis of wireless channel impairments in DSRC vehicular communications. In Proceedings of the 2008 IEEE International Conference on Communications, Beijing, China, 19–23 May 2008; pp. 4882–4888.
24. Zemen, T.; Bernadó, L.; Czink, N.; Molisch, A.F. Iterative time-variant channel estimation for 802.11 p using generalized discrete prolate spheroidal sequences. *IEEE Trans. Veh. Technol.* **2012**, *61*, 1222–1233. [CrossRef]
25. Zhuang, Y.; Hua, J.; Wen, H.; Meng, L. An iterative Doppler shift estimation in vehicular communication systems. *Procedia Eng.* **2012**, *29*, 4129–4134. [CrossRef]
26. Ghanavati, A.Z.; Pareek, U.; Muhaidat, S.; Lee, D. On the performance of imperfect channel estimation for vehicular ad-hoc networks. In Proceedings of the 2010 IEEE 72nd Vehicular Technology Conference-Fall, Ottawa, ON, Canada, 6–9 September 2010; pp. 1–5.
27. Sutar, M.B.; Patil, V.S. LS and MMSE estimation with different fading channels for OFDM system. In Proceedings of the 2017 International conference of Electronics, Communication and Aerospace Technology (ICECA), Coimbatore, India, 20–22 April 2017; Volume 1, pp. 740–745.
28. Kukačka, J.; Golkov, V.; Cremers, D. Regularization for deep learning: A taxonomy. *arXiv* **2017**, arXiv:1710.10686.
29. Abadi, M.; Barham, P.; Chen, J.; Chen, Z.; Davis, A.; Dean, J.; Devin, M.; Ghemawat, S.; Irving, G.; Isard, M.; et al. Tensorflow: A system for large-scale machine learning. In Proceedings of the 12th USENIX Symposium on Operating Systems Design and Implementation (OSDI 16), Savannah, GA, USA, 2–4 November 2016; pp. 265–283.
30. Kingma, D.P.; Ba, J. Adam: A method for stochastic optimization. *arXiv* **2014**, arXiv:1412.6980.
31. He, K.; Zhang, X.; Ren, S.; Sun, J. Identity mappings in deep residual networks. 2016. *arXiv* **2016**, arXiv:1603.05027.
32. Chollet, F. Xception: Deep learning with depthwise separable convolutions. In Proceedings of the IEEE Conference on Computer Vision and Pattern Recognition, Honolulu, HI, USA, 21–26 July 2017; pp. 1251–1258.
33. Szegedy, C.; Vanhoucke, V.; Ioffe, S.; Shlens, J.; Wojna, Z. Rethinking the inception architecture for computer vision. In Proceedings of the IEEE Conference on Computer Vision and Pattern Recognition, Las Vegas, NV, USA, 27–30 June 2016; pp. 2818–2826.
34. Szegedy, C.; Ioffe, S.; Vanhoucke, V.; Alemi, A. Inception-v4, inception-resnet and the impact of residual connections on learning. *arXiv* **2016**, arXiv:1602.07261.
35. Huang, G.; Liu, Z.; Van Der Maaten, L.; Weinberger, K.Q. Densely connected convolutional networks. In Proceedings of the IEEE Conference on Computer Vision and Pattern Recognition, Honolulu, HI, USA, 21–26 July 2017; pp. 4700–4708.
36. Sandler, M.; Howard, A.; Zhu, M.; Zhmoginov, A.; Chen, L.C. Mobilenetv2: Inverted residuals and linear bottlenecks. In Proceedings of the IEEE Conference on Computer Vision and Pattern Recognition, Salt Lake City, UT, USA, 18–23 June 2018; pp. 4510–4520.
37. Elwekeil, M.; Wang, T.; Zhang, S. Deep Learning for Environment Identification in Vehicular Networks. *IEEE Wirel. Commun. Lett.* **2019**, *9*, 576–580. [CrossRef]
38. Dixit, V.V.; Chand, S.; Nair, D.J. Autonomous vehicles: Disengagements, accidents and reaction times. *PLoS ONE* **2016**, *11*, e0168054. [CrossRef] [PubMed]

electronics

Article

Automatic Modulation Classification with Deep Neural Networks

Clayton A. Harper *, Mitchell A. Thornton and Eric C. Larson

Computer Science, Lyle School of Engineering, Southern Methodist University, Dallas, TX 75205, USA; mitch@smu.edu (M.A.T.); eclarson@smu.edu (E.C.L.)
* Correspondence: caharper@smu.edu

Abstract: Automatic modulation classification is an important component in many modern aeronautical communication systems to achieve efficient spectrum usage in congested wireless environments and other communications systems applications. In recent years, numerous convolutional deep learning architectures have been proposed for automatically classifying the modulation used on observed signal bursts. However, a comprehensive analysis of these differing architectures and the importance of each design element has not been carried out. Thus, it is unclear what trade-offs the differing designs of these convolutional neural networks might have. In this research, we investigate numerous architectures for automatic modulation classification and perform a comprehensive ablation study to investigate the impacts of varying hyperparameters and design elements on automatic modulation classification accuracy. We show that a new state-of-the-art accuracy can be achieved using a subset of the studied design elements, particularly as applied to modulation classification over intercepted bursts of varying time duration. In particular, we show that a combination of dilated convolutions, statistics pooling, and squeeze-and-excitation units results in the strongest performing classifier achieving 98.9% peak accuracy and 63.7% overall accuracy on the RadioML 2018.01A dataset. We further investigate this best performer according to various other criteria, including short signal bursts of varying length, common misclassifications, and performance across differing modulation categories and modes.

Keywords: automatic modulation classification; machine learning; convolutional neural networks

Citation: Harper, C.A.; Thornton, M.A.; Larson, E.C. Automatic Modulation Classification with Deep Neural Networks. *Electronics* **2023**, *12*, 3962. https://doi.org/10.3390/electronics12183962

Academic Editors: Yichuang Sun, Haeyoung Lee and Oluyomi Simpson

Received: 14 August 2023
Revised: 15 September 2023
Accepted: 18 September 2023
Published: 20 September 2023

1. Introduction

Automatic modulation classification (AMC) holds particular significance in aerospace applications, specifically in radio frequency (RF) signal analysis and modern software-defined radios. It serves a multitude of crucial tasks including "spectrum interference monitoring, radio fault detection, dynamic spectrum access, opportunistic mesh networking, and numerous regulatory and defense applications" [1]. Upon detection of an RF signal with unknown characteristics, AMC is a crucial initial procedure in order to demodulate the signal for receivers supporting a variety of standard and non-standard modulation schemes. Efficient AMC allows for maximal usage of transmission mediums and can enhance resilience in modern cognitive radios. Systems capable of adaptive modulation schemes can monitor current channel conditions with AMC and adjust exercised modulation schemes to maximize usage across the transmission medium.

Moreover, for receivers that have a versatile demodulation capability, AMC is a requisite task. The correct demodulation scheme must be applied, as a first step, to recover the modulated message within a detected signal. Aerospace communication systems, such as those employed in satellites, unmanned aerial vehicles (UAVs), and aircraft often operate in dynamic and congested environments [2]. AMC is critical in these applications to ensure efficient spectrum utilization. In systems where the modulation scheme is unknown a priori, AMC allows for efficient prediction of the employed modulation scheme.

Higher performing AMC can increase the throughput and accuracy of these systems; therefore, AMC is currently an important research topic in the fields of machine learning and communication systems, specifically for software-defined radios.

Common benchmarks are formulated with the underlying assumption that the AMC model needs to perform classification for both the modulation mode (e.g., QAM) and the specific variant within that mode (e.g., 32QAM as opposed to 64QAM). While many architectures have proven to be effective at high signal-to-noise ratios (SNRs), performance degrades significantly at lower SNRs that often occur in real-world applications. Other works have investigated increasing classification performance at lower SNR levels through the use of SNR-specific modulation classifiers [3] and clustering based on SNR ranges [4]. For the purpose of classification, various signal characteristics have been explored. Traditionally, AMC has made use of statistical moments and higher-order cumulants derived from the received signal [5,6]. Recently, direct employment of the raw in-phase (I) and quadrature (Q) components in the time domain have been embraced [1,7–9]. Additionally, alternative studies have investigated supplementary attributes, including I/Q constellation plots [10–13].

Upon the selection of signal input features, the subsequent step involves the utilization of machine learning models to discern statistical patterns within the data for classification. Classifiers such as support vector machines, decision trees, K-nearest neighbors, and neural networks are commonly used for this application [1,4,7–10,14–17]. Residual neural networks (ResNets), along with convolutional neural networks (CNNs), have been shown to achieve high classification performance for AMC [1,4,7–10,18–21]. Thus, deep learning-based methods in AMC have become more prevalent due to their promising performance and their ability to generalize to large, complex datasets comprising a variety of standard and non-standard modulation schemes.

While other works have contributed to increased AMC performance, the importance of many design elements for AMC remains unclear and a number of architectural elements have yet to be investigated. Therefore, in this work, we aim to formalize the impact of a variety of architectural changes and model design decisions on AMC performance. Numerous modifications to architectures from previous works, including our own [7], and novel combinations of elements applied to AMC are considered. After an initial investigation, we provide a comprehensive ablation study in this work to investigate the performance impact of various architectural modifications. Additionally, we achieve new state-of-the-art classification performance on the RadioML 2018.01A dataset [22] that benefits from the results of the ablation study. Using the best-performing model, we provide additional analyses that characterize its performance across modulation modes and signals with variable duration bursts.

2. Related Work

The area of AMC has been investigated by several research groups. We provide a summary of recent results in AMC to provide context and motivation for our contributions to AMC and the corresponding ablation study described in this paper. The results of the ablation study are then used to determine a new AMC architecture that demonstrates increased performance.

Corgan et al. demonstrate that deep convolutional neural networks exhibit notable classification efficacy, particularly under low SNRs, evidenced by their study on a dataset encompassing 11 distinct modulation types [8]. It was found that CNNs exceeded performance over expertly crafted features. Comparing results with architectures in [1,8], Liu et al. improved AMC performance utilizing self-supervised contrastive learning [23]. First, an encoder is pre-trained in a self-supervised manner through creating contrastive pairs with data augmentation. By creating different views of the input data through augmentation, contrastive loss is used to maximize the cosine similarity between positive pairs (augmented views of the same input). Once converged, the encoder is frozen (i.e., the weights are set to fixed values) and two fully-connected layers are added following the encoder to form the classifier. The classifier is trained using supervised learning to predict the 11 different modulation schemes. Chen et al. applied a novel architecture to the same dataset where the input signal is sliced and transformed into a square matrix and applies a residual network to predict the modulation schemes [24]. A multidimensional CNN-LSTM architecture was utilized in [25], where the CNN performed feature extraction that would later be processed by LSTM (long short-term memory) [26] and classification layers. Other work has investigated empirical and variational mode decomposition to improve few-shot learning for AMC [27]. In our work, we utilize a larger, more complex dataset consisting of 24 modulation schemes, as well as modeling improvements.

Spectrograms and I/Q constellation plots in [28] were found to be effective input features to a traditional CNN achieving nearly equivalent performance as the baseline CNN network in [1], which used raw I/Q signals. Furthermore, Refs. [10–12] employed I/Q constellations as input features in their machine learning models, focusing on a more constrained context involving four or eight modulation types. Additionally, other approaches have been explored for AMC. For instance, Refs. [29,30] utilized statistical features in conjunction with support vector machines, while [31,32] integrated fusion methodologies into CNN classifiers. Mao et al. utilized various constellation diagrams at varying symbol timings, alleviating symbol timing synchronization concerns [33]. A squeeze-and-excitation-inspired [34] architecture was used as an attention mechanism to focus on the most important diagrams.

Although spectrograms and constellation plots have shown promise, they require additional processing overhead and have had comparable performance to raw I/Q signals. In addition, models that use raw I/Q signals could be more adept at handling varying-length signals than constellation plots because they are not limited by periodicity constraints for short-duration signals (i.e., burst transmissions). Consequently, we utilize raw I/Q signals in our work.

Expanding upon these investigations, Tridgell's dissertation [35] explores the application of these architectures within the context of resource-limited Field Programmable Gate Arrays (FPGAs). His research underscores the significance of parameter reduction for modulation classifiers, given their typical deployment in embedded systems characterized by resource constraints. Addressing this concern, Mendis et al. proposed the use of multiplierless deep belief networks that map directly to binary circuits [36].

In [1], Oshea et al. created a dataset with 24 different types of modulation, known as RadioML 2018.01A, and achieved high classification performance using convolutional neural networks, specifically using residual connections (see Figure 1) within the network (ResNet). A total of six residual stacks were used in the architecture. A residual stack is defined as a series of a convolutional layers, residual units, and a max pooling operation as shown in Figure 1. The ResNet employed by [1] attained approximately 95% classification accuracy at high SNR values. Wang et al. also made use of residual connections along with depthwise separable convolutions for feature extraction [37]. This architecture was able to achieve a maximum performance of 97% accuracy and an average of 53.85% accuracy across all signal-to-noise ratios while greatly reducing model complexity.

Harper et al. proposed the use of X-Vectors [38] to increase classification performance using CNNs [7]. X-Vectors are traditionally used in speaker recognition and verification systems making use of aggregate statistics. X-Vectors utilize statistical moments, specifically the mean and variance, computed over convolutional filter outputs. It can be postulated that computing the mean and variance of the embedding layer contributes to the removal of signal-specific details, leaving broader modulation-specific characteristics. Figure 2 illustrates the X-Vector architecture, where statistics are computed over the activations from a convolutional layer producing a fixed-length vector.

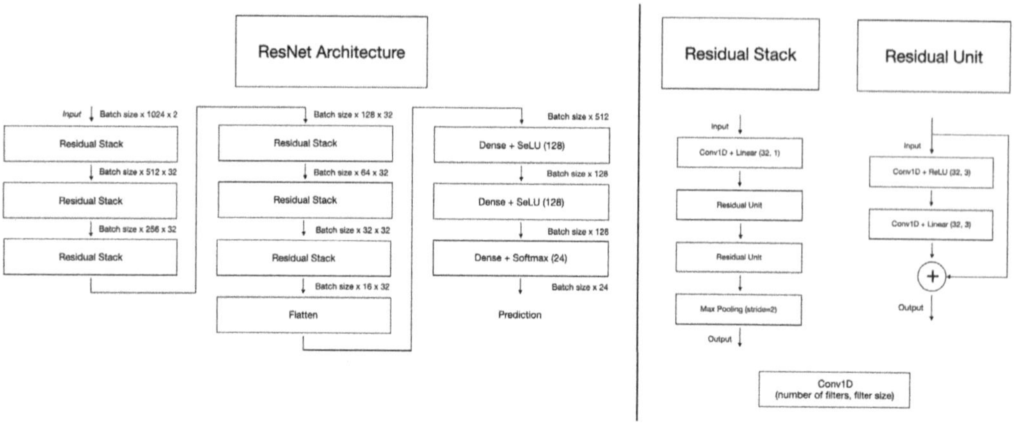

Figure 1. ResNet architecture used in [1]. Each block represents a unit in the network, which may be composed of several layers and connections as shown on the right of the figure. Dimensions of the tensors on the output of each block are also shown where appropriate.

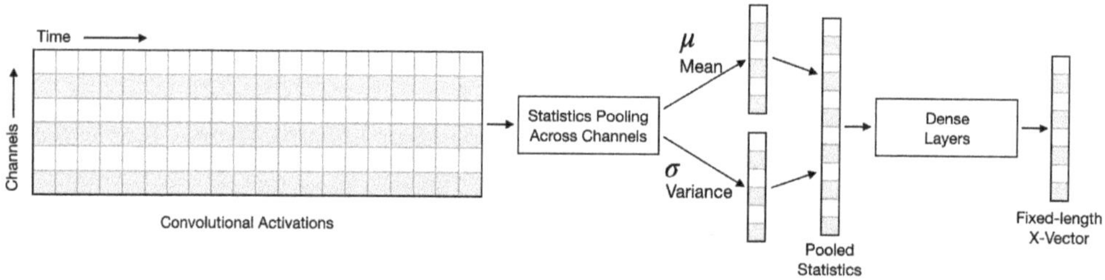

Figure 2. X-Vector architecture overview. The convolutional activations immediately before pooling are shown. These activations are fed into two statistical pooling layers that collapse the activations over time, creating a fixed-length tensor that can be further processed by fully connected dense layers.

Additionally, this architecture upholds a completely convolutional framework, enabling adaptability to inputs of varying sizes within the network. The utilization of statistical aggregations capitalizes on this characteristic. With statistical aggregations, the input to the initial dense layer becomes contingent upon the quantity of filters in the final convolutional layer. The number of filters is a hyperparameter that remains distinct from the temporal length of the input signal fed into the neural network.

In the absence of statistical aggregations, input signals for a conventional CNN or ResNet would require resampling, cropping, or padding to attain a consistent temporal length for the subsequent dense layers. While the dataset used in this work has uniformly sized signals in terms of duration, (1024×2), this is an architectural advantage in our deployment, as received signals may vary in duration. Instead of modifying the inputs to the network via sampling, cropping, padding, etc., the X-Vector architecture can directly operate with variable-length inputs without modifications to the network or input signal. Work by Li et al. [39] utilizes LSTMs while highlighting this desirable characteristic.

Figure 3 outlines the employed X-Vector architecture in [7] where $F = [f_1, f_2, ..., f_7] = 64$ and $K = [k_1, k_2, ..., k_7] = 3$. Mean and variance pooling are performed on the final convolutional outputs, concatenated, and fed through a series of dense layers creating the fixed-length X-Vector. A maximum of 98% accuracy was achieved at high SNR levels.

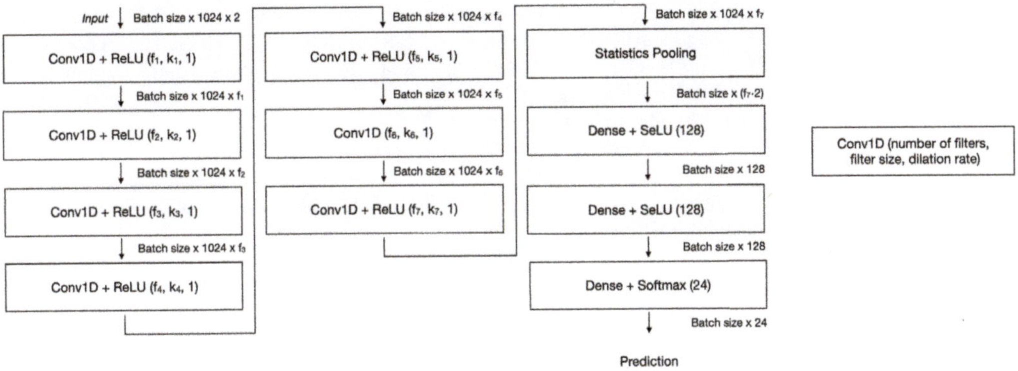

Figure 3. Proposed CNN Architecture in [7]. This is the first work to employ an X-Vector-inspired architecture for AMC showing strong performance. This architecture is used as a baseline for the modifications investigated in this paper. The f and k variables shown designate the number of kernels and size of each kernel, respectively, in each layer. These parameters are investigated for optimal sizing in our initial investigation.

The work of [7] replicated the ResNet architecture from [1] and compared the results with the X-Vector architectures as seen in Figure 4. Harper et al. [7] were able to reproduce this architecture, achieving a maximum of 93.7% accuracy. The authors attribute the difference in performance to differences in the train and test set separation that they used, since these parameters were unavailable.

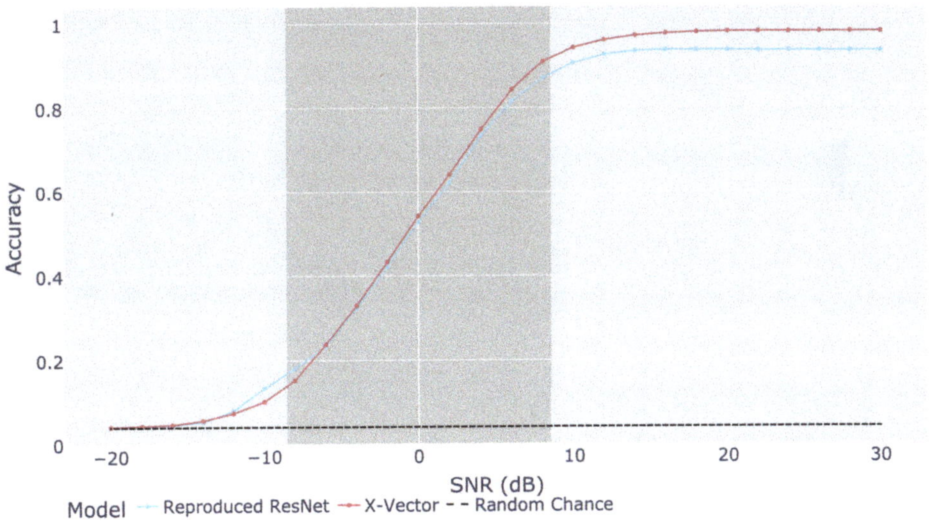

Figure 4. Accuracy comparison of the ResNet reproduced in [1] and the X-Vector-inspired model from [7] over varying SNRs. This accuracy comparison shows the superior performance of the X-Vector architecture, especially at higher SNRs, and supports using this architecture as a baseline for the improvements investigated in this paper.

As expected, the classifiers perform with a higher accuracy as the SNR value increases. At low SNR values, the classification task becomes more difficult due to the increased presence of noise. High SNR values are not invariably guaranteed in software-defined radios. However, notable enhancements are evident when compared to random chance, even under conditions of diminished SNR. In time-critical classification scenarios, this factor gains heightened significance, potentially leading to a pivotal advantage, as fewer demodulation schemes would require trial-and-error application to ascertain the correct scheme, thus streamlining the process.

One challenge of AMC is that it is desirable for performance to work well across a large range of SNRs. For instance, Figure 4 illustrates that modulation classification performance reached a plateau beyond +8 dB SNR, and approached chance-level classification performance when the SNR dipped below −8 dB on the RadioML 2018.01A dataset. This range is denoted by the shaded region. Harper et al. investigated methods to improve classification performance in this range by employing an SNR regression model to aid separate modulation classifiers (MCs). While other works have trained models to be robust across diverse SNR scenarios, Harper et al. employed SNR-specific MCs [3].

Six MCs were created by discretizing the SNR range to ameliorate performance between −8 dB and +8 dB SNR. These groupings were chosen in order to provide sufficient training data to avoid overfitting the MCs and provide enough resolution, so that combining MCs provided more value than a single classifier.

Firstly, by predicting the SNR of the received signal with a regression model, an SNR-specific MC that was trained on signals with the predicted SNR is applied to make the final prediction. While the dataset's SNR values are discretized, the SNR is measured on a continuous scale in practical deployment scenarios, subject to temporal fluctuations. Consequently, a regression approach is adopted instead of classification. By employing this methodology, various classifiers can tailor their feature processing to accommodate distinct SNR ranges. Each MC in this approach uses the same architecture as that proposed in [7]; however, each MC is trained with signals within each MC's SNR training range (see Table 1).

Table 1. SNR-specific modulation classifiers (MCs) groupings during training and inference phases, adapted from [3].

AMC Model	Training Range (dB)	Employed during Inference (dB)
MC 1	$[-20, -8]$	$(-\infty, -8)$
MC 2	$[-8, -4]$	$[-8, -4)$
MC 3	$[-4, 0]$	$[-4, 0)$
MC 4	$[0, 4]$	$[0, 4)$
MC 5	$[4, 8]$	$[4, 8)$
MC 6	$[8, 30]$	$[8, \infty)$

Illustrating enhancements across diverse SNR levels, Figure 5 presents the performance improvement (expressed as percentage accuracy) achieved through the employment of the SNR-informed architecture, contrasted with the baseline classification architecture detailed in [7]. While a marginal decline in performance was evident at −8 dB and a more substantial reduction at −2 dB, discernible enhancement is observable across most SNR conditions, with a pronounced emphasis on the desired range, spanning from −8 dB to +8 dB.

Declined performance at specific SNRs could be attributed to the optimization of a specific modulation classifier (MC), which led to an enhanced performance for a specific SNR grouping at the cost of lower performance for an individual value within the same group. To elaborate, the MC designed for the $[-4, 0)$ dB range may have bolstered the overall performance by accurately classifying signals at −4 dB and 0 dB, potentially at the expense of −2 dB accuracy. Given the substantial size of the testing dataset, these marginal percentage gains hold significance, as they result in thousands of additional correct classifications. Importantly, all outcomes achieve statistical significance according to McNemar's test [40], consequently achieving new state-of-the-art performance at the time.

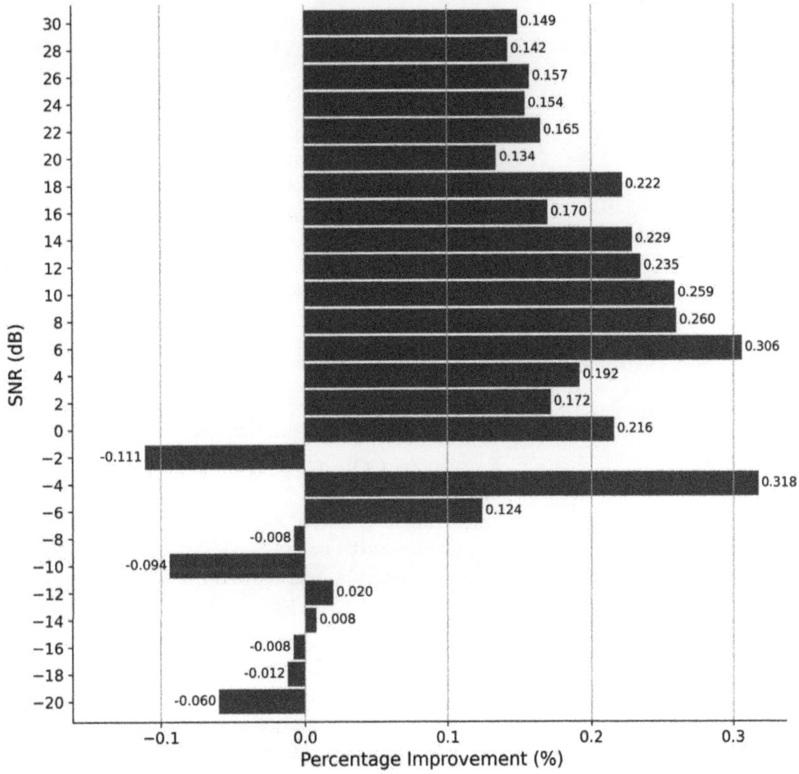

Figure 5. Summary of percentage improvement in accuracy over [7] seen in [3]. This work showed how the baseline architecture could be tuned to specific SNR ranges. Positive improvement is observed for most SNR ranges.

Soltani et al. found that SNR regions of $[-10, -2]$ dB, $[0, 8]$ dB, and $[10, 30]$ dB had similar classification patterns [4]. Instead of predicting exact modulation variants, the authors grouped commonly confused variants into a more generic, coarse-grained label. This grouping increases the performance of AMC by combining modulation variants that are commonly confused. However, it also decreases the sensitivity of the model to the numerous possible variants.

Cai et al. utilized a transformer-based architecture to aid performance at low SNR levels with relatively few training parameters (approximately 265,000 parameters) [41]. Ren et al. proposed *ResSwinT-SwinT*, making use of transformers to denoise signals under low SNR conditions prior to classification [17]. A multi-scale network along with center loss [42] was used in [43]. It was found that larger kernel sizes improved AMC performance. We further explore kernel size performance impacts in this work. Zhang et al. proposed a high-order attention mechanism using the covariance matrix achieving a maximum accuracy of 95.49% [44].

Although many discussed works use the same RadioML 2018.01A dataset, there is a lack of a uniform dataset split to establish a benchmark for papers to report performance. In an effort to make AMC more reproducible and comparable across publications, we have made our dataset split and accompanying code available on GitHub (https://github.com/caharper/Automatic-Modulation-Classification-with-Deep-Neural-Networks).

While numerous works have investigated architectural improvements, we aim to improve upon these works by introducing additional modifications, as well as a comprehensive ablation study that illustrates the improvement of each modification. With the new modifications, we achieve new state-of-the-art AMC performance.

3. Dataset

In order to assess different machine learning architectures, we employ the RadioML 2018.01A dataset, which encompasses a collection of 24 distinct modulation types [1,22]. Due to the complexity and variety of modulation schemes in the dataset, it is fairly representative of typically encountered modulation schemes. Moreover, this variety increases the likelihood that AMC models will generalize to more exotic or non-existing modulation schemes in the training data that are derived from these traditional variants.

There are a total of 2.56 million labeled signals, $S(T)$, each consisting of 1024 time domain digitized intermediate frequency (IF) samples of in-phase (I) and quadrature (Q) signal components where $S(T) = I(T) + jQ(T)$. The data were collected at a 900 MHz IF with an assumed sampling rate of 1MS/sec, such that each 1024 time domain digitized I/Q sample is 1.024 ms [8]. The 24 modulation types and the representative groups that we chose for each are listed as follows:

- **Amplitude**: OOK, 4ASK, 8ASK, AM-SSB-SC, AM-SSB-WC, AM-DSB-WC, and AM-DSB-SC.
- **Phase**: BPSK, QPSK, 8PSK, 16PSK, 32PSK, and OQPSK.
- **Amplitude and Phase**: 16APSK, 32APSK, 64APSK, 128APSK, 16QAM, 32QAM, 64QAM, 128QAM, and 256QAM.
- **Frequency**: FM and GMSK.

Each modulation type has a total of 106,496 observations ranging from −20 dB to +30 dB SNR in 2 dB increments. In total, there are 26 different SNR values. The SNR is assumed to be consistent over the same window length as the I/Q sample window.

The dataset was partitioned into 1 million training observations and 1.5 million testing observations through a random shuffle split, as carried out in [3,7]. This division was performed in a stratified manner, taking into account modulation type and the SNR. As a result of this balanced approach, the anticipated performance for a classifier employing random chance is 1/24 or approximately 4.2%. Considering the dataset's incorporation of diverse SNR levels, it is reasonable to expect that the classifier's accuracy would increase with the SNR. For consistency, each model investigated in this work was trained and evaluated on the same train and test set splits.

4. Initial Investigation

In this work, we use the architecture described in [7] as the baseline architecture. We note that [3] improved upon the baseline; however, each individual MC used the baseline architecture, except each is trained on specific SNR ranges. Therefore, the base architectural elements were similar to [7], but separated for different SNRs. In this work, our focus is to improve upon the employed CNN architecture for an individual MC rather than the use of several MCs. Therefore, we use the architecture from [7] as our baseline.

Before exploring an ablation study, we make a few notable changes from the baseline architecture in an effort to increase AMC performance. This initial exploration is for clarity as it reserves the ablation study that follows from requiring an inordinate number of models. It also introduces the general training procedures that assist and orient the reader in following the ablation study—the ablation study mirrors these procedures. We first provide an initial investigation exploring these notable changes.

We train each model using the Adam optimizer [45] with an initial learning rate $lr = 0.0001$, and a decay factor of 0.1, if the validation loss does not decrease for 12 epochs, and a minimum learning rate of 1×10^{-7}. If the validation loss does not decrease after 20 epochs, training is terminated and the models are deemed converged. For all experiments, mini-batches of size 32 are used. As has been established in most programming packages for neural networks, we refer to fully connected neural network layers as *dense* layers, which are typically followed by an activation function.

4.1. Architectural Changes

A common property of neural networks is using fewer but larger kernels in the early layers of the network, and an increased number of smaller kernels is used in the later layers, compared to the baseline architecture. This is commonly referred to as the information distillation pipeline [46]. By utilizing a smaller number of large kernels in early layers, we are able to increase the temporal context of the convolutional features without dramatically increasing the number of trainable parameters. Numerous, but smaller kernels are used in later convolutional layers to create more abstract features. Configuring the network in this manner is especially popular in image classification, where later layers represent more abstract, class-specific features.

We investigate this modification in three stages, using the baseline architecture described in Figure 3 [7]. We denote the number of filters in the network and the filter sizes as $F = [f_1, f_2, ..., f_7]$ and $K = [k_1, k_2, ...k_7]$ in Figure 3. The baseline architecture used $f = 64$ (for all layers) and $k = 3$ (consistent kernel size for all layers). Our first modification to the baseline architecture is $F = [32, 48, 64, 72, 84, 96, 108]$, but keeping $k = 3$ for all layers. Second, we use the baseline architecture, but change the size of filters in the network where $f = 64$ (the same as the baseline) and $K = [7, 5, 7, 5, 3, 3, 3]$. Third, we make both modifications and compare the result to the baseline model where $F = [32, 48, 64, 72, 84, 96, 108]$ and $K = [7, 5, 7, 5, 3, 3, 3]$. These modifications are not exhaustive searches; rather, these modifications are meant to guide future changes to the network by understanding the influence of filter quantity and filter size in a limited context.

4.2. Initial Investigation Results

As shown in Table 2, increasing the size of the filters in earlier layers increases both average and maximum test accuracy over [7], but at the cost of additional parameters. A possible explanation for the increase in performance is the increase in temporal context due to the larger kernel sizes. Increasing the number of filters without increasing temporal context decreases performance. This is possibly because it increases the complexity of the model without adding additional signal context.

Table 2. Initial investigation performance overview. All architectures employ the baseline with varying numbers of kernels and kernel sizes.

Notes	# Params	Avg. Accuracy	Max Accuracy
Reproduced ResNet [1]	165,144	59.2%	93.7%
X-Vector [7]	110,680	61.3%	98.0%
More Filters (Same Filter Sizes)	149,168	61.0%	96.1%
Larger Filter Sizes (Same # Filters)	143,960	62.6%	98.2%
Combined	174,000	62.9%	98.6%

Figure 6 illustrates the change in accuracy with varying SNR. The combined model, utilizing various kernel sizes and numbers of filters, consistently outperforms the other architectures across changing SNR conditions.

Although increasing the number of filters decreases performance alone, combining the approach with larger kernel sizes yields the best performance in our initial investigation. Increasing the temporal context may have allowed additional filters to better characterize the input signal. Because increased temporal context improves AMC performance, we are inspired to investigate additional methods, such as squeeze-and-excitation blocks and dilated convolutions, that can increase global and local context [34,47].

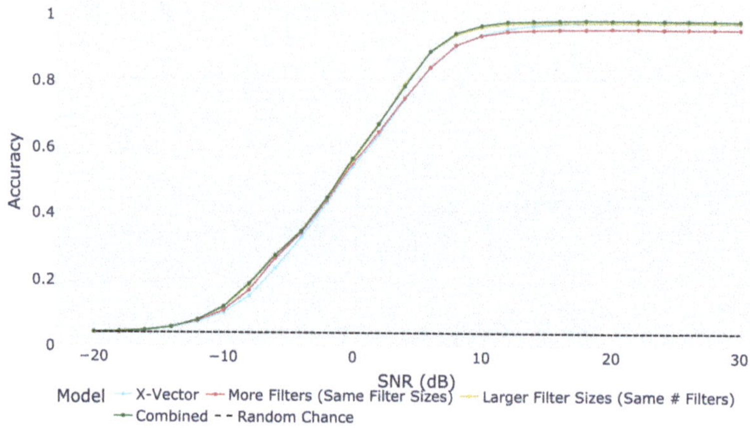

Figure 6. SNR vs. accuracy comparison of the initial investigation using the X-Vector baseline architecture [7]. Noticeable improvements can be observed across all SNRs.

5. Ablation Study Architecture Background

Building upon our findings from our initial investigation, we make additional modifications to the baseline architecture. For the MCs, we introduce dilated convolutions, squeeze-and-excitation blocks, self-attention, and other architectural changes. We also investigate various kernel sizes and the quantity of kernels employed from the initial investigation. Our goal is to improve upon existing architectures while investigating the impact of each modification on classification accuracy through an ablation study. In this section, we describe each modification performed.

5.1. *Squeeze-and-Excitation Networks*

Squeeze-and-excitation (SE) blocks introduce a channel-wise attention mechanism, first proposed in [34]. Due to the limited receptive field of each convolutional filter, SE blocks propose a recalibration step based on global statistics across channels (average pooling) to provide global context. Although initially utilized for image classification tasks [34,48,49], we argue the use of SE blocks can provide meaningful global context to the convolutional network used for AMC over the time domain.

Figure 7 depicts an SE block. The squeeze operation is defined as temporal global average pooling across convolutional filters. For an individual channel, c, the squeeze operation is defined as:

$$z_c = F_{sq}(x_c) = \frac{1}{T}\sum_{i=1}^{T} x_{i,c} \tag{1}$$

where $X \in \mathbb{R}^{T\times C} = [x_1, x_2, ..., x_C]$, $Z \in \mathbb{R}^{1\times C} = [z_1, z_2, ..., z_C]$, T is the number of samples in time, and C is the total number of channels. To model nonlinear interactions between channel-wise statistics, Z is fed into a series of dense layers followed by nonlinear activation functions:

$$s = F_{ex}(z, W) = \sigma(g(z, W)) = \sigma(W_2\delta(W_1 z)) \tag{2}$$

where δ is the rectified linear (ReLU) activation function, $W_1 \in \mathbb{R}^{\frac{C}{r}\times C}$, $W_2 \in \mathbb{R}^{C\times\frac{C}{r}}$, r is a dimensionality reduction ratio, and σ is the sigmoid activation function. The sigmoid function is chosen, as opposed to the softmax function, so that multiple channels can be accentuated and are not mutually exclusive. That is, the normalization term in the softmax can cause dependencies among channels, so the sigmoid activation is preferred. W_1 imposes a bottleneck to improve generalization performance and reduce parameter counts, while W_2 increases the dimensionality back to the original number of channels for the recalibration operation. In our work, we use $r = 2$ for all SE blocks to ensure a reasonable number of trainable parameters without over-squashing the embedding size.

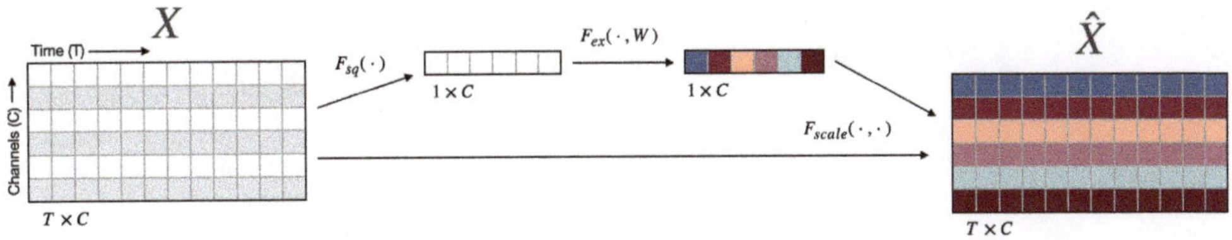

Figure 7. Squeeze-and-excitation block proposed in [34]. One SE block is shown applied to a single layer convolutional output activation. Two paths are shown: a scaling path and an identity path. The scaling vector is applied across channels to the identity path of the activations.

The final operation in the SE block, scaling or recalibration, is obtained by scaling the the input X by s:

$$\hat{x_c} = F_{scale}(x_c, s_c) = s_c x_c \tag{3}$$

where $\hat{X} \in \mathbb{R}^{T\times C} = [\hat{x_1}, \hat{x_2}, ..., \hat{x_C}]$.

5.2. Dilated Convolutions

As proposed in [47], Figure 8 depicts dilated convolutions, where the convolutional kernels are denoted by the colored components. In a traditional convolution, the dilation rate is equal to 1. Dilated convolutions build temporal context by increasing the receptive field of the convolutional kernels without increasing parameter counts, as the number of entries in the kernel remains the same.

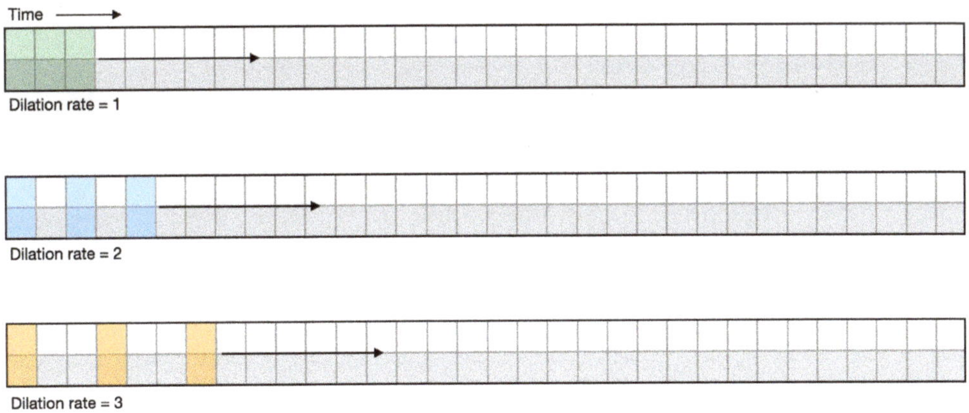

Figure 8. Dilated convolutions diagram. The top shows a traditional kernel applied to sequential time series points. The middle and bottom diagrams illustrate dilation rates of two and three, respectively. These dilations serve to increase the receptive field of the filter without increasing the number of trainable variables in the kernel.

Also, dilated convolutions do not downsample the signals like strided convolutions. Instead, the output of a dilated convolution can be the exact size of the input after properly handling edge effects at the beginning and end of the signal.

5.3. Final Convolutional Activation

We also investigate the impact of using an activation function (ReLU) after the last convolutional layer, just before statistics pooling. Because ReLU transforms the input sequence to be non-negative, the distribution characterized by the pooling statistics may become skewed. In [3,7], no activation was applied after the final convolutional layer, as shown in Figure 3. We investigate if this transformation impacts classification performance.

5.4. Self-Attention

Self-attention allows the convolutional outputs to interact with one another, enabling the network to learn to focus on important outputs. Self-attention before statistics pooling essentially creates a weighted summation over the convolutional outputs, weighting their importance similarly to [50–52].

We use the attention mechanism described by Vaswani et al. in [53], where each output element is a weighted sum of the linearly transformed input, where the dimensionality of K is d_k, as seen in Equation (4).

$$Attention(Q, K, V) = softmax\left(\frac{QK^T}{|\sqrt{d_k}|}\right)V \quad (4)$$

In the case of self-attention, Q, K, and V are equal. A scaling factor of $\frac{1}{|\sqrt{d_k}|}$ is applied to counteract vanishing gradients in the softmax output when d_k is large.

6. Ablation Study Architecture

Applying the specified modifications to the architecture in [7], Figure 9 illustrates the proposed architecture with every modification included in the graphic. Each colored block represents an optional change to the architecture that will be investigated in the ablation study. That is, each combination of network modifications is analyzed to aid understanding of each modification's impact on the network.

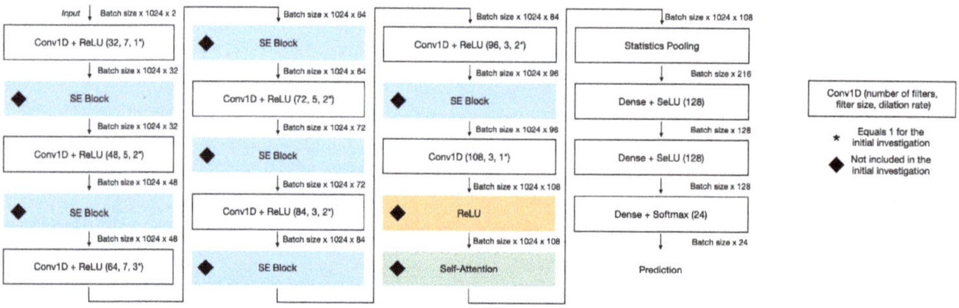

Figure 9. Proposed architecture with modifications including SENets, dilated convolutions, optional ReLU activation before statistics pooling, and self-attention. The output tensor sizes are also shown for each unit in the diagram. * denotes where the sizes differ from the baseline architecture.

Each convolutional layer has the following parameters: number of filters, kernel size, and dilation rate. The asterisk next to each dilation rate represents the changing of dilation rates in the ablation study. If dilated convolutions are used, then the dilation rate value in the graphic is used. If dilated convolutions are not used, each dilation rate is set to 1. That is, a traditional convolution is applied. All convolutions use a stride of 1, and the same training procedure from the initial investigation is used.

7. Evaluation Metrics

We present several evaluation metrics to compare the different architectures considered in the ablation study. In this section, we will discuss each evaluation technique used in the results section.

Due to the varying levels of SNRs in the employed dataset, we plot classification accuracy over each true SNR value. This allows for a visualization of the trade-off in performance, as noise becomes more or less dominant in the received signals. Additionally, we report average accuracy and maximum accuracy across the entire test set for each model. While we note that average accuracy is not indicative of the model's performance, as accuracy is highly correlated to the SNR of the input signal, we share this result to give other researchers the ability to reproduce and compare works.

As discussed in [35], AMC is often implemented on resource-constrained devices. In these systems, using larger models in terms of parameter counts may not be feasible. We report the number of parameters for each model in the ablation study to examine the trade-off in AMC performance and model size.

Additional analyses are also carried out. However, due to the large number of models investigated in this study, we will select the best-performing model from the ablation study for brevity and analyze the performance of this model in greater detail. For example, confusion matrices for the best-performing model from the ablation study are provided to show common misclassifications for each modulation type. Additionally, there exist several use-cases where relatively short signal bursts are received. For example, a wide-band scanning receiver may only detect a short signal burst. Therefore, signal duration in the time domain versus AMC performance is investigated to determine the robustness of the best-performing model when short signal bursts are received.

8. Ablation Results

8.1. Overall Performance

Table 3 lists the maximum and average accuracy performance for each model in the ablation study. A binary naming convention is used to indicate the various methods used for each architecture. Similarly to the result found in Section 4, increasing the temporal context typically results in increased performance. Models that incorporate dilated convolutions tended to have higher average accuracies than models without dilated convolutions.

The best-performing model, in terms of average accuracy across all SNR conditions, included SE blocks, dilated convolutions, and a ReLU activation, prior to statistics pooling (model 1110), with an average accuracy of approximately 63.7%. This model also achieved the highest maximum accuracy of about 98.9% at a 22 dB level. Both values achieve new state-of-the-art performance on the RadioML 2018.01A dataset. In terms of overall accuracy, model 1110 outperforms the results reported in prior work [1,3,7,37,54] (all between 52.47% and 61.3%) and all other models investigated in this work. In terms of peak accuracy, model 1110 outperforms the methods proposed in [1,3,4,7,37,39,41,43,44,54]—each with a reported peak accuracy between 80% and 98%.

SE blocks did not increase performance compared to model 0000, with the exception of models 1110 and 1111. However, SE blocks were incorporated in the best-performing model, 1110. Self-attention was not found to aid classification performance in general with the proposed architecture. Self-attention introduces a large number of trainable parameters, possibly forming a complex loss space.

Table 4 lists the performances of single modification (from baseline) architectures. Each component of the ablation study, with the exception of dilated convolutions, decreased performance when applied individually. When combined, however, the best-performing model was found. Therefore, we conclude that each component could possibly aid the optimization of each other—and, in general, dilated convolutions tend to have the most dramatic performance increases.

Table 3. Ablation study performance overview.

Model Name	Notes	SENet	Dilated Convolutions	Final Activation	Attention	# Params	Avg. Accuracy	Max Accuracy
—	Reproduced ResNet [1]	—	—	—	—	165,144	59.2%	93.7%
—	X-Vector [7]	—	—	—	—	110,680	61.3%	98.0%
0000	Best-performing model from the initial investigation	—	—	—	—	174,000	62.9%	98.6%
0001		—	—	—	✓	221,088	62.3%	97.6%
0010		—	—	✓	—	174,000	62.8%	98.6%
0011		—	—	✓	✓	221,088	62.3%	97.5%
0100		—	✓	—	—	174,000	63.2%	98.9%
0101		—	✓	—	✓	221,088	63.1%	97.9%
0110		—	✓	✓	—	174,000	63.2%	98.9%
0111		—	✓	✓	✓	221,088	63.0%	98.0%
1000		✓	—	—	—	202,880	62.9%	98.5%
1001		✓	—	—	✓	249,968	62.6%	98.2%
1010		✓	—	✓	—	202,880	62.6%	98.3%
1011		✓	—	✓	✓	249,968	62.8%	98.1%
1100		✓	✓	—	—	202,880	62.8%	98.2%
1101		✓	✓	—	✓	249,968	63.0%	97.7%
1110	Overall best performing model	✓	✓	✓	—	202,880	63.7%	98.9%
1111		✓	✓	✓	✓	249,968	63.0%	97.8%

Table 4. Individual network modification performance overview. Entries are repeated from Table 3 for clarity.

Model Name	Notes	SENet	Dilated Convolutions	Final Activation	Attention	# Params	Avg. Accuracy	Max Accuracy
—	X-Vector [7]	—	—	—	—	110,680	61.3%	98.0%
0000		—	—	—	—	174,000	62.9%	98.6%
0001		—	—	—	✓	221,088	62.3%	97.6%
0010		—	—	✓	—	174,000	62.8%	98.6%
0100		—	✓	—	—	174,000	63.2%	98.9%
1000		✓	—	—	—	202,880	62.9%	98.5%
1110	Best-performer	✓	✓	✓	—	202,880	63.7%	98.9%

8.2. Accuracy over Varying SNR

Figure 10 summarizes the ablation study in terms of classification accuracy over varying SNR levels. We add this figure for completeness and reproducibility for other researchers. The accuracy within each SNR band is shown along with the modifications used, similar to Table 3. The coloring in the figure denotes the accuracy in each SNR band. The performance follows a trend similar to that of a sigmoid function, where the rate at which peak classification accuracy is achieved is the most distinguishing feature between the different models. With the improved architectures, a maximum of 99% accuracy is achieved at high SNR levels (starting around 12 dB SNR).

Model Name	Notes	SENet	Dilated Convolutions	Final Activation	Attention
–	Reproduced ResNet	–	–	–	–
–	X-Vector	–	–	–	–
0000	–	–	–	–	–
0001	–	–	–	–	✓
0010	–	–	–	✓	–
0011	–	–	–	✓	✓
0100	–	–	✓	–	–
0101	–	–	✓	–	✓
0110	–	–	✓	✓	–
0111	–	–	✓	✓	✓
1000	–	✓	–	–	–
1001	–	✓	–	–	✓
1010	–	✓	–	✓	–
1011	–	✓	–	✓	✓
1100	–	✓	✓	–	–
1101	–	✓	✓	–	✓
1110	Overall best performing model	✓	✓	✓	–
1111	–	✓	✓	✓	✓

Modulation Classification Results

Model Name	-20	-18	-16	-14	-12	-10	-8	-6	-4	-2	0	2	4	6	8	10	12	14	16	18	20	22	24	26	28	30
Reproduced ResNet	0.04	0.04	0.05	0.05	0.08	0.13	0.18	0.23	0.33	0.42	0.53	0.62	0.73	0.82	0.87	0.91	0.93	0.93	0.94	0.94	0.94	0.94	0.94	0.94	0.93	0.93
X-Vector	0.04	0.05	0.05	0.06	0.07	0.10	0.15	0.24	0.33	0.43	0.54	0.64	0.75	0.84	0.91	0.94	0.96	0.97	0.98	0.98	0.98	0.98	0.98	0.98	0.98	0.98
0000	0.04	0.04	0.05	0.06	0.08	0.12	0.19	0.28	0.35	0.45	0.57	0.67	0.79	0.89	0.95	0.97	0.98	0.98	0.98	0.99	0.99	0.99	0.99	0.99	0.99	0.99
0001	0.04	0.05	0.05	0.06	0.08	0.12	0.19	0.27	0.35	0.45	0.56	0.66	0.78	0.88	0.94	0.96	0.97	0.97	0.97	0.97	0.98	0.98	0.98	0.98	0.98	0.98
0010	0.04	0.05	0.05	0.06	0.08	0.12	0.18	0.26	0.35	0.44	0.56	0.67	0.79	0.89	0.95	0.97	0.98	0.98	0.98	0.99	0.99	0.99	0.99	0.99	0.98	0.99
0011	0.04	0.04	0.05	0.06	0.08	0.12	0.19	0.27	0.35	0.45	0.56	0.67	0.78	0.89	0.94	0.96	0.97	0.97	0.97	0.97	0.97	0.97	0.97	0.98	0.98	0.97
0100	0.04	0.04	0.05	0.06	0.08	0.12	0.18	0.25	0.35	0.45	0.57	0.68	0.81	0.92	0.97	0.98	0.99	0.99	0.99	0.99	0.99	0.99	0.99	0.99	0.99	0.99
0101	0.04	0.05	0.05	0.06	0.09	0.13	0.19	0.28	0.35	0.45	0.57	0.68	0.81	0.92	0.96	0.97	0.98	0.98	0.98	0.98	0.98	0.98	0.98	0.98	0.98	0.98
0110	0.04	0.04	0.05	0.05	0.07	0.11	0.17	0.25	0.35	0.46	0.57	0.68	0.81	0.92	0.97	0.98	0.99	0.99	0.99	0.99	0.99	0.99	0.99	0.99	0.99	0.99
0111	0.04	0.05	0.05	0.06	0.08	0.12	0.18	0.26	0.35	0.46	0.57	0.69	0.81	0.92	0.96	0.97	0.98	0.98	0.98	0.98	0.98	0.98	0.98	0.98	0.98	0.98
1000	0.04	0.05	0.05	0.06	0.08	0.12	0.19	0.28	0.36	0.46	0.57	0.67	0.78	0.88	0.94	0.97	0.98	0.98	0.98	0.99	0.98	0.98	0.98	0.98	0.98	0.98
1001	0.04	0.05	0.05	0.06	0.08	0.12	0.19	0.28	0.35	0.45	0.56	0.67	0.78	0.87	0.93	0.96	0.97	0.98	0.98	0.98	0.98	0.98	0.98	0.98	0.98	0.98
1010	0.04	0.05	0.05	0.06	0.08	0.12	0.18	0.27	0.35	0.46	0.57	0.67	0.78	0.88	0.94	0.97	0.98	0.98	0.98	0.98	0.98	0.98	0.98	0.98	0.98	0.98
1011	0.04	0.05	0.05	0.06	0.08	0.12	0.20	0.28	0.35	0.46	0.57	0.67	0.79	0.89	0.94	0.96	0.98	0.98	0.98	0.98	0.98	0.98	0.98	0.98	0.98	0.98
1100	0.04	0.05	0.05	0.06	0.08	0.12	0.19	0.28	0.35	0.45	0.55	0.67	0.79	0.91	0.96	0.97	0.98	0.98	0.98	0.98	0.98	0.98	0.98	0.98	0.98	0.98
1101	0.04	0.05	0.05	0.06	0.09	0.13	0.20	0.28	0.36	0.46	0.58	0.69	0.81	0.91	0.95	0.97	0.97	0.98	0.98	0.98	0.98	0.98	0.98	0.98	0.98	0.98
1110	0.04	0.05	0.05	0.06	0.08	0.12	0.19	0.28	0.36	0.46	0.58	0.69	0.82	0.92	0.97	0.98	0.99	0.99	0.99	0.99	0.99	0.99	0.99	0.99	0.99	0.99
1111	0.04	0.05	0.05	0.06	0.09	0.13	0.20	0.28	0.36	0.46	0.57	0.69	0.81	0.91	0.96	0.97	0.97	0.98	0.98	0.98	0.98	0.98	0.98	0.98	0.98	0.98

SNR (dB)

Figure 10. Ablation study results in terms of classification accuracy across SNR ranges. The reproduced ResNet [1] and X-Vector baseline [7] architectures are included. The best-performing model is in the second-to-last row and displays strong performance across SNR values.

While the proposed changes to the architectures generally improve performance at higher SNR levels, the largest improvements occur between −12 dB and 12 dB, compared to the baseline model in [7]. For example, at 4 dB, the performance increases from 75% up to 82%. Incorporating these modifications to the network may prove to be critical in real-world situations, where noisy signals are likely to be obtained. Improving AMC performance at lower SNR ranges (<−12 dB) is still an open research topic, with accuracies at near-chance level.

A receiver using model 1110 within its demodulator achieves a notable 91% classification accuracy at 6 dB SNR, which is an improvement compared to previous work [1], which achieved a similar accuracy of around 10 dB, and [7], which achieved a similar accuracy of around 8 dB. Wireless communications systems employed in various applications can suffer from poor reception in low SNR environments due to environmental conditions, such as complex channel characteristics, multipath interference, fading, and man-made conditions, such as congested channels, among other factors. Therefore, any improvement that can increase performance at low SNR is desirable. Because AMC can directly impact decision-making algorithms, in situations where reliable communications are essential, such as emergency response systems, military operations, or autonomous vehicle networks, the ability to accurately classify modulations under challenging SNR conditions becomes a pivotal determinant of system effectiveness and safety.

One observation is that the best-performing model can vary with the SNR. In systems that have available memory and processing power, an approach similar to [3] may be used to utilize several models and intelligently choose predictions based on estimated SNR conditions. That is, if the SNR of the signal of interest is known, a model can be tuned to increase performance slightly, as shown in [3]. Using the results presented here, researchers could also choose the architecture differences that perform best for a given SNR range (although performance differences are subtle).

8.3. Parameter Count Trade-Off

An overview of each model's complexity and overall performance across the entire testing set is shown in Table 3. This information is also shown graphically in Figure 11 for the maximum accuracy over SNR and the average accuracy across all SNRs. Whether looking at the maximum or the average measures of performance, the conclusions are similar. The previously described binary model name also appears in the figure. We found a slight correlation between the number of model parameters and overall model performance; however, with the architectures explored, there was a general parameter count where performance peaked. Models with parameter counts between approximately 170 k and 205 k generally performed better than smaller and larger models. We note that the models with more than 205 k parameters included self-attention, which was found to decrease model performance with the proposed architectures. This implies that one possible reason self-attention did not perform as well as other modifications is because of the increase in parameters, resulting in a more difficult loss space, from which to optimize.

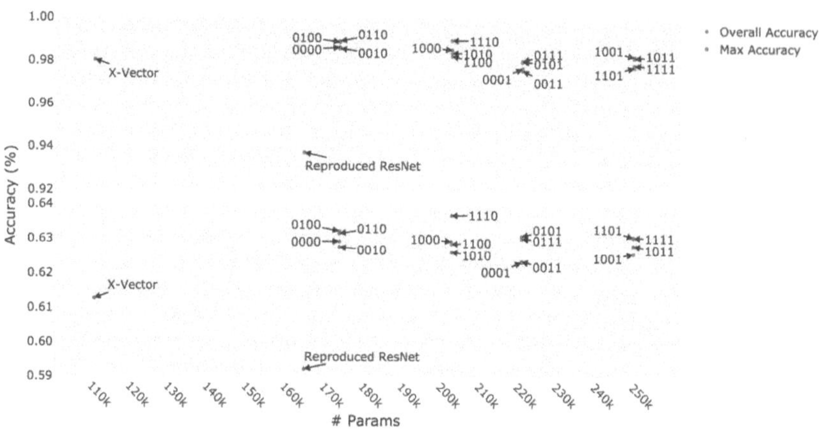

Figure 11. Ablation study parameter count trade-off including the reproduced ResNet [1] and X-Vector baseline [7]. The x-axis shows the number of trainable variables in each model and the y-axis shows max or average accuracy. The callout for each point denotes the model name, as shown in Table 3.

9. Best-Performing Model Investigation

Due to the large volume of models, we focus upon the best-performing model, model 1110, for the remainder of this work. As previously mentioned, this model employs all modifications except self-attention.

9.1. Top-K Accuracy

As discussed, in systems where the modulation schemes must be classified quickly, it is advantageous to apply fewer demodulation schemes in a trial-and-error fashion. This is particularly significant at lower SNR values, where accuracy is mediocre. Top-k accuracy allows an in-depth view of the expected number of trials before finding the correct modulation scheme. Although traditional accuracy (top-1 accuracy) characterizes the performance of the model in terms of classifying the exact variant, top-k accuracy characterizes the percentage of the classifier predicting the correct variant among the top-k predictions (sorted by descending class probabilities). We plot the top-1, top-2, and top-5 classification accuracy over varying SNR conditions for each modulation grouping, as defined in Section 3 in Figure 12.

Although performance decays to approximately random chance for the overall (all modulation schemes) performance curves for each top-k accuracy, it is notable that some modulation group performances drop below random chance. The models are trained to maximize the overall model performance. This could explain why certain modulation groups dip below random chance but the overall performance and other modulation groups remain at or above random chance.

Using the proposed method greatly reduces the correct modulation scheme search space. While high performance in top-1 accuracy is increasingly difficult to achieve with low

SNR signals, top-2 and top-5 accuracies converge to higher values at a much faster rate. This indicates that our proposed method greatly reduces the search space from 24 modulation candidates to fewer candidate types when employing trial-and-error methods to determine the correct modulation scheme. Further, if the group of modulation is known (e.g., FM), one can view a more specific trade-off curve in terms of SNR and top-k accuracy, as given in Figure 12.

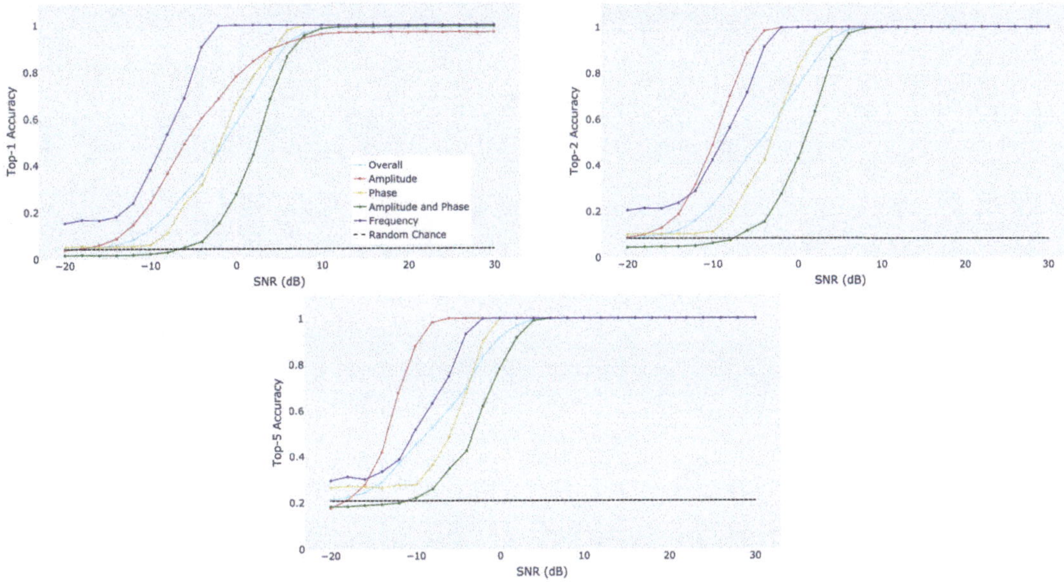

Figure 12. Top-1 (**top left**), top-2 (**top right**), and top-5 (**bottom**) accuracy over varying SNR conditions for model 1110. Random chance for each is defined as 1/24, 2/24, and 5/24, respectively.

9.2. Short-Duration Signal Bursts

Due to the rapid scanning characteristic of some modern software-defined radios, we investigate the performance trade-off of varying signal duration and AMC performance. This analysis is meant to emulate the situation wherein a receiver only detects a short RF signal burst. We investigate signal burst durations of 1.024 ms (full length signal from original dataset), 512 µs, 256 µs, 128 µs, 64 µs, 32 µs, and 16 µs. We assume the same 1 MS/sec sampling rate, as in the previous analyses, such that the 16 µs burst is captured in 16 I/Q samples.

In this section, we use the same test set as our other investigations; however, a uniformly random starting point is determined for each signal such that a contiguous sample of the desired duration, starting at the random point, is chosen. Thus, the chosen segment from a test set sample is randomly assigned.

We also note that, although the sample length for the evaluation is changed, the best-performing model is the same architecture with exactly the same trained weights, because this model uses statistics pooling from the X-Vector-inspired modification. A significant benefit of the X-Vector-inspired architecture is its ability to handle variable-length inputs without the need of padding, retraining, or other network modifications. This is achieved by taking global statistics across convolutional channels, producing a fixed-length vector, regardless of signal duration. Due to this flexibility, the same model (model 1110) weights are used for each duration experiment. This fact also emphasizes the desirability of using X-vector-inspired AMC architectures for receivers that are deployed in an environment where short-burst and variable duration signals are anticipated to be present.

For each signal duration in the time domain, we plot the overall classification accuracy over varying SNR conditions, as well as the accuracy for each modulation grouping defined in Section 3 in Figure 13, which demonstrates the trade-off for various signal durations, where n is the number of samples from the time domain I/Q signal. The first observation is, as we would expect, that classification performance degrades with decreased signal duration, similarly to [39]. For example, the maximum accuracy begins to degrade at 256 µs and is more noticeable at 128 µs. This is likely a result of using sample statistics that result in unstable or biased estimates for short signal lengths, since the number of received signal data points are insufficient to characterize the sample statistics used during training. Random classification accuracy is approximately 4% and is shown in the black dotted line in Figure 13. Although classification performance decreases with decreased duration, we are still able to achieve significantly higher classification accuracy than random chance, down to 16 µs of signal capture.

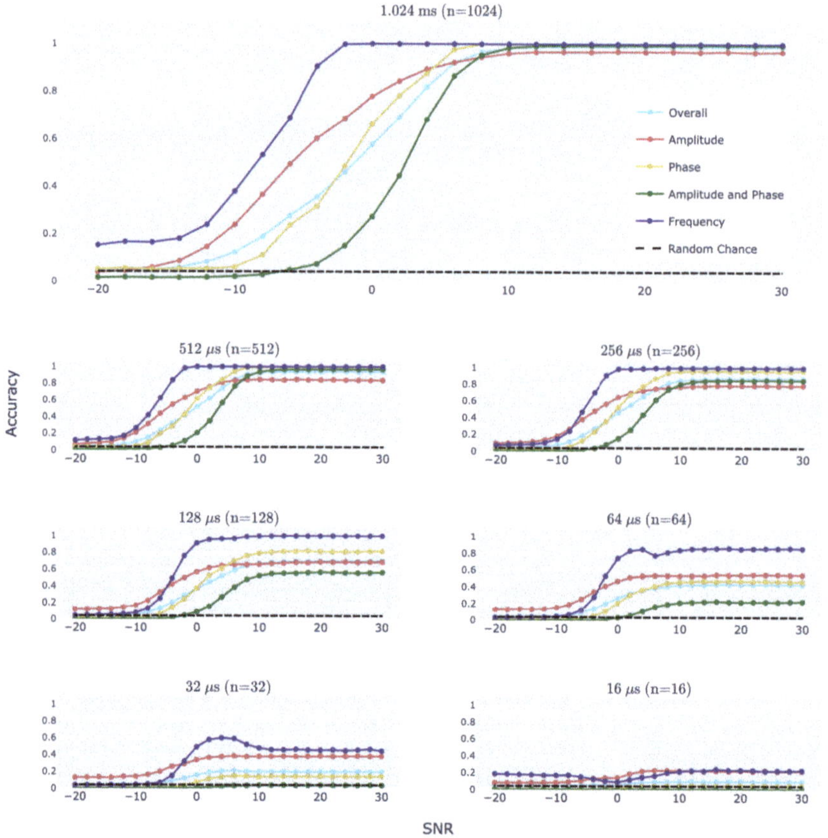

Figure 13. Trade-off in accuracy for various signal lengths across the SNR, grouped by modulation category for the best-performing model, 1110. The top plot shows the baseline performance using the full sequence. Subsequent plots show the same information using increasingly smaller signal lengths for classification.

FM (frequency modulation) signals were typically more resilient to noise interference than AM (amplitude modulation) and AM–PM (amplitude and phase modulation) signals in our AMC. This was observed across all signal burst durations and our top-k accuracy analysis. This behavior indicates that the performance of our AMC for short bursts, in the presence of increasing amounts of noise, is more robust for signals modulated by changes in the carrier frequency and is more sensitive to signals modulated by varying the carrier amplitude. We attribute this behavior to our AMC architecture, the architecture of the receiver, or a combination of both of the AMC and receiver.

9.3. Confusion Matrices

While classification accuracy provides a holistic view of model performance, it lacks the granularity to investigate where misclassifications are occurring. Confusion matrices are used to analyze the distribution of classifications for each given class. For each true label, the proportion of correctly classified samples is calculated along with the proportion of incorrect predictions for each opposing class. In this way, we can see which classes the model is struggling to distinguish from one another. A perfect classifier would be the identity matrix where the diagonal values represent the true class, and which match the predicted class. Each matrix value represents the percentage of classifications for the true label and each row sums to 1 (100%).

Figure 14 illustrates the class confusion matrices for SNR levels greater than or equal to 0 dB for models 1110, the reproduced ResNet architecture from [1], and the baseline X-Vector architecture from [7], respectively. Shown in [7], the X-Vector architecture was able to distinguish PSK and AM-SSB variants to a higher degree and performed better overall than [1]. Both architectures struggled to differentiate QAM variants.

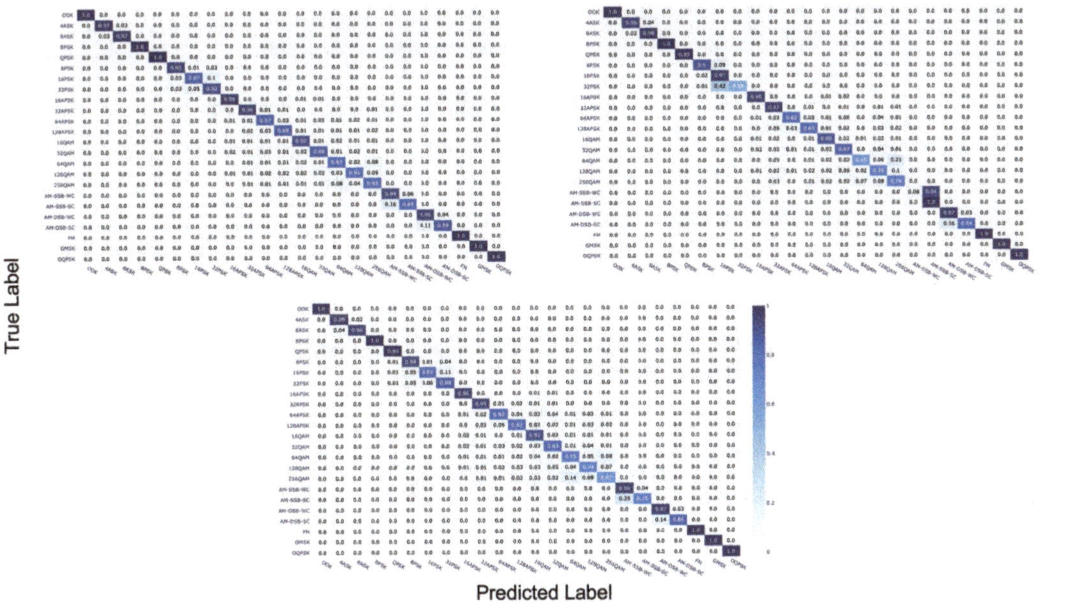

Figure 14. Confusion matrices for model 1110—the best-performing model from this work (**top left**), the reproduced ResNet model from [1] (**top right**), and the X-Vector-inspired model from [7] (**bottom**) with SNR $\geq$ 0 dB.

Model 1110 improved upon these prior results for QAM signals and, in general, has higher diagonal components than the other architectures. This, again, supports a conclusion that model 1110 achieves a new state of the art in AMC performance.

10. Conclusions

A comprehensive ablation study was carried out with regard to AMC architectural features using the extensive RadioML 2018.01A dataset. This ablation study built upon a strong performance of a new baseline model that was also introduced in the initial investigation of this study. This initial investigation informed the design of a number of AMC architecture modifications—specifically, the use of X-Vectors, dilated convolutions, and SE blocks. With the combined modifications, we achieved a new state of the art in AMC accuracy, improving upon prior work by approximately 2.5% overall accuracy on the RadioML 2018.01A dataset. We also achieve a new state of the art in peak performance with 98.9% accuracy at high SNR values. Among these modifications, dilated convolutions were found to be the most critical architectural feature for model performance. Self-attention was also investigated, but was not found to increase performance—although increased temporal context improved upon prior works. Additionally, the best-performing model was found to be robust against signals of varying duration, down to 128 μs of signal capture.

Author Contributions: Conceptualization, C.A.H.; Software, C.A.H.; Supervision, M.A.T. and E.C.L.; Visualization, C.A.H.; Writing—original draft, C.A.H.; Writing—review and editing, C.A.H., M.A.T. and E.C.L. All authors have read and agreed to the published version of the manuscript.

Funding: This research received no external funding.

Data Availability Statement: Publicly available datasets were analyzed in this study. This data can be found here: https://www.deepsig.ai/datasets (accessed on 1 February 2020).

Conflicts of Interest: The authors declare no conflict of interest.

References

1. O'Shea, T.; Roy, T.; Clancy, T.C. Over-the-Air Deep Learning Based Radio Signal Classification. *IEEE J. Sel. Top. Signal Process.* **2018**, *12*, 168–179. [CrossRef]
2. Jacob, P.; Sirigina, R.P.; Madhukumar, A.S.; Prasad, V.A. Cognitive Radio for Aeronautical Communications: A Survey. *IEEE Access* **2016**, *4*, 3417–3443. [CrossRef]
3. Harper, C.A.; Sinha, A.; Thornton, M.A.; Larson, E.C. SNR-Boosted Automatic Modulation Classification. In Proceedings of the 2021 55th Asilomar Conference on Signals, Systems, and Computers, Pacific Grove, CA, USA, 31 October–3 November 2021; pp. 372–375.
4. Soltani, N.; Sankhe, K.; Ioannidis, S.; Jaisinghani, D.; Chowdhury, K. Spectrum awareness at the edge: Modulation classification using smartphones. In Proceedings of the 2019 IEEE International Symposium on Dynamic Spectrum Access Networks (DySPAN), Newark, NJ, USA, 11–14 November 2019; pp. 1–10.
5. Swami, A.; Sadler, B. Hierarchical Digital Modulation Classification using Cumulants. *IEEE Trans. Commun.* **2000**, *48*, 416–429. [CrossRef]
6. Abdelbar, M.; Tranter, W.H.; Bose, T. Cooperative Cumulants-Based Modulation Classification in Distributed Networks. *IEEE Trans. Cogn. Commun. Netw.* **2018**, *4*, 446–461. [CrossRef]
7. Harper, C.A.; Lyons, L.; Thornton, M.A.; Larson, E.C. Enhanced Automatic Modulation Classification Using Deep Convolutional Latent Space Pooling. In Proceedings of the 2020 54th Asilomar Conference on Signals, Systems, and Computers, Pacific Grove, CA, USA, 1–4 November 2020.
8. O'Shea, T.; Corgan, J.; Clancy, T. Convolutional Radio Modulation Recognition Networks. In Proceedings of the International Conference on Engineering Applications of Neural Networks, Aberdeen, UK, 2–5 September 2016; pp. 213–226.
9. Huynh-The, T.; Hua, C.H.; Kim, J.W.; Kim, S.H.; Kim, D.S. Exploiting a Low-Cost CNN with Skip Connection for Robust Automatic Modulation Classification. In Proceedings of the 2020 IEEE Wireless Communications and Networking Conference (WCNC), Seoul, Republic of Korea, 25–28 May 2020; pp. 1–6. [CrossRef]
10. Peng, S.; Jiang, H.; Wang, H.; Alwageed, H.; Yao, Y. Modulation Classification using Convolutional Neural Network Based Deep Learning Model. In Proceedings of the 2017 26th Wireless and Optical Communication Conference (WOCC), Newark, NJ, USA, 7–8 April 2017; pp. 1–5. [CrossRef]
11. Peng, S.; Jiang, H.; Wang, H.; Alwageed, H.; Zhou, Y.; Sebdani, M.M.; Yao, Y. Modulation Classification Based on Signal Constellation Diagrams and Deep Learning. *IEEE Trans. Neural Netw. Learn. Syst.* **2019**, *30*, 718–727. [CrossRef]
12. Wang, F.; Wang, Y.; Chen, X. Graphic Constellations and DBN Based Automatic Modulation Classification. In Proceedings of the 2017 IEEE 85th Vehicular Technology Conference (VTC Spring), Sydney, NSW, Australia, 4–7 June 2017; pp. 1–5. [CrossRef]
13. Dulek, B. A Sparse Approach for Identification of Signal Constellations Over Additive Noise Channels. *IEEE Trans. Aerosp. Electron. Syst.* **2020**, *56*, 817–822. [CrossRef]
14. Hassan, K.; Dayoub, I.; Hamouda, W.; Nzeza, C.N.; Berbineau, M. Blind Digital Modulation Identification for Spatially-Correlated MIMO Systems. *IEEE Trans. Wirel. Commun.* **2012**, *11*, 683–693. [CrossRef]
15. Abdelbar, M.; Tranter, B.; Bose, T. Cooperative Modulation Classification of Multiple Signals in Cognitive Radio Networks. In Proceedings of the 2014 IEEE International Conference on Communications (ICC), Sydney, NSW, Australia, 10–14 June 2014; pp. 1483–1488. . [CrossRef]

16. Zhang, Q.; Guan, Y.; Li, H.; Song, Z. Distributed Cooperative Automatic Modulation Classification Using DWA-ADMM in Wireless Communication Networks. *Electronics* **2023**, *12*, 3002. [CrossRef]
17. Ren, B.; Teh, K.C.; An, H.; Gunawan, E. Automatic Modulation Recognition of Dual-Component Radar Signals Using ResSwinT-SwinT Network. *IEEE Trans. Aerosp. Electron. Syst.* **2023**, pp. 1–13. [CrossRef]
18. Alzaq-Osmanoglu, H.; Alrehaili, J.; Ustundag, B.B. Low-SNR Modulation Recognition based on Deep Learning on Software Defined Radio. In Proceedings of the 2022 5th International Conference on Advanced Communication Technologies and Networking (CommNet), Marrakech, Morocco, 12–14 December 2022; pp. 1–6.
19. Wang, Y.; Liu, M.; Yang, J.; Gui, G. Data-Driven Deep Learning for Automatic Modulation Recognition in Cognitive Radios. *IEEE Trans. Veh. Technol.* **2019**, *68*, 4074–4077. [CrossRef]
20. Meng, F.; Chen, P.; Wu, L.; Wang, X. Automatic Modulation Classification: A Deep Learning Enabled Approach. *IEEE Trans. Veh. Technol.* **2018**, *67*, 10760–10772. [CrossRef]
21. Zhang, L.; Liu, H.; Yang, X.; Jiang, Y.; Wu, Z. Intelligent Denoising-Aided Deep Learning Modulation Recognition With Cyclic Spectrum Features for Higher Accuracy. *IEEE Trans. Aerosp. Electron. Syst.* **2021**, *57*, 3749–3757. [CrossRef]
22. DeepSig Incorporated. RF Datasets for Machine Learning. 2018. Available online: https://www.deepsig.ai/datasets (accessed on 29 April 2021).
23. Liu, D.; Wang, P.; Wang, T.; Abdelzaher, T. Self-Contrastive Learning based Semi-Supervised Radio Modulation Classification. In Proceedings of the MILCOM 2021-2021 IEEE Military Communications Conference (MILCOM), San Diego, CA, USA, 29 November–2 December 2021; pp. 777–782.
24. Chen, Z.; Cui, H.; Xiang, J.; Qiu, K.; Huang, L.; Zheng, S.; Chen, S.; Xuan, Q.; Yang, X. SigNet: A Novel Deep Learning Framework for Radio Signal Classification. *IEEE Trans. Cogn. Commun. Netw.* **2022**, *8*, 529–541. [CrossRef]
25. Wang, N.; Liu, Y.; Ma, L.; Yang, Y.; Wang, H. Multidimensional CNN-LSTM network for automatic modulation classification. *Electronics* **2021**, *10*, 1649. [CrossRef]
26. Hochreiter, S.; Schmidhuber, J. Long short-term memory. *Neural Comput.* **1997**, *9*, 1735–1780. [CrossRef] [PubMed]
27. Chen, T.; Gao, S.; Zheng, S.; Yu, S.; Xuan, Q.; Lou, C.; Yang, X. EMD and VMD Empowered Deep Learning for Radio Modulation Recognition. *IEEE Trans. Cogn. Commun. Netw.* **2022**, *9*, 43–57. [CrossRef]
28. Uppal, A.J.; Hegarty, M.; Haftel, W.; Sallee, P.A.; Cribbs, H.B.; Huang, H.H. High-Performance Deep Learning Classification for Radio Signals. In Proceedings of the 2019 53rd Asilomar Conference on Signals, Systems, and Computers, Pacific Grove, CA, USA, 3–6 November 2019.
29. Park, C.; Choi, J.; Nah, S.; Jang, W.; Kim, D.Y. Automatic Modulation Recognition of Digital Signals using Wavelet Features and SVM. In Proceedings of the 2008 10th International Conference on Advanced Communication Technology, Gangwon, Republic of Korea, 17–20 February 2008; Volume 1, pp. 387–390. [CrossRef]
30. Teng, X.; Tian, P.; Yu, H. Modulation Classification Based on Spectral Correlation and SVM. In Proceedings of the 2008 4th International Conference on Wireless Communications, Networking and Mobile Computing, Dalian, China, 12–14 October 2008; pp. 1–4. [CrossRef]
31. Zhang, Z.; Wang, C.; Gan, C.; Sun, S.; Wang, M. Automatic Modulation Classification Using Convolutional Neural Network With Features Fusion of SPWVD and BJD. *IEEE Trans. Signal Inf. Process. Netw.* **2019**, *5*, 469–478. [CrossRef]
32. Zheng, S.; Qi, P.; Chen, S.; Yang, X. Fusion Methods for CNN-Based Automatic Modulation Classification. *IEEE Access* **2019**, *7*, 66496–66504. [CrossRef]
33. Mao, Y.; Dong, Y.Y.; Sun, T.; Rao, X.; Dong, C.X. Attentive Siamese Networks for Automatic Modulation Classification Based on Multitiming Constellation Diagrams. *IEEE Trans. Neural Netw. Learn. Syst.* **2021**, *34*, 5988–6002. [CrossRef]
34. Hu, J.; Shen, L.; Sun, G. Squeeze-and-excitation networks. In Proceedings of the IEEE Conference on Computer Vision and Pattern Recognition, Salt Lake City, UT, USA, 18–22 June 2018; pp. 7132–7141.
35. Tridgell, S. Low Latency Machine Learning on FPGAs. Ph.D. Thesis, The University of Sydney Australia, Sydney, NSW, Australia, 2019.
36. Mendis, G.J.; Wei-Kocsis, J.; Madanayake, A. Deep Learning Based Radio-Signal Identification With Hardware Design. *IEEE Trans. Aerosp. Electron. Syst.* **2019**, *55*, 2516–2531. [CrossRef]
37. Wang, Z.; Sun, D.; Gong, K.; Wang, W.; Sun, P. A Lightweight CNN Architecture for Automatic Modulation Classification. *Electronics* **2021**, *10*, 2679. [CrossRef]
38. Snyder, D.; Garcia-Romero, D.; Sell, G.; Povey, D.; Khudanpur, S. X-vectors: Robust DNN Embeddings for Speaker Recognition. In Proceedings of the 2018 IEEE International Conference on Acoustics, Speech and Signal Processing (ICASSP), Calgary, AB, Canada, 15–20 April 2018; pp. 5329–5333.
39. Li, L.; Dong, Z.; Zhu, Z.; Jiang, Q. Deep-Learning Hopping Capture Model for Automatic Modulation Classification of Wireless Communication Signals. *IEEE Trans. Aerosp. Electron. Syst.* **2023**, *59*, 772–783. [CrossRef]
40. McNemar, Q. Note on the Sampling Error of the Difference Between Correlated Proportions or Percentages. *Psychometrika* **1947**, *12*, 153–157. [CrossRef]
41. Cai, J.; Gan, F.; Cao, X.; Liu, W. Signal Modulation Classification Based on the Transformer Network. *IEEE Trans. Cogn. Commun. Netw.* **2022**, *8*, 1348–1357. [CrossRef]

42. Wen, Y.; Zhang, K.; Li, Z.; Qiao, Y. A discriminative feature learning approach for deep face recognition. In Proceedings of the European Conference on Computer Vision, Amsterdam, The Netherlands, 11–14 October 2016; Springer: Berlin/Heidelberg, Germany, 2016; pp. 499–515.
43. Zhang, H.; Zhou, F.; Wu, Q.; Wu, W.; Hu, R.Q. A Novel Automatic Modulation Classification Scheme Based on Multi-Scale Networks. *IEEE Trans. Cogn. Commun. Netw.* **2022**, *8*, 97–110. [CrossRef]
44. Zhang, D.; Lu, Y.; Li, Y.; Ding, W.; Zhang, B. High-Order Convolutional Attention Networks for Automatic Modulation Classification in Communication. *IEEE Trans. Wirel. Commun.* **2022**, *22*, 4600–4610. [CrossRef]
45. Kingma, D.P.; Ba, J. Adam: A Method for Stochastic Optimization. *arXiv* **2015**, arXiv:1412.6980.
46. Chollet, F. *Deep Learning with Python*; Manning Publications: Shelter Island, NY, USA, 2021.
47. Yu, F.; Koltun, V. Multi-Scale Context Aggregation by Dilated Convolutions. In Proceedings of the International Conference on Learning Representations (ICLR), San Juan, Puerto Rico, 2–4 May 2016.
48. Zhang, X.; Zhou, X.; Lin, M.; Sun, J. Shufflenet: An extremely efficient convolutional neural network for mobile devices. In Proceedings of the IEEE Conference on Computer Vision and Pattern Recognition, Salt Lake City, UT, USA, 18–22 June 2018; pp. 6848–6856.
49. Tan, M.; Le, Q. Efficientnet: Rethinking model scaling for convolutional neural networks. In Proceedings of the International Conference on Machine Learning. PMLR, Long Beach, CA, USA, 9–15 June 2019; pp. 6105–6114.
50. Okabe, K.; Koshinaka, T.; Shinoda, K. Attentive Statistics Pooling for Deep Speaker Embedding. In Proceedings of the Interspeech 2018, Hyderabad, India, 2–6 September 2018; pp. 2252–2256. [CrossRef]
51. Safari, P.; India, M.; Hernando, J. Self-Attention Encoding and Pooling for Speaker Recognition. In Proceedings of the Interspeech 2020, Shanghai, China, 25–29 October 2020; pp. 941–945. [CrossRef]
52. Sammit, G.; Wu, Z.; Wang, Y.; Wu, Z.; Kamata, A.; Nese, J.; Larson, E.C. Automated prosody classification for oral reading fluency with quadratic kappa loss and attentive x-vectors. In Proceedings of the ICASSP 2022–2022 IEEE International Conference on Acoustics, Speech and Signal Processing (ICASSP), Singapore, 23–27 May 2022.
53. Vaswani, A.; Shazeer, N.; Parmar, N.; Uszkoreit, J.; Jones, L.; Gomez, A.N.; Kaiser, Ł.; Polosukhin, I. Attention is all you need. In Proceedings of the Advances in Neural Information Processing Systems, Long Beach, CA, USA, 4–9 December 2017; Volume 30.
54. Huynh-The, T.; Pham, Q.V.; Nguyen, T.V.; Nguyen, T.T.; Costa, D.B.d.; Kim, D.S. RanNet: Learning Residual-Attention Structure in CNNs for Automatic Modulation Classification. *IEEE Wirel. Commun. Lett.* **2022**, *11*, 1243–1247. [CrossRef]

electronics

MDPI

Article

A Cascade Network for Blind Recognition of LDPC Codes

Xiang Zhang and Wei Zhang *

School of Microelectronics, Tianjin University, Tianjin 300072, China; zhangxiang2012@tju.edu.cn
* Correspondence: tjuzhangwei@tju.edu.cn

Abstract: Coding blind recognition plays a vital role in non-cooperative communication. Most of the algorithm for coding blind recognition of Low Density Parity Check (LDPC) codes is difficult to apply and the problem of high time complexity and high space complexity cannot be solved. Inspired by deep learning, we propose an architecture for coding blind recognition of LDPC codes. This architecture concatenates a Transformer-based network with a convolution neural network (CNN). The CNN is used to suppress the noise in real time, followed by a Transformer-based neural network aimed to identify the rate and length of the LDPC codes. In order to train denoise networks and recognition networks with high performance, we build our own datasets and define loss functions for the denoise networks. Simulation results show that this architecture is able to achieve better performance than the traditional method at a lower signal-noise ratio (SNR). Compared with the existing methods, this approach is more flexible and can therefore be quickly deployed.

Keywords: coding blind recognition; low density parity check codes; deep learning; denoise

Citation: Zhang, X.; Zhang, W. A Cascade Network for Blind Recognition of LDPC Codes. *Electronics* **2023**, *12*, 1979. https://doi.org/10.3390/electronics12091979

Academic Editors: Yichuang Sun, Haeyoung Lee and Oluyomi Simpson

Received: 15 March 2023
Revised: 8 April 2023
Accepted: 10 April 2023
Published: 24 April 2023

1. Introduction

Forward error correcting codes counteract the random errors over the noisy channel by inserting redundant bits into code words [1]. In order to balance the quality and rate of communication, different coding schemes have been proposed over the past few decades, and the corresponding decoding scheme has increasingly attracted the attention of many researchers. In the traditional communication system, only the decoder knows the encoding parameters it can decode accurately. However, under the conditions of non-cooperative communication [2], such as cognitive radio, it is impossible for the non-cooperative receiver to decode without prior knowledge of the code parameters. Hence, coding blind recognition is urgently required, and has attracted extensive research interest [3].

Among the existing coding schemes, Low Density Parity Check (LDPC) code, which was first proposed by Gallager in the 1960s [4], has been widely used in modern communication systems, and has been identified as the long code coding scheme in the 5G enhanced mobile broadband scene due to its long code length, rich combination, and sparse check matrix. Since LDPC codes have these basic characteristics, it also poses a challenge to decoding and coding blind recognition. In practical terms, LDPC codes are usually too long to reconstruct the parity-check matrix directly.

In order to solve the problem of high time complexity and high computational complexity, most of the existing methods of LDPC coding blind recognition proposed in recent years use closed set identification. The identification methods based on the closed set utilize a known set which contains all probable parameters [5–8]. The log likelihood ratio (LLR) used in [5,6] for coding blind recognition performs well at a low SNR. In [7], LDPC code is identified by the average likelihood difference (LD) of parity-checks. Since the LLR of syndrome a posterior probability is widely used in these methods, it is often limited by channel conditions. Wu proposed to calculate the average cosine conformity (CC) for recognition, which not only has an explicit probability density, but also has low computational complexity [8]. The code parameters within a given closed set can be recognized using the methods above. However, the identification methods without a candidate set are more universal, and take a longer time to calculate. The minimum Hamming weight vector is used in the search algorithm proposed by Valembois [9], but the fault tolerance capability is weak. Cluzeau used iterative coding [10] to improve the tolerance of the method in [9] followed by the problem of longer time consumption for finding sparse parity-check vectors.

Deep learning as an emerging technique has been widely applied in the many fields, such as image classification [11] and natural language processing [12]. The best solutions are found in specific problems through the connection of multi-layer networks. Different network models have been proposed by scholars in different fields—e.g., the Transformer model proposed by A. Vaswani [13] is a well-known sequence to sequence (seq2seq) architecture that performs well by treating a sentence as a sequence of words in the field of natural language processing (NLP). The self-attention layer structure in the Transformer greatly reduces computation time using the trick of parallel calculation.

In recent years, the combination of neural networks and coding blind recognition has progressed rapidly [14–17]. It has been proven that both of the coding schemes and the coding parameter are able to be identified using neural networks [14]. Two types of LDPC codes can be identified in [15], while a 2-dimensional convolution neural network is used to identify the parity-check matrix of LDPC codes with the help of a candidate set [16]. Moreover, a joint modulation and channel coding recognition framework is proposed in [17] for the practical 5G-PDSCH protocols using the novel Res-Inception convolutional NN and the algorithm based on maximal ALLR. Inspired by the deep learning model used in NLP, we propose a novel method based on Transformer for recognizing the coding parameter of LDPC codes.

Furthermore, channel condition is the key factor in coding blind recognition. Almost all of the traditional coding blind recognition methods of LDPC codes adequately utilize the channel condition. On the contrary, most existing recognition methods based on deep learning pay little attention to the channel condition, which made these deep learning methods identify accurately at high SNR but fare badly when the channel condition is not good enough. Channel noise is similar to image noise in some areas. The deep learning method for two-dimensional image denoising has been widely studied in the field of computer vision [18], and has made considerable progress over the last couple of years [19,20]. In addition to this denoising, the design aims for two-dimensional data. Denoising networks are also widely used in other fields. In the area of decoding, a novel receiver architecture [21] concatenates a belief propagation (BP) decoder for decoding with a convolution neural network (CNN) for denoising. The iteration between BP decoding and CNN will gradually improve the SNR, and achieve better decoding performance. In [22], the double-CNN denoiser is designed to surpass the noise by estimating the channel state information (CSI) under the Rayleigh fading channel. However, these two methods can hardly be applied in practical terms due to the constraint of the information bit length [23], which can cause the dimension explosion of the neural network.

In this paper, we design a deep-learning-based architecture for coding blind recognition of LDPC codes under an additive white Gaussian noise channel with different SNRs. Briefly, the main contributions of this paper include: The code words are treated as a sequence of words which is sent to the proposed Transformer-based neural network for coding blind recognition. A denoising network and new loss functions are proposed in order to get reliable recovery of the bit stream for recognition. Besides, the denoising network and blind recognition network are cascaded, and the simulation results show that the accuracy of the cascade structure is better than that of the non-cascade structure. Furthermore, we compared the proposed method with traditional methods. When the SNR is low, our method performs better than the traditional one.

2. Related Work

In this section, the communication principles of typical digital systems and the location of blind recognition in the communication system are briefly introduced. In order to illustrate our method clearly, the encoding method of the LDPC code and the basic knowledge of the neural network related to this design are introduced.

2.1. Communication System

The typical digital system communication block diagram is shown in Figure 1. The binary information sequence after the source coding is converted into code word information by means of channel coding. The code words are then converted into a waveform signal that can be transmitted on the channel after modulation. During the transmission, the signal is interfered with by various noises, so that the waveform information received by the demodulator may be wrong. When the information from the channel is transmitted to the receiving end, it will firstly be demodulated, and then enter the channel decoding module.

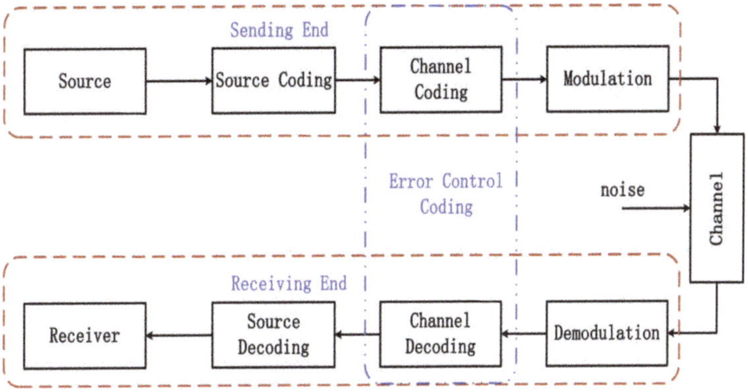

Figure 1. The typical digital system communication block.

In cooperative communication, the decoding module knows the channel conditions and encoding parameters used for real-time communication. In contrast, coding parameters need to be identified in non-cooperative communication before decoding. The block diagram of a typical non-cooperative communication system is shown in Figure 2.

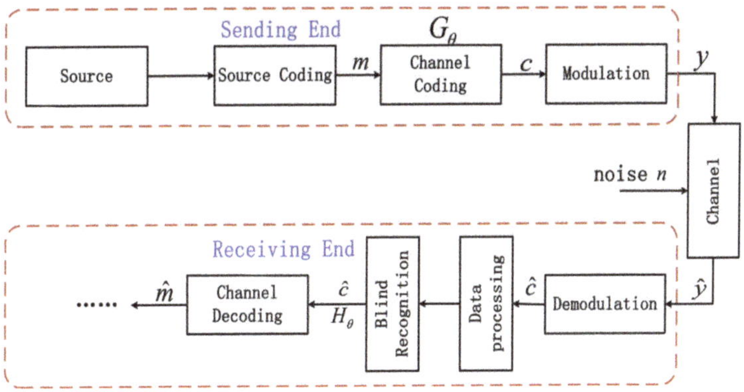

Figure 2. The digital system communication block diagram with blind recognition.

2.2. Basic Theory of LDPC Code

LDPC code is a linear block code whose parity-check matrix contains only a few non-zero elements. The selection of check matrix is very important for LDPC codes, which can not only affect the error correction ability, but also can affect the complexity of LDPC encoding and decoding. The theoretical analysis and formula derivation in this section are all based on *Error Control Coding* [24].

As shown in Figure 2, $G^{\ominus} = \{G^1, G^2, \cdots, G^N\}$ represents the set of encoders, where N is the number of generator matrix, and G represents the generation matrix of size (k, n). The encoder is switched by the transmitter according to the channel conditions. At the sending end, original bits are divided into A groups, and each group has a length of k. The ith information $m_i = [m_{i,1}, m_{i,2}, \cdots, m_{i,k}]$ is sent into encoder with n-bits code word $c_i = [c_{i,1}, c_{i,2}, \cdots, c_{i,n}]$ output, where $m_{i,j}, c_{i,j} \in GF(2)$.The encoding formula is shown in (1):

$$c_i = m_i \cdot G \tag{1}$$

The parity check matrix H of size $(n-k, n)$ satisfies $c_i \cdot H = 0$. Since H is a dual matrix of G, the reconstruction of H and G carries the same meaning for blind recognition of LDPC code. In binary phase-shift keying (BPSK) modulation, the result can be expressed as

$$y = 2c_i - 1 \tag{2}$$

After modulation, channel transmission, and demodulation, the code word $\widehat{c}$ is sent to the receiving end for decoding. The additive white Gaussian noise (AWGN) of n with zero-mean and variance of σ^2 is considered in this paper. $\widehat{c}$ can then be represented by

$$\widehat{c} = y + n \tag{3}$$

Traditional soft decision for decoding is to calculate the LLR using the prior information of channel condition.

$$\text{LLR}_i = ln\left(\frac{P(c[i] = 0|\widehat{c}[i])}{P(c[i] = 1|\widehat{c}[i])}\right) \tag{4}$$

where $P(c[i] = 0|\widehat{c}[i])$ and $P(c[i] = 1|\widehat{c}[i])$ donate the probability of $c[i]$ to be considered as 0 and 1 at the receiver with a known channel condition. When channel is AWGN, we get:

$$P(c[i] = 0|\widehat{c}[i]) = \frac{e^{-2\cdot\widehat{c}[i]/\sigma^2}}{1 + e^{-2\cdot\widehat{c}[i]/\sigma^2}} \tag{5}$$

Substitute (5) for (4). The LLR is represented as follows:

$$\text{LLR}_i = -2 \cdot \frac{\widehat{c}[i]}{\sigma^2} \tag{6}$$

LLR is then used for decoding or blind recognition. Furthermore, LLR can be also used for describing the relationship H and $\widehat{c}$.

2.3. Deep Learning Method

In this subsection, we introduce the basic theory of CNN and Transformers, and then give some examples of the applications of CNN in the area of coding blind recognition.

2.3.1. Convolution Neural Networks

The CNN is mainly composed of convolution layers, pooling layers, and fully connected layers. The features of adjacent data are extracted by convolution layer. Pooling layer retains the main feature, and reduces the number of parameters. The cascade of convolution layer and pooling layer transforms local features into global features. Finally, the fully connected layer takes the global feature as the input and the prediction result as the output. The CNN is widely used in computer vision.

In [16], authors use CNN for LDPC code blind recognition with the help of closed set. It firstly receives the output sequence. The output sequence $\widehat{c}$ processed by LLR generators is described by (6). In order to make better use of LLR, this method utilizes LLR of syndrome a posteriori probability (SPP) [7],

$$\phi_{i,j}^{\theta} = 2arctanh\left(\prod_{\lambda \in \{\Lambda\}} tanh(L(\widehat{c}[\lambda])/2)\right) \tag{7}$$

and then get the set of SPP vectors $\Phi = \{\phi^1, \phi^2, \cdots, \phi^N\}$. The elements $\phi = \{\phi_1^{\theta}, \phi_2^{\theta}, \cdots, \phi_{(n-k)}^{\theta}\}$ in Φ are generated by parity-check matrix H and different channel conditions. Since the size of parity-check matrices H is different, ϕ is of different length correspondingly. In order to get feature matrix F, each ϕ picks α elements randomly, where α is the smallest length of ϕ. Obviously, F is of size $\alpha \times N$. Finally, F is sent to blind recognition network for recognition. Since it considers channel condition within known data sets, the calculation of SPP with different LLR will cause a waste of time. This method cannot applied in the real scene.

2.3.2. The Model Architecture of the Transformer

Figure 3 describes the model architecture of the Transformer in [13]. The encoder stack and decoder stack are two critical components with most of the trainable parameters and the most complex computations. The encoder layers and the decoder layers are composed of the multi-head attention (MHA) ResBlock and the position-wise feed-forward network (FFN) ResBlock.

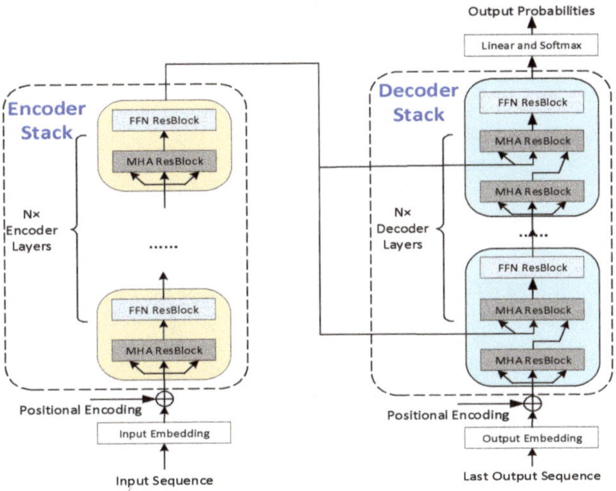

Figure 3. The model architecture of the Transformer.

According to the description in [13], 1an MHA ResBlock has h Attention Heads. For each Head, the input is the same, including three tensors: Queries(Q), Keys(K), and values(V). Figure 4 shows the detail of the MHA ResBlock. The Attention function in the MHA of one Head can be expressed as follows:

$$Attention(Q_i, K_i, V_i) = softmax\left(Mask\left(\frac{Q_i K_i^T}{\sqrt{d_k}}\right)\right)V_i \tag{8}$$

The output tensors from different Heads are put together as tensor P. (9) represents the output of the MHA ResBlock:

$$\mathrm{MHA}(x) = LayerNorm(PW_G + x) \tag{9}$$

The FFN ResBlock is composed of two linear sublayers and a *ReLU* activation function between them. The output of this block is expressed as:

$$\mathrm{FFN}(x) = LayerNorm(x + ReLU(xW_1 + b_1)W_2 + b_2) \tag{10}$$

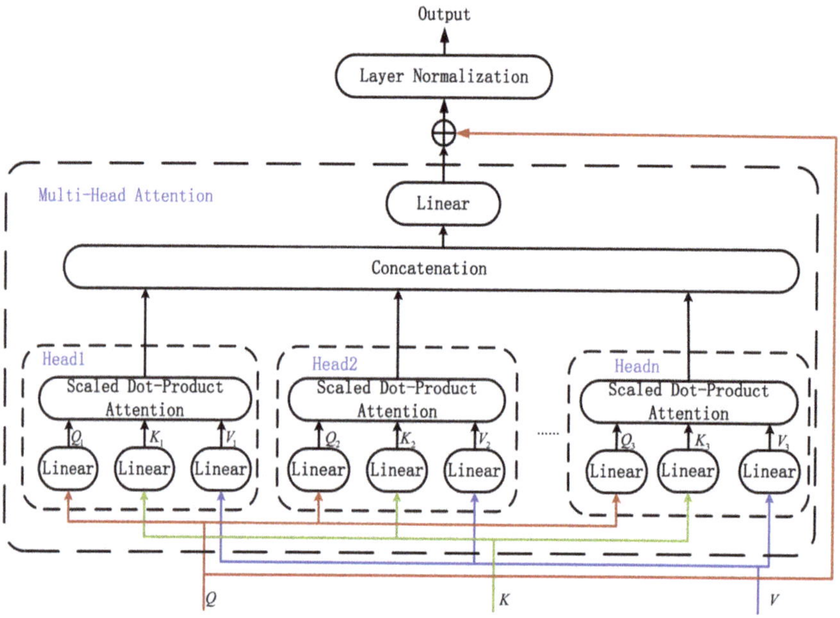

Figure 4. The MHA ResBlock with h Heads.

3. Proposed Cascade Neural Network

This section contains the main innovation of this work: a cascade neural network for LDPC coding recognition. In order to better present the proposed design, we firstly describe the system framework. Then, the CNN structure and the coding blind recognition network will be introduced, and their functions will be explained specifically.

The input of the denoising CNN is a 1-D vector x, while the output vector is y. Both x and y are of size $(1, N)$. As shown in Figure 5, the received code words $\widehat{c}$ are uniformly distributed as N bits, and then sent into CNN. After denoising, y are used to recalculate $LLR(L)$. $\widehat{L}$ denotes the concatenation of L with of size $(1, 15N)$. The blind recognition network based on the Transformer takes $\widehat{L}$ as input. The output of the recognition network is the label, corresponding to the parity-check matrix H.

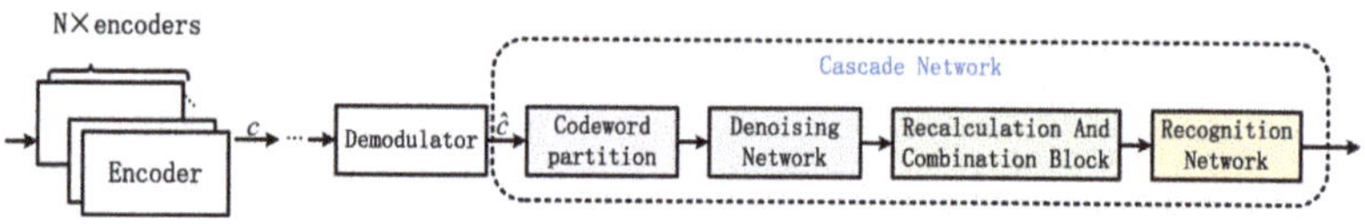

Figure 5. The system framework of the proposed network.

3.1. Denoising Networks

Figure 6 shows the denoising networks using CNN. The input of our networks is a 1-D vector. The feature map at the ith layer o_i can be expressed in (11) as

$$o_{i,j} = ReLU(w_{i,j} * o_{i-1} + b_{i,j}) \tag{11}$$

where $*$ represents the convolution operation, and $w_{i,j}$ is the jth convolution kernel in layer i. $b_{i,j}$ represents the corresponding bias. The activation function is ReLU. In addition, the structure parameters of the proposed denoising network are given in Table 1.

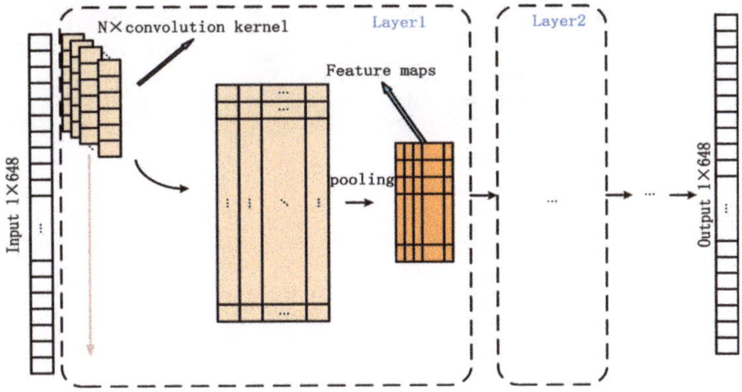

Figure 6. The details of the denoising network.

Table 1. The structure of the denoising network.

Layers	Numbers of Filters	Filter Size	Output Size
Conv2d1	64	(9, 1)	(128, 64, 648)
Conv2d2-8	64	(9, 1)	(128, 64, 648)
Conv2d9	1	(9, 1)	(128, 1, 648)

In addition to the network framework, the loss function is also important for designing the network. It guides the training in the correct direction. L_2 loss, which is also called the mean squared error (MSE), is widely used in the area of image denoising. Suppose that $f(x)$ is the forward calculation output of the network, and y is the expected output. L_2 can be represented by

$$L_2 = \frac{\sum_{i=1}^{n} (f(x) - y)^2}{n} \tag{12}$$

Inspired by the image denoising, $L2$ loss is also used in this design. Furthermore, in [22], the authors propose a method for recalculating the LLR. It obtains the empirical probability distribution function (EPDF) F of the residual noise through histogram statistics. In order to recalculate the LLR more easily, we consider that the residual noise preserves Gaussian distribution with various values of $\widehat{\sigma}^2$. Thus, L_K [25] is used to evaluate whether the residual noise meets Gaussian distribution. Residual noise is defined as $\tilde{n}$, and the numerical expectation as $E[\tilde{n}]$. L_K is expressed by:

$$\begin{aligned} L_K &= \left(S^2 + \frac{1}{4}(C-3)^2\right) \\ S &= \frac{\frac{1}{D}\sum_{d=1}^{D}(\tilde{n}_d - E[\tilde{n}])^3}{\left(\frac{1}{D}\sum_{d=1}^{D}(\tilde{n}_d - E[\tilde{n}])^2\right)^{3/2}} \\ C &= \frac{\frac{1}{D}\sum_{d=1}^{D}(\tilde{n}_d - E[\tilde{n}])^4}{\left(\frac{1}{D}\sum_{d=1}^{D}(\tilde{n}_d - E[\tilde{n}])^2\right)^{2}} \end{aligned} \tag{13}$$

S is called skewness, and C is called kurtosis. For a Guassian distribution, $S = 0, C = 3$. Thus, the loss function is

$$Ł = \alpha L_2 + \beta L_K \tag{14}$$

where $\alpha + \beta = 1$.

In this paper, the candidate sets of LDPC codes have the length $L = [648, 1296, 1944]$, and rate $R = [1/2, 2/3, 3/4, 5/6]$. IEEE802.11 elaborates on these LDPC codes specifically, and is used for Wireless Fidelity(Wi-Fi). The H matrix of the LDPC code with $n = 648$, $R = 2/3$ is shown in Figure 7. The H matrix is constituted by two basic elements, i.e., a zero matrix represented by -1 and an identity matrix. The elements in the H matrix which are not equal to -1 represent the digits of rotating right of the identity matrix. In order to train denoising networks, we consider the AWGN with different SNR $S = [0, 0.5, 1, 2, 4, 6, 8]$ dB. Three code words $[c_1, c_2, c_3]$ with different lengths are sent to the channel. In order to prevent the dimension explosion of the neural network and reduce training time, the appropriate input size of the network is chosen, which is 648 bits. Thus, these code words are divided into 648 bits c, which is the shortest code word length. After the process of the sending end and channel, we get $\widehat{c}$ with white Gaussian noise added. Denoising networks use c and $\widehat{c}$ for training.

$$H = \begin{bmatrix} 25 & 26 & 14 & -1 & 20 & -1 & 2 & -1 & 4 & -1 & -1 & 8 & -1 & 16 & -1 & 18 & 1 & 0 & -1 & -1 & -1 & -1 & -1 & -1 \\ 10 & 9 & 15 & 11 & -1 & 0 & -1 & 1 & -1 & -1 & 18 & -1 & 8 & -1 & 10 & -1 & -1 & 0 & 0 & -1 & -1 & -1 & -1 & -1 \\ 16 & 2 & 20 & 26 & 21 & -1 & 6 & -1 & 1 & 26 & -1 & 7 & -1 & -1 & -1 & -1 & -1 & -1 & 0 & 0 & -1 & -1 & -1 & -1 \\ 10 & 13 & 5 & 0 & -1 & 3 & -1 & 7 & -1 & -1 & 26 & -1 & -1 & 13 & -1 & 16 & -1 & -1 & -1 & 0 & 0 & -1 & -1 & -1 \\ 23 & 14 & 24 & -1 & 12 & -1 & 19 & -1 & 17 & -1 & -1 & -1 & 20 & -1 & 21 & -1 & 0 & -1 & -1 & -1 & 0 & 0 & -1 & -1 \\ 6 & 22 & 9 & 20 & -1 & 25 & 17 & -1 & 8 & -1 & -1 & 14 & -1 & 18 & -1 & -1 & -1 & -1 & -1 & -1 & -1 & 0 & 0 & -1 \\ 14 & 23 & 21 & 11 & 20 & -1 & 24 & -1 & 18 & -1 & 19 & -1 & -1 & -1 & -1 & 22 & -1 & -1 & -1 & -1 & -1 & -1 & 0 & 0 \\ 17 & 11 & 11 & 20 & -1 & 21 & -1 & 26 & -1 & 3 & -1 & -1 & 18 & -1 & 26 & -1 & 1 & -1 & -1 & -1 & -1 & -1 & -1 & 0 \end{bmatrix}$$

Figure 7. H matrix of LDPC code with $n = 648$, $R = 2/3$ in 802.11n.

3.2. *Recalculation of LLR*

As mentioned in Section 3.1, the residual noise preserves the Gaussian distribution with various values of $\widehat{\sigma}^2$. It is easy to recalculate the $\widetilde{LLR}$:

$$\widetilde{LLR} = -2 \cdot \frac{\widehat{c}[i]}{\widehat{\sigma}^2} \tag{15}$$

3.3. *Recognition Networks*

After the above operation, the $\widetilde{LLR}$s are calculated at sizes of (1, 648). Then, 15 $\widetilde{LLR}$s are concatenated into 1-D data of size (1, 9720) defined as I_R, which is the input of the recognition network. However, because of the positional encoding layer in Transformer, the input of the network must be a fixed size. In order to identify these three different lengths of LDPC code and reduce the number of parameters, the ideas in the Swin Transformer are used in the recognition network.

Figure 8 shows the architecture of the network. This network takes I_R as input. Since the token size of this network is 324, I_R is divided into 30×324 by Patch Partition layer. The window size is of 10 tokens and the relative position encoding (RPE) B is a learnable variable of size $(1, 19)$. B is calculated by substituting its index into an RPE matrix of size $(1, 19)$. The size of the RPE matrix is determined by the number of tokens in the window, and the relationship between tokens and windows is shown in Figure 9. The Swin Transformer Block contains two layers: one for MHA calculation, and another for shifted window MHA(SW-MHA) calculation. Figure 10 shows the detail of the Swin Transformer block. The window calculates MHA firstly, which is introduced in Section 2.3. In order to calculate SW-MHA, the feature map must shift circularly as shown in Figure 9. Then, the SW-MHA can be calculated the same way as the MHA.

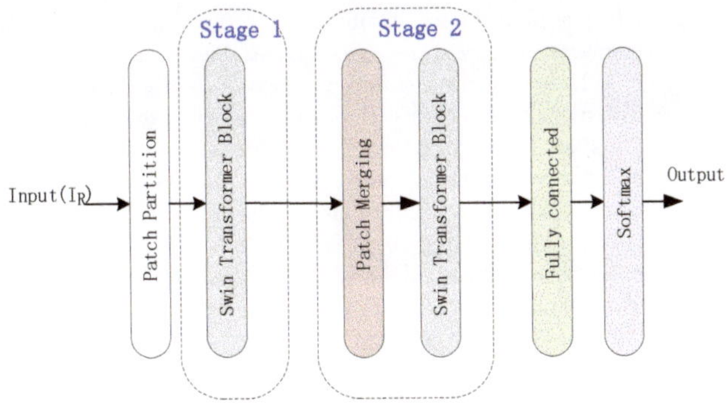

Figure 8. The architecture of the recognition network.

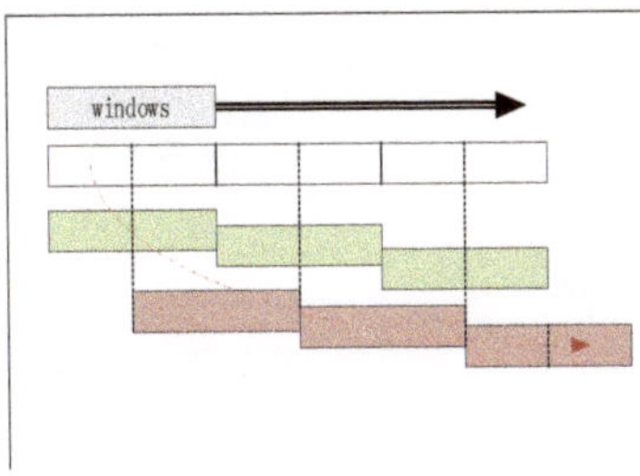

Figure 9. The relationship between patches and windows.

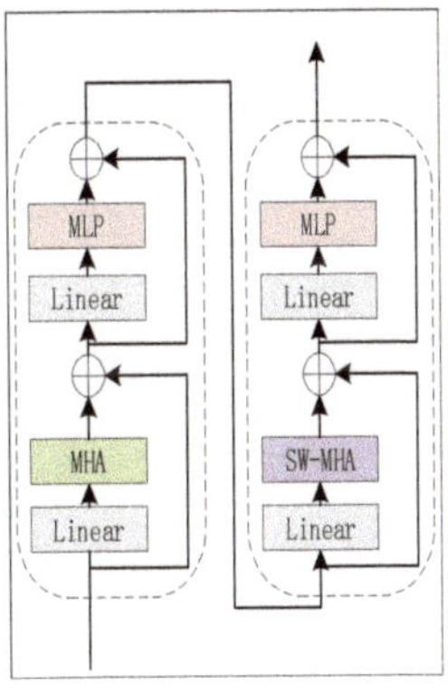

Figure 10. The Swin Transformer block.

It should be noted that the formula is a little different from (8) because of the difference in the strategy of linear embedding.

$$Attention(Q_i, K_i, V_i) = softmax\left(\frac{Q_i K_i^T}{\sqrt{d_k}} + B\right) V_i \tag{16}$$

B is the RPE mentioned before, and a mask is useless in this task.

The function of the patch merging layer is similar to the max-pooling layer in CNN, which is used to expand the receptive field. The vector size of stage 1's output is the the same as its input with the size of $(30, 324)$. Then, the outputs are resized into the size of $(10, 324)$ by taking one of every three pieces of data from the output of stage 1. After a normalization in each row and a fully connected layer, the output becomes $(10, 324)$. The Swin Transformer block in stage 2 is similar to stage 1. The difference is only the dimension, which is 10 in stage 2. Finally, the outputs of stage 2 of the size $(10, 324)$ are sent to the fully connected layer and processed using the softmax function for classification.

There are three different types of datasets for the recognition network. The main difference between these three datasets is that the impact of noise on their data inside varies. The data in dataset 1 are noiseless, the data in dataset 2 are the output of the recalculation and combination block, which retain the residual noise, and the data in dataset 3 come directly from the demodulator without denoising. The label of these datasets is one-hot codes of 12 bits, which corresponds to each parity-check matrix H. The details of the datasets are given in Table 2. Each dataset included in Table 2 is randomly divided into 3 groups, i.e., the train set, validation set and test set. The train set accounts for 80%, while the validation set and test set account for 10%. Note that there is no other mechanism for accessing channel information, so dataset 1 and dataset 3 are sent to the network for training directly. The parameters of the recognition network are given in Table 3.

Table 2. The details of the datasets.

Attributes	Dataset 1	Dataset 2	Dataset 3
Numbers of H	12	12	12
SNRs for generation	∞	—	$[0, 0.5, 1, 2, 4, 6, 8]$
Modulation	BPSK	BPSK	BPSK
Numbers of training data	24,000	168,000	168,000
Numbers of validation data	7200	50,400	50,400
Whether channel soft information was used	-	YES	NO

Table 3. The parameters of the recognition network.

Layers	Input	Output	d_k	Number of Heads
Patch Partition	$(128, 1, 9720)$	$(128, 30, 324)$	-	-
Swin Transformer Block	$(128, 30, 324)$	$(128, 30, 324)$	36	9
Patch Merging	$(128, 30, 324)$	$(128, 10, 324)$	-	-
Swin Transformer Block	$(128, 10, 324)$	$(128, 10, 324)$	36	9
Fully connected	$(128, 10, 324)$	$(128, 12)$	-	-
Softmax	$(128, 12)$	$(128, 12)$	-	-

4. Experiment and Result

In this section, the function of the proposed network is verified in three steps. Firstly, the denoising network is trained and evaluated. Dataset 2 will also be built at this stage. Then, the datasets 1, 2, and 3 mentioned in Section 3.3 are used to train the recognition network respectively and evaluate the accuracy of this network. Finally, we analyse the cascade neural of the proposed network, and compare with the method in [7,16].

4.1. Denoising Network

As shown in Table 1, the batch size for training is 128. Furthermore, hyperparameters for training are given in Table 4. The first-order moment estimation and second-order moment estimation of gradients are calculated using Adaptive Moment Estimation (Adam) to set the independent adaptive learning rates of different parameters. Kaiming initializers [26] are widely used in convolutional networks and perform well. After extensive experiments, we found that the convergence speed is the fastest when the learning rate is 1E-4, and the training will stop until the validation loss doesn't drop in 1E+4 epochs.

Table 4. Hyperparameters for training.

Optimizer	Initializer	Learning Rate	Stop Strategy
Adam	Kaiming	$1e{-4}$	Val loss not drop in $1e+4$ epoch

In order to intuitively demonstrate that if the function of this network is as expected, the loss functions will be L_2 and L_K, this is shown in Figure 11.

Moreover, the residual noise is exported and fitted to the Gaussian function in Figure 12. The blue line indicates the distribution of the residual noise from −5 dB to 5 dB, while the red one is the Gaussian fitting of these data. Both the Jarque-Bera test (JB-test) and Figure 12 indicate that the residual noise preserves the Gaussian distribution.

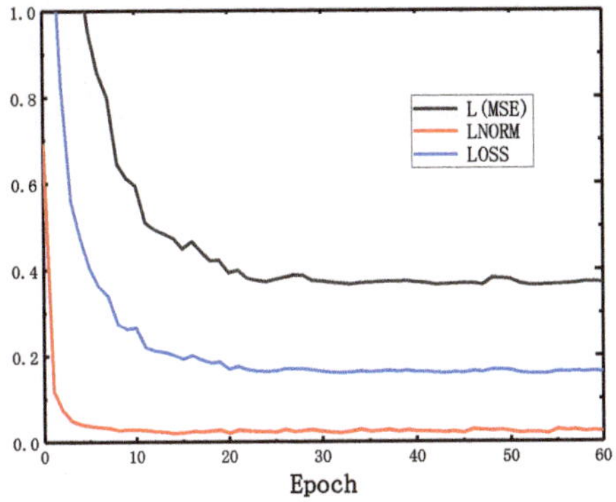

Figure 11. Loss value at the offline training stage.

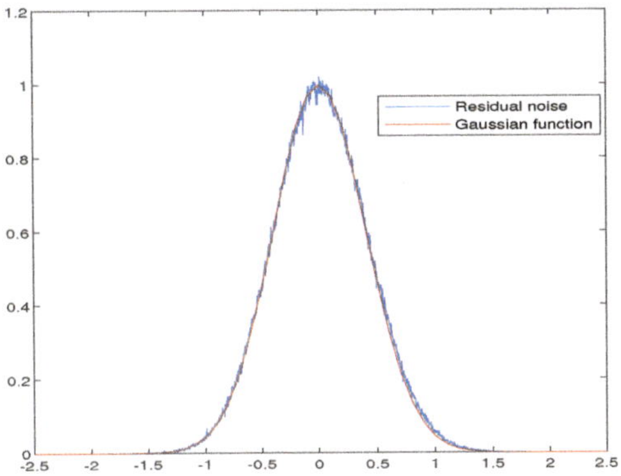

Figure 12. Gaussian function and the residual noise.

As mentioned in Section 3.3, dataset 2 is the output of this network with the corresponding labels. The variance $\widehat{\sigma}$ of the residual noise generated by the denoising network with different SNR inputs is counted, and the statistical results show that $\widehat{\sigma}^2$ is around 0.4, which means that the SNR of the output is about 4 dB, according to the formula below.

$$SNR = 10log_{10}\left(\frac{1}{\widehat{\sigma}^2}\right) \tag{17}$$

It leads to good results when SNR is smaller than 4 dB. Even if the channel condition is good enough, this network can also provide approximate channel information, which can be used for the following work.

4.2. Recognition Network

Similar to the denoising network, the recognition network is trained using the three datasets mentioned above. The hyperparameters of these three networks are all the same, which are given in Table 5. The Optimizer is the same as the denoising network, while the initializer used is Xavier, which initializes the mean value of the weights and bias to 0 and of various other values to 1. The learning rate is initially set to 0.001 and adjusted dynamically during training. In order to prevent overfitting, the drop out value we set is 0.5.

Table 5. Hyperparameters for training.

Optimizer	Initializer	Learning Rate	Drop Out
Adam	Xavier	0.001 at the beginning; gradually decreased while training	0.5

In order to verify the accuracy and robustness of the network, twelve types of LDPC codes with $[0, 0.5, 1, 2, 4, 6, 8, 10, \infty]$ dB are used as the test sets. Figure 13 shows the performance of these three networks trained by different datasets.

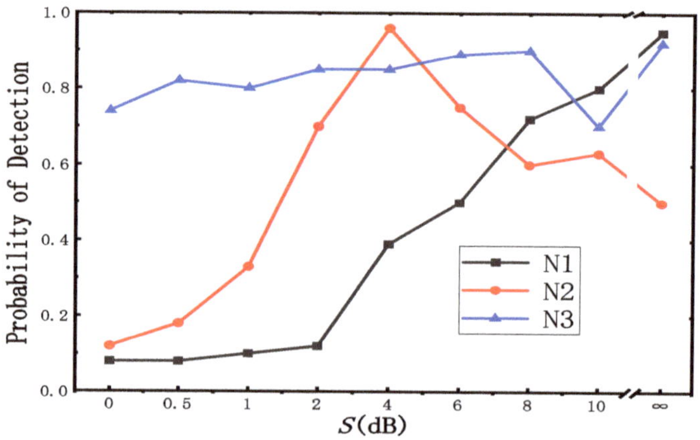

Figure 13. The performance of the recognition network trained by different datasets.

$[N1, N2, N3]$ are the networks trained by $[dataset1, dataset2, dataset3]$, respectively, for convenience. As shown in Figure 13, $N1$ and $N2$ perform fairly well when the SNR(S) of the test set is the same as that of the training set. This network is robust, since it is able to recognize the LDPC codes with a SNR that has not appeared in the training set. However, the performance of $N1$ and $N2$ gradually deteriorates as the gap in the SNR between the test sets and training sets becomes larger. The accuracy of $N1$ is the worst when the S of the test set is 0 dB. On the contrary, $N3$ has more flexibility and is more adaptable to multiple situations due to the large variety of data in its training set. But it is less accurate than $N1$ and $N2$ for a particular SNR The above discussion shows that it is possible to obtain a highly accurate blind recognition network, as long as we ensure that the SNR deviation of the data after the denoising network is as small as possible.

Furthermore, it is noticeable that the accuracy is quite low when the S is less than 0.5 dB. We consider the trade-off of the accuracy and flexibility by rebuilding the dataset for the denoising network. The addition of new data with a low SNR will inevitably cause the degradation of denoising network performance, resulting in a decline in the accuracy of the recognition network. In order to improve the accuracy of the prediction of $\widehat{\sigma}$ from the denoising network, we rebuild datasets with a low SNR ($[-3, -2, -1, 0, 1, 2, 4]$ dB) for the denoising network. After removing the low-SNR data, the recognition network using only high-SNR data can be further optimized. Hence, the proposed optimized structure is shown in Figure 14. The accuracy of this structure is shown in Table 6. It is noteworthy that this system performs better with SNRs from -3 dB to 4 dB than with other SNRs due to the use of the channel information. Its accuracy is around 90% when $-3\text{ dB} \leq S \leq 4\text{ dB}$, while the accuracy is around 87% when $5\text{ dB} \leq S \leq 10\text{ dB}$.

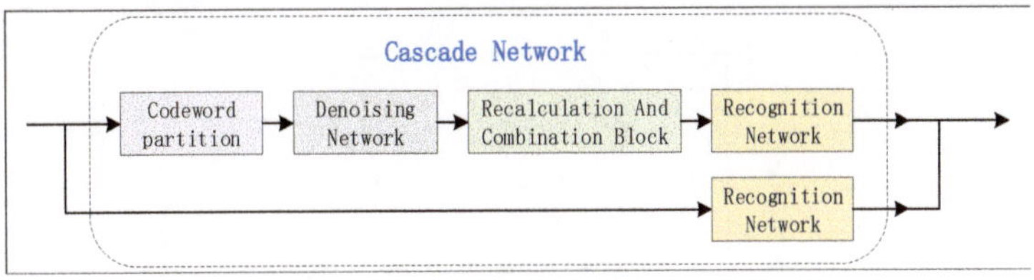

Figure 14. The optimized structure for blind recognition.

Table 6. The accuracy of the optimized structure.

SNR/*dB*	−3	−2	−1	0	1	2	4	6	8	10
Accuracy	0.866	0.871	0.895	0.920	0.925	0.951	0.911	0.874	0.862	0.867

4.3. Experimental Setup

P_d represents the probability of correct recognition, which can be described as

$$P_d = \frac{N_d}{N_t} \tag{18}$$

N_d is the number of correctly recognized code words, and N_t is the total number of the code words.

The method in [7] firstly estimates the signal amplitude and noise variance for the construction of the LLRs. It shows that P_d is higher when they collect multiple blocks jointly for blind encoder identification. They assume that each encoder θ lasts for M consecutive blocks. As the value of M increases, the reliability of the estimated LLR and the accuracy of blind recognition will be better. Since a test sample in datasets we built contains at least five consecutive code words, the P_d for this network is compared with that of the method in [7] under the condition of $M = 5$.

The P_d for different code-rates$[1/2, 2/3, 3/4, 5/6]$ is shown in Figure 15a, when the code word length is fixed at 648. Statistical methods are used in [7] for coding blind recognition. EM (expectation-maximization) represents the method in [7]. CN (cascade-network) represents the method we proposed. According to Figure 15a, the probability of detection increases as the code-rate becomes lower [7], while the P_d of these four code words is mainly consistent using our method. In addition, it is easy to find that the method we proposed performs better when the S is between 0 dB and 5 dB. Although the accuracy of our method is not good enough when S is larger than 5 dB, it can also precisely recognize H in most cases.

In order to estimate P_d for different code word lengths, the code-rates R are fixed at the same value of 5/6. Figure 15b demonstrates the P_d of these test samples. The overall trend of these curves is similar to that of Figure 15.

In addition to comparing with traditional blind recognition algorithms, another blind recognition method based on deep learning [16] is also compared. Figure 16 shows the accuracy of these two methods. Since the dataset mentioned in [16] is difficult to build, we can only roughly compare their accuracy. The results also show that our proposed network performs better when $-4 \text{ dB} \leq S \leq -2 \text{ dB}$. Since our design does not need to generate LLR information for each SNR, our network is more generic.

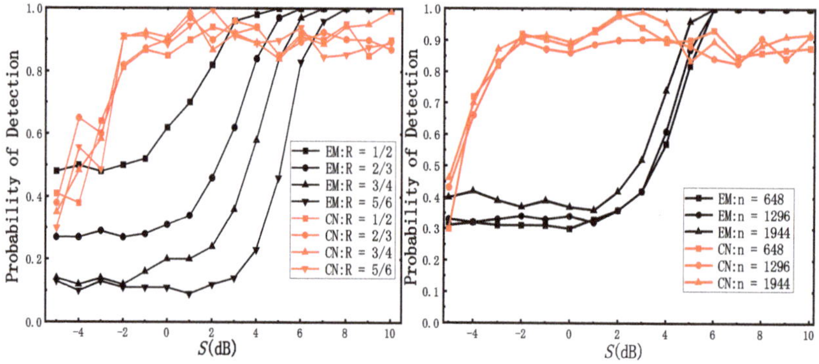

(a) (b)

Figure 15. (**a**) Recognition performance of these two methods with the fixed code length $n = 648$. (**b**) Recognition performance of these two methods with the fixed rate $R = 5/6$.

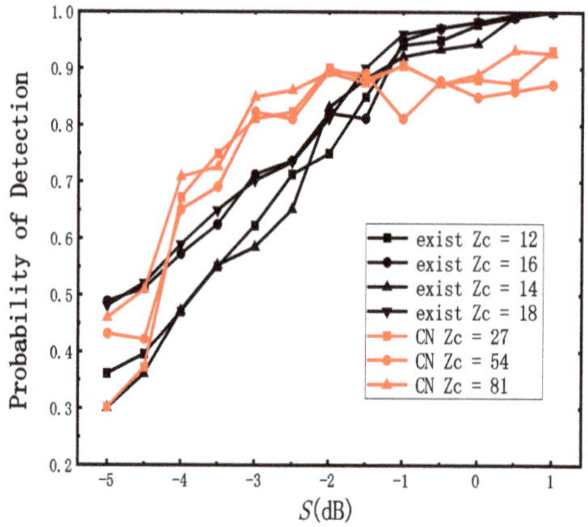

Figure 16. The results compared with another method based on deep learning.

5. Conclusions and Discussion

In this paper, we proposed a cascade network structure for LDPC coding blind recognition. Compared with the traditional blind recognition algorithm, it performs better under the condition of low SNR.

The original intention of our design of the denoising network was to filter part of the noise and obtain the code word, which has residual noise in the Gaussian distribution. The results show that the denoising network can provide accurate channel information when the SNR gap in datasets is not large. In addition, it performs better when the SNR is small. For the recognition network, training for various different SNRs is not reliable. We prefer focalization training, which coincides with the function of the noise reduction network.

The generation matrix of LDPC codes in datasets is described in IEEE802.11n, which is widely used in short-range wireless communications technology. The proposed cascade network can identify the parity-check matrix of LDPC codes in datasets with a probability of 90% when the S is larger than 0 dB. Furthermore, this network achieves about 80% accuracy when the S is around -2 dB, which is better than other traditional methods.

After the division of the dataset and the optimization of the structure, this architecture performed better than the traditional method and another deep learning method. However, due to the limitation of the cascade structure, the training cost of the network is high, and the data division is also critical. Hence, the fusion of this structure and traditional algorithms may achieve better results. Here, we suppose a hypothetical structure with two different data paths, $PathA$ and $PathB$. The demodulated data is sent to two different data paths, and the final output is the one with higher confidence of these two data paths.

$PathA$ is for data with low SNR, while $PathB$ is on the contrary. Both paths have a recognition network, but the training sets of the two networks are different. Recognition networks on $PathA$ use the output of the denoising network for training. The datasets of the denoising network are of a low SNR, such as $[-5, -4, -3, -2, -1, 0]$ dB or even lower, aimed to handle the situation of a bad channel condition. On the contrary, $PathB$ is for data with a high SNR. Firstly, the LLR is evaluated blindly with the help of the traditional method. The soft information is subsequently used for training. Note that the recognition network can also be replaced by the traditional blind identification method, since the traditional method has already given good results when the SNR is high. At the end of these two paths, a discriminant function must be designed to determine whether the final output is from $PathA$ or $PathB$.

In short, our work confirmed that it is feasible to use the pure deep learning method for LDPC blind recognition, and that the performance is better after using the CNN for denoising. In the subsequent work, we will rebuild the dataset with a different channel condition and different parity check matrix of LDPC codes, which is widely used in modern communication, and verify the accuracy of the existing network on the rebuild dataset. In addition, the hardware acceleration architecture for the network is also being designed synchronously.

Author Contributions: Conceptualization, W.Z.; Methodology, X.Z.; Software, X.Z.; Data curation, X.Z.; Writing—review and editing, W.Z.; Writing—original draft preparation, X.Z.; Visualization, X.Z.; Formal analysis, X.Z.; Supervision, W.Z. All authors have read and agreed to the published version of the manuscript.

Funding: This research received no external funding.

Data Availability Statement: Data is unavailable due to privacy.

Conflicts of Interest: The authors declare no conflict of interest.

References

1. Perrins, E. FEC systems for aeronautical telemetry. *IEEE Trans. Aerosp. Electron. Syst.* **2013**, *49*, 2340–2352. [CrossRef]
2. Bonvard, A.; Houcke, S.; Marazin, M.; Gautier, R. Order statistics on minimal Euclidean distance for blind linear block code identification. In Proceedings of the 2018 IEEE International Conference on Communications (ICC), Kansas City, MO, USA, 20–24 May 2018; pp. 1–5.
3. Yu, P.; Peng, H.; Li, J. On blind recognition of channel codes within a candidate set. *IEEE Commun. Lett.* **2016**, *20*, 736–739. [CrossRef]
4. Gallager, R. Low-density parity-check codes. *IRE Trans. Inf. Theory* **1962**, *8*, 21–28. [CrossRef]
5. Moosavi, R.; Larsson, E.G. Fast blind recognition of channel codes. *IEEE Trans. Commun.* **2014**, *62*, 1393–1405. [CrossRef]
6. Xia, T.; Wu, H.-C. Blind identification of nonbinary LDPC codes using average LLR of syndrome a posteriori probability. *IEEE Commun. Lett.* **2013**, *17*, 1301–1304.
7. Xia, T.; Wu, H.-C. Novel blind identification of LDPC codes using average LLR of syndrome a posteriori probability. *IEEE Trans. Signal Process.* **2014**, *62*, 632–640. [CrossRef]
8. Wu, Z.; Zhang, L.; Zhong, Z.; Liu, R. Blind Recognition of LDPC Codes Over Candidate Set. *IEEE Commun. Lett.* **2020**, *24*, 11–14. [CrossRef]
9. Valembois, A. Detection and Recognition of a Binary Linear Code. *Discret. Appl. Math.* **2001**, *111*, 199–218. [CrossRef]
10. Cluzeau, M. Block Code Reconstruction Using Iterative Decoding Techniques. In Proceedings of the 2006 IEEE International Symposium on Information Theory, Seattle, WA, USA, 9–14 July 2006; pp. 2269–2273.
11. Redmon, J.; Divvala, S.; Girshick, R.; Farhadi, A. You only look once: Unified, real time object detection. In Proceedings of the IEEE Conference on Computer Vision and Pattern Recognition (CVPR), Las Vegas, NV, USA, 27–30 June 2016; pp. 779–786.
12. Young, T.; Hazarika, D.; Poria, S.; Cambria, E. Recent Trends in Deep Learning Based Natural Language Processing. *IEEE Comput. Intell. Mag.* **2018**, *13*, 55–75. [CrossRef]
13. Vaswani, A.; Shazeer, N.; Parmar, N.; Uszkoreit, J.; Jones, L.; Gomez, A.N.; Kaiser, Ł.; Polosukhin, I. Attention is all you need. In Proceedings of the 31st International Conference on Neural Information Processing Systems, Long Beach, CA, USA, 4–9 December 2017; pp. 6000–6010.

14. Shen, B.; Wu, H.; Huang, C. Blind Recognition of Channel Codes via Deep Learning. In Proceedings of the 2019 IEEE Global Conference on Signal, Information Processing (GlobalSIP), Ottawa, ON, Canada, 11–14 November 2019; pp. 1–5.
15. Ni, Y.; Peng, S.; Zhou, L.; Yang, X. Blind Identification of LDPC Code Based on Deep Learning. In Proceedings of the 6th International Conference on Dependable Systems and Their Applications (DSA), Harbin, China, 3–6 January 2020; pp. 460–464.
16. Li, L.; Huang, Z.; Liu, C.; Zhou, J.; Zhang, Y. Blind Recognition of LDPC Codes Using Convolutional Neural Networks. In Proceedings of the 2021 IEEE 4th International Conference on Electronics Technology (ICET), Chengdu, China, 7–10 May 2021; pp. 696–700.
17. Chen, X.; Wang, X.; Zhao, H.; Fei, Z. Deep Learning-Based Joint Modulation and Coding Scheme Recognition for 5G New Radio Protocols. In Proceedings of the 2022 IEEE 22nd International Conference on Communication Technology (ICCT), Nanjing, China, 11–14 November 2022; pp. 1411–1416.
18. Zhang, K.; Zuo, W.; Chen, Y.; Meng, D.; Zhang, L. Beyond a Gaussian Denoiser: Residual Learning of Deep CNN for Image Denoising. *IEEE Trans. Image Process.* **2017**, *26*, 3142–3155. [CrossRef] [PubMed]
19. Zhang, K.; Zuo, W.; Zhang, L. FFDNet: Toward a Fast and Flexible Solution for CNN-Based Image Denoising. *IEEE Trans. Image Process.* **2018**, *27*, 4608–4622. [CrossRef] [PubMed]
20. Guo, S.; Yan, Z.; Zhang, K.; Zuo, W.; Zhang, L. Toward Convolutional Blind Denoising of Real Photographs. In Proceedings of the 2019 IEEE/CVF Conference on Computer Vision and Pattern Recognition (CVPR), Long Beach, CA, USA, 15–20 June 2019; pp. 1712–1722.
21. Liang, F.; Shen, C.; Wu, F. An Iterative BP-CNN Architecture for Channel Decoding. *EEE J. Sel. Top. Signal Process.* **2018**, *12*, 144–159. [CrossRef]
22. Li, J.; Zhao, X.; Fan, J.; Shu, F.; Jin, S.; Qian, Y. A double-CNN BP decoder on fast fading channels using correlation information. In Proceedings of the 2019 IEEE/CIC International Conference on Communications in China (ICCC), Changchun, China, 11–13 August 2019; pp. 53–58.
23. Gruber, T.; Cammerer, S.; Hoydis, J.; Brink, S.T. On deep learning based channel decoding. In Proceedings of the 2017 51st Annual Conference on Information Sciences and Systems (CISS), Baltimore, MD, USA, 22–24 March 2017.
24. Lin, S.; Costello, D.J. *Error Control Coding*; Prentice-Hall: Hoboken, NJ, USA, 2007.
25. Thadewald, T.; Büning, H. Jarque-Bera test and its competitors for testing normality: A power comparison. *J. Appl. Stat.* **2007**, *34*, 87–105. [CrossRef]
26. He, K.; Zhang, X.; Ren, S.; Sun, J. Delving Deep into Rectifiers: Surpassing Human-Level Performance on ImageNet Classification. In Proceedings of the 2015 IEEE International Conference on Computer Vision (ICCV), Santiago, Chile, 7–13 December 2015; pp. 1026–1034.

sensors

MDPI

Communication

Optical Encoding Model Based on Orbital Angular Momentum Powered by Machine Learning

Erick Lamilla [1,2,*], Christian Sacarelo [1], Manuel S. Alvarez-Alvarado [3], Arturo Pazmino [1] and Peter Iza [1,4]

[1] Escuela Superior Politécnica del Litoral, ESPOL, Departamento de Física, Campus Gustavo Galindo, Km 30.5 Vía Perimetral, P.O. Box 09-01-5863, Guayaquil 090150, Ecuador
[2] Facultad de Ciencias Matemáticas y Físicas, Universidad de Guayaquil, Guayaquil 090514, Ecuador
[3] Escuela Superior Politécnica del Litoral, ESPOL, Facultad de Ingeniería en Electricidad y Computación(FIEC), Campus Gustavo Galindo, Km 30.5 Vía Perimetral, P.O. Box 09-01-5863, Guayaquil 090150, Ecuador
[4] Center of Research and Development in Nanotechnology, CIDNA, Escuela Superior Politécnica del Litoral, ESPOL, Campus G. Galindo, Km 30.5 víA Perimetral, Guayaquil 090150, Ecuador
* Correspondence: ealamill@espol.edu.ec

Abstract: Based on orbital angular momentum (OAM) properties of Laguerre–Gaussian beams LG(p, ℓ), a robust optical encoding model for efficient data transmission applications is designed. This paper presents an optical encoding model based on an intensity profile generated by a coherent superposition of two OAM-carrying Laguerre–Gaussian modes and a machine learning detection method. In the encoding process, the intensity profile for data encoding is generated based on the selection of p and ℓ indices, while the decoding process is performed using a support vector machine (SVM) algorithm. Two different decoding models based on an SVM algorithm are tested to verify the robustness of the optical encoding model, finding a BER $= 10^{-9}$ for 10.2 dB of signal-to-noise ratio in one of the SVM models.

Keywords: machine learning; LG-beams; OAM-beams; optical encoding model

Citation: Lamilla, E.; Sacarelo, C.; Alvarez-Alvarado, M.S.; Pazmino, A.; Iza, P. Optical Encoding Model Based on Orbital Angular Momentum powered by Machine Learning. *Sensors* **2023**, *23*, 2755. https://doi.org/10.3390/s23052755

Academic Editors: Yichuang Sun, Haeyoung Lee and Oluyomi Simpson

Received: 19 January 2023
Revised: 18 February 2023
Accepted: 22 February 2023
Published: 2 March 2023

1. Introduction

Since the study of Allen et al. [1], optical beams with orbital angular momentum (OAM) have aroused growing interest from researchers around the world due to their wavefront helical shape properties that provide a new degree of freedom for exploration of new applications in particle manipulation [2,3], image processing [4,5] and optical communications [6,7]. In this context, optical communications systems have found a possibility of exploring vortex beams properties in multiplexing [8,9] and data encoding [10,11] pathways. Concerning data encoding, OAM states to encode different data symbols are evidenced by Fang et al. [12], where OAM holography is performed by OAM selectivity in a spatial-frequency domain without a theoretical helical mode index limit. In the area of holographic encryption, Xiao et al. [13] propose a two-coding information metasurface to achieve OAM-encrypted holography. OAM encoding has also been explored in multicasting links; for instance, Shiyao Fu et al. [14] encode digital signals through the OAM free space one-to-many multicasting link. Within the same research line, data coding has also been explored experimentally as demonstrated by Willner et al. in [15], where data encoding at 20 Gb/s, using 4 possible OAM modes, is performed. High-dimensional data encoding through a hybrid OAM-radial index is also demonstrated in [14]. Optical encoding and multiplexing techniques in OAM channels for highly dispersive media have also been implemented [11], where a novel scattering-matrix-assisted retrieval technique was proposed to demultiplex OAM channels from highly scattered optical fields.

There is a lot of evidence of OAM applications, for instance, in the data encoding field in free space and fiber-based transmission channels [16,17], polarization-based [18,19] and intensity and vortices in phase-based channels [20]. However, implementation of an

OAM-based encoding system requires overcoming several challenges from the point of view of information medium propagation and system detection implementation. Due to the nature of information propagation, some effects can be induced in the medium, such as absorption, scattering and turbulence, spatial distortion (amplitude and phase), modal coupling and crosstalk. Some of these effects, as in the case of turbulence and modal crosstalk, have been potentially suppressed in coding and multiplexing systems through mitigation methods [21], but for the most part, these effects constitute a great challenge [11,22]. Such challenges have captured the attention of the scientific community to focus their studies on designing more robust and flexible optical encoders and encryptors based on coding techniques that minimize noise and information distortion, while correctly maximizing the amount of data coded. In this way, the efforts to improve optical encoding systems are reflected in image recognition methods for encoded data, as in the case of [23], where an index modulation is implemented for OAM states with a uniform scheme circular array (OAM-UCA) to build low-intensity parity coding to improve error performance and transmit additional bits of information. Incoherent detection methods have also been implemented for data decoding [24], where an image information transfer method based on petal-like beam lattices for coding is used. In this case, a decoding system works directly with the identification of the intensity patterns captured. Another example of an image-based method is presented in [20] that employs the amplitude and the phase of an optical field into a phase-only hologram to control spatial transverse modes for data symbol mapping. A similar study can be found in [25] that uses an OAM array for a free-space communication encoding/decoding link with 625 states. A proposal for OAM light encoding in magnets has also been developed in [26], where the possible sub-wavelength magnetic phenomena induced by a vortex beam and their applications in the generation of topological defects in chiral magnets is discussed. Although the aforementioned studies show the feasibility of encoding systems based on OAM modes, OAM does not increase the amount of information, nor does it exceed the multiple-input multiple-output (MIMO) transmission of current standards in optical communications [7,27]. In fact, the number of spatial modes available for data encoding is limited by the space-bandwidth product of a given optical system [27,28]. A solution to this problem is to use all spatial degrees of freedom offered by OAM modes. A commonly used OAM beam for this purpose is a Laguerre–Gaussian (LG) beam [7,17], which provides eigen-modes dependent on both radial (p) and azimuthal (ℓ) indices, being able to use the superposition of modes to increase the number of encoding data in a limited system. On the other hand, a decoding system (which is generally based on image detection and classification) can present strong signal distortion (both in the intensity profile and in the phase distribution) due to optical alignment, turbulence and scattering [29]. Recently, convolutional neural networks (CNNs) and machine learning techniques have been implemented in optical coding systems as an alternative solution for image detection and classification [30–32]. High-resolution recognition techniques based on deep learning to encode data in spatial modes have already been implemented [33]. The deep-learning-based approach has also been used to recover the sparse data from multiplexed OAM channels independent of phase information [34]. Although these studies demonstrate the feasibility of encoding systems based on OAM modes as well as various methods implemented for data decoding, there is still a gap concerning image detection and classification methods in decoding due to degradation effects that the medium induces in the transmitted signal, which brings the motivation for this research work.

Motivated by previous statements, this paper proposes a comprehensive optical encoding–decoding system based on the intensity profile generated by a coherent superposition of two OAM-carrying Laguerre–Gaussian (LG) modes and a machine learning detection method. In the encoding process, an intensity profile for data encoding is generated based on the selection of p and ℓ indices of LG beams, while the decoding process is performed using support vector machine (SVM). Different from other existing encoding systems that require the additional extraction of phase information, this paper proposes a novel optical encoder based on the number of spatial modes carrying data symbols increased in a limited optical system. Moreover, the proposed optical encoding model opens a pathway to a stable image detection and classification system based on machine learning that only uses the intensity profile for target modes. As a result, the main contributions of this paper are: (1) a comprehensive design of a coherent optical encoding system based on the superposition of LG modes carrying OAM that is independent of phase information and (2) a robust decoding system based on intensity profile recognition using the machine learning SVM method. Section 2 presents the concept and operating principle of the optical encoder. In Section 3, the SVM-based decoding method for image recognition and classification is explained in detail. In Section 4, a case study for a 4-bit coding system with different types of noise is considered to validate the robustness of the proposed encoder. Finally, Sections 5 and 6 exhibit the results and conclusion, respectively.

2. Concept and Principle of the Optical Encoding Model

The schematic setup of the conceptual art of this proposed optical encoding model is illustrated in Figure 1. On the transmitter side, an optical system based on a Mach–Zehender interferometer is used to generate a coherent combination of two Laguerre–Gaussian (LG) beams carrying orbital angular momentum (OAM). A laser source provides a coherent fundamental Laguerre–Gaussian (LG00) beam in free space that is launched to a polarization beam splitter (PBS) to control relative power between the reference and the selector arm. Both arms will go through an OAM generator to convert a fundamental LG($p = 0, \ell = 0$) mode to a higher-order LG(p, ℓ) mode carrying OAM. The reference arm is converted to an LG($p, \ell = 1$) mode (via OAM Generator 1 in Figure 1), while selector arm is converted to an LG(p, ℓ) mode carrying OAM (via OAM Generator 2 in Figure 1). Since the topological charge ℓ and the radial index p at the LG(p, ℓ) selector beam can be properly selected to generate the intensity pattern for the optical encoder, this mode index will be the code-key numbers associated with the data symbol. The reference and selector arms are combined through a beam splitter (BS1), and the intensity profile of this superposition will be the pattern corresponding to a unique data symbol associated with (p, ℓ) combination. After encoding, the transmitted output beam is transferred to a communication channel in which different noise sources will be added in order to affect the signal. On the receiver side, the received beam is decoded by a machine learning process using an SVM-based method for image recognition and classification.

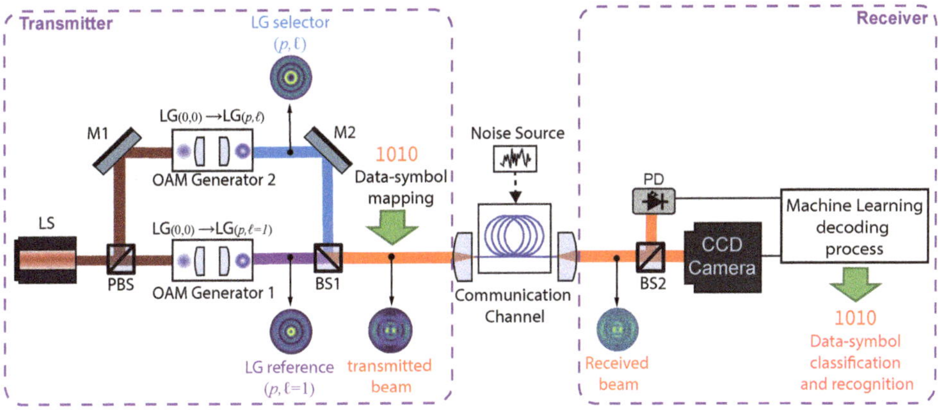

Figure 1. Concept and proposed setup of an optical encoding model. LS: laser source; PBS: polarization beam splitter, M1,2: mirror; BS1,2: beam splitter; PD: photodetector.

For the generation stage of the intensity pattern that will be used on the transmitter side, the mathematical formulation of the Laguerre–Gaussian beams [35,36] has been used, which is characterized by two indices (p, ℓ) corresponding to radial and azimuthal distribution, respectively. The optical field of an LG(p, ℓ) mode can be represented by:

$$
\begin{aligned}
&LG_{p,\ell}(r,\theta,z) \\
&= \sqrt{\frac{2p!}{\pi(p+|\ell|)!}}\frac{1}{w(z)}\left(\frac{r\sqrt{2}}{w(z)}\right)^{|\ell|} L_p^{|\ell|}\left(\frac{2r^2}{w^2(z)}\right) \\
&\exp\left(\frac{-r^2}{w^2(z)}\right)\exp\left(\frac{ik_o r^2 z}{2(z^2+z_R^2)}\right)\exp(\Phi(z))\exp(i\ell\theta)
\end{aligned}
\tag{1}
$$

where $w(z)$ is the beam width, z_R is the Rayleigh range and $\Phi(z)$ is the Gouy phase. $L_p^{|\ell|}$ are the generalized Laguerre polynomials, and (r, θ, z) represents the cylindrical coordinate. Then, the superposition of two LG modes carrying OAM [37] can be expressed as:

$$
u(r,\theta,z) = LG_{p',\ell'}(r,\theta,z) + LG_{p,\ell}(r,\theta,z) \tag{2}
$$

The first term in Equation (2) describes the reference field, while the second term represents the optical field that acts as a selector. As mentioned, for the optical encoder presented in this work, the $LG(p, \ell = 1)$ mode will be used as the $LG_{p',\ell'}(r, \theta, z)$ reference beam, while the selector beam $LG_{p,\ell}(r, \theta, z)$ will be a previously selected $LG_{p,\ell}$ mode. The same radial index p has been chosen for both reference and selector beam in order to simplify the design of the encoder. An intensity profile of $u(r, \theta, z)$ is associated with a data-bit sequence according to the (p, ℓ) parameters used in the selector beam generation. Since OAM beams have twisted helical phase fronts, often characterized by the azimuthal index ℓ (also named topological charge), while propagating[11], the intensity profile will be most affected in rotations along the propagation axis, without significant changes in the intensity pattern. Additionally, the property of orthogonality between LG modes allows the resulting intensity pattern to be unique for each data symbol [38].

3. SVM-Based Decoding Method for Image Recognition and Classification

The proposed optical encoding model takes as input a 4-bit code defined by the variable X. In addition, a signal noise ratio (SNR) is used to emulate the noise in the communication channel that is given in decibels. The encoding starts with the definition of the variables ℓ_1, ℓ_2 and p to establish the intensity profile, which is executed by the function *selectCode*. Then, two different intensity profiles are generated using the mathematical formulation given in Equations (1) and (2) and declared in $functionLG$. This is followed by the representation of the intensity profile in terms of Cartesian coordinates x, y and the intensity of the resulting beam profile declared in variable I. Next, with a view to emulate a real communication, signal noise is added to the transmitted intensity profile stated in the function addNoise. Later, the $extractHOGFeaturesFromIntensity$ function is used to extract useful patterns for information recognition through histogram of oriented gradients (HOG) detection [14]. Finally, the function *predict*, which is based on a linear regression model, is used as a 4-bit classifier through a multiclass error-correcting output codes (ECOC) model using SVM binary learners. For more details about the followed process, Algorithm 1 is presented. As the process involves training procedures, the decoding processing at the receiver side of the encoder is based on SVM. More details on this SVM algorithm can be found in the Appendix A.

Algorithm 1 Pseudocode for decoding processing using an SVM–ECOC model

Input X, SNR
Transmission side:
$[\ell_1, \ell_2, p]$=*selectCode*(X) ▷ *Give* ℓ_1, ℓ_2, *and* p *values according to Code Table*
$[x_1, y_1, z_1] = functionLG(\ell_1, \theta_o = 0, \lambda, z = 0, p)$ ▷ *Generate 1st Intensity Profile accord. Equation1*
$[x_2, y_2, z_2] = functionLG(\ell_2, \theta_o = 0, \lambda, z = 0, p)$ ▷ *Generate 2nd Intensity Profile accord. Equation1*
$x \leftarrow x_1$
$y \leftarrow y_1$ ▷ *Generate Value for Cartesian coordinates*
$I \leftarrow z_1 + z_2$ ▷ *Superposition of Intensity Profile accord. Equation2*
Communication Channel:
$n = addNoise(I, SNR)$ ▷ *Add Noise to the intensity in order to simulate real communication signal*
Receiver Side:
$TestFeatures = extractHOGFeaturesFromIntensity(n)$ ▷ *Extract HOG features for the Intensity profile with noise*
$Y = predict(classifier, testFeatures)$ ▷ *Use model from SVM-ECOC Multiclass Training*
Output Y

4. Case Study

As mentioned in the operating principle of the proposed optical encoding model, each data symbol is mapped to a corresponding $u(r, \theta)$ profile according to the selected modal indices ℓ and p in the selector beam. Since the reference beam is restricted to the $LG(p', \ell' = 1)$ mode, the alphabet for possible data symbols within a discrete time window can be calculated as $log_2 N$ with $N = n_\ell n_p$, where $n_{\ell,p}$ represents the number of ℓ and p indices used in the selector arm, and N represents the different data symbols that can be encoded as N -ary numbers: $0, 1, \ldots (N-1)$ [14]. For validation purposes, a data symbol code based on a 4-bit data symbol $(N = 16)$ is designed, which is associated with the resulting intensity profile according to the selection of the (p, ℓ) combination, as shown in Figure 2. In the simulations presented in this work, LG beams with wavelength $\lambda = 1550$ nm, fundamental beam width $w_0 = 100\lambda$ and a propagation distance $z = 200\lambda$ have been considered. This table shows all possible combinations of data symbols and the normalized intensity profiles of the reference, selector and the transmitted beam for data mapping.

Since the proposed optical encoding model operates based on the Mach–Zehnder [39,40] interferometric method on the transmitter side, and an image-based detection system on the receiver side, the following OAM generation methods must be considered for an experimental implementation: For the experimental generation of OAM modes, it is common to use mode converters composed of several cylindrical lenses, which can convert high-order Hermite–Gaussian beams into high-order Laguerre–Gaussian beams. However, mode converters are limited to a specific order Hermite–Gaussian beam, which needs to be generated by certain technical means as presented in [2]. The size of the mode converter is large, which presents strict requirements for the relative position and angle of the cylindrical lens. Typical mode converter configurations can be founded in [41]. Another alternative of mechanism for the OAM generation mode is the employment of a spatial light modulator (SLM) [11] that uses configurations based on changing modulation patterns loaded into the spatial light modulator. This can be achieved with a laser that can achieve various OAM beams with different output degrees. However, it is important to consider that under current technical conditions, the reflectivity of liquid crystal spatial light modulators is from 60 % to 90% [11].

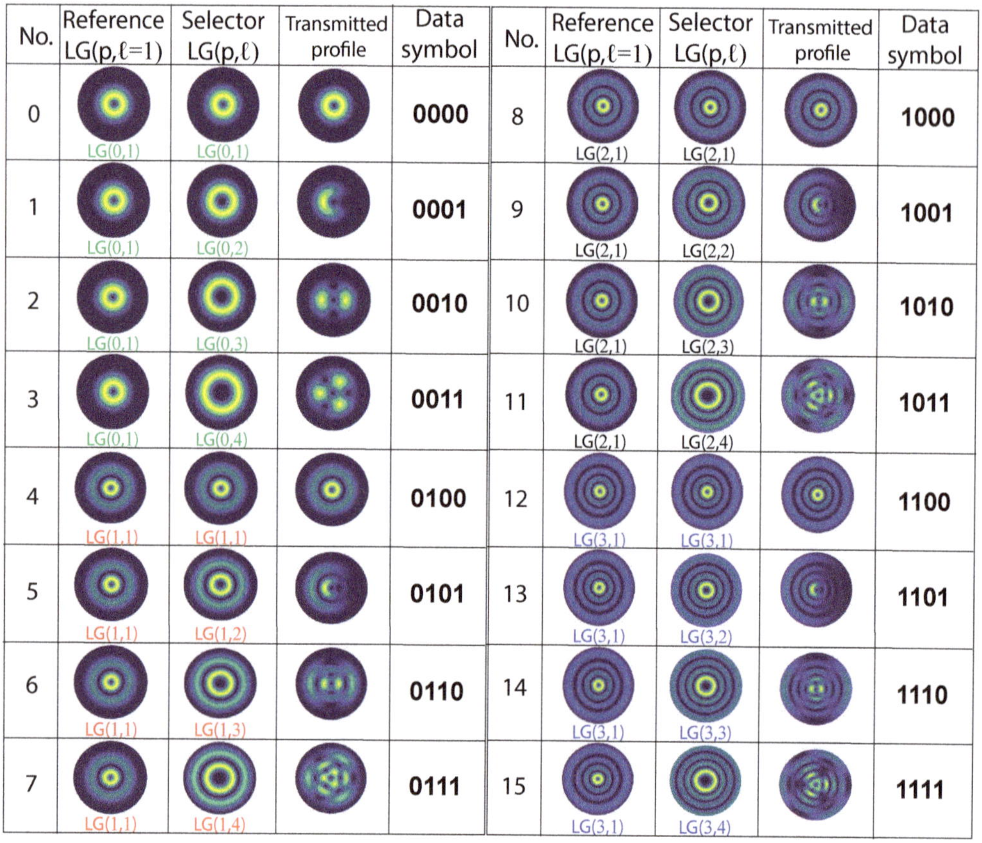

Figure 2. Data symbol set based on a 4-bit data symbol for the case study presented.

5. Results

The performance of the proposed optical encoding model is measured in terms of signal degradation due to the addition of noise and the bit error rate (BER) presented by the system. It is known that accuracy and precision of an optical encoder depend on detection method and robustness of the SVM training algorithm used [42,43]. In this context, to validate the influence of noise on transmission and therefore measure the degree of degradation and signal detection, a combination of RIN and AWGN noises has been used as channel noise for all detection and classification cases. The value of α for RIN has been established by a factor of 0.5 of the uniform random distribution, while for AWGN the mean $\mu = 0$ and signal–noise ratio (SNR) levels have been established at 36 dB (low), 30 dB (medium) and 24 dB (high) that are typical noise levels in optical communication systems [44,45].

To understand how the combination of these noises affects the transmitted signal, three different data symbols are presented in Figure 3: 0011 (Figure 3a.i), 0110 (Figure 3b.i) and 1011 (Figure 3c.i) with their corresponding 2D linear transformations of 200 $\times$ 200 pixels (computational burden), Figure 3a.ii, b.ii and c.ii. With a view to show the impact of the noises in the transmitted signal, the horizontal position arrangement for pixel 50 of the vertical position (x,50) has been chosen for display purposes, which is presented as a yellow dotted line in Figure 3a.ii, b.ii and c.ii. As a result, the normalized intensity curve for such an array is presented in Figure 3a.iii, b.iii and c.iii.

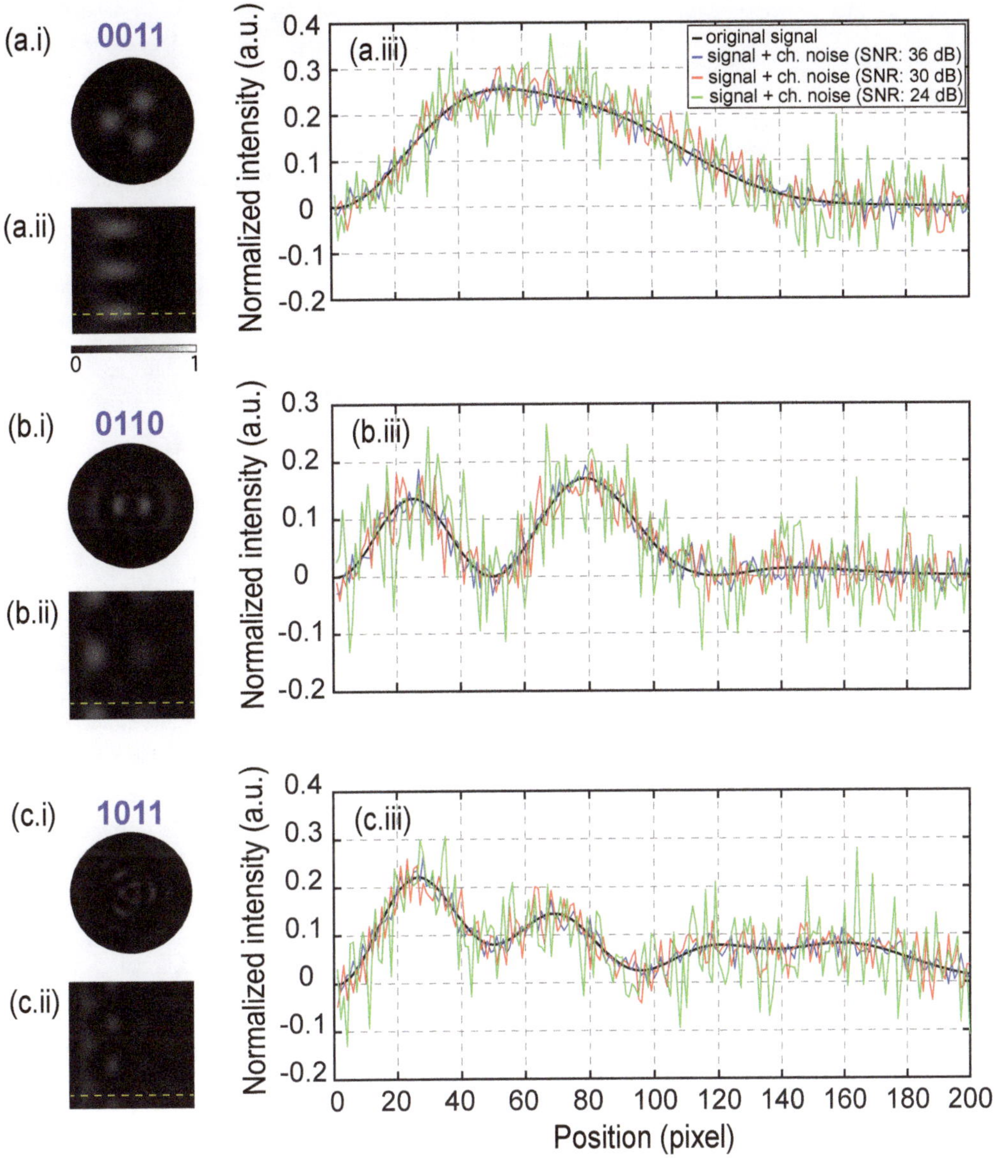

Figure 3. Different data symbols and their corresponding normalized intensity curves: (**a.i**) data symbol 0011; (**b.i**) data symbol 0110; (**c.i**) data symbol 1011; (**a.ii,b.ii,c.ii**) linear transformation of (**a.i,b.i,c.i**); (**a.iii,b.iii,c.iii**) normalized intensity curve corresponding to a pixel array of a 2D image (dotted yellow line) with different channel noise levels.

The normalized intensity curves show the original transmitted signal in a black curve, the received signal with the same α for each case (α = 0.5 for RIN) and with low level channel noise in a blue curve, medium level in a red curve and high level in a green curve. The results of Figure 3 reveal that despite observing distortion in the signal due to the addition of noise for levels greater than 24 dB of SNR, the SVM–ECOC model allows each image to be correctly classified and recognized, with a percentage of 100 % recovery for each data symbol. For this reason, a computed BER measurement for values less than 24 dB of SNR is necessary to validate the robustness of the optical encoding model at much more critical noise levels.

An end-to-end performance measure for data transmission is BER, which quantifies the reliability of an entire coding system from "input bits" to "output bits", including the behaviour of all components and elements between the transmitted signal and the received signal in addition to considering the path of the signal in the middle [46,47]. BER is mathematically defined as the relation between the number of bit errors and the total number of bits [48], which expresses the probability of a bit error. The machine learning model for prediction, recognition and classification of images on the receiver side of the proposed optical encoder is based on the SVM–ECOC multicast algorithm, which can be modelled with binary combinations of each class (one-vs.-one) or with binary combinations of one to multiple classes (one-vs.-all) [49,50], so the BER measurement for each machine learning model becomes a reliable metric of confidence level at the receiving point. To calculate the BER as a function of SNR at the receiver end using the SVM-ECOC model, the training model (one-vs.-one or one-vs.-all) is first created based on the data set of 4-bit symbols (see Figure 2). Then, the algorithm is trained with 750 images at different SNR levels (from 12 dB to 36 dB in steps of 6 dB) in the received signal, to test the functionality of the model at these noise levels. Once the functionality of the SVM model has been verified through the previous training, the images are processed with the model. For image processing, a database consisting of 10,000 images for each 4-bit data symbol combination (between 0000 and 1111) was used, resulting in a total of 160,000 processed images. The HOG features are extracted from each of these images to predict the combination of bits corresponding to each image, using the model. Finally, after each prediction, the acquired combination is compared with the original combination, and then the BER is calculated. Figure 4 shows the computed BER points as a function of signal-to-noise ratio (SNR) from 0 to 14 dB in steps of 1 dB for the two proposed SVM-ECOC models: the multicast one-vs.-one algorithm (Model 1) in the red curve and the Multicast one-vs.-all algorithm (Model 2) in the blue curve.

Since the standard maximum BER for most optical systems is 10^{-9} [51], and for applications in optical communications the maximum BER range is in the range 10^{-9} to $10^{-12)}$ [52], the adjustment curve for each model is also shown in Figure 4 in order to predict noise levels for these values. The BER curve for Model 1 reaches BER $= 10^{-9}$ for 12.8 dB of SNR (see green line in Figure 4), and BER $= 10^{-12}$ for 13.4 dB of SNR. For the case of Model 2 (blue curve in Figure 4), BER $= 10^{-9}$ for 10.2 dB of SNR (see green line in Figure 4), while for a BER $= 10^{-12}$ for 10.9 dB of SNR. Additionally, for comparison purposes, a BER estimation curve assuming a probability of error with a Gaussian random variable [53] is also shown in a black curve. It is observed that both SVM–ECOC models have better performance compared to the simplified Gaussian BER model in terms of noise levels, highlighting that Model 2 has a better probability of error compared to Model 1, with a difference of 2.65 dB of noise level for BER $= 10^{-9}$. Note that since the channel noises used in the simulation are AWGN and RIN, the bit errors generated in this case study are directly due to signal degradation by these types of noise. This fact is observed in the results of Figure 4 for each model, indicating that for an SNR level greater than 9 dB, the bit error probability is below 10% for Model 1 and below 0.0001% for Model 2.

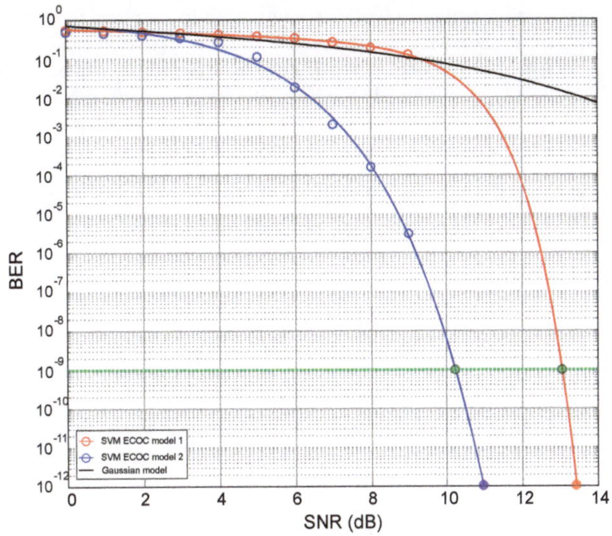

Figure 4. Computed BER for each SVM-ECOC model as a function of SNR for critical noise levels (from 0 to 14 dB).

6. Conclusions

A comprehensive design of an optical encoding model based on the coherent superposition of two *LG* beams with OAM is proposed for the generation of a coding system independent of phase information. The proposed approach employs an SVM-ECOC algorithm machine learning that enables image prediction, recognition and classification. To verify the robustness of the proposed optical encoding model, a data symbol code based on a 4-bit data symbol is designed, which is associated with the intensity profile according to the (p, ℓ) combination. A channel noise made up of the RIN and AWGN is added to the images generated in the encoding stage to emulate a real environment. In order to identify each data symbol, two different algorithms based on an SVM-ECOC model are used. The efficacy of the proposed approach is validated through BER measurements. The results reveal that the proposed algorithms are able to recognize the data symbol set with a degree of confidence greater than 90 % for noise levels up to 9 dB in both models. Even though both models present high efficiency, the Multicast one-vs.-all model (Model 2) presents the best BER curve between the two models studied, with a BER $= 10^{-9}$ for 10.2 dB of SNR.

The proposed encoding model can be employed on optical free-space (OFS) data links, which according to the state of the art, such encoding potentially increases data capacity for wireless systems and satellite communication systems [25,54]. These systems present typical link distances between 1 km and 143 km (verified experimentally), for 532 nm, 633 nm and 1550 nm of operating wavelength and a range between 150 Mbps and 200 Gbps of data rate [11], which complicates the data transmission. However, the proposed optical encoding model can be a solution as it can be used over optical fiber links as evidenced in [25], in which an optical encoding system based on OAM beams has been implemented for data transmission at 80 Gbps using 5 km few mode fibers (FMF) to data transmission at 640 Gbps using 18 km of ring-core-fiber. On the other hand, some constraints must be considered when choosing the type of encrypted data transmission channel. For free-space links, atmospheric turbulence can cause a random phase and intensity distortion on the transversal beam profile [55], which can be quantified by the refractive index structure constant C_n^2 that has typical values between 10^{-17} $m^{-2/3}$ and 10^{-13} $m^{-2/3}$. According to Allen et al. [1], the Rytov variance is an adequate indicator to quantify turbulence fluctuations in OFS links, since this is related to C_n^2 and the propagation distance. Also demonstrated in [56], for a link with low power fluctuations (low Rytov variance), the recommended propagation length should be less than 10 km, and for greater distances, the use of mitigation methods such as adaptive optics beam shaping is recommended [57]. Focusing on optical fiber links, the fundamental limitations lie in the type of fiber used for the transmission channel. The use of few mode fibers (FMFs) or the use of micro-structured fibers is necessary to excite OAM modes within the fiber as evidenced in [25].

For future research, it is relevant to mention that the number of circular fringes in the intensity profile of an LG mode are directly related to the index p, while the spatial distribution of these fringes is related to the index ℓ; therefore, the number of bits can be extended to more than 4-bits for the case where $\ell \geq 5$. This fact opens new opportunities for the development of advanced encoding systems.

Author Contributions: Conceptualization, E.L. and C.S.; methodology, M.S.A.-A.; software, C.S.; validation, A.P., E.L. and P.I.; formal analysis, E.L.; investigation, C.S; writing—review and editing, E.L. and M.S.A.-A. All authors have read and agreed to the published version of the manuscript.

Funding: This study is financially supported by the Decanato de Investigación from the Escuela Superior Politécnica del Litoral (ESPOL) under the project FCNM-210-2020.

Institutional Review Board Statement: Not applicable.

Informed Consent Statement: Not applicable.

Data Availability Statement: Not applicable.

Conflicts of Interest: The authors declare no conflict of interest.

Appendix A. Support Vector Machine Algorithm

As the process involves training procedures, the decoding processing on the receiver side of the encoder is based on support vector machine (SVM). SVM is a machine learning algorithm that employs the concept of the kernel function to map the data in a different dimensional space, such that the information is grouped according to similar attributes. The algorithm takes as input the raw data, which is classified depending on the kernel function. Then, the data are saved and compared with the original figure to identify similar patterns that are used for image identification. This process is repeated until the maximum number of iterations n is reached, as shown in Figure A1a. As a result, a simplification of complex nonlinear decision boundaries is obtained to derive in a linear dimensional space [58]. Mathematically, the characterization is driven by the kernel that can take the form as presented in Table A1. For a better understanding, a flowchart of SVM is presented in Figure A1b. The kernel used in the SVM algorithm for the proposed optical encoding model is the basis function (Gaussian).

Table A1. Brief description of the kernels that are used in the different types of SVM algorithms.

Type of SVM	Kernel	Description
Base function (Gaussian)	$K(x_1, x_2) = e^{-\frac{\|x_1 - x_2\|^2}{2\sigma^2}}$	Learning of one class, where σ represents the width of the kernel
Linear	$K(x_1, x_2) = x_1^T x_2$	Learning of two classes
Polynomial	$K(x_1, x_2) = (x_1^T x_2 + 1)^\rho$	ρ is the polynomial degree
Sigmoid	$K(x_1, x_2) = \tanh(\beta_0 x_1^T x_2 + \beta_1)$	The kernel is determined by specific β_0 and β_1

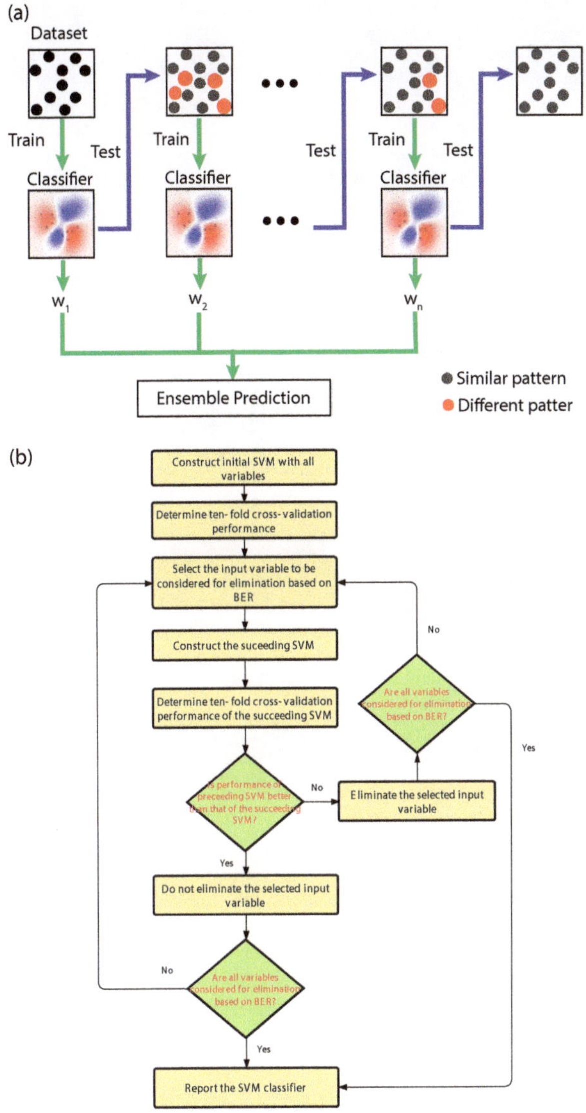

Figure A1. (**a**) Flow diagram of an SVM algorithm. The ensemble classifiers consist of a set of weak classifiers. The weights (wn) of the incorrectly predicted points are increased in the next classifier. The final decision is based on the weighted average of the individual predictions; (**b**) flowchart of the application of the support vector machine (SVM) algorithm in the decoding processing.

References

1. Allen, L.; Beijersbergen, M.W.; Spreeuw, R.; Woerdman, J. Orbital angular momentum of light and the transformation of Laguerre–Gaussian laser modes. *Phys. Rev. A* **1992**, *45*, 8185. [CrossRef] [PubMed]
2. Lian, Y.; Qi, X.; Wang, Y.; Bai, Z.; Wang, Y.; Lu, Z. OAM beam generation in space and its applications: A review. *Opt. Lasers Eng.* **2022**, *151*, 106923. [CrossRef]
3. Zhao, J.; Chremmos, I.D.; Song, D.; Christodoulides, D.N.; Efremidis, N.K.; Chen, Z. Curved singular beams for three-dimensional particle manipulation. *Sci. Rep.* **2015**, *5*, 12086. [CrossRef] [PubMed]

4. Liu, K.; Cheng, Y.; Gao, Y.; Li, X.; Qin, Y.; Wang, H. Super-resolution radar imaging based on experimental OAM beams. *Appl. Phys. Lett.* **2017**, *110*, 164102. [CrossRef]
5. Wang, J.; Liu, K.; Cheng, Y.; Wang, H. Vortex SAR imaging method based on OAM beams design. *IEEE Sensors J.* **2019**, *19*, 11873–11879. [CrossRef]
6. Wang, J. Twisted optical communications using orbital angular momentum. *Sci. China Phys. Mech. Astron.* **2019**, *62*, 34201. [CrossRef]
7. Trichili, A.; Rosales-Guzmán, C.; Dudley, A.; Ndagano, B.; Ben Salem, A.; Zghal, M.; Forbes, A. Optical communication beyond orbital angular momentum. *Sci. Rep.* **2016**, *6*, 27674. [CrossRef]
8. Gong, L.; Zhao, Q.; Zhang, H.; Hu, X.Y.; Huang, K.; Yang, J.M.; Li, Y.M. Optical orbital-angular-momentum-multiplexed data transmission under high scattering. *Light Sci. Appl.* **2019**, *8*, 27. [CrossRef]
9. Zhao, Q.; Yu, P.P.; Liu, Y.F.; Wang, Z.Q.; Li, Y.M.; Gong, L. Light field imaging through a single multimode fiber for OAM-multiplexed data transmission. *Appl. Phys. Lett.* **2020**, *116*, 181101. [CrossRef]
10. Kai, C.; Feng, Z.; Dedo, M.I.; Huang, P.; Guo, K.; Shen, F.; Gao, J.; Guo, Z. The performances of different OAM encoding systems. *Opt. Commun.* **2019**, *430*, 151–157. [CrossRef]
11. Willner, A.E.; Pang, K.; Song, H.; Zou, K.; Zhou, H. Orbital angular momentum of light for communications. *Appl. Phys. Rev.* **2021**, *8*, 041312. [CrossRef]
12. Fang, X.; Ren, H.; Gu, M. Orbital angular momentum holography for high-security encryption. *Nat. Photonics* **2020**, *14*, 102–108. [CrossRef]
13. Xiao, Q.; Ma, Q.; Yan, T.; Wu, L.W.; Liu, C.; Wang, Z.X.; Wan, X.; Cheng, Q.; Cui, T.J. Orbital-angular-momentum-encrypted holography based on coding information metasurface. *Adv. Opt. Mater.* **2021**, *9*, 2002155. [CrossRef]
14. Fu, S.; Zhai, Y.; Zhou, H.; Zhang, J.; Wang, T.; Yin, C.; Gao, C. Demonstration of free-space one-to-many multicasting link from orbital angular momentum encoding. *Opt. Lett.* **2019**, *44*, 4753–4756. [CrossRef]
15. Willner, A.J.; Ren, Y.; Xie, G.; Zhao, Z.; Cao, Y.; Li, L.; Ahmed, N.; Wang, Z.; Yan, Y.; Liao, P.; et al. Experimental demonstration of 20 Gbit/s data encoding and 2 ns channel hopping using orbital angular momentum modes. *Opt. Lett.* **2015**, *40*, 5810–5813. [CrossRef]
16. Li, S.; Xu, Z.; Liu, J.; Zhou, N.; Zhao, Y.; Zhu, L.; Xia, F.; Wang, J. Experimental demonstration of free-space optical communications using orbital angular momentum (OAM array encoding/decoding. In Proceedings of the 2015 Conference on Lasers and Electro-Optics (CLEO), Busan, Republic of Korea, 24–28 August 2015; pp. 1–2.
17. Trichili, A.; Salem, A.B.; Dudley, A.; Zghal, M.; Forbes, A. Encoding information using Laguerre Gaussian modes over free space turbulence media. *Opt. Lett.* **2016**, *41*, 3086–3089. [CrossRef]
18. Wang, J.; Yang, J.Y.; Fazal, I.M.; Ahmed, N.; Yan, Y.; Huang, H.; Ren, Y.; Yue, Y.; Dolinar, S.; Tur, M.; et al. Terabit free-space data transmission employing orbital angular momentum multiplexing. *Nat. Photonics* **2012**, *6*, 488–496. [CrossRef]
19. Zhou, H.; Sain, B.; Wang, Y.; Schlickriede, C.; Zhao, R.; Zhang, X.; Wei, Q.; Li, X.; Huang, L.; Zentgraf, T. Polarization-encrypted orbital angular momentum multiplexed metasurface holography. *ACS Nano* **2020**, *14*, 5553–5559. [CrossRef]
20. Bolduc, E.; Bent, N.; Santamato, E.; Karimi, E.; Boyd, R.W. Exact solution to simultaneous intensity and phase encryption with a single phase-only hologram. *Opt. Lett.* **2013**, *38*, 3546–3549. [CrossRef] [PubMed]
21. Willner, A.E.; Song, H.; Liu, C.; Zhang, R.; Pang, K.; Zhou, H.; Hu, N.; Song, H.; Su, X.; Zhao, Z.; et al. Causes and mitigation of modal crosstalk in OAM multiplexed optical communication links. In *Structured Light for Optical Communication*; Elsevier: Amsterdam, The Netherlands, 2021; pp. 259–289.
22. Ouyang, X.; Xu, Y.; Xian, M.; Feng, Z.; Zhu, L.; Cao, Y.; Lan, S.; Guan, B.O.; Qiu, C.W.; Gu, M.; et al. Synthetic helical dichroism for six-dimensional optical orbital angular momentum multiplexing. *Nat. Photonics* **2021**, *15*, 901–907. [CrossRef]
23. Zhu, F.; Jiang, J.; Li, Y.; Zhou, C.; Tang, L.; Lai, Z. Index Modulation of OAM-UCA with LDPC Transmission. In Proceedings of the 2021 IEEE 21st International Conference on Communication Technology (ICCT), Tianjin, China, 13–16 October 2021; pp. 1300–1303.
24. Li, Y.; Zhang, Z. Image information transfer with petal-like beam lattices encoding/decoding. *Opt. Commun.* **2022**, *510*, 127931. [CrossRef]
25. Du, J.; Li, S.; Zhao, Y.; Xu, Z.; Zhu, L.; Zhou, P.; Liu, J.; Wang, J. Demonstration of M-ary encoding/decoding using visible-light Bessel beams carrying orbital angular momentum (OAM) for free-space obstruction-free optical communications. In Proceedings of the Optical Fiber Communication Conference, Los Angeles, CA, USA, 22–26 March 2015; p. M2F-4.
26. Fujita, H.; Sato, M. Encoding orbital angular momentum of light in magnets. *Phys. Rev. B* **2017**, *96*, 060407. [CrossRef]
27. Zhao, N.; Li, X.; Li, G.; Kahn, J.M. Capacity limits of spatially multiplexed free-space communication. *Nat. Photonics* **2015**, *9*, 822–826. [CrossRef]
28. Chen, M.; Dholakia, K.; Mazilu, M. Is there an optimal basis to maximise optical information transfer? *Sci. Rep.* **2016**, *6*, 22821. [CrossRef]
29. Trichili, A.; Park, K.H.; Zghal, M.; Ooi, B.S.; Alouini, M.S. Communicating using spatial mode multiplexing: Potentials, challenges, and perspectives. *IEEE Commun. Surv. Tutorials* **2019**, *21*, 3175–3203. [CrossRef]
30. Bartkiewicz, K.; Gneiting, C.; Černoch, A.; Jiráková, K.; Lemr, K.; Nori, F. Experimental kernel-based quantum machine learning in finite feature space. *Sci. Rep.* **2020**, *10*, 12356. [CrossRef]
31. Zhou, J.; Huang, B.; Yan, Z.; Bünzli, J.C.G. Emerging role of machine learning in light-matter interaction. *Light Sci. Appl.* **2019**, *8*, 84. [CrossRef]

32. Neary, P.L.; Watnik, A.T.; Judd, K.P.; Lindle, J.R.; Flann, N.S. Machine learning-based signal degradation models for attenuated underwater optical communication OAM beams. *Opt. Commun.* **2020**, *474*, 126058. [CrossRef]
33. Kirchner, T.; Gröhl, J.; Maier-Hein, L. Context encoding enables machine learning-based quantitative photoacoustics. *J. Biomed. Opt.* **2018**, *23*, 056008. [CrossRef]
34. Zhou, L.; Chen, X.; Chen, W. Deep learning based attack on phase-truncated optical encoding. In Proceedings of the 2020 IEEE MTT-S International Conference on Numerical Electromagnetic and Multiphysics Modeling and Optimization (NEMO), Hangzhou, China, 7–9 December 2020; pp. 1–4.
35. Doster, T.; Watnik, A.T. Laguerre–Gauss and Bessel–Gauss beams propagation through turbulence: Analysis of channel efficiency. *Appl. Opt.* **2016**, *55*, 10239–10246. [CrossRef]
36. Paufler, W.; Böning, B.; Fritzsche, S. High harmonic generation with Laguerre–Gaussian beams. *J. Opt.* **2019**, *21*, 094001. [CrossRef]
37. Litvin, I.A.; Burger, L.; Forbes, A. Angular self-reconstruction of petal-like beams. *Opt. Lett.* **2013**, *38*, 3363–3365. [CrossRef] [PubMed]
38. Guo, Z.; Wang, Z.; Dedo, M.I.; Guo, K. The orbital angular momentum encoding system with radial indices of Laguerre–Gaussian beam. *IEEE Photonics J.* **2018**, *10*, 7906511. [CrossRef]
39. Zetie, K.; Adams, S.; Tocknell, R. How does a Mach–Zehnder interferometer work? *Phys. Educ.* **2000**, *35*, 46. [CrossRef]
40. Schimpf, D.; Barankov, R.; Ramachandran, S. Cross-correlated (C 2) imaging of fiber and waveguide modes. *Opt. Express* **2011**, *19*, 13008–13019. [CrossRef]
41. Beijersbergen, M.W.; Allen, L.; Van der Veen, H.; Woerdman, J. Astigmatic laser mode converters and transfer of orbital angular momentum. *Opt. Commun.* **1993**, *96*, 123–132. [CrossRef]
42. Zheng, G.; Qian, Z.; Yang, Q.; Wei, C.; Xie, L.; Zhu, Y.; Li, Y. The combination approach of SVM and ECOC for powerful identification and classification of transcription factor. *BMC Bioinform.* **2008**, *9*, 282. [CrossRef]
43. Liu, M.; Zhang, D.; Chen, S.; Xue, H. Joint binary classifier learning for ECOC-based multi-class classification. *IEEE Trans. Pattern Anal. Mach. Intell.* **2015**, *38*, 2335–2341. [CrossRef]
44. Binh, L.N. *Noises in Optical Communications and Photonic Systems*; CRC Press LLC: Boca Raton, FL, USA, 2016.
45. Kareem, F.Q.; Zeebaree, S.; Dino, H.I.; Sadeeq, M.; Rashid, Z.N.; Hasan, D.A.; Sharif, K.H. A survey of optical fiber communications: Challenges and processing time influences. *Asian J. Res. Comput. Sci.* **2021**, *7*, 48–58. [CrossRef]
46. Sheng, M.; Jiang, P.; Hu, Q.; Su, Q.; Xie, X.x. End-to-end average BER analysis for multihop free-space optical communications with pointing errors. *J. Opt.* **2013**, *15*, 055408. [CrossRef]
47. Freude, W.; Schmogrow, R.; Nebendahl, B.; Winter, M.; Josten, A.; Hillerkuss, D.; Koenig, S.; Meyer, J.; Dreschmann, M.; Huebner, M.; et al. Quality metrics for optical signals: Eye diagram, Q-factor, OSNR, EVM and BER. In Proceedings of the 2012 14th International Conference on Transparent Optical Networks (ICTON), Coventry, UK, 2–5 July 2012; pp. 1–4.
48. Hayal, M.R.; Yousif, B.B.; Azim, M.A. Performance enhancement of DWDM-FSO optical fiber communication systems based on hybrid modulation techniques under atmospheric turbulence channel. *Photonics* **2021**, *8*, 464. [CrossRef]
49. Hernández, J.A.; Martín, I.; Camarillo, P.; de Arcaute, G.M.R. Applications of Machine Learning Techniques for What-if Analysis and Network Overload Detection. In Proceedings of the 2022 18th International Conference on the Design of Reliable Communication Networks (DRCN), Vilanova i la Gelrú, Spain, 28–31 March 2022; pp. 1–7.
50. Zhang, L.; Li, X.; Tang, Y.; Xin, J.; Huang, S. A survey on QoT prediction using machine learning in optical networks. *Opt. Fiber Technol.* **2022**, *68*, 102804. [CrossRef]
51. Walsh, D.; Moodie, D.; Mauchline, I.; Conner, S.; Johnstone, W.; Culshaw, B. Practical bit error rate measurements on fibre optic communications links in student teaching laboratories. In Proceedings of the 9th International Conference on Education and Training in Optics and Photonics (ETOP), Marseille, France, 24–27 October 2005.
52. Balsells, J.M.G.; López-González, F.J.; Jurado-Navas, A.; Castillo-Vázquez, M.; Notario, A.P. General closed-form bit-error rate expressions for coded M-distributed atmospheric optical communications. *Opt. Lett.* **2015**, *40*, 2937–2940. [CrossRef]
53. Keiser, G. *Fiber Optic Communications*; Springer: Berlin/Heidelberg, Germany, 2021.
54. Chan, V.W. Free-space optical communications. *J. Light. Technol.* **2006**, *24*, 4750–4762. [CrossRef]
55. Andrews, L.C.; Phillips, R.L. Laser beam propagation through random media. In *Laser Beam Propagation Through Random Media: Second Edition*; SPIE Publications: Bellingham, WA, USA, 2005.
56. Bhatnagar, M.R.; Ghassemlooy, Z. Performance analysis of gamma–gamma fading FSO MIMO links with pointing errors. *J. Light. Technol.* **2016**, *34*, 2158–2169. [CrossRef]
57. Li, L.; Xie, G.; Ren, Y.; Ahmed, N.; Huang, H.; Zhao, Z.; Liao, P.; Lavery, M.P.; Yan, Y.; Bao, C.; et al. Orbital-angular-momentum-multiplexed free-space optical communication link using transmitter lenses. *Appl. Opt.* **2016**, *55*, 2098–2103. [CrossRef]
58. Cervantes, J.; Garcia-Lamont, F.; Rodriguez, L.; López, A.; Castilla, J.R.; Trueba, A. PSO-based method for SVM classification on skewed data sets. *Neurocomputing* **2017**, *228*, 187–197. [CrossRef]

applied sciences

MDPI

Communication

High Speed Decoding for High-Rate and Short-Length Reed–Muller Code Using Auto-Decoder

Hyun Woo Cho and Young Joon Song *

Department of Electronic Engineering, Kumoh National Institute of Technology, Gumi 39177, Korea
* Correspondence: yjsong@kumoh.ac.kr

Abstract: In this paper, we show that applying a machine learning technique called auto-decoder (AD) to high-rate and short length Reed–Muller (RM) decoding enables it to achieve maximum likelihood decoding (MLD) performance and faster decoding speed than when fast Hadamard transform (FHT) is applied in additive white Gaussian noise (AWGN) channels. The decoding speed is approximately 1.8 times and 125 times faster than the FHT decoding for $\mathfrak{R}(1,4)$ and $\mathfrak{R}(2,4)$, respectively. The number of nodes in the hidden layer of AD is larger than that of the input layer, unlike the conventional auto-encoder (AE). Two ADs are combined in parallel and merged together, and then cascaded to one fully connected layer to improve the bit error rate (BER) performance of the code.

Keywords: Reed–Muller (RM) code; machine learning; auto-decoder; auto-encoder

1. Introduction

Machine learning (ML) techniques are widely used in many fields, such as image recognition, natural language processing, and autonomous driving [1–3]. Auto-encoder (AE) unsupervised ML techniques are known for their capacity to extract important features of data while reducing unwanted noise, and thus are useful in generating new images with key features [4]. AEs perform roles such as dimensionality reduction, image denosing, image generation, and abnormality detection, and are used in various fields such as medical care, autonomous driving, and image recognition [5–8]. In addition, various studies are being undertaken to apply machine learning technology to communication systems, such as channel coding, massive multi-input and multi-output, multiple access, resource allocation, and network security [9,10]. In this study, we modify an AE to a new model called the auto-decoder (AD), which is suitable for reducing the noise that corrupts the transmitted information signal in channel coding. The proposed AD is used to decode Reed–Muller (RM) code of high-rate and short length, which is used in many communication systems, such as long-term evolution (LTE) and fifth-generation wireless (5G) cellular systems [11,12], where the minimum latency delay is 5 ms. The requirement is to further reduce this delay in sixth-generation (6G) wireless systems [13,14]. Since we consider high-rate Reed–Muller code of short length, such as $\mathfrak{R}(2,4)$ with code rate 0.6875, we use a fast Hadamard transform (FHT) decoding method [15] for performance comparison instead of the recursive decoding of [16,17] which is useful for low-rate RM code. Because the RM code has an extremely simple structure and can be decoded with maximum likelihood decoding (MLD) performance using FHT, it is especially useful in control channels in wireless communication systems. We first illustrate the key differences between the conventional AE and the proposed AD, and then show how to construct the decoder for the RM code using it. After training the AD model for the code, we found that the proposed method showed similar performance to the conventional FHT method with faster decoding speed. For improved performance, we present a parallel auto-decoder (PAD) that combines a couple of ADs in parallel.

Citation: Cho, H.W.; Song, Y.J. High Speed Decoding for High-Rate and Short-Length Reed–Muller Code Using Auto-Decoder. *Appl. Sci.* **2022**, *12*, 9225. https://doi.org/10.3390/app12189225

Academic Editor: Valentino Santucci

Received: 2 August 2022
Accepted: 13 September 2022
Published: 14 September 2022

2. RM Decoder Based on AD

This section explains the RM decoder using AD and compares the performance of FHT decoding. Table 1 shows the notations used in this paper.

Table 1. Summarizes the notation used in the paper.

Symbol	Description
n	length of codeword (bits)
r, m	parameters of RM code ($0 \leq r \leq m$)
k	length of message (bits)
N	number of nodes in FC layer
$\mathbf{y}$	one-hot encoding vector
$\mathbf{m}$	message vector
$\mathbf{z}$	output of FC layer
S	number of validation sets
$\rho_t, \rho_{v,s}$	SNRs for the training set and the s-th validation set

2.1. Auto-Decoder

The AD, which plays a central role in decoding of the RM code, is modified from a conventional AE. Figure 1 shows the basic structures of the conventional AE and the proposed AD, highlighting the difference between the number of nodes in two different hidden layers. The number of nodes in the hidden layer of the AE in Figure 1a is less than that of the input layer. In comparison to the AE, the number of nodes in the hidden layer of the AD in Figure 1b is larger than that of the input layer. Figure 2 shows the typical structure of an AD that is composed of three hidden layers for the decoder; the number of nodes in each layer is presented in Table 2. The number of nodes in the input layer is the same as the length of the codeword n. The number of nodes in the first hidden layer is $2n$, in the second hidden layer is $4n$, in the last hidden layer is again $2n$, and finally the output layer becomes n again. In general, a neural network with a multilayer perceptron shows much better performance than a single-layer perceptron. From this perspective, we can speculate that the hidden layer structure of the AD is more suitable for decoding short length codes, such as RM code, than the AE, in which the number of nodes of the hidden layer becomes smaller as the depth of the hidden layer increases, which may result in a smaller number of nodes, especially in the middle of the hidden layer, resulting in poor decoding performance.

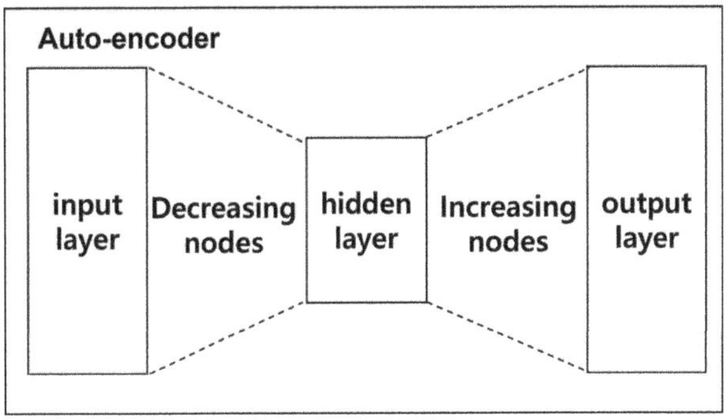

(a)

Figure 1. *Cont.*

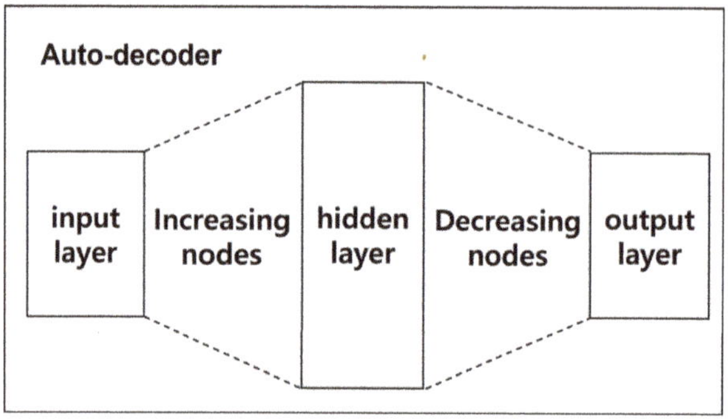

(b)

Figure 1. Structure of (**a**) auto-encoder and (**b**) auto-decoder.

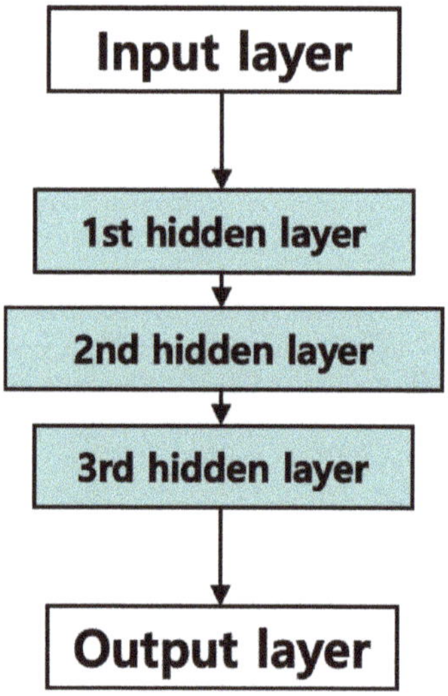

Figure 2. Structure of auto-decoder with 3 hidden layers.

Table 2. Number of nodes for layers in AD.

Layer	Number of Nodes
input	n
1st hidden	$2n$
2nd hidden	$4n$
3rd hidden	$2n$
output	n

2.2. RM Decoding Model

The RM code of $n = 2^m$ bits with the minimum Hamming distance of 2^{m-r} used to encode $k = \sum_i^r \binom{m}{i}$ bits of message is denoted as $\mathfrak{R}(r, m)$ [15,18]. To illustrate the proposed decoding model, we consider two cases of RM code, $\mathfrak{R}(1,4)$ and $\mathfrak{R}(2,4)$ of code length 16, because the model training for longer code consumes more time. Figure 3 shows the proposed RM decoder structure constructed in such a way that the AD of Figure 2 is followed by one fully connected (FC) layer with the number of nodes defined as $N = 2^k$, and, thus, $N = 2^5$ for $\mathfrak{R}(1,4)$ and $N = 2^{11}$ for $\mathfrak{R}(2,4)$. Because $N = 2^k$ is equivalent to the number of all possible messages transmitted, we use the FC layer as the output layer.

Moreover, a softmax activation function is used to normalize the output of the model to a probability distribution over all possible transmitted messages. Figure 4 shows the method used to estimate the transmitted message bits of the (16,11) code of $\mathfrak{R}(2,4)$ from the output layer. The output node index of the FC layer is the decimal value corresponding to the message bits, and the transmitted message bits are estimated by converting the index of the maximum output node value into $k = 11$ binary bits.

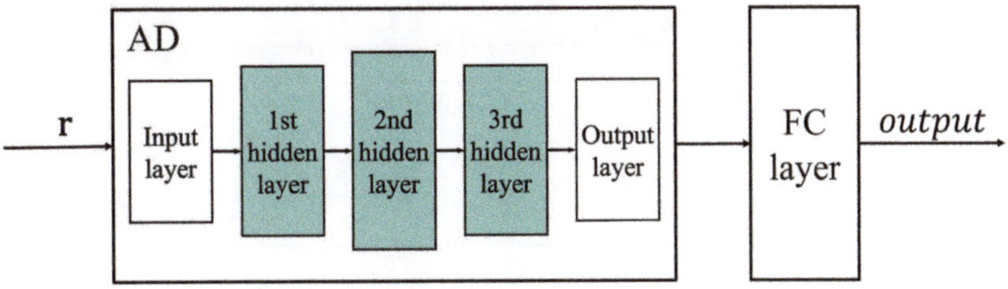

Figure 3. Structure of RM decoder based on AD.

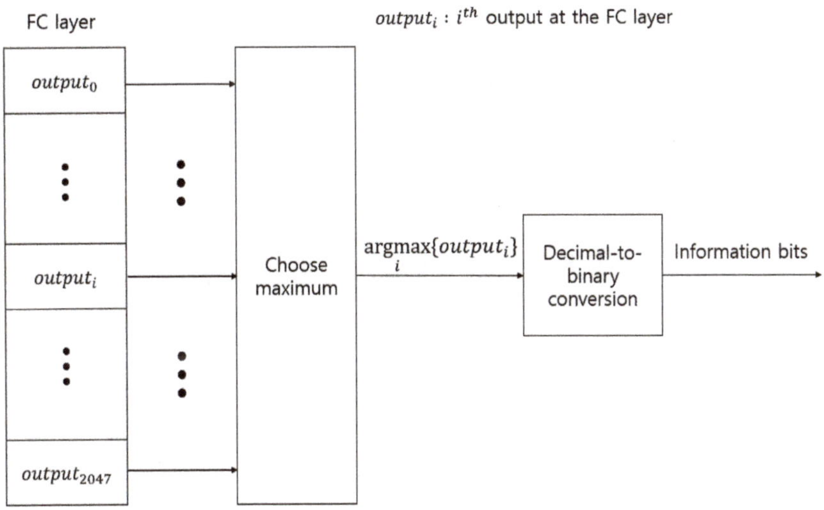

Figure 4. Message estimation of $\mathfrak{R}(2,4)$ from the output of FC layer.

2.3. Hyperparameters

Table 3 lists the hyperparameters used for training the RM decoder. Let $\mathbf{y} = \{0,1\}^N = (y_0, y_1, \cdots, y_j, \cdots, y_{N-1}) = (0, 0, \cdots, 1, \cdots, 0)$ be the one-hot encoded vector for the message $\mathbf{m} = (m_0, m_1, \cdots, m_k - 1)$, where $j = \sum_{l=0}^{k-1} m_l 2^l$ is the decimal value for $\mathbf{m}$, and all bits of $\mathbf{y}$ are zeros, except the value of one at the j-th bit. Let z_i be the i-th output of the FC layer. The cross-entropy is given by:

$$L(\mathbf{y}, \mathbf{z}) = -\sum_{i=0}^{N-1} [y_i \log z_i + (1 - y_i) \log(1 - z_i)] \tag{1}$$

is used as the loss function, and the Adam optimizer is used for model training. The training data set was $2^k \times 10^5$, the epoch was 10^2, and the batch size was set to 10^4. Normalized validation error (NVE) of [19]

$$\mathrm{NVE} = \frac{1}{S} \sum_{s=1}^{S} \frac{BER_{AD}(\rho_t, \rho_{v,s})}{BER_{FHT}(\rho_{v,s})} \tag{2}$$

is used to select the appropriate signal-to-noise (SNR) ratio for model training, where ρ_t and $\rho_{v,s}$ are the SNRs for the training set and the s-th validation set, respectively, and S is the number of all possible different validation sets. $BER_{AD}(\rho_t, \rho_{v,s})$ is the bit error rate when the RM decoder with AD is trained at ρ_t and evaluated at $\rho_{v,s}$. Table 4 shows the NVE values at different training SNRs, from 0 dB to 7 dB with 1 dB intervals, and we observe that NVE = 0.945 at $\rho_t = 1$ dB is the lowest value among them. Thus, the training SNR is set to 1 dB.

Table 3. Hyperparameters for model training.

loss function	cross-entropy
optimizer	Adam
training data set	$2^k \times 10^5$
epoch	10^2
batch size	10^4

Table 4. NVE for different training SNRs.

Training SNR (ρ_t)	0	1	2	3
NVE	0.972	0.945	0.982	1.138
Training SNR (ρ_t)	4	5	6	7
NVE	0.981	1.320	1.529	2.489

2.4. *Performance Evaluation*

To evaluate the performance of the proposed decoder using AD, we consider two cases of RM code, $\mathfrak{R}(1,4)$ and $\mathfrak{R}(2,4)$ whose code structure is simple, with a short message length. Figure 5 shows the BER of RM coding with the AD compared with FHT decoding. The BER at each SNR was calculated when the maximum number of error bits was 500, or the number of codewords generated reached 10^5. From the graph, we can see that the two methods show almost the same BER performance. Table 5 shows the decoding times for $\mathfrak{R}(1,4)$ and $\mathfrak{R}(2,4)$ using FHT decoding and the AD. A computer with a central processing unit (CPU) of Intel i9-7920, graphics processing unit (GPU) of Nvidia Titan XP, and 64 GB random access memory (RAM) was used for the evaluation. If a GPU is used for decoding, the proposed method can have an advantage over the method using FHT, and for a fair comparison, we measure the decoding time using a CPU without a GPU. The decoding time of the proposed method was 1.8 times faster than the decoding time of FHT decoding for $\mathfrak{R}(1,4)$ and 125 times faster for $\mathfrak{R}(2,4)$. This derives from the fact that (16,11) $\mathfrak{R}(2,4)$ decoding using FHT needs 2^6 FHTs, considering six masking bits, whereas (16,5) $\mathfrak{R}(1,4)$ decoding needs only one FHT operation [15]. The main difference between the RM decoding using the AD for $\mathfrak{R}(1,4)$ and $\mathfrak{R}(2,4)$ is the number of nodes in the FC layer, which increases the number of parameters.

Table 5. Decoding time of RM code using FHT and AD.

RM Code	Method	Time (ms)
$\mathfrak{R}(1,4)$	FHT	0.6012
	AD	0.3327
$\mathfrak{R}(2,4)$	FHT	46.625
	AD	0.3704

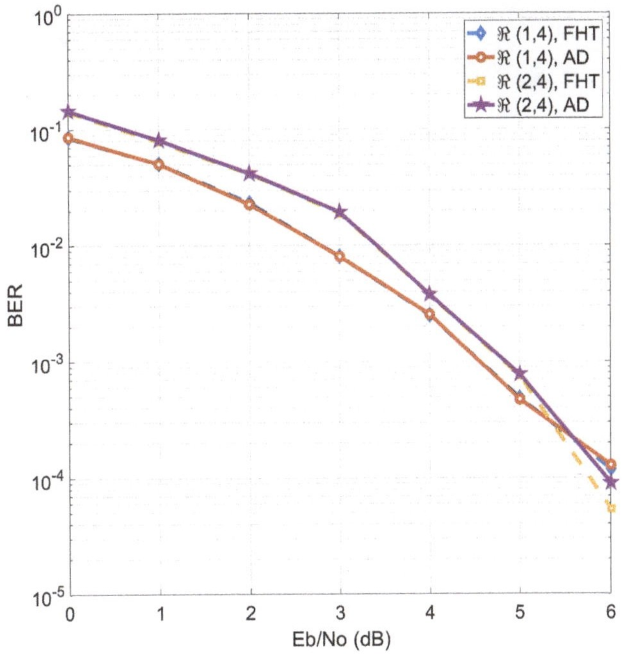

Figure 5. BER of RM decoding with FHT and AD.

3. PAD

To improve the BER performance of the RM decoder using a single AD, we present a PAD composed of multiple ADs. We call an AD in a PAD a constituent auto-decoder (CAD). To illustrate the structure of a PAD, we set the number of CADs in the PAD to five, as shown in Figure 6, where the output layers of all CADs are simply added at the merge layer with $n = 2^m$ nodes. Each CAD has the same structure, except for the different number of nodes at the first, second, and third hidden layers, as described in Table 6, where the first and the third hidden layers have the same number of nodes; thus, we find the symmetry structure of the PAD based on the second hidden layer or middle layer. As in the case of the AD, the activation functions for the hidden and output layers of CADs in the PAD are exponential linear unit (ELU) and tanh functions, respectively [20,21]. The PAD is followed by the FC layer, and the hyperparameters for the decoder model using PAD are the same as in the case of the AD. Figure 7 shows the BER performance for $\mathfrak{R}(1,4)$ and $\mathfrak{R}(2,4)$ using the conventional FHT decoder, and the proposed decoder model denoted as PAD-i, where i CADs are used. The specifications of the computer and the conditions for the BER calculation in Figure 7 for PAD are the same as those used in Figure 5 for the AD. Figure 7a shows the BER for $\mathfrak{R}(1,4)$ in the range from 0 to 6 dB with 1 dB steps, and (b) in the range of 1.5 dB to 2.5 dB, which provides a better discrimination between different cases where no significant difference can be observed in terms of BER. However, PAD-2 shows the best performance. Figure 7c shows the BER for $\mathfrak{R}(2,4)$ in the range from 0 to 6 dB with a 1 dB step, and (d) in the range of 1.5 dB to 2.5 dB for better observation. PAD-3 demonstrates the best performance, but there is no significant improvement as we increase the number of CADs in PAD. The comparison of decoding process times and the number of parameters for $\mathfrak{R}(1,4)$ and $\mathfrak{R}(2,4)$ using FHT and PADs is described in Table 7, where the higher the number of CADs in PAD, the higher the number of parameters and the complexity, and the longer the decoding process time. In $\mathfrak{R}$(1,4), the number of parameters of PAD-1 and PAD-5 are about 10-fold different, but the decoding time is only about 1.3-fold different. This is because the computation of CADs is performed in parallel.

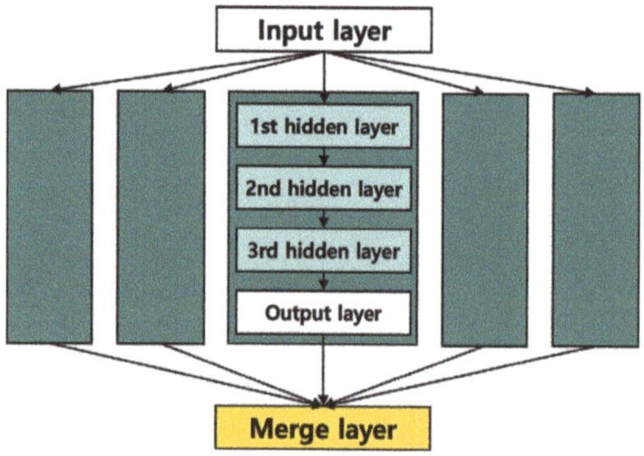

Figure 6. Structure of PAD with 5 CADs.

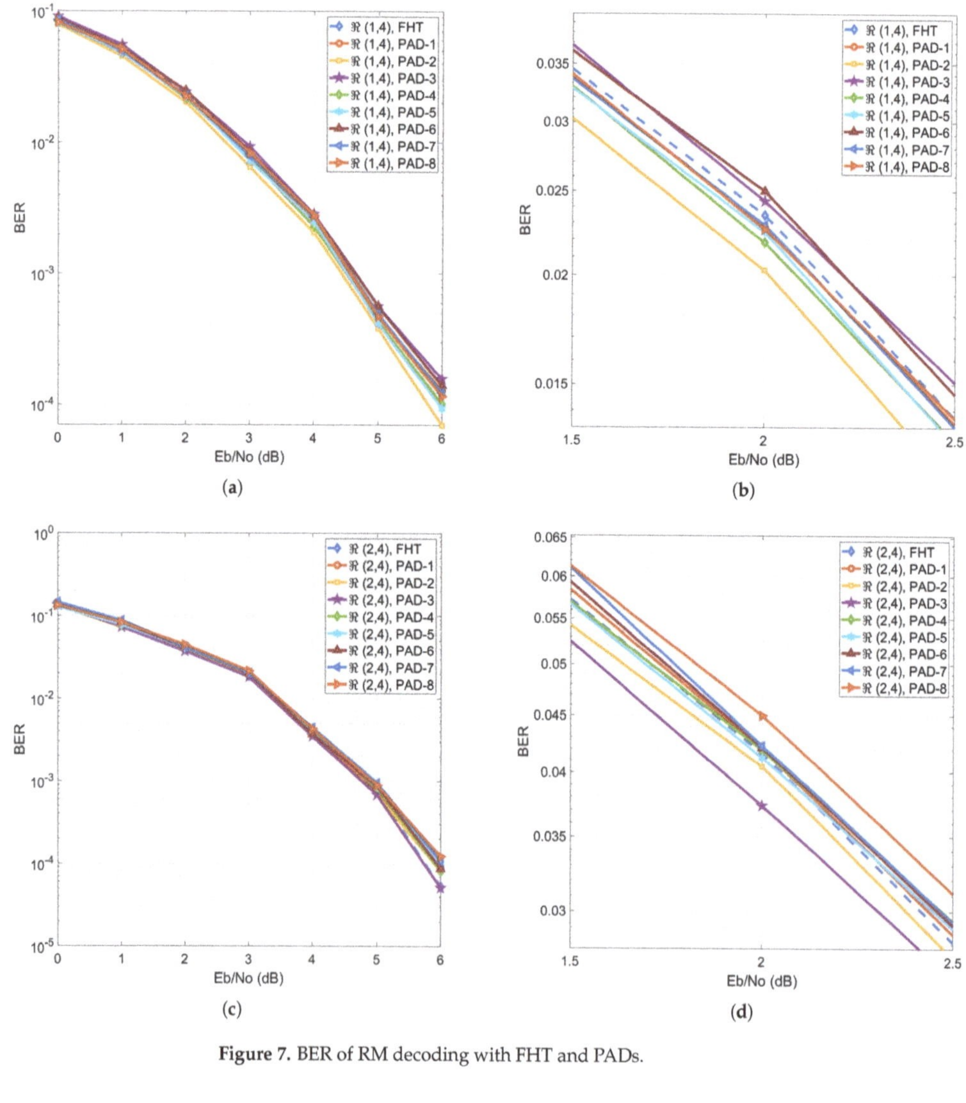

Figure 7. BER of RM decoding with FHT and PADs.

electronics

MDPI

Article

Sightless but Not Blind: A Non-Ideal Spectrum Sensing Algorithm Countering Intelligent Jamming for Wireless Communication

Ziming Pu [1], Yingtao Niu [1,*], Peng Xiang [2] and Guoliang Zhang [1]

[1] The Sixty-Third Research Institute, National University of Defense Technology, Nanjing 210007, China
[2] College of Communications Engineering, Army Engineering University of PLA, Nanjing 210007, China
* Correspondence: niuyingtao78@hotmail.com

Abstract: Aiming at the existing intelligent anti-jamming communication methods that fail to consider the problem that sensing is inaccurate, this paper puts forward an intelligent anti-jamming method for wireless communication under non-ideal spectrum sensing (NISS). Under the malicious jamming environment, the wireless communication system uses Q-learning (QL) to learn the change law of jamming, and considers the false alarm and missed detection probability of jamming sensing, and selects the channel with long-term optimal reporting in each time slot for communication. The simulation results show that under linear sweep jamming and intelligent blocking jamming, the proposed algorithm converges faster than QL with the same decision accuracy. Compared with wide-band spectrum sensing (WBSS), an algorithm which failed to consider non-ideal spectrum sensing, the decision accuracy of the proposed algorithm is higher with the same convergence rate.

Keywords: non-ideal spectrum sensing; intelligent anti-jamming; Q-learning; wide-band spectrum sensing

Citation: Pu, Z.; Niu, Y.; Xiang, P.; Zhang, G. Sightless but Not Blind: A Non-Ideal Spectrum Sensing Algorithm Countering Intelligent Jamming for Wireless Communication. *Electronics* **2022**, *11*, 3402. https://doi.org/10.3390/electronics11203402

Academic Editors: Yichuang Sun, Haeyoung Lee and Oluyomi Simpson

Received: 22 September 2022
Accepted: 18 October 2022
Published: 20 October 2022

1. Introduction

In the past 20 years, wireless communication technology was widely used. However, due to the openness of the wireless channel, the challenge of artificial interference in wireless communication is becoming more and more serious. Artificial interference mainly includes two types. One is unintentional interference [1], and the other one is intentional jamming. Intentional jamming refers to the jamming behavior taken for the purpose of destroying the information transmission process of a wireless communication system. According to whether the strategy is fixed or not, artificial intentional jamming can be divided into fixed-strategy jamming and dynamic strategy jamming. Fixed-strategy jamming mainly includes multi-tone jamming, partial-band jamming, periodic-pulse jamming, linear-sweep jamming, etc., and its strategy is fixed and the law is easily perceived. Dynamic strategy jamming mainly includes dynamic probability jamming, intelligent blocking jamming, etc., and its jamming law is not easily obtained through simple observation or sensing.

1.1. Related Works

In recent years, the continuous development of machine learning algorithms provided new intelligent ideas for communication anti-jamming. In the frequency domain, the author in [2] modeled the problem of multi-channel jamming and anti-jamming as a Markov decision process (MDP). The best defense strategy is obtained through value iteration under the channel transition probability, and rewards are completely known. Similarly, in [3], the author modeled the game problem between the secondary user and the jammer in the cognitive radio system as MDP, using Q-learning and maximum likelihood estimation to obtain attacker parameters and obtain the optimal channel switching strategy. In [4], the author also modeled the jamming and anti-jamming process as MDP, and proposed a game theory anti-jamming scheme (GTAS), which achieved higher returns. In [5,6], the authors also studied the anti-jamming problem of user channel selection by using the method of game theory. In [7], the author used Q-learning to solve the channel allocation problem, and proposed a channel allocation strategy with the lowest failure rate. In [8], the author used MDP to model dynamic and complex spectrum environment, and used Q-learning to obtain the optimal communication strategy. Because Q-learning has the problem of disaster maintenance, deep learning is introduced to solve the complex anti-jamming strategy learning problem. In [9], the author proposed a hierarchical deep reinforcement learning algorithm without a jamming mode and channel model, which solved the problem of selecting many optional frequencies in the jamming environment. In the power domain, the author in [10] proposed a power control strategy based on reinforcement learning for the case of unknown jamming patterns and channel parameters, which improved the communication efficiency. Aiming at multi-domain joint jamming in the frequency domain, time domain, and power domain, the authors in [11] proposed a multi-parameter intelligent anti-jamming method based on 3D Q-learning. The proposed algorithm has a lower jamming collision rate than the traditional Q-learning algorithm. The authors in [12] proposed the CAAQ algorithm to solve the problem of multi-user cooperative anti-jamming. By increasing the distance threshold, the problem of mutual jamming between users was well avoided. The author in [13] proposed an anti-jamming power control strategy based on Q-learning, which has better performance in the condition of the game model unknown. However, the above mentioned algorithms require the acknowledgment character (ACK) fed back from the receiver after data packets are successfully received as a means to sense whether the communication channel is subjected to jamming, which increases the operation overhead and can induce additional risk of ACK frame jamming. Furthermore, those strategies can only update the Q value of the selected channel one by one, and their convergence rates are low.

To address these issues, the author in [14] used the wide-band spectrum sensing technology to sense the jamming state of multiple channels simultaneously, which significantly improved the convergence rate of the algorithms without the need of an ACK feedback channel. However, the strategy proposed in [14] assumed that the observed jamming state by the system was the actual jamming state of the system, which neglected the non-ideality of the jamming state observation. In practice, due to the existence of many non-ideality factors, such as false alarm and the missed jamming states in the observed results, the observed jamming state is not necessarily equivalent to the actual jamming state of the system, which may lead to large uncertainties in the algorithm. Therefore, if we ignore these undesirable characteristics of perception, we will not be able to correctly learn the behavior of jamming, which will lead to large disturbances in the algorithm. Aiming at the problem that channel quality varies with probability, reference [15] studied the time-constrained downlink scheduling strategy for the actual channel observation environment, proposed a simplified partially observable Markov decision process (POMDP) modeling method for downlink transmission, and proposed a low-complexity suboptimal strategy method based on finite time domain Q_MDP algorithm.

Inspired by the work in [14], this paper proposes an intelligent anti-jamming method for wireless communication in the case of non-ideal spectrum sensing. Compared with the traditional Q-learning algorithms in Refs. [3–13], this paper uses the idea of broadband sensing technology to update the Q value of multiple channels at a time; compared with WBSS algorithms in Ref. [14], this paper considers the non-ideal sensing, which is closer to the actual electromagnetic environment. Firstly, the algorithm models the problem of communication channel selection in a jamming environment with false alarm and missed detection as an improved Markov decision process (IMDP) combined with false alarm and missed detection. Then, it is solved by the NISS algorithm, obtaining the selection of the optimal communication channel. The advantages for decision accuracy and convergence rate of the proposed algorithm are proved comparison to the traditional Q-learning [16] and the conventional wide-band spectrum sensing algorithm [14] through the MATLAB simulation results.

1.2. Contribution and Structure

The contribution of this paper is as follows:

- This paper proposes a NISS algorithm, which combines the advantages of Q-learning and the WBSS algorithm. The proposed algorithm has a fast convergence rate and high decision accuracy.
- This paper takes the probability of false alarm and missed detection into account in anti-jamming communication for the first time, which is closer to the actual electromagnetic environment and fills the blank of intelligent anti-jamming wireless communication in the case of non-ideal sensing.

The remainder of this paper is organized as follows. Section 2 presents the system model and problem formulation. In Section 3, we introduce the detailed derivation of the NISS algorithm. The simulation results and analysis are presented in Section 4. Our concluding remarks are given in Section 5.

2. System Model and Problem Formulation

2.1. System Model

Figure 1 shows the model of the communication system. There is one communication transmitter, one communication receiver, and one jammer. The communication transmitter and receiver can use the communication channel to communicate. There are M possible channels in the wireless communication system, which are recorded as $\{1, 2, 3, \ldots, M\}$, selecting one channel at a time for communication. The jammer has a non-intelligent mode and an intelligent mode. In the non-intelligent mode, the jammer performs conventional jamming, such as linear sweep; in the intelligent mode, the jammer has a greater probability of jamming with high-communication frequency channels according to the communication frequency of each channel detected in advance.

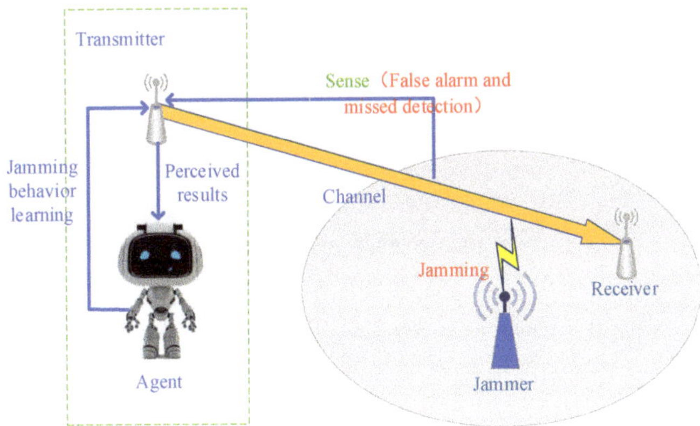

Figure 1. Communication system model.

Figure 2 shows the structure of communication time slot. Each jamming time slot corresponds to a communication time slot, which can be divided into transmission sub-slot, sensing sub-slot and learning sub-slot. In the transmission sub-slot, the transmitter selects an undisturbed channel to transmit information to the receiver according to the judgment of the previous time slot on the jamming channel. In the sensing sub-slot, the receiver senses each channel, and transmits the sensing results to the agent for learning. The agent obtains the judgment of the available channels of the same time slot in the next jamming period. The transmitter selects the optimal channel for the communication transmission of the next time slot according to the judgment.

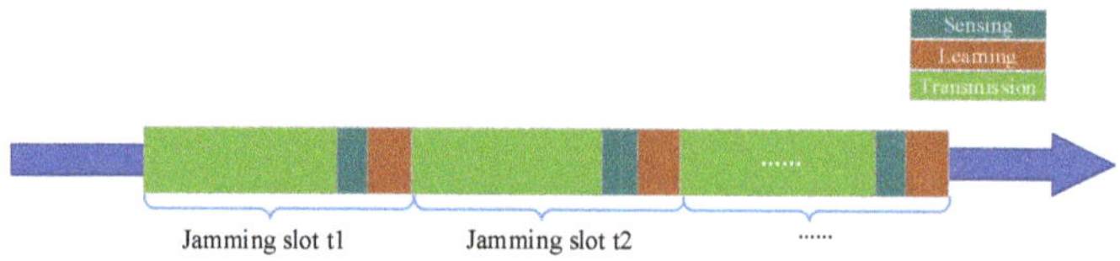

Figure 2. Structure of the communication time slot.

Figure 3 is a state transition diagram between the actual state and the observed (sensing) state for channel i due to the existence of false alarm and missed detection. Where p_i is the missed detection probability, indicating that the observation state is no jamming, while the actual state is jamming. q_i is the false alarm probability, indicating that the observation state is jamming, while the actual state is no jamming.

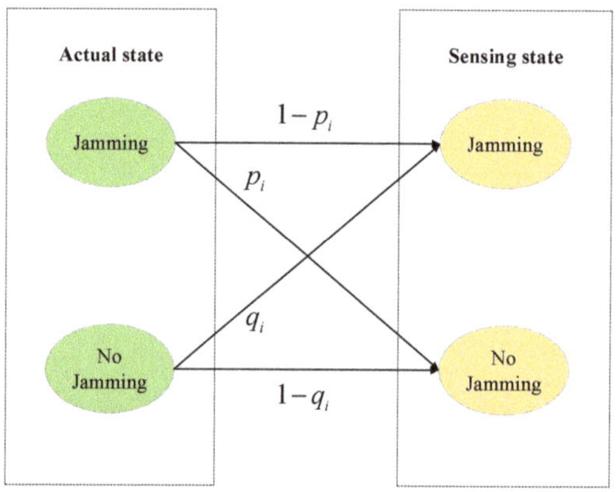

Figure 3. Transition diagram between actual state and observation state caused by false alarm and missed detection.

2.2. Problem Formulation

To simplify the research, the following assumptions are made:

- The communication frequency band is divided into N channels with the same bandwidth, and there is no frequency overlap between the channels, and the fading characteristics of each channel are the same and flat fading.
- The sensing result is only affected by false alarm and missed detection, which leads to inaccuracy, and there is no inaccuracy caused by other factors.
- In the same time slot, the channel of jamming does not change.

Based on the above assumptions, since the agent cannot accurately perceive the state of the system, we use the improved Markov decision process (IMDP) for modeling and solving. IMDP can be expressed as a five tuple $\langle S, A, P, O, r\rangle$, in which, in addition to the state space S, action space A, state transition probability P, and real-time reward function r of the general MDP, there is also observation space O.

The state space S can be expressed as:

$$S \triangleq \{n_1, n_2, \ldots, n_i : n_k \in \{1, 2, 3, \ldots, N\}\} \tag{1}$$

where $n_k = j$ indicates that the k_{th} channel to be interfered is channel j, and N channels can be interfered at most in the same time slot.

Action space A can be expressed as:

$$A \triangleq \{a : a \in \{1,2,3,\ldots,N\}\} \quad (2)$$

where $a = i$ indicates that the transmitter chooses to communicate on channel i.

Observation space O can be expressed as:

$$O \triangleq \left\{o_t = \left(o_t^1, o_t^2, \ldots, o_t^M\right) : o_t^i \in \{0, i\}\right\} \quad (3)$$

where $o_t^i = 0$ and $o_t^i = i$, respectively, indicate that the observation status of channel i is undisturbed and disturbed in time slot t.

The reward function r is defined as follows:

$$r_t^i = \begin{cases} 0 & a \neq i; \\ E & a = i \& i \notin s_t \\ -L & a = i \& i \in s_t \end{cases} \quad (4)$$

where r_t^i represents the reward function of channel i in time slot t, and $-L$ represents the loss in case of message transmission failure. E represents the return of successful message transmission. s_t represents the jamming status of each channel at time t, and $s_t \in S$.

The observed state and actual state transition diagram of each time slot are shown in Figure 4.

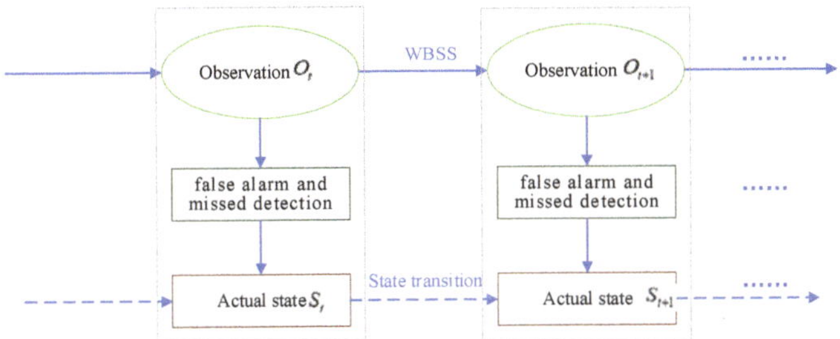

Figure 4. The observed state and actual state transition diagram of each time slot.

Where the observed states in time slot t and time slot $t+1$ can be obtained through wide-band spectrum sensing. At the same time, the false alarm probability P_f and missed detection probability P_d can be calculated according to the energy detection theory [17].

$$P_d = \Pr(Y > \lambda | H_1) = Q_u\left(\sqrt{2\gamma}, \sqrt{\lambda}\right) \quad (5)$$

$$P_f = \Pr(Y > \lambda | H_0) = \frac{\Gamma(u, \lambda/2)}{\Gamma(u)} \quad (6)$$

where $u = TW$ is time domain bandwidth product, and γ is jamming to noise ratio (JNR), and λ is decision threshold of energy detection.

3. Detailed Derivation of Algorithm

Q-learning is a form of typical model-free learning. Its basic idea is to establish a Q table. The values in the table represent the long-term cumulative rewards of executing the current strategy after the state s_t selects the action a_t. The long-term cumulative reward can be expressed as follows:

$$V = E\left[\sum_{\tau=0}^{+\infty} \gamma^\tau r_{t+\tau}\right] \quad (7)$$

where γ is the discount factor, indicating the importance of future returns, and r_t is the immediate reward value obtained in step t. The goal of Q learning is to find a strategy π to maximize the long-term cumulative rewards under this strategy.

To solve the optimal strategy, the state value function V and the state action value function Q are defined as follows:

$$V^{\pi}(s) = E_{\pi}\left[\sum_{\tau=0}^{+\infty} \gamma^{\tau} r_{t+\tau} | s_t = s\right] \tag{8}$$

$$Q^{\pi}(s,a) = E_{\pi}\left[\sum_{\tau=0}^{+\infty} \gamma^{\tau} r_{t+\tau} | s_t = s, a_t = a\right]. \tag{9}$$

Since the MDP model is satisfied, it can be converted into a recursive form as follows:

$$\begin{aligned} V^{\pi}(s) &= E_{\pi}\{r_t + \gamma[r_{t+1} + \gamma(r_{t+2} + \ldots)]|s_t = s\} \\ &= \sum_{a \in A} \pi(a|s) \sum_{s' \in S}\{P(s'|s,a)[r(s'|s,a) + \gamma V^{\pi}(s')]\} \end{aligned} \tag{10}$$

$$Q^{\pi}(s,a) = \sum_{s' \in S}\{P(s'|s,a)[r(s'|s,a) + \gamma V^{\pi}(s')]\} \tag{11}$$

where $P(s'|s,a)$ represents the probability of taking action a represents the probability of taking action in state s and transferring the state to s', and $r(s'|s,a)$ represents the corresponding reward.

According to the Bellman optimization principle, the optimal value $Q^{\pi^*}(s,a)$ can be obtained as follows [18]:

$$Q_*^{\pi}(s,a) = \sum_{s' \in S}\left\{P(s'|s,a)\left[r(s'|s,a) + \gamma \max_{\pi} Q^{\pi}(s,a)\right]\right\} \tag{12}$$

$$Q^{\pi^*}(s,a) = \max_{\pi} Q_*^{\pi}(s,a). \tag{13}$$

Therefore, the optimal strategy $\pi*$ can be obtained as follows:

$$\pi* = \underset{\pi}{\operatorname{argmax}}\{Q^{\pi}(s,a)\} \tag{14}$$

Since the Q-learning algorithm does not need prior knowledge, such as state transition probability, its update formula is as follows:

$$Q_{t+1}(s,a) = (1-\alpha)Q_t(s,a) + \alpha\left(r_t + \gamma \max_{a'} Q_t(s,a')\right) \tag{15}$$

where α is the learning rate. The reference [19] and [20] proved that if α meets the conditions:

$$\alpha_t \in [0,1), \sum_t^{\infty} \alpha_t = \infty, \text{ and } \sum_t^{\infty} (\alpha_t)^2 < \infty \tag{16}$$

then the Q-learning algorithm can converge after finite iterations. When the Q table converges, the action corresponding to the maximum Q value in each state is the optimal action in that state.

Wide-band spectrum sensing algorithm (WBSS) senses multiple channels in the same time slot, obtains the actual state of each channel (whether it is jammed) according to the sensing, and updates the Q value of each channel at the same time slot. Therefore, compared with the conventional Q-learning algorithm [16], which only updates the Q value of the selected channel in one time slot, the convergence rate of the WBSS algorithm will be greatly improved.

The Q value of the WBSS algorithm is updated as follows:

$$Q_{t+1}(s,n_i) = (1-\alpha)Q_t(s,n_i) + \alpha\left[r_t + \gamma \max_{n_j} Q_t(s,n_j)\right] \tag{17}$$

where n_i refers to different channels, and its value is $\{1,2,3,\ldots,M\}$. That is, for any time slot, the Q values of the M channels are updated at the same time. r_t represents the instant benefit of the time slot $t+1$ selecting channel n_i for communication.

It can be seen from Equations (11) and (12) that to update the Q value and obtain the optimal strategy, it is necessary to know the state of the current time slot. However, in the actual electromagnetic environment, due to the inaccuracy of observation, the state of the current time slot is not a completely determined state, but there are multiple possible states related to the probability of false alarm and missed detection. Therefore, different from the conventional WBSS algorithm, the proposed algorithm takes the false alarm and missed detection probability into account when calculating the Q value and making decisions, obtaining the NISS algorithm.

Since there are different states, such as $s_t^1 = \{1,0,0,\ldots,0\}$ (only channel 1 is jammed) and $s_t^2 = \{1,2,0,\ldots,0\}$ (only channel 1 and channel 2 are jammed) for actions such as $a=1$ (select channel 1 for communication), the communication results are the same and the benefits are the same. Therefore, we change the state from the set of all channel states as one state to the time slot as the state. When calculating the immediate return r, we only need to consider the actual state of the selected communication channel, not the actual state of other channels.

Then, for the time slots n_{t-1} and n_t, where the system is located, $o_{t-1}, o_t \in \mathbf{O}$ is observed. For each $a_t \in \mathbf{A}$, the Q value is calculated as follows:

If the observation of the selected channel is jamming, that is $a_t \in o_{t-1}$, update the Q value as Equation (18).

$$Q(n_{t-1},a_t) = Q(n_{t-1},a_t) + \alpha\left[p_i r_1 + (1-p_i)r_2 + \gamma \max_{a_{t+1}\in A} Q(n_t,a_{t+1}) - Q(n_{t-1},a_t)\right] \tag{18}$$

If the observation of the selected channel is no jamming, that is $a_t \notin o_{t-1}$, update the Q value as Equation (19).

$$Q(n_{t-1},a_t) = Q(n_{t-1},a_t) + \alpha\left[(1-q_i)r_1 + q_i r_2 + \gamma \max_{a_{t+1}\in A} Q(n_t,a_{t+1}) - Q(n_{t-1},a_t)\right] \tag{19}$$

For the time slot n_t, the observation is o_t. We obtain the actions as Equation (20).

$$a_{t+1} = \arg\max_{a_{t+1}\in A} Q(n_t,a_{t+1}) \tag{20}$$

Algorithm 1 is the flow of intelligent anti-jamming communication decision algorithm based on NISS.

Algorithm 1: Intelligent anti-jamming communication decision algorithm based on NISS.

1. **Initialization:** Learning factor α, Discount factor γ and other parameters in Table 1.
The Q table is initialized as a zero matrix with N_T rows and M columns, that is, for any n_T and a, let $Q(n_T,a)=0$.
2. **for** $t = 1,2,\ldots T$ **do**
3. In the current transmitter state s_t, the transmitter performs the optimal policy selection action a_t obtained in the last timeslot or the initial action a_0.
4. The transmitter detects the energy of each channel.
5. Calculate the probability of false alarm p_f and missed detection P_m according to the detection results.
6. According to the detection results, false alarm, and missed detection, the real-time reward r is calculated and the next state s_{t+1} is predicted to obtain the optimal communication channel a_{t+1}.
7. The agent updates the Q value according to (18) and (19).
8. The agent obtains the optimal strategy $\pi*$ according to (20) and instructs the transmitter to transmit in the next time slot.
9. $t = t+1$
10: **end for**

First, initialize the system. Second, according to the last decision result, the optimal communication channel is selected. Third, the transmitter detects the energy of all channels. Fourthly, the false alarm probability and missed detection probability are calculated according to the energy detection results, the reward of each channel is obtained, and the Q table is updated. Finally, the optimal communication channel of the next slot is selected according to the Q table, and an iteration is completed.

4. Simulation Result and Analysis

4.1. Parameter Settings

Table 1 shows the simulation parameters.

Table 1. Simulation parameter settings.

Parameters	Value
Communication timeslot length T_s	0.6 ms
Transmission timeslot length T_{trans}	0.5 ms
Perception timeslot length $T_{sensing}$	0.04 ms
Learning timeslot length $T_{learning}$	0.06 ms
Total transmission timeslots T_a	10,000
Number of available channels N	10
Transmission power of transmitter P	30 dBm
Fading of communication signal B_S	−130 dB
Transmission power of jamming J	30 dBm
Fading of jamming B_J	−134 dB
Power spectral density of ambient noise	−174 dBm/Hz
Channel bandwidth B_W	1 MHz
Learning rate factor α_m	0.1
The discount factor γ	0.5
Transmission success reward E	1
Transmission failure loss L	−3

According to the parameters in Table 1, the jamming noise ratio (JNR) can be calculated as 10 dB, and the time bandwidth product is 5, so we can calculate the false alarm probability as $P_f = 0.0549$ and missed detection probability as $P_m = 0.1021$ according to Equations (5) and (6). To evaluate the performance of this algorithm, this algorithm is compared with the traditional QL algorithm and WBSS algorithm.

In this paper, the effectiveness and universality of the algorithm will be verified by simulation from both fixed-strategy jamming and dynamic strategy jamming. The first is the fixed-strategy jamming, and the linear sweep jamming is selected as the research object. Figure 5 shows the time-frequency distribution of linear sweep jamming, and the red background indicates jamming. The jamming channel changes linearly at any time slot, and 10 time slots are a jamming period. In the same time slot, the jamming channel does not change.

The second is the dynamic strategy jamming, and the intelligent blocking jamming is selected as the research object. Intelligent blocking jamming refers to a jamming strategy in which the jammer selects the channels with the highest number of communications to interfere with the relative number of communications in each channel in the previous period of time.

Figure 6 shows the probability distribution of intelligent blocking jamming. The number represents the jamming probability of the channel in that time slot, which is determined by the proportion of the communication times of each channel in the total communication time of the previous period. A jamming period is 10 time slots.

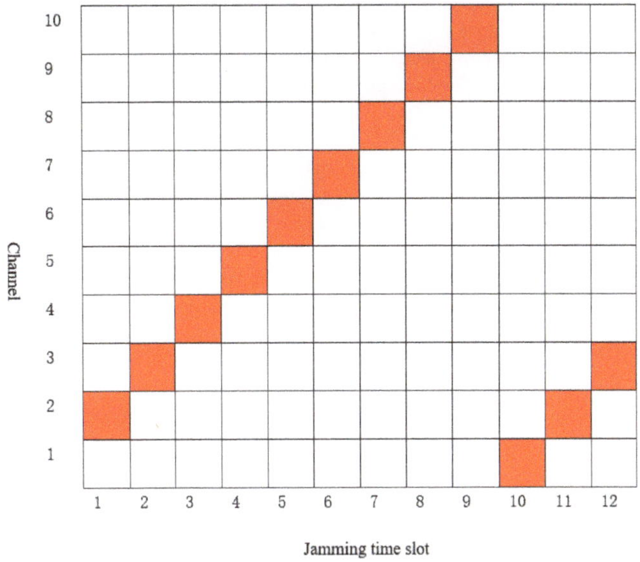

Figure 5. Linear sweep jamming distribution.

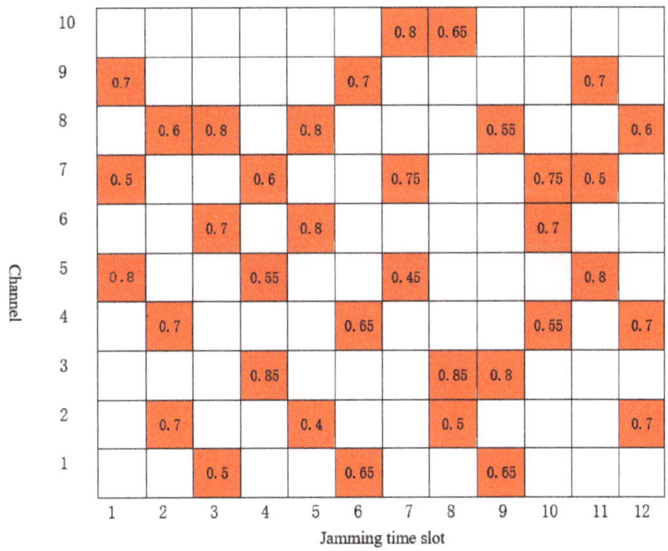

Figure 6. Probability distribution of intelligent blocking jamming.

4.2. Result Analysis

Figure 7 is a comparison of the decision accuracy (ratio of successful transmission times to total communication time) of the proposed algorithm anti-linear sweep jamming with traditional Q-learning and wide-band spectrum sensing algorithms. The decision accuracy of traditional Q-learning converges to 100% after about 30 rounds of algorithm iteration. To accelerate the convergence of the algorithm, aiming at the problem that Q-learning only updates one Q value of the state action pair at a time, the WBSS algorithm senses the jamming states of all channels in each time slot, and updates the Q value of all actions at the same time in each state, which greatly accelerates the convergence rate of the algorithm. However, due to false alarm $P_f = 0.0549$ and missed detection $P_m = 0.1021$, the sensing results cannot be completely accurate. For example, in a certain time slot, the channel state perceived is free of jamming, but the sensing result may be caused by missed detection, and the actual channel may be jammed. Therefore, the WBSS algorithm takes the sensing result directly as the actual state of the system and does not consider the impact of false alarm and missed detection on the sensing result. Its accuracy after convergence is only 90%, which is lower than the Q-learning algorithm. By taking the false alarm and missed detection probability into account, the inaccuracy of the sensing results and the decision of the optimal channel is more accurate and reasonable. Therefore, the decision accuracy of the NISS algorithm for the channel of linear sweep jamming can also reach 100%.

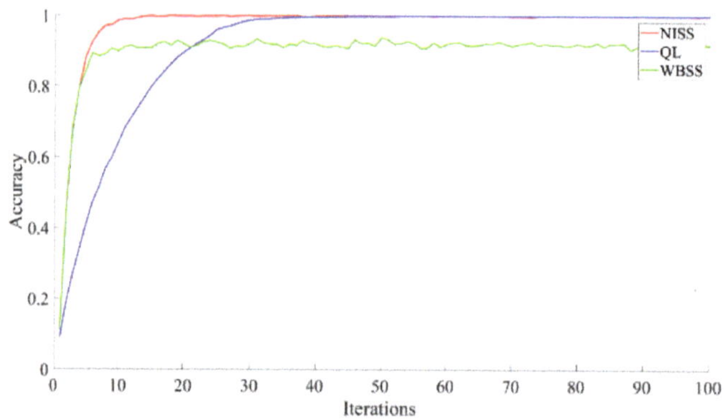

Figure 7. Decision accuracy of three anti-linear sweep jamming algorithms.

Since the actual jamming cannot be such regular linear sweep jamming, to verify the applicability of the algorithm under complex jamming patterns, the simulation verification against intelligent blocking jamming is carried out, and the results are shown in Figure 8.

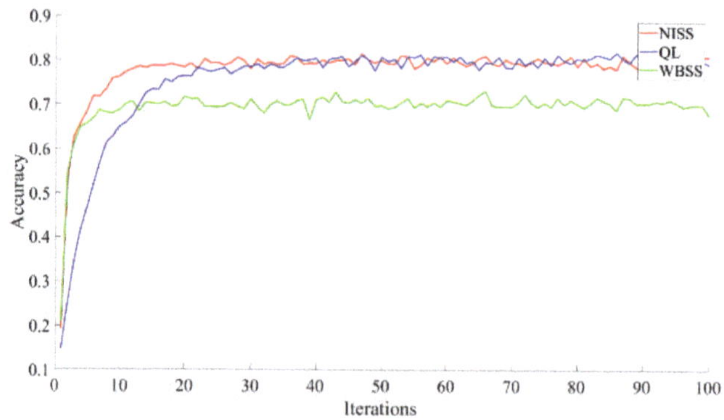

Figure 8. Decision accuracy of three anti-intelligent blocking jamming algorithms.

As can be seen from Figure 8, since the intelligent blocking jamming interferes with different channels in each time slot with probability, it is impossible to fully predict the jamming channel of the next time slot. Therefore, even with Q-learning, the decision accuracy is only 80%.

From the results of Figures 7 and 8, we can see that the NISS algorithm can converge faster than Q-learning with the same decision accuracy, and has higher decision accuracy than the WBSS algorithm with the same convergence rate.

Figure 9 is a comparison diagram of the change rule of successful transfer rate with JNR when the false alarm probability of the NISS algorithm and the WBSS algorithm is $P_f = 0.0549$. It can be seen from Figure 9 that with the increase in JNR, the decision accuracy of both algorithms increases. When JNR is taken from 2dB to 12dB, the decision accuracy of the NISS algorithm is significantly higher than that of the WBSS algorithm, which proves that the performance of the NISS algorithm is better than that of the WBSS algorithm. It should be noted that when JNR is low, the reason why the decision accuracy of the two algorithms is similar is that it is difficult to distinguish whether the channel contains jamming due to the low JNR. When JNR is very high, both decision accuracies reach the same maximum. The reason is that the energy of the jamming signal is very strong, the probability of missed detection is almost negligible, and the observation accuracy is only related to the probability of false alarm. The probability of false alarm is equal, so the decision accuracy is equal. The results show that the performance of NISS is greater than WBSS at the medium JNR.

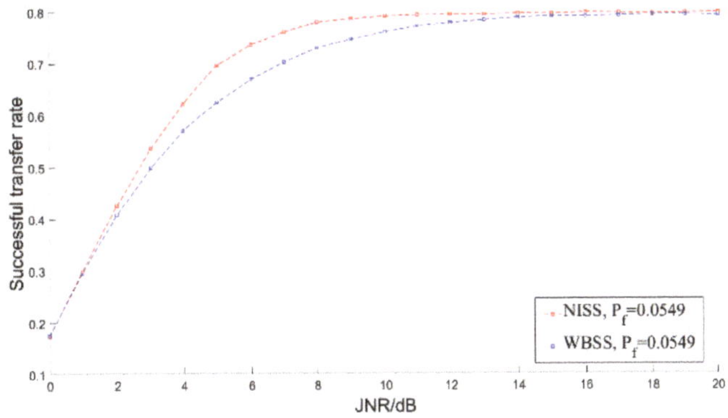

Figure 9. Variation diagram of decision accuracy with JNR of two anti-intelligent blocking algorithms.

Figure 10 is a channel decision diagram of the WBSS algorithm against intelligent blocking jamming. The vertical coordinate indicates the channel serial number, the horizontal coordinate indicates the current communication slot. The green area indicates that the current channel is predicted to be jammed, but there is no jamming in the actual channel. The light red area indicates that the current channel is jammed but was not predicted. The dark red area indicates that the current channel has jamming and was successfully predicted. For the WBSS algorithm, when the algorithm converges, some channels will be jammed, but the agent was not successfully predicted. If the channel is selected for communication at this time, the communication will fail and cause great losses.

Figure 11 is a channel decision diagram of the proposed algorithm against intelligent blocking jamming. Compared with Figure 10, after the algorithm converges, the NISS algorithm in this paper can accurately judge those channels that are actually jammed, and there is no missed detection. Although there will be no jamming but predicted jamming, the loss caused by this false alarm is very small. Therefore, the algorithm proposed in this paper can effectively solve the problem of wireless communication intelligent anti-jamming in the case of non-ideal spectrum sensing.

Comparing the performance of the algorithm in the anti-jamming channel decision under the two jamming patterns, we can see that: Although the accuracy of the Q-learning algorithm is high, the convergence rate of the algorithm is slow. For jammers with dynamic jamming patterns, they may not be able to learn the jamming rules in a short time and make effective anti-jamming decisions. In the anti-jamming channel decision of the WBSS

sensors

Article

Fast-Convergence Reinforcement Learning for Routing in LEO Satellite Networks

Zhaolong Ding [1,2], Huijie Liu [2,*], Feng Tian [2], Zijian Yang [2] and Nan Wang [1]

1 School of Information Science and Technology, ShanghaiTech University, Shanghai 201210, China
2 Innovation Academy for Microsatellites of CAS, Shanghai 201304, China
* Correspondence: liuhj@microsate.com

Abstract: Fast convergence routing is a critical issue for Low Earth Orbit (LEO) constellation networks because these networks have dynamic topology changes, and transmission requirements can vary over time. However, most of the previous research has focused on the Open Shortest Path First (OSPF) routing algorithm, which is not well-suited to handle the frequent changes in the link state of the LEO satellite network. In this regard, we propose a Fast-Convergence Reinforcement Learning Satellite Routing Algorithm (FRL–SR) for LEO satellite networks, where the satellite can quickly obtain the network link status and adjust its routing strategy accordingly. In FRL–SR, each satellite node is considered an agent, and the agent selects the appropriate port for packet forwarding based on its routing policy. Whenever the satellite network state changes, the agent sends "hello" packets to the neighboring nodes to update their routing policy. Compared to traditional reinforcement learning algorithms, FRL–SR can perceive network information faster and converge faster. Additionally, FRL–SR can mask the dynamics of the satellite network topology and adaptively adjust the forwarding strategy based on the link state. The experimental results demonstrate that the proposed FRL–SR algorithm outperforms the Dijkstra algorithm in the performance of average delay, packet arriving ratio, and network load balance.

Keywords: LEO satellite networks; satellite routing; multi-agent reinforcement learning; distributed routing

Citation: Ding, Z.; Liu, H.; Tian, F.; Yang, Z.; Wang, N. Fast-Convergence Reinforcement Learning for Routing in LEO Satellite Networks. *Sensors* **2023**, *23*, 5180. https://doi.org/10.3390/s23115180

Academic Editors: Yichuang Sun, Haeyoung Lee and Oluyomi Simpson

Received: 17 April 2023
Revised: 23 May 2023
Accepted: 26 May 2023
Published: 29 May 2023

1. Introduction

Terrestrial communications already meet much of our daily communication needs, but for users in remote areas and at sea, their communications needs cannot be met. Unlike terrestrial communication, satellite communication has a longer range and better communication quality [1]. It meets the communication needs of users in remote areas and at sea, and can also be used to complement terrestrial communications to serve urban users. Therefore, satellite communication research has attracted much attention.

Based on the distance of the orbital plane from the Earth's surface, satellite systems can be classified as geostationary Earth orbit (GEO), medium Earth orbit (MEO), and low Earth orbit (LEO) satellite systems. Unlike GEO and MEO, LEO satellites are usually located in the orbital plane, at an altitude of 500 to 2000 km [2]. Because of low latency, low path loss, and low launch costs, LEO satellite networks are a major research direction for the industry. At present, there are many large LEO satellite constellations in operation, such as Starlink and Oneweb. The link status of the satellite network LEO is volatile due to the complex space environment and frequent satellite laser failures. In addition, satellites move at high speeds, resulting in frequent changes in satellite topology. These two characteristics are the main differences between satellite networks and terrestrial networks. Therefore, traditional routing algorithms are not suitable for direct use for satellite networks. New routing algorithms need to be designed to solve the LEO satellite routing problem.

Typically, the routing problem of the LEO satellite is mainly divided into two parts: how to obtain the network topology, and how to generate the route paths based on the network topology [3]. Mauger [4] first proposed the concept of virtual node. When the satellite moves to the virtual node location, its logical address becomes the logical address of that virtual node. The routing method is to give priority to the path with the highest latitude. The authors in [5] proposed the dynamic virtual topology routing (DV–DVTR), which divided the system period into time slices. For each time slot, the network topology was considered to be fixed [6], and then packets were forwarded according to the shortest path first algorithm. Although DV–DVTR is easy to implement, the division of time intervals is a very difficult task. Smaller time intervals require more storage space, and larger time intervals affect the performance of the algorithm. The authors in [7] proposed the Temporal Netgrid Model (TNM) to portray the time-varying topology of LEO satellite networks, in which the Dijkstra algorithm was used to generate forwarding policies. The Dijkstra algorithm needs to obtain global network information to operate, which can cause an increase in the communication load on the satellite network. Also, the link state of the satellite network changes very quickly and, by the time the node has collected the global information, much of it may be invalid. The authors in [8] proposed a distributed routing method, in which the surface was divided into several spaces with corresponding logical areas. In [9], each node sent congestion information to neighbor nodes, including queue length and available resources. Therefore, satellite nodes could route packages based on congestion information to achieve load balancing.

There has been a significant amount of research on the application of Software Defined Network (SDN) technology to address satellite routing challenges. For example, a study conducted in [10] investigated a LEO satellite routing algorithm in a software-defined space terrestrial integrated network, where SDN technology was applied to the LEO satellite network. The lattice-like topology of the satellite network created a shared bottleneck problem, which was addressed in the study. To tackle these issues, ref. [11] proposed an SDN-coordinated Multipath TCP (scMPTCP) architecture and related algorithms. In [12], the authors aimed to introduce an SDN-based terrestrial satellite network architecture and estimate the mean time required to transport data of a new traffic flow from the source to the destination while considering the time required to transfer SDN control actions. It should be noted that the proposed algorithms in these studies were centralized routing algorithms, which require frequent communication with the ground station and may cause some latency, presenting a significant disadvantage.

The above approaches mainly consider how to shield network dynamics and then run traditional routing algorithms for static network topology. These methods take up a certain amount of storage space and do not yield accurate satellite network topology. In recent years, a lot of research has started to use machine learning methods to solve routing problems. The difference between machine learning methods and traditional methods is that the former are data-driven, while the latter are model-driven. If the model does not describe the problem accurately, the performance of the model-driven method will be poor. Recently, machine learning has been applied in network areas, such as regulating congestion at the transport layer [13], optimizing video stream transmission [14], allocating resources [15], and so on.

The most suitable machine learning method for solving routing problems is the reinforcement learning technique. In [16], they used the Q-routing method to decide how to forward packages in the LEO satellite network. In [17], deep reinforcement learning was used to solve the routing problem; they used the neural network to replace Q-tables to store Q values. They both use centralized routing algorithms that viewed all satellite nodes as the agent that learned packet forwarding policies as it interacted with the network. The disadvantages of this approach are the need to collect global link state information and the high signaling overhead. The authors in [18] proposed a dynamic distributed routing scheme based on reinforcement learning; they considered each satellite node as an agent, and agents trained and executed routing operations. However, this did not consider the problem of the convergence speed of the algorithm, but simply provided the approximate running process of the algorithm.

Even though many researchers have applied reinforcement learning to routing problems [19], few of them have improved the convergence speed of the algorithm to face the dynamically changing link state. In this paper, we propose a distributed reinforcement learning method named FRL–SR, which not only learns routing and forwarding policies by distributed reinforcement learning, but also accelerates the convergence speed of the algorithm and senses the satellite network state faster. Our main contributions can be summarized as follows:

- We propose a distributed reinforcement learning method named FRL–SR; it has lower end-to-end latency and lower signaling overhead than traditional algorithms;
- We propose a learning mechanism that allows the algorithm to converge faster in the face of changes in the network link state;
- We compare the impact of different reward functions on the performance of reinforcement learning algorithms. The experimental results show that the design of reward functions is crucial for reinforcement learning algorithms.

The remainder of this paper is organized as follows. In Section 2, we describe the system model and Q-routing algorithm. In Section 3, we give the details of our FRL–SR method. We discuss the experimental results in Section 4. Finally, Section 5 concludes our work.

2. System Model

In this section, we first give a model of the LEO satellite constellation and its characteristics, based on which research scenario is depicted. After that, we describe the traditional Q learning algorithm and its application in routing problems. The definition of the routing problem is given in the last sub-section.

2.1. LEO Satellite Constellation

Currently, LEO satellite constellations can be divided into two types based on the number of satellite orbital planes: single-layer and multi-layer constellations. The Iridium system is representative of single-layer constellations, while Starlink is a multi-layer constellation with satellite orbital planes mainly distributed between 300 and 600 km. To facilitate the analysis, this paper uses the Iridium system as the satellite environment.

As shown in Figure 1, the Iridium constellation consists of multiple satellite orbits that are evenly distributed which intersect at the pole position. The area at the ends of the constellation is known as the polar region, where satellites are prohibited from communicating. The direction of motion of the satellite changes as it passes the pole, and the relative positions between the satellites change, which leads to periodic changes in the topology of the satellite. There are two satellite-to-satellite links within the satellite network. The link with an orbiting satellite is an intra-satellite link, and the link between adjacent orbiting satellites is an inter-satellite link [20]. Thus, each satellite has up to four neighbor nodes with which to communicate, leaving little decision space for the satellite when considering packet forwarding.

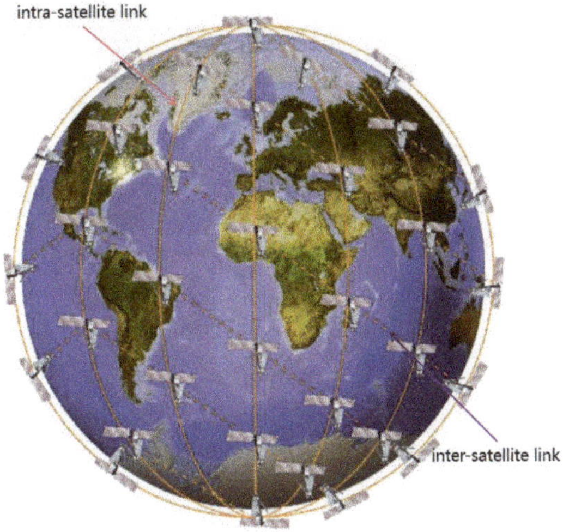

Figure 1. LEO satellite constellation.

2.2. Q-routing Algorithm

Reinforcement learning is one of the areas in machine learning [21]. Inspired by behavioral psychology, it focuses on what actions an agent should perform to maximize cumulative rewards when confronted with a given environment. Reinforcement learning consists of agent, environment, state, action, and reward, where the environment consists of some states. The agent performs an action based on the current state of the environment, after which the environment moves to the next state based on the action performed and provides the agent with a reward value to evaluate the action. In the long run, the agent learns how to perform the optimal action in a given state [22].

Q learning is a traditional reinforcement learning algorithm; it provides the method by which the intelligence chooses actions in a given state by maintaining a Q-table. Each Q value in the Q-table represents the total benefit of taking a certain action in a certain state. The update of the Q value is mainly realized through the Bellman Equation:

$$Q(s,a) = (1-\alpha)Q(s,a) + \alpha[r + \gamma \max_{d \in A} Q(s^{'},a^{'})] \tag{1}$$

where s represents the current state of the agent, a is the action performed, α is the learning rate, r is the reward value by performing action a under state s, γ represents the discount factor, and A is the action space in state $s^{'}$; α and γ both are in the interval [0, 1]. Therefore, $\max Q(s^{'},a^{'})$ represents the max Q value of state $s^{'}$.

Q routing is the application of the Q learning algorithm to the routing problem. In the Q routing algorithm, each communication node is treated as an agent which can independently learn the forwarding policy and forward packets to the next port [23]. As shown in Table 1, each node maintains a Q-table, which records the Q value of all actions and states. The agent looks up the Q-table based on the destination node of the packet, finds the action corresponding to the maximum Q value, and then executes it, which is a packet forwarding process. Similar to Equation (1), the update of Q-table is as follows:

$$Q_i(s,a) = (1-\alpha)Q_i(s,a) + \alpha[r + \gamma \max_{d \in A} Q_j(s^{'},a^{'})] \tag{2}$$

where α is the learning rate which determines the updating rate, γ is discount factor, and i, j represent the index of different nodes. This equation is the essence of agent learning.

Table 1. Q-table of node I.

State	Neighbor			
	N_1	N_2	N_3	...
(i, D_1)	$Q_i((i, D_1), N_1)$	$Q_i((i, D_1), N_2)$	$Q_i((i, D_1), N_3)$	...
(i, D_2)	$Q_i((i, D_2), N_1)$	$Q_i((i, D_2), N_2)$	$Q_i((i, D_2), N_3)$	...
(i, D_3)	$Q_i((i, D_3), N_1)$	$Q_i((i, D_3), N_2)$	$Q_i((i, D_3), N_3)$	...
$(i, \ldots)$	...	...	...	...

2.3. Problem Statement

The LEO satellite network can be represented as graph G=(V,E), where V represents the set of satellite nodes and E represents the set of satellite links. Consider an Iridium-like system with M number of orbits and N satellites per orbit; we use (i, j) to represent the position of a satellite, where i represents the satellite's orbit number and j represents the satellite's number in orbit $(1 \leq i \leq M, 1 \leq j \leq N)$. There are intra-satellite links between satellites in the same orbit and inter-satellite links between satellites in different orbits, which means that each satellite can communicate directly with up to four satellites. For clarity, we list the notations and definitions in Table 2.

Table 2. Notations of variables.

Notations	Definition
G	graph of the LEO satellite network
V	set of satellite nodes
E	set of satellite links
M	number of orbits
N	number of satellites per orbit
D_{ij}	transmission delay between node i and node j
Ng_i	the set of neighbors of node i

In this article, we only consider the process of packet transmission between satellites. The packet starts from the initial node N_i; the node first finds the next hop node N_j from the set of neighbor nodes N_{gi}, and then sends the packet out. The transmission delay of the packet is D_{ij}, and then the next hop node repeats the previous action, sending the packet to its neighbor node, and then updating the transmission delay of the packet D according to the delay accumulation rule. The above steps are repeated until the data packet is forwarded to the destination node. The problem is planning a route which minimizes D. In a real-world scenario, thousands of packets are passed through the network, so the algorithm needs to consider the congestion of the link. Firstly, the algorithm must be able to plan a feasible path from the source node to the destination node, and secondly, the algorithm should minimize the delay of this path, including propagation delay and queuing delay. Therefore, the ultimate goal of the algorithm is to minimize the transmission delay of packets while ensuring the packet arrival rate.

3. Proposed FRL–SR Algorithm

In this section, we first discuss the design of the reinforcement learning model, including states, reward functions, and actions in Section 3.1. Then, we briefly introduce the neighborhood discovery and explain the propagation range of 'hello' packets in reinforcement learning in Section 3.2. The training approach and algorithm are proposed in Sections 3.3 and 3.4. In Section 3.5, we analyze the time complexity and space complexity of the FRL–SR algorithm and the Dijkstra algorithm.

3.1. Reinforcement Learning Model Setting

In solving the satellite routing problem using multi-agent reinforcement learning, we consider each satellite node as an agent. When the data packet arrives at the satellite node, it observes the current state and forwards the packet based on the present situation. The node adjusts the forwarding strategy according to the current state of the network. Once a link is broken or there is congestion present, the reward of this port is decreased, and packets are forwarded to another path. This is the main advantage of reinforcement learning compared with traditional routing algorithms.

The entire packet forwarding process can be viewed as a finite-state Markov decision process (MDP) whose final state occurs when the packets have arrived at the destination node. We use (S, A, P, R) to represent a state of the MDP, where S is the state of the current system, A is the action space, P is the probability of state transition, and R is the reward. Each satellite node only forwards packets to its neighbor nodes, which means that the action space is up to four. Therefore, we chose reinforcement learning rather than deep reinforcement learning to achieve this.

When using reinforcement learning to solve problems, it is crucial to design state, action, and reward functions [24]. For different scenarios, we should follow different design principles. In our satellite network routing scenario, the states, actions, and rewards are defined as follows:

- States: Each state $s_t \in S = \{N_c, N_d, q_1^t, q_2^t, \dots, q_P^t\}$ indicates the present situation in the satellite network environment, where N_c, N_d represent current node and destination node for packet, respectively. The parameter q_p^t represents the current queue length of the p-th node for $p = 1$ to $p = P$. In multi-agent reinforcement learning, each agent observes a different state, and they make independent routing decisions based on the current state;
- Actions: The action $a_t \in A = \{p_1, p_2, \dots, p_P\}$ represents the agent choosing a node from its neighborhood nodes for each upcoming packet, where p_p represents the p-th satellite node. In satellite networks, each satellite has up to four neighbor nodes, so the length of A is up to four [25];
- Reward: The reward function is designed according to the goal we want to achieve, such as the minimum time delay and the maximum packet arrival rate. In this paper, the transmission delay consists of two parts, propagation time and waiting time, so the reward function consists of the following three parts:
    - Propagation delay. In satellite networks, the star link often fails and becomes unavailable, which requires frequent inter-satellite reconnections. For convenience, we consider both reconnection time and propagation time as propagation time delay. D_{ij} represents propagation delay between node N_i and node N_j;
    - Queue length. Each satellite node maintains the receiving queue q_r and the transmitting queue q_t. Queuing delay occurs when the number of packets received by the satellite is larger than the length of the satellite receiving queue. The agent learns to forward data packets to satellite nodes with small queue length to reduce the waiting time;
    - Load growth. To avoid packets being forwarded centrally to individual satellite nodes, causing network congestion, we record the receiving queue length in the previous stage as the load growth of the satellite, which is recorded as g_i. Therefore, g_i could be seen as the congestion level of nodes N_i. This avoids the situation that everyone sends data to "high-quality" nodes at the same time.

Equation (3) gives the expression of reward function, where q_{max} represents the maximum queue length, and w_1 and w_2 represent the growth and delay coefficients, respectively. When the next hop is the destination node, we set the reward to $20N$, which ensures that the packets are transmitted to the destination node as soon as possible.

$$reward_j = \begin{cases} 20N & N_j \text{ is the destination} \\ q_{max} - (q_r + q_t) - w_1 * g_j - w_2 * D_j & \text{Otherwise} \end{cases} \quad (3)$$

As shown in Figure 2, the system framework consists of three parts: Neighbor Node Discovery, Offline Training, and Online Training. In multi-agent reinforcement learning, the learning process of the agent needs the information of the neighboring agents, which includes Q-table, connection status, and available resources. Therefore, we develop the neighborhood discovery part for agents to share information. During the offline training phase, we perform the initial training of the agents in a ground-based network environment. By randomly generating packets, the agents act by observing the state of the packets and the state of the neighboring nodes. To avoid local optimal solutions, we use the ϵ-greedy strategy to explore a larger unknown space. Static network training results are not fully suitable for dynamic LEO satellite networks, so online training is required to fine-tune the Q-table. Agents make routing decisions for satellite networks based on the pre-trained model and update the Q-table with the feedback results. It is important to note that the ϵ-greedy strategy is not used at this stage, as we only need to fine-tune the Q-table. The advantage of using the pre-trained method is that it saves onboard resources and improves the performance of the initial phase of the algorithm.

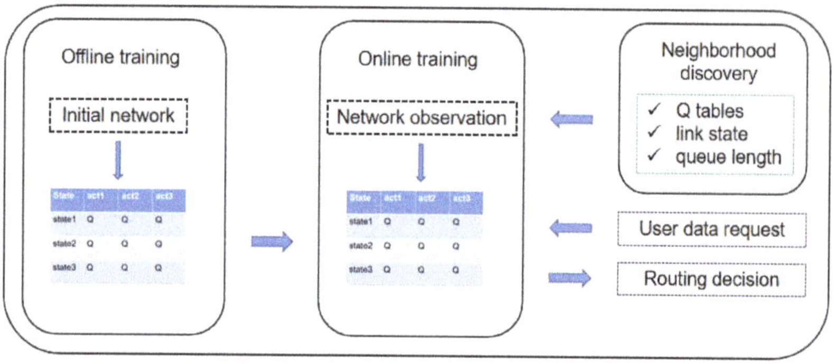

Figure 2. Model framework of FRL–SR.

3.2. *Neighborhood Discovery*

Since the topology of the satellite network changes dynamically and the links between satellites are unstable, the neighbor nodes of the satellite change. Therefore, agents must periodically check the neighbor nodes so that they know where to forward packets when they arrive.

Satellite nodes receive information about their neighbors by receiving 'hello' packets from neighboring nodes [26]. In addition, the "hello" packet contains a Q-table, link information, and available resources that nodes can use to calculate reward values and update their own Q-table. If a node does not receive "hello" packets from a neighboring node for a long time, it assumes that the neighboring node is down and will not send packets to it later. In this paper, nodes only send 'hello' packets to their neighbors. Compared to flooding 'hello' packets in terrestrial networks, this method saves energy cost for the satellite and does not burden the network.

3.3. *Offline Training*

The algorithm proposed in this paper needs offline training before being applied to satellite nodes. The purpose of this process is to reduce the training time of the algorithm online and improve its initial performance. The network G_{t0} at t_0 time is input as the initial state, and the output is that each satellite node receives a Q-table.

To simulate the random nature of packet generation, the initial and destination nodes of packets are randomly selected from the set of satellite nodes. To reduce training time, if a packet is forwarded to its destination node, a new packet is generated, and its initial node and destination node are also randomly generated.

To better explore the unknown space, the ϵ-greedy strategy is used in the offline training phase. As shown in Equation (4), the agent randomly chooses an action with probability ϵ, and, with probability $1-\epsilon$, it chooses the action with the maximum q-value. This strategy prevents local optimal solutions, but the convergence speed is low. To solve the above problem, the value of ϵ gradually decreases as reinforcement learning progresses, which can speed up the convergence of the algorithm without affecting the learning effect.

$$a_t = \begin{cases} \text{random action} & \text{w.p.}\epsilon \\ argmax_a Q_{t+1} & \text{w.p.}1-\epsilon \end{cases} \tag{4}$$

Algorithm 1 is the process of offline training. First we need to initialize the training parameters: num_{step} is the total number of steps trained, ϵ is the probability value of the greedy strategy, l_r is the learning rate of reinforcement learning, and γ is the discount factor of the Q value update function. At each step, the algorithm first cleans up the remaining packets in the network and then randomly generates a preset number of packets. For each satellite node, we first determine whether its sending queue q_t is empty, and the top packet pops up when it is not empty. Then, we select the next hop node m according to the ϵ-greedy strategy. If the receive queue q_r of node m is not full, the packet is forwarded to m, and the current node will receive the Q-table and reward value of node m. After that, the Q-table of the current node should be updated according to Equation (2). Otherwise, the network is congested and packets will be inactive until the next round arrives.

Algorithm 1: FRL–SR offline training algorithm.

Input: $G_{t0} =< N, E >; num_{step}; \epsilon; l_r; \gamma$
Output: Q-tables for each satellite node

initial env, Q-tables;
for *k=1 to* num_{step} **do**
  Clear all data packages;
  Randomly generate x_t data packages;
  for *n=1 to N* **do**
    if q_t *in current node* $n \neq \varnothing$ **then**
      Pop a package p from q_t;
      Select an action m according to *ϵ-greedy*;
      if q_r *of node m is not full* **then**
        Forward packets p to node m;
        Node m send Q-table and reward to node n;
        Update the state and Q-table of node n;
        Update the state of node m;
        if *m is the destination node of package p* **then**
          reward = 20;
          Randomly generate a package;
        end
      else
        Wait a round;
      end
    end
  end
end

3.4. Online Training

In the online training phase, the pre-trained parameter in Section 3.3 is used as the initial routing decision strategy. The online training algorithm carries out two things: one is to make routing decisions based on the pre-trained model, and the other is to fine-tune the Q-table based on the real satellite network environment.

Unlike offline training, agents in online training do not make decisions according to the ϵ-greedy strategy, since agents avoided the local optimal solution in the previous step. The main purpose of this process is to fine-tune the routing strategy. Moreover, the agents are already making routing decisions at this stage, and the algorithm must ensure that these decisions are sound.

We simplify the satellite network routing problem to finding the smallest delay path. The delay of a packet consists of two parts: propagation delay and queuing delay. The propagation delay depends on the inter-satellite link distance and link status, and the queuing delay depends on the available resources. According to Equation (3), we know that the reward value is linearly related to the delay of the packet. Equation (5) is the definition of the Q value in reinforcement learning, from which we can conclude that the Q value represents the estimated total return of an action in the current state, and the larger the Q value, the smaller the delay. Therefore, the path with the maximum Q is the path with the minimum delay. According to the greedy algorithm, if each agent chooses the action with the largest Q value, the latency of the obtained packet is close to the optimal choice.

As shown in Equation (6), we use Q_{sum} to represent the goal of algorithm optimization, which is a linear combination of the Q values of each agent. Combined with Equation (5), we can derive the relationship between Q_{sum} and each reward value. If we know that the reward value and delay are linear, then Q_{sum} and delay are also linear, so we only need to maximize Q_{sum} to obtain the minimum delay path.

$$Q_i(s,a) = R_i + \gamma(R_{i+1} + R_{i+2} + \ldots + R_n) \tag{5}$$

$$Q_{sum} = \sum_{i=1}^{n} w_i Q_i(s_i, a) \tag{6}$$

If we suppose that the link state of a satellite changes, it is obtained first by the two satellite nodes of this link, followed by the neighboring nodes of the two satellite nodes. Therefore, the state of links is serial propagation, which causes certain difficulties for the convergence of the reinforcement learning algorithm. Especially, the state of the satellite network is prone to change; it is possible that the convergence of the previous stage is not yet complete and the link state has changed again.

As shown in Figure 3, there are five orbits, each with five satellites. Each satellite is represented by a dot with a different color. The red node satellite transmits a message outwards, and the first to receive that message are the yellow nodes around it. The yellow node then passes the message to the green node, and the white node does not receive the message until the fourth round of message dissemination. This indicates that the message transmission in the satellite network is linear. In the traditional Q-routing algorithm, the agent receives the Q-table and link state information feedback from the neighbor nodes when and only when it sends a packet to its neighbor nodes. Then, the node updates its Q-table, which is a learning process. This paper proposes a method named empty packet convergence method to accelerate the perception of the network state by agents. Neighboring nodes do not only send status information after receiving a packet, but also broadcast its message by period t. The traditional learning process only updates a certain item of the two-dimensional Q-table at a time, while the empty packet convergence method updates the entire content of a node's action space at a time. The smaller the t, the more often agents perceive the network. Therefore, we designed t to be inversely proportional to the node traffic density; the higher the traffic density, the smaller the t, and the faster agents perceive the network. This ensures faster awareness of the state of the network without increasing the network overhead too much.

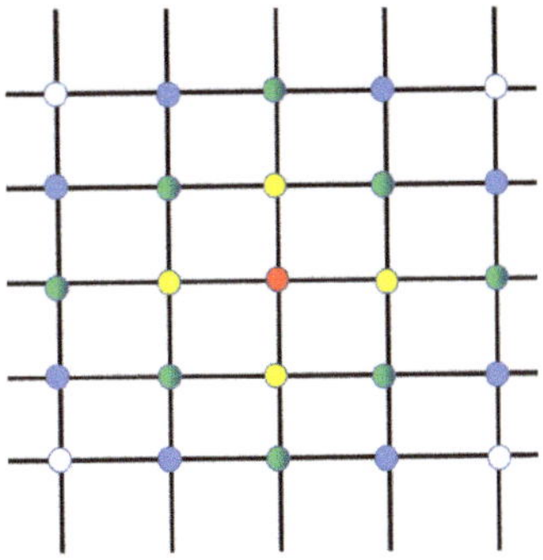

Figure 3. Schematic diagram of link state propagation.

Algorithm 2 is the pseudocode of the online learning process of the algorithm. It differs from the traditional reinforcement learning algorithm in that this paper proposes an empty packet convergence method, which accelerates the obtaining of network status and is more suitable for the state-variant satellite network environment.

Algorithm 2: FRL–SR online training algorithm.

Input: Trained routing strategy model; user data request; l_r; γ
Output: Real time routing decision

```
while not done do
    if time mod t == 0 then
        Neighborhood discovery;
        Update the Q-table according to Equation (2);
    end
    while user data request received do
        Choose the next hop according to Q-table;
        if q_r of the next hop is not full then
            Transmit data packets;
            Calculate the reward;
            Update the Q-table according to Equation (2);
        else
            Wait a round;
        end
    end
end
```

3.5. Complexity Analysis

In this paper, we compare FRL–SR with the Dijkstra algorithm in terms of algorithm complexity. The time complexity of the Dijkstra algorithm is $O(n^2)$: it has to obtain global state information and then compute the shortest path from the current node to any node through two layers of loops. The FRL–SR algorithm proposed in this paper only queries the Q-table stored in the satellite when making routing decisions, so the time complexity is constant. Therefore, FRL–SR is more suitable for solving routing problems for large satellite constellations. In terms of spatial complexity, both of them store a two-dimensional array. The spatial complexity of the Dijkstra algorithm is O(E +4N) when using adjacency tables for data storage, where E is the number of edges in the network diagram. The maximum action space of the FRL–SR algorithm is 4, so the space complexity is O(4N), which does not take up more storage space than the Dijkstra algorithm.

Another important indicator is the communication overhead during the execution of the algorithm. In the Dijkstra algorithm, each satellite node must send its own status information in a flood so that each satellite node can receive the global information. The communication overhead is high and also has some negative impact on the network load. In the FRL–SR algorithm, each satellite node only needs to send the status information to its neighboring nodes without forwarding it to the whole network. Even if we increase the number of communications between neighbor nodes to improve the convergence speed, it is much lower than the communication overhead of the Dijkstra algorithm.

4. Simulation Results

In this section, we present the simulation results of the FRL–SR algorithm. The experiment was mainly divided into two parts: First, we simulated different reward functions in reinforcement learning and found out which one could minimize the delay for our subsequent experiments. Then, we compared the performance differences between the FRL–SR algorithm proposed in this paper and the traditional Dijkstra algorithm, both running in the same network environment and with the same user data requirements. We compared these two algorithms in terms of average delay, packet arriving ratio, and network load balance, and explained the reasons for their performance differences. To make the experimental results more accurate, we repeated all the experiments three times and took the average of the results as the final result.

In this paper, we chose the Dijkstra algorithm as the comparison algorithm because it is the most mature and widely used algorithm in satellite routing algorithm, and through experiments, we verified that the proposed algorithm can effectively improve the satellite routing strategy and better support communication services [27].

The parameters of the experiment are given in Table 3. The network had a total of 7 satellite orbits, each containing 7 satellites, for a total of 49 satellites [28]. Because interstellar links fail frequently and link recovery varies, to simulate a satellite network more realistically, we set the propagation delay to vary according to a sinusoidal curve. In the offline training phase, the algorithm ran for 30 episodes—the step for each episode was 200, which ensured each agent could learn the routing strategy in a limited number of training steps. In the online training phase, we observed the delivery of packets in the network every 1 s and recorded it. In order to ensure the stability of the network, we adjusted the learning rate of this stage to 0.2, and the corresponding learning rate of the offline training stage was 0.8.

Table 3. Simulation parameters.

Parameters	Values
Number of satellites	49
Delay type	sinusoidal
Trials	3
Offline training network load	3000
Initial network load for online training	3000
Max queue length	150
Max transmit packages at one time	10
Number of episodes	30
Number of steps peer episode	200
discount factor	0.9
Learning rate for offline training	0.8
Learning rate for online training	0.2

The performance of different rewards is shown in Figure 4. The reward1 function was designed to be related only to the length of the link between the two nodes, in which D_{ij} was the distance between node i and node j.

$$reward1 = \begin{cases} 980 & \text{Next node is the destination} \\ -0.1 * D_{ij} & \text{Otherwise} \end{cases} \tag{7}$$

The *reward2* function was inversely relative to the queue length of the node and the distance between the two nodes.

$$reward2 = \begin{cases} 980 & \text{Next node is the destination} \\ 300 - (q_r + q_t) - 5 * g_j - 0.1 * D_{ij} & \text{Otherwise} \end{cases} \quad (8)$$

Based on the simulation results, we can conclude that the design of the reward function has a great influence on the performance of the algorithm. The goal of the algorithm was to obtain the path with the least delay, so we chose the second reward function for subsequent simulations.

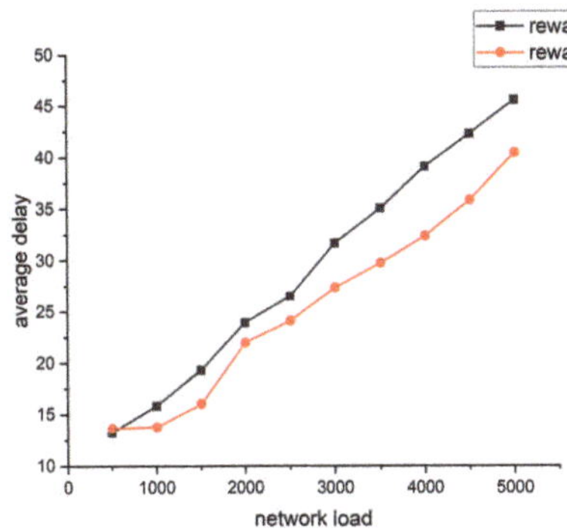

Figure 4. Comparison of different reward performances.

Both Figure 5 and Figure 6 show the relationship between the average delay and time in operation for the FRL–SR algorithm and the Dijkstra algorithm. To better demonstrate the actual effect of the algorithm, we took the study time after the network was relatively stable, rather than when the system was first put into operation. The initial number of packages in Figure 5 is 3000, and the initial number of packages in Figure 6 is 5000. We can see that the FRL–SR algorithm showed consistent performance in environments with different initial packet counts.

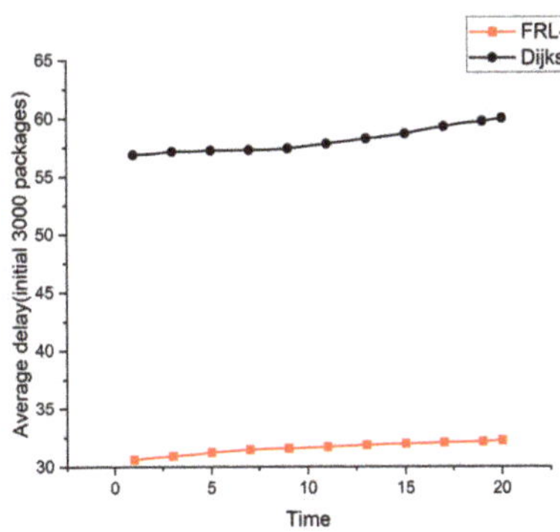

Figure 5. Comparison of the average delay of the FRL–SR algorithm and the Dijkstra algorithm with 3000 initial packages.

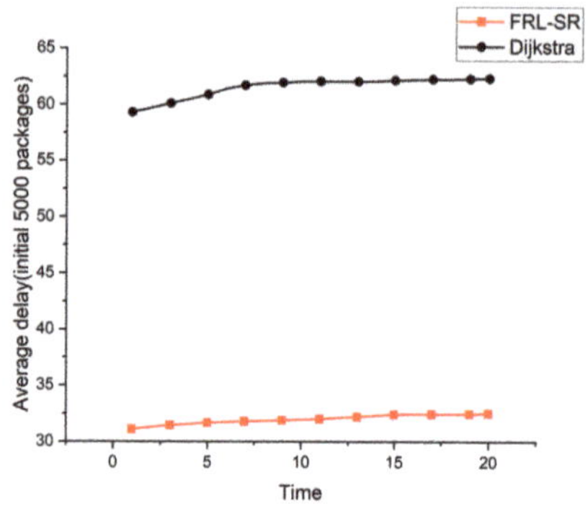

Figure 6. Comparison of the average delay of the FRL–SR algorithm and the Dijkstra algorithm with 5000 initial packages.

The FRL–SR algorithm outperforms the Dijkstra algorithm in terms of average delay in the same network environment. This is because the FRL–SR algorithm adjusts its forwarding policy based on the current state of the network and then forwards the packets to the optimal port, while the Dijkstra algorithm makes forwarding decisions based on the previously collected global information. The link state of the satellite network changes rapidly, and the information collected by Dijkstra cannot accurately reflect the state of the satellite network at that time. As can be observed from the above graph, the average latency increases slowly with time because the inter-star link is prone to failure, resulting in packet loss. The algorithm in this paper does not have a data retransmission function, which means the delay of lost packets will keep increasing, resulting in a rising average delay.

The relationship between cumulative number of packets and time is shown in Figure 7. We can see that, over time, the advantages of the FRL–SR algorithm over the Dijkstra algorithm gradually become apparent. Each satellite node has limited resources, so the communication capacity of the whole network is also limited. The FRL–SR algorithm considers the resource utilization of satellite nodes as an optimization objective. When a satellite node selects the next hop, it takes the queue length of neighboring nodes as a consideration parameter and forwards the packets to the satellite with sufficient communication resources first, which improves the resource utilization of the network and forwards more packets per unit time.

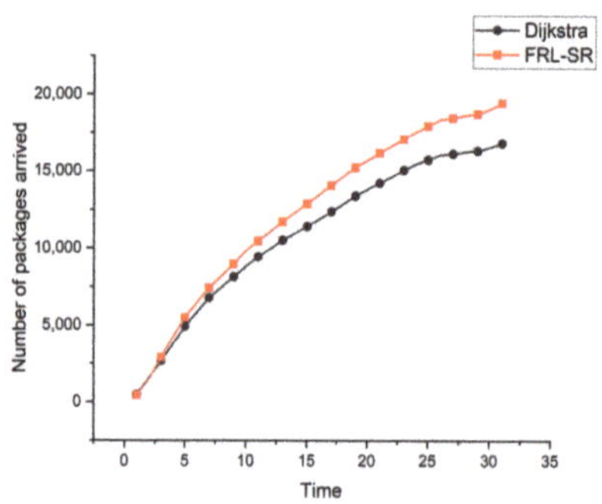

Figure 7. Comparison of the number of successfully delivered packets by FRL–SR algorithm and Dijkstra algorithm.

Considering Figures 5–7 together, we observe that the FRL–SR algorithm transmits a higher number of successful packets in the same amount of time with a smaller average delay per packet. This is enough to show the advantages of the FRL–SR algorithm proposed in this paper over the traditional satellite routing algorithm.

$$s = \sqrt{\frac{\sum(x - M)^2}{n}} \quad (9)$$

Figure 8 shows a comparison of the load balance of nodes in the network. We use the population standard deviation to express the load balance of the network, as shown in Equation (9), where M is the mean of the data and n is the total number of nodes in the network, which is a commonly used parameter to describe the degree of discreteness of the system. By observing the simulation results, we can observe that, under the same user request conditions, the FRL–SR algorithm has a better load balancing effect than the Dijkstra algorithm. The former can make full use of the resources of each node for routing, while the Dijkstra algorithm is more likely to cause network congestion.

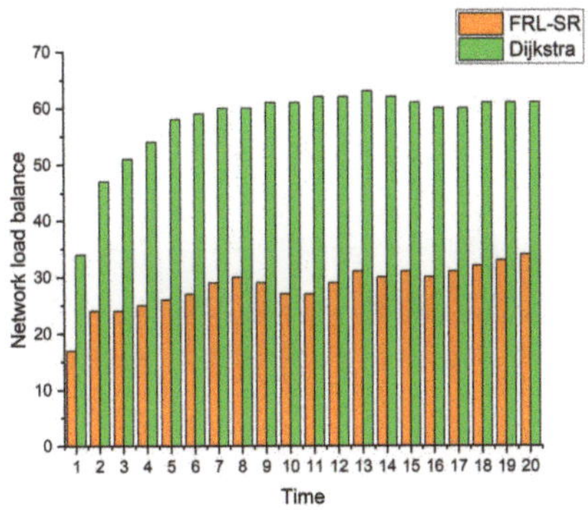

Figure 8. Measure the balance of node load in the network.

Based on the above simulation results, we conclude that the FRL–SR algorithm has lower network latency and higher data throughput, which is more suitable for satellite networks with variable network status. In addition, the FRL–SR algorithm also has good load balancing characteristics, and it considers both satellite link status and satellite load when making data forwarding decisions, avoiding network congestion. The Dijkstra algorithm only blindly forwards packets to ports with good link status, causing congestion on some network links.

5. Conclusions

In this paper, we proposed a fast-convergence reinforcement learning algorithm to construct the routing issue in the LEO constellation. Aiming at addressing the characteristics of large dynamic satellite network status and unstable inter-satellite links, we designed a routing method named the fast-convergence reinforcement learning satellite routing (FRL–SR) for online decision-making. This method is always aware of the network link status and dynamically adjusts its routing strategy. The FRL–SR algorithm includes three parts: neighbor node discovery, offline training, and online training. By shortening the cycle time of agent obtaining network states, we accelerated the convergence speed of the algorithm, so that the routing decision was more suitable for the current network state. In addition, we also performed a complexity analysis, and the FRL–SR algorithm was superior to the Dijkstra algorithm in both time complexity and spatial complexity.

The simulation results showed that the FRL–SR algorithm had a lower average delay and higher packet arriving ratio compared with the Dijkstra algorithm. In addition, the FRL–SR algorithm also had a good performance with respect to load balancing. It made full use of the resources of each node and reduced the probability of network congestion. Multi-agent cooperation is a promising method to solve the problem of large-scale satellite network routing. In future work, we will continue to work on multi-agent reinforcement learning algorithms to better solve the problem of satellite network routing.

Author Contributions: Conceptualization, Z.D.; Formal analysis, Z.D.; Funding acquisition, H.L.; Investigation, Z.D.; Methodology, Z.D.; Project administration, H.L.; Resources, Z.Y.; Software, Z.D.; Supervision, N.W.; Validation, F.T.; Writing—original draft, Z.D.; Writing—review & editing, Z.D. and F.T. All authors have read and agreed to the published version of the manuscript.

Funding: This research received no external funding.

Institutional Review Board Statement: Not applicable.

Informed Consent Statement: Not applicable.

Data Availability Statement: The relevant data of this paper can be contacted by dingzhl@shanghaitech.edu.cn.

Conflicts of Interest: The authors declare no conflict of interest.

References

1. Tsai, K.-C.; Fan, L.; Wang, L.-C.; Lent, R.; Han, Z. Multi-Commodity Flow Routing for Large-Scale LEO Satellite Networks Using Deep Reinforcement Learning. In Proceedings of the 2022 IEEE Wireless Communications and Networking Conference (WCNC), Austin, TX, USA, 10–13 April 2022; pp. 626–631.
2. Su, Y.; Liu, Y.; Zhou, Y.; Yuan, J.; Cao, H.; Shi, J. Broadband LEO Satellite Communications: Architectures and Key Technologies. *IEEE Wirel. Commun.* **2019**, *26*, 55–61. [CrossRef]
3. Liu, J.; Zhao, B.; Xin, Q.; Su, J.; Ou, W. DRL-ER: An Intelligent Energy-Aware Routing Protocol With Guaranteed Delay Bounds in Satellite Mega-Constellations. *IEEE Trans. Netw. Sci. Eng.* **2021**, *8*, 2872–2884. [CrossRef]
4. Mauger, R.; Rosenberg, C. QoS guarantees for multimedia services on a TDMA-based satellite network. *IEEE Commun. Mag.* **1997**, *35*, 56–65. [CrossRef]
5. Werner, M. A dynamic routing concept for ATM-based satellite personal communication networks. *IEEE J. Sel. Areas Commun.* **1997**, *15*, 1636–1648. [CrossRef]
6. Cao, X.; Li, Y.; Xiong, X.; Wang, J. Dynamic Routings in Satellite Networks: An Overview. *Sensors* **2022**, *22*, 4552. [CrossRef] [PubMed]
7. Li, J.; Lu, H.; Xue, K.; Zhang, Y. Temporal Netgrid Model-Based Dynamic Routing in Large-Scale Small Satellite Networks. *IEEE Trans. Veh. Technol.* **2019**, *68*, 6009–6021. [CrossRef]
8. Ekici, E.; Akyildiz, I.F.; Bender, M.D. A distributed routing algorithm for datagram traffic in LEO satellite networks. *IEEE/ACM Trans. Netw.* **2001**, *9*, 137–147. [CrossRef]
9. Liu, X.; Yan, X.; Jiang, Z.; Li, C.; Yang, Y. A low-complexity routing algorithm based on load balancing for LEO satellite networks. In Proceedings of the 2015 IEEE 82nd Vehicular Technology Conference (VTC2015-Fall), Boston, MA, USA, 6–9 September 2015; pp. 1–5.
10. Guo, A.; Zhao, C.; Xu, F.; Qiu, C. LEO satellite routing algorithm in software defined space terrestrial integrated network. In Proceedings of the 2017 17th International Symposium on Communications and Information Technologies (ISCIT), Cairns, Australia, 25–27 September 2017; pp. 1–6.
11. Jiang, Z.; Wu, Q.; Li, H.; Wu, J. scMPTCP: SDN Cooperated Multipath Transfer for Satellite Network With Load Awareness. *IEEE Access* **2018**, *6*, 19823–19832. [CrossRef]
12. Boero, L.; Marchese, M.; Patrone, F. The impact of delay in software-defined integrated terrestrial-satellite networks. *China Commun.* **2018**, *15*, 11–21. [CrossRef]
13. Jay, N.; Rotman, N.; Godfrey, B.; Schapira, M.; Tamar, A. A deep reinforcement learning perspective on internet congestion control. In Proceedings of the International Conference on Machine Learning, PMLR, Long Beach, CA, USA, 9–15 June 2019; pp. 3050–3059.
14. Mao, H.; Netravali, R.; Alizadeh, M. Neural adaptive video streaming with pensieve. In Proceedings of the Conference of the ACM Special Interest Group on Data Communication, Los Angeles, CA, USA, 21–25 August 2017; pp. 197–210.
15. Wei, Y.; Yu, F.R.; Song, M.; Han, Z. User scheduling and resource allocation in hetnets with hybrid energy supply: An actor-critic reinforcement learning approach. *IEEE Trans. Wireless Commun.* **2018**, *17*, 680–692. [CrossRef]
16. Wang, X.; Dai, Z.; Xu, Z. LEO Satellite Network Routing Algorithm Based on Reinforcement Learning. In Proceedings of the 2021 IEEE 4th International Conference on Electronics Technology (ICET), Chengdu, China, 7–10 May 2021; pp. 1105–1109.
17. Xu, L.; Huang, Y.-C.; Xue, Y.; Hu, X. Deep Reinforcement Learning-Based Routing and Spectrum Assignment of EONs by Exploiting GCN and RNN for Feature Extraction. *J. Light. Technol.* **2022**, *40*, 4945–4955. [CrossRef]
18. Huang, Y.-X.; Wu, S.-F.; Kang, Z.-Y. Reinforcement learning based dynamic distributed routing scheme for mega LEO satellite networks. *Chin. J. Aeronaut.* **2023**, *36*, 284–291. [CrossRef]

19. Fadlullah, Z.M.; Tang, F.; Mao, B.; Kato, N.; Akashi, O.; Inoue, T.; Mizutani, K. State-of-the-art deep learning: Evolving machine intelligence toward tomorrow's intelligent network traffic control systems. *IEEE Commun. Surv. Tutor.* **2017**, *19*, 2432–2455. [CrossRef]
20. Chang, H.S.; Kim, B.W.; Lee, C.G.; Choi, Y.; Min, S.L.; Yang, H.S.; Kim, C.S. Topological design and routing for low-Earth orbit satellite networks. In Proceedings of the GLOBECOM'95, Singapore, 14–16 November 1995; Volume 1, pp. 529–535.
21. Sutton, R.S.; Barto, A.G. *Reinforcement Learning: An Introduction*; MIT Press: Cambridge, UK, 2018.
22. Zhuang, Z.; Wang, L.C.; Wang, J.; Qi, Q.; Han, Z.; Lent, R. Tensor-based reinforcement learning for network routing. *IEEE J. Sel. Top. Signal Process.* **2021**, *15*, 617–629.
23. Li, X.; Hu, X.; Li, W.; Hu, H. A Multi-Agent Reinforcement Learning Routing Protocol for Underwater Optical Sensor Networks. In Proceedings of the ICC 2019—2019 IEEE International Conference on Communications (ICC), Shanghai, China, 20–24 May 2019; pp. 1–7.
24. Liu, Y.; Jiang, Z.; Zhang, S.; Xu, S. Deep Reinforcement Learning-Based Beam Tracking for Low-Latency Services in Vehicular Networks. In Proceedings of the ICC 2020—2020 IEEE International Conference on Communications (ICC), Dublin, Ireland, 7–11 June 2020.
25. Lowe, R.; Wu, Y.; Tamar, A.; Harb, J.; Abbeel, P.; Mordatch, I. Multi-agent actor-critic for mixed cooperative-competitive environments. In Proceedings of the 31st International Conference on Neural Information Processing Systems, Long Beach, CA, USA, 4–9 December 2017; pp. 6379–6390.
26. Jung, W.S.; Yim, J.; Ko, Y.B. QGeo: Q-learning-based geographic ad hoc routing protocol for unmanned robotic networks. *IEEE Commun. Lett.* **2017**, *21*, 2258–2261. [CrossRef]
27. Zheng, F.; Wang, C.; Zhou, Z. LEO laser microwave hybrid inter-satellite routing strategy based on modified Q-routing algorithm. *EURASIP J. Wirel. Commun. Netw.* **2022**, *2022*, 1–18. [CrossRef]
28. Tang, F.; Zhang, H.; Yang, L.T. Multipath Cooperative Routing with Efficient Acknowledgement for LEO Satellite Networks. *IEEE Trans. Mob. Comput.* **2019**, *18*, 179–192. [CrossRef]

sensors

MDPI

Article

Joint Optimization of Bandwidth and Power Allocation in Uplink Systems with Deep Reinforcement Learning

Chongli Zhang [1], Tiejun Lv [1], Pingmu Huang [2], Zhipeng Lin [3,*], Jie Zeng [4] and Yuan Ren [5]

1 School of Information and Communication Engineering, Beijing University of Posts and Telecommunications (BUPT), Beijing 100876, China; chonglizhang@bupt.edu.cn (C.Z.); lvtiejun@bupt.edu.cn (T.L.)

2 School of Artificial Intelligence, Beijing University of Posts and Telecommunications (BUPT), Beijing 100876, China; pmhuang@bupt.edu.cn

3 Key Laboratory of Dynamic Cognitive System of Electromagnetic Spectrum Space, College of Electronic and Information Engineering, Nanjing University of Aeronautics and Astronautics (NUAA), Nanjing 211106, China

4 School of Cyberspace Science and Technology, Beijing Institute of Technology, Beijing 100081, China; zengjie@bit.edu.cn

5 Shaanxi Key Laboratory of Information Communication Network and Security, School of Communications and Information Engineering, Xi'an University of Posts and Telecommunications, Xi'an 710121, China; renyuan@xupt.edu.cn

* Correspondence: linlzp@nuaa.edu.cn

Abstract: Wireless resource utilizations are the focus of future communication, which are used constantly to alleviate the communication quality problem caused by the explosive interference with increasing users, especially the inter-cell interference in the multi-cell multi-user systems. To tackle this interference and improve the resource utilization rate, we proposed a joint-priority-based reinforcement learning (JPRL) approach to jointly optimize the bandwidth and transmit power allocation. This method aims to maximize the average throughput of the system while suppressing the co-channel interference and guaranteeing the quality of service (QoS) constraint. Specifically, we de-coupled the joint problem into two sub-problems, i.e., the bandwidth assignment and power allocation sub-problems. The multi-agent double deep Q network (MADDQN) was developed to solve the bandwidth allocation sub-problem for each user and the prioritized multi-agent deep deterministic policy gradient (P-MADDPG) algorithm by deploying a prioritized replay buffer that is designed to handle the transmit power allocation sub-problem. Numerical results show that the proposed JPRL method could accelerate model training and outperform the alternative methods in terms of throughput. For example, the average throughput was approximately 10.4–15.5% better than the homogeneous-learning-based benchmarks, and about 17.3% higher than the genetic algorithm.

Keywords: uplink; multi-cell multi-user system; joint-priority-based reinforcement learning (JPRL); prioritized replay buffer; throughput

Citation: Zhang, C.; Lv, T.; Huang, P.; Lin, Z.; Zeng, J.; Ren, Y. Joint Optimization of Bandwidth and Power Allocation in Uplink Systems with Deep Reinforcement Learning. *Sensors* **2023**, *23*, 6822. https://doi.org/10.3390/s23156822

Academic Editors: Yichuang Sun, Haeyoung Lee and Oluyomi Simpson

Received: 2 July 2023
Revised: 28 July 2023
Accepted: 28 July 2023
Published: 31 July 2023

1. Introduction

The fifth generation (5G) and beyond fifth generation (B5G) era is boosting a mega growth in the number of mobile devices [1], thereby resulting in explosive increasing demand that prompts people to explore new technologies to ease the demand strains. Recently, the large-scale dense network is gradually developing as a trend for the next-generation communication networks [2,3] due to its advantages traffic capacity and diversified services [4]. The densification of the network [5] is one of the key features of the 5G wireless network architecture, which not only contributes to increasing the system capacity of 5G networks, but also is closely related to user experience enhancement. As an important technique for improving the efficiency and quality of communications, dense networks still suffer from extremely complex interference problems [6]. In the dense multi-cell multi-user system, explosive rising users in different cells have an interplay due to the reuse of resources, which leads to increased co-channel interference and scarce resources. Furthermore, it is not conducive to deliver high throughput and a good quality of service (QoS) [7]. As a result, reasonable radio resource management [8] is imperative to improve the performance of future communications.

As pointed out in [9,10], whether resource allocation is rational or not determines the throughput performance of the system. Consider the multi-cell multi-user system where multiple resources (e.g., the bandwidth and transmitted power) are allocated to each user. As one user who interferes other users improving individual throughput causes serious interference, the coordination of resource allocation can avoid this situation efficiently. Therefore, bandwidth assignment and power allocation are the essential components of radio resource management, which can effectively suppress co-channel interference and conserve frequency resources. For the challenges of bandwidth and power allocations in the multi-cell multi-user network, a variety of methods have been proposed to increase throughput. Xu et al. [11] improved the throughput by selecting mobile relay and assigning subcarriers in the existence of various interferences. Liu et al. [12] increased the throughput by means of fast power allocation while guaranteeing stringent latency and reliability. The authors in [13] proposed a metaheuristic algorithm to solve the power control problem, which relied on discrete power allocation schemes. For the network cost problem of the large-scale heterogeneous system, Cao et al. [14] improved the network coverage using an adaptive seagull algorithm. In addition, various joint allocation methods have been proposed to maximize the rate, energy efficiency, and spectral efficiency [15–17]. The above-mentioned research works are based on traditional methods, such as the genetic algorithm [18], game theory [19], water-filling method [20], graph theory [21], and so on. These approaches are usually able to achieve the goals for different optimizations and application scenarios. Nevertheless, all of them experience dilemmas in exponentially growing the search space for the large-scale system, which are unsuitable for addressing high-dimensional joint optimization problems.

Reinforcement learning (RL) has been an efficient tool to solve optimization problems with a large number of data. It relies on uncharted exploitation with available samples for good reward feedback, which has been widely applied in large-scale scenarios [22,23]. Han et al. [24] proposed a State-Action-Reward-State-Action (SARSA) algorithm for power control to improve throughput. By taking advantage of machine learning, deep RL (DRL) is more effective for multi-user systems with large action spaces, which speeds the training process. The deep Q network (DQN) combines deep neural networks with Q-learning to approximate the value function with the help of maximizing the Q value [25], which has been deployed in many studies [26–28]. In [26], the authors developed a DQN-based method to allocate resource blocks in order to reduce the collision ratio and improve the throughput. Instead of directly using the maximum Q value, the double DQN (DDQN) selects the action by de-coupling the maximum Q value, which can avoid the overestimation of the Q value and speed up the convergence. Iqbal et al. [29] designed a DDQN method for power allocation to minimize the total power consumption. Nevertheless, many optimization variables, such as power allocation, are continuous in practice and are not applicable to the DQN and DDQN due to the discrete nature of actions. Furthermore, although the DQN and DDQN can transform continuous ranges into actions with different discrete granularities, they are impractical because of the limited granularity. For problems with infinite choices (e.g., power allocation), continuous action-selection-based algorithms such as the deep deterministic policy gradient (DDPG) [30] can overcome the disadvantages of discretization. Meng et al. [31] customized a DDPG to maximize the sum rate in a downlink cellular communication system. The authors in [32] optimized the long-term throughput using the adjusted DDPG extended from the DDPG, which is valid for two absolutely different action spaces. However, a centralized method such as the above works is feasible but inefficient and unsuitable for large-scale systems [33]. Multi-agent DRL (MADRL) is an advanced

RL method that can outperform the single agent in resource allocation, especially in the multi-cell multi-user system [34,35]. In [36], a joint resource allocation problem is settled by a MADRL relying on the independent Q-learning method [37]. Similarly, Tian et al. [38] presented a DDPG-based MADRL method to allocate the channel and power by optimizing the QoS in vehicular networks.

Though MADRL contributes a great progress in the filed of joint resource allocation, it still continues to have the following limitations typically: (1) It generally ignoring the importance of the transition replay in sampling a mini-batch. In the traditional MADRL, since the complex communication environments usually contain a large amount of information, uniform experience replay leads to poor stability and the slow convergence of neural networks; (2) It weakens the interconnectivity between agents, especially in the system where the agent plays a direct role with the other agents (for example, an agent promotes individually and hinders others). Therefore, the traditional MADRL, which uses a distributed training process to explore solutions, is unsuitable for finding the action characteristics of each agent; and (3) It is not realistic to simplify the channel with a free-space propagation model, since some test scenarios are neglected in different channel models [39], including the urban macro-cell (UMa), rural macro0cell (RMa), and rural micro-cell (RMi) in IMT-2020.

Inspired by the success of DRL and the above research, the joint-priority-based RL (JPRL) method has been proposed to maximize the average throughput, which considers the co-channel interference between different cells. Unlike the traditional DRL algorithm that optimizes multiple variables, we selected different algorithms to optimize variables according to the problem property and deployed a distributed learning and centralized training framework. The main contributions of this paper are summarized as follows:

- We proposed a joint bandwidth and power allocation framework based on the JPRL method to maximize the average throughput of the uplink large-scale system, which considered the co-channel interference between different cells with the assurance of the QoS. For the joint optimization problem, since the bandwidth assignment is a discrete problem, while the power allocation is continuous, we decomposed the joint problem into two sub-problems and used different algorithms to solve them.
- We proposed a priority experience replay mechanism for power allocation. By analyzing the characteristics of the optimization sub-problems, the proposed experience replay mechanism was applied to a multi-agent DDPG (MADDPG), which was named the prioritized MADDPG (P-MADDPG), which trained valuable experiences to improve the throughput in the training process, thereby surpassing the issue of infinite power action space.
- The proposed JPRL method is shown in Figure 1. It consists of a multi-agent DDQN (MADDQN) algorithm and the P-MADDPG algorithm, where MADDQN was developed to solve the bandwidth assignment sub-problem, and the P-MADDPG was employed to solve the transmit power allocation. Besides, both the MADDQN and P-MADDPG used a centralized training framework with a joint action-value function.

The remainder of this paper is organized as follows. Section 2 introduces the system model and optimization problem. The proposed JPRL method is described in Section 3. Section 4 demonstrates the simulation results, and the conclusions are presented in Section 5.

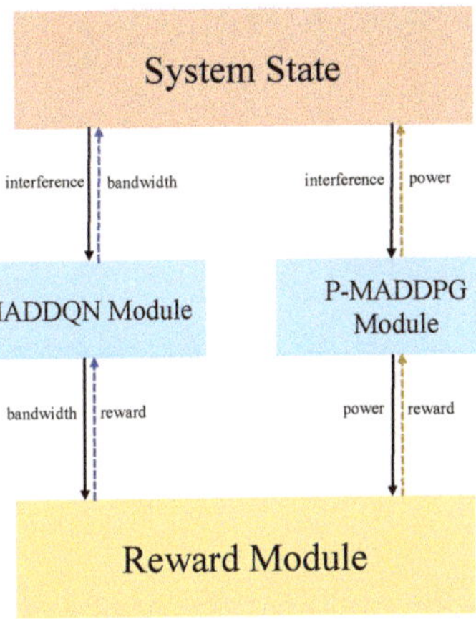

Figure 1. Joint bandwidth and power allocations scheme.

2. System Model and Problem Formulation

2.1. System Model

Consider a large-scale uplink multi-cell multi-user network, where M single-antenna base stations (BSs) are collected by the set $\mathcal{M} = \{1, 2, \ldots, M\}$ are deployed at the center of M cells, respectively. Assume that there are N users collecting by the set $\mathcal{L}_m = \{l_{m,1}, l_{m,2}, \ldots, l_{m,N}\}$ in each cell m, where $l_{m,n}$ denotes the index of the n-th user in the m-th BS. The total users of the considered system are collected by the set $\mathcal{K} = \{1, 2, \ldots, K\}$, where $K = MN$. The total bandwidth of the considered system is denoted as W and is divided into three widths, which are collected by the set $\mathcal{B} = \{B_i\} = \{15\text{kHz}, 30\text{kHz}, 60\text{kHz}\}$, where $i \in \{1, 2, 3\}$ [40]. Let $\mathcal{X}_i = \{1, 2, \ldots, X_i\}$ denote the set of the sub-bands of the width i of the bandwidth, where X_i is the total number of allocated bandwidth of width i.

Since users in different cells would occupy the same frequency band when transmitting their uplink signals, there exists interference between these users. This interference is called co-channel interference [41]. In this paper, each cell occupies the same frequency band and serves the same number of users N. For each user $l_{m,n}$, some users in the neighboring cells can cause co-channel interference. In other words, users in the same cell can use different frequency band sub-carriers, and, thus, each user is subject to co-channel interference from users in other cells. Let $\mathcal{M}' = \{l_{m',n} | m' \in \mathcal{M}, m' \neq m\}$ denote the set of interfering users. Thus, these users from different cells belonging to the set $\mathcal{M}'$ will interfere with user $l_{m,n}$. The channel gain between user $l_{m',n}$ and BS m at the slot t is represented by the following: $g(d_{l_{m',n},m}) = h_{l_{m',n}} [\beta(d_{l_{m',n},m})]^{\frac{1}{2}}$, where $\beta(d_{l_{m',n},m}) = 10^{\frac{PL_{l_{m',n}} + \sigma_\beta z_\beta}{10}}$ is the large scale fading corresponding to the distance $d_{l_{m',n},m}$ between user $l_{m',n}$, BS m, $PL_{l_{m',n}}$ is the path loss of user $l_{m',n}$, σ_β is the standard deviation of shadow fading, $z_\beta \sim \mathcal{N}(0,1)$ is a Gaussian random variable, and $h_{l_{m',n}} \sim \mathcal{CN}(0, \sigma_h^2)$ is the small-scale fading with variance σ_h^2. Then, the power of co-channel interference on user $l_{m,n}$ is expressed as follows:

$$I_{l_{m,n}} = \sum_{m' \in \mathcal{M}'} g\left(d_{l_{m',n},m}, t\right) p_{l_{m',n}}, \tag{1}$$

where $p_{l_{m',n}}$ denotes the transmit power for user $l_{m',n}$.

The signal $y_{l_{m,n}}$ received by BS m from user $l_{m,n}$ can be written as

$$y_{l_{n,m}} = x_{l_{n,m}} + I_{l_{m,n}} + n_{l_{m,n}}, \tag{2}$$

where $x_{l_{m,n}} = b_{l_{m,n}} |g(d_{l_{m,n},m}, t)| p_{l_{m,n}}$ denotes the transmitted signal by user $l_{m,n}$, $b_{l_{m,n}}$ is the transmitted symbol from user $l_{m,n}$ to BS m, and $n_0 \sim \mathcal{CN}\left(0, \sigma^2_{l_{m,n}}\right)$ is the additive white complex Gaussian noise. As a result, the received signal-to-interference-plus-noise ratio (SINR) at BS m of user $l_{m,n}$ is given by

$$\xi_{l_{m,n}}\left(\mathbf{p_{l_{m,n}}}, B_{i,l_{m,n}}\right) = \frac{p_{l_{m,n}} g(d_{l_{m,n},m}, t)}{\sigma^2_{l_{m,n}} + I_{l_{m,n}}}, \tag{3}$$

where $\sigma^2_{l_{m,n}} = n_f B_{i,l_{m,n}}$ indicates the variance of the Gaussian white noise, and n_f is the power spectral density of noise. $\mathbf{p_{l_{m,n}}}$ is the power vector that includes the power of user $l_{m,n}$ and its interfering users, and $B_{i,l_{m,n}}$ is the i-th width of the bandwidth allocated to the user $l_{m,n}$. Then, by considering the normalized rate [42], the achievable throughput of user $l_{m,n}$ at BS m is

$$TH_{l_{m,n}} = \log_2\left(1 + \xi_{l_{m,n}}\left(\mathbf{p_{l_{m,n}}}, B_{i,l_{m,n}}\right)\right). \tag{4}$$

2.2. Problem Formulation

This paper mainly focuses on maximizing the average throughput of the considered large-scale multi-cell multi-user system subject to QoS of all users by jointly optimizing the transmit power and bandwidth allocation of all the users. Denote the average throughput of all the users by $\overline{TH}$; then, the joint resource allocation problem is formulated as follows:

$$\begin{aligned} \text{P1}: \ & \max_{\mathbf{p_{l_{m,n}}}, B_{i,l_{m,n}}} \ \overline{TH} \triangleq \frac{1}{K} \sum_{m=1}^{M} \sum_{n=1}^{N} TH_{l_{m,n}} \\ s.t.\ & \text{C1}: P_{\min} \le p_{l_{m,n}} \le P_{\max}, \forall l_{m,n} \in \mathcal{L}_m, m \in \mathcal{M}, \\ & \text{C2}: \sum_{i=1}^{3} B_i X_i \le W, \\ & \text{C3}: TH_{l_{m,n}} \ge TH^{\text{th}}, \forall l_{m,n} \in \mathcal{L}_m, m \in \mathcal{M}, \end{aligned} \tag{5}$$

where $P_{\min}$ and $P_{\max}$ are the minimum and maximum transmit power of each user, respectively. Constraint C1 limits the transmit power budget per user; C2 indicates that the allocated bandwidth cannot exceed the total bandwidth of the system; and C3 ensures the QoS of each user. TH^{th} denotes the required minimum throughput. Note that $p_{l_{m,n}}$ and $B_{i,l_{m,n}}$ are the decision variables associated with user $l_{m,n}$, where $p_{l_{m,n}}$ is the allocated power of the user $l_{m,n}$, and $B_{i,l_{m,n}}$ denotes the bandwidth assigned to the user $l_{m,n}$ of width i. This paper aims at obtaining better throughput by jointly optimizing the two variables.

Problem P1 is non-convex; it is difficult to solve using traditional methods due to the high computational complexity. Furthermore, owing to the intricacy of the co-channel interference relationship in large-scale systems and the interaction between users in different cells, it is challenging to find the effective solution for joint transmission power and bandwidth allocation directly. To tackle these challenges, we proposed the JPRL method, which is excellent for the multi-cell multi-user system. In the proposed method, the MAD-DQN algorithm was used to allocate the bandwidth, and the P-MADDPG algorithm was developed to optimize the transmit power.

3. JPRL-Based Joint Resource Allocation Approach

The detailed structure of the joint uplink bandwidth and transmit power allocation is shown in Figure 2. Joint resource allocation often optimizes multiple variables consistently. However, for the problem of the joint allocation of the bandwidth and transmit power, there exist infinite combinations of joint assignment schemes that are influenced by the users interactions, thereby leading to unfortunate performance. In addition, the bandwidth assignment with limited choices is a discrete assignment scheme, rather than the continuous range such as for the power allocation. Thus, we de-coupled problem P1 into two sub-problems and designed an efficient JPRL method to solve the joint resource allocation problem in the considered large-scale multi-cell multi-user system. Specifically, the MADDQN algorithm was developed to solve the bandwidth allocation sub-problem with a discrete action space, and the P-MADDPG algorithm was designed to solve the transmit power allocation subproblem in the continuous domain. This resource assignment procedure satisfies the decentralized partially observable Markov decision process. Therefore, the proposed JPRL based on the RL method employed each user as an agent to model the optimization, which could solve large-scale resource allocation while meeting QoS constraints.

The RL can be described as a stochastic game, which is defined by a tuple $\langle \mathcal{K}, \mathcal{S}, \mathcal{A}, R, P \rangle$, where $\mathcal{K}$ is the set of agents, and $\mathcal{S}$ and $\mathcal{A}$ denote the set of states and the joint actions the space of all agents, respectively. The R is the reward function, and P is the state transition probability. The game is generally concerned with the interaction between the environment and one or more agents in a series of iterations. In each iteration, the agent observes the environmental state $\mathcal{S}$ to take action from action space $\mathcal{A}$. Thenm the agent receives an immediate reward R_t to reflect the quality of this iteration and observes a new state to the next step. Our goal was to maximize of the long-term rewards over various iterations. The details of the proposed framework are illustrated as follows.

- Agent: All users K.
- State space: The state $s_k(t)$ of agent k is denoted as its co-channel interference, and the global environment state is thus defined as a set including the state of all agents, i.e.,

$$\begin{aligned} \mathcal{S}_t &= \{s_1(t), \ldots, s_k(t), \ldots, s_K(t)\}, \\ &= \{I_{l_{1,1}}(t), \ldots, I_{l_{1,n}}(t), \ldots, I_{l_{M,N}}(t)\}. \end{aligned} \quad (6)$$

- Actions space: The actions of each agent consist of the bandwidth and power allocation and can be expressed as

$$\mathcal{A}_t = \{\left(a_1^b(t), a_1^p(t)\right), \ldots, \left(a_K^b(t), a_K^p(t)\right)\}, \quad (7)$$

  where $\mathcal{A}_t^b = \{a_1^b(t), \ldots, a_K^b(t)\}$ is defined as the bandwidth allocation, and $\mathcal{A}_t^p = \{a_1^p(t), \ldots, a_K^p(t)\}$ is defined as the power allocation of all agents.
- Reward function: Since the whole performance is influenced by all users in the considered system, the sparse reward is a serious issue. Inspired by the entire long-term evaluation mechanism, in the learning process, previous lessons are indicative of the current learning. Therefore, a novel reward function is defined as

$$R_t = \overline{TH}_t - \widetilde{TH}_{t,\tau} - c, \quad (8)$$

  where $\overline{TH}_t$ denotes the average throughput of the current step t, τ denotes the moving step, and $\widetilde{TH}_{t,\tau} = \frac{1}{\tau}\sum_{\tau=1}^{\tau}\left(\overline{TH}_{t-\tau+1}\right)$ is the moving average of $\overline{TH}_t$. c is a non-negative value. Especially, $c = 0$ if constraint C3 of Problem P1 is satisfied for all users; otherwise, $c > 0$. Unlike the typical reward functions that evaluate the single-step target by setting a threshold, the proposed reward function employs a long short-term criterion that varies autonomously as the performance over time, which allows agents to perform more stable exploration in the multi-cell multi-user system.

In the proposed JPRL method, we developed a distributed learning and centralized training framework, as shown in Figure 3, which promised to explore the entire action space fully and encourage each agent to leverage the experience of other agents. Specifically, all agents are guided by the harmonized loss feedback value of the MADDQN and P-MADDPG when learning the bandwidth and power individually. The details of the proposed JPRL method are given as follows, its structure is illustrated in Algorithm 1, and the flow chart is shown in Figure 4. In the learning phase, the state of each agent is input into the the MADDQN and P-MADDPG algorithms synchronously, and then each agent individually performs the bandwidth allocation and power allocation actions. Based on the actions, rewards and new states are generated and stored in the replay buffers of the two algorithms. Note that the reward is calculated by Equation (8), which corresponds to all actions of the bandwidth and power. In the training phase, the values in the buffer are randomly selected to compute correlation values to guide the intelligence in the direction of increasing throughput. The details are described as follows.

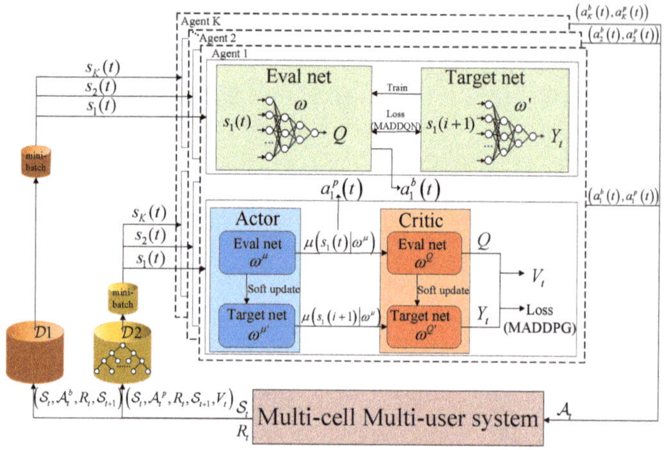

Figure 2. System model of the JPRL-based bandwidth and power allocations.

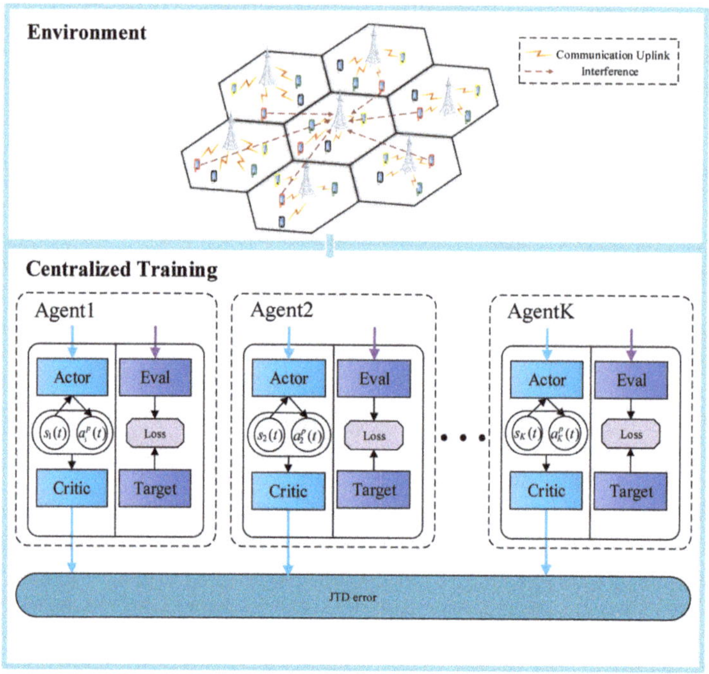

Figure 3. Framework of centralized training.

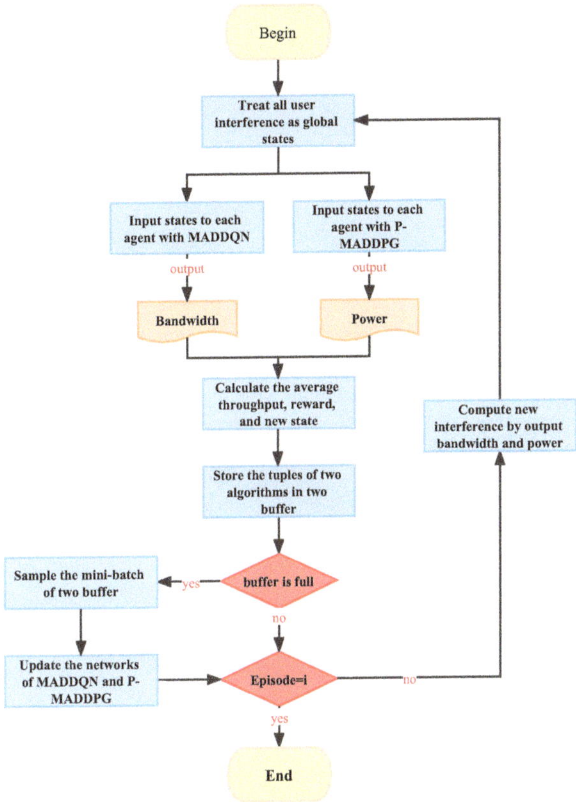

Figure 4. Flow chart of the JPRL method.

3.1. Bandwidth Allocation of MADDQN

Bandwidth allocation is a non-convex problem with discrete space; there are finite choices. The size of the action space grows exponentially with the number of users. Therefore, a MADDQN algorithm with centralized training was presented to achieve sufficient exploration of the actions, which had good performance in large-scale discrete action spaces.

A MADDQN model consists of a target Q network and an evaluated Q network, which creates a copy of neural network for the two networks, respectively. For multiple agents, an arbitrary agent taking actions to improve its performance could lead to the degradation of the overall performance as the agents are interacting with each other. Therefore, the effect of mutual synergy between agents cannot be ignored. A centralized training architecture, to this end, denotes a joint action-state function Q^b_{sum} that composes the action-state functions from different agents to promote cooperation between agents. The concrete formula is defined as

$$Q^b_{\text{sum}}\left(\mathcal{S}_t, \mathcal{A}^b_t\right) = \sum_{k=1}^{K} Q^b_k(s_k(t), a^b_k(t) \mid \omega), \tag{9}$$

where ω is the parameter of the evaluated Q network. Q^b_k is the k-th user's action-state function based on its own state. In the training phase, the joint action-state function is used for back propagation to promote cooperation, and a mini-batch sample is randomly sampled from the replay memory $\mathcal{D}1$ that stores the states, actions, next states, and rewards of all the agents (note that all the agents have the same reward value) to minimize the loss function, which is written as

$$L = \mathop{\mathbb{E}}_{(\mathcal{S}_t,\mathcal{A}_t^b,\mathcal{S}_{t+1},R_t)\sim\mathcal{D}1}\left[\left(Y_t - Q_{\text{sum}}^b\left(\mathcal{S}_t,\mathcal{A}_t^b \mid \omega\right)\right)^2\right], \tag{10}$$

where $\mathbb{E}[\bullet]$ denotes the mathematical expectation and

$$Y_t = R_t + \gamma_1 \underset{\mathcal{A}}{\text{argmax}} Q_{\text{sum}}^b\left(\mathcal{S}_{t+1},\mathcal{A}_{t+1}^b \mid \omega\right); \tag{11}$$

γ_1 is the discount ratio. For each agent, the soft updating is given by

$$\omega' \longleftarrow \eta\omega + (1-\eta)\omega', \eta \in (0,1), \tag{12}$$

where ω' are the parameters (including the weights and biases) of target Q network.

In the multi-cell multi-user system, the MADDQN model of agent k chooses the bandwidth assignment action according to its own state $s_k(t)$ in step t. Note that the agents can share their past training process (state, the influence based on training). Then, all the agents are centralized trained to minimize the loss value by Q_{sum}^b.

3.2. P-MADDPG-Based Uplink Power Allocation

For power allocation, a huge action space is not helpful for exploitation. In addition, although the discrete DRL algorithms can quantize power, they ignore the diversity of power choices. To this end, a novel P-MADDPG algorithm was proposed to solve the transmit power allocation subproblem. This is an enhancement of the DDPG with a prioritized replay buffer. In contrast to the power quantization, the P-MADDPG directly outputs the power of all the users in a continuous domain with infinite choices. Furthermore, by applying the prioritized replay buffer, it is more sensitive to the negative effect of the bad actions than the general MADDPG algorithms.

Similar to DDPG, an actor-critic architecture [43] applies for learning and training; both the actor and critic networks of each agent contain two identical neural networks, which are named the online network and target network, respectively. For a multi-agent system, the actor network of agent k outputs the power allocation under the current state through a policy π, i.e., $a_k^p(t) = \pi(s_k(t))$. However, the inherent exploration–exploitation dilemma in the DRL is prevalent for an inflexible action policy. By taking advantage of the DQN, it is balanced by a stochastic noise whose function is similar to the $\epsilon - \textit{greedy}$ mechanism. Consequently, the actions of all agents are written as

$$\mathcal{A}_t^p = [\pi(\mathcal{S}_t \mid \omega^\mu) + \Sigma_t]_{P_{\min}}^{P_{\max}}, \tag{13}$$

where ω^u is the weight of the actor network, and Σ_t follows a Gaussian distribution $\mathcal{N}(0,\varrho)$; ϱ is the variance of Gaussian noise and decreases linearly to zero as the iteration proceeds. Similarly, applying the individual action-value function to each agent ignores the features of others, which reduces the learning stability and weakens agent interaction. To this end, the critic network uses the joint action-value function $Q_{\text{sum}}^p(\mathcal{S}_t,\mathcal{A}_t)$ to evaluate all actions. The specific Q_{sum}^p is defined as

$$Q_{\text{sum}}^p(\mathcal{S}_t,\mathcal{A}_t) = \mathbb{E}_{R_t,\mathcal{S}_t\sim\mathcal{D}2}[R_t + \gamma_2 Q_{\text{sum}}^p(\mathcal{S}_{t+1},\pi(\mathcal{S}_{t+1}))], \tag{14}$$

where $\mathcal{D}2$ is the experience replay buffer, and $\gamma_2 \in (0,1]$ is a discount factor. According to the deterministic policy gradient theorem, the action-value function Q_{sum}^p is used to update the actor parameters ω^{μ} in the direction of increasing the cumulative discounted reward with D samples of a mini-batch, that is

$$\begin{aligned}\nabla_{\omega^{\mu}}\pi \approx&\mathbb{E}_{\pi'}[\nabla_{\omega^{\mu}}Q^{p}_{sum}(\mathcal{S},\mathcal{A}\mid\omega^{Q})\mid_{\mathcal{S}=\mathcal{S}_t,\mathcal{A}=\pi(\mathcal{S}_t|\omega^{\mu})}],\\ =&\mathbb{E}_{\pi'}[\nabla_{\omega^{\mu}}Q^{p}_{sum}(\mathcal{S},\mathcal{A}\mid\omega^{Q})\mid_{\mathcal{S}=\mathcal{S}_t,\mathcal{A}=\pi(\mathcal{S}_t)}\nabla_{\omega^{\mu}}\pi(\mathcal{S}\mid\omega^{\mu})\mid_{\mathcal{S}=\mathcal{S}_t}],\\ =&\frac{1}{D}\sum_{k}\nabla_{a^{p}_{k}(t)}Q^{p}_{\text{sum}}\left(\mathcal{S}_t,\mathcal{A}^{p}_{t}\mid\omega^{Q}\right)|\nabla_{\omega^{\mu}}\pi(s_k(t)\mid\omega^{\mu})|_{s_k(t)},\end{aligned} \tag{15}$$

where ω^{Q} is the weight of critic network.

A common method for training neural networks is to randomly and uniformly sample mini-batches from the buffer D2, which often results in a high probability of selecting bad actions among the vast combinations of different actions, thereby lowering performance. This method is inefficient and poorly helpful for guiding the networks to update in the correct direction. Considering the transition samples of all agents, we designed the P-MADDPG algorithm to enhance the MADDPG by customizing a prioritized experience replay technique, where the more important transition samples have a higher probability of being replayed to participate in network updating. Specifically, in each step t, the transition samples of all agents are measured by the corresponding importance denoted by V_t, which is combined with $\mathcal{S}_t$, $\mathcal{A}^{p}_{t}$, R_t, and $\mathcal{S}_{t+1}$ to form a tuple $\left(\mathcal{S}_t,\mathcal{A}^{p}_{t},R_t,\mathcal{S}_{t+1},V_t\right)$ being stored in $\mathcal{D}2$. Similar to the MADDQN, the goal of P-MADDPG updating is to minimize the magnitude between the joint Q-value and target joint Q-value, i.e., joint temporal-difference (JTD) error. The transitions with the large JTD error contain more information and are more necessary to the update of neural networks. Thus, the JTD error is a reasonable proxy measure of important value, and V_t is written as

$$V_t = |Y_t - Q^{p}_{\text{sum}}\left(\mathcal{S}_t,\mathcal{A}^{p}_{t}\mid\omega^{Q}\right)|. \tag{16}$$

However, in the sampling process, initially stored transition samples with small JTDs may not be sampled to replay if the sampling only relies on the importance. This can result in over-fitting, since the system lacks the sampling diversity of transitions. To avoid the issue, a probability associated with the importance is assigned to each transition sample, which can overcome the above issues effectively. The probability of the arbitrary transition sample φ at the step t is expressed as

$$P^{\varphi}_{t} = \frac{\left(V^{\varphi}_{t}\right)^{\alpha}}{\left(\sum_{\varphi=1}^{|\mathcal{D}2|}(V^{\varphi}_{t})\right)^{\alpha}}, \tag{17}$$

where $\alpha \in [0,1]$ is a contribution factor that controls the impact of importance. In particular, $\alpha = 0$ means that all samples are equally distributed, i.e., no contribution is made according to importance (uniform sampling). Original samples are equally probability-distributed in the replay buffer, but prioritized experience replay introduces bias, since it changes the original distribution by assigning different probabilities to the transitions. The compensation weight is thus introduced to correct this bias, which is expressed as

$$\lambda^{\varphi}_{t} = \left(\frac{1}{D}\frac{1}{P^{\varphi}_{t}}\right)^{\beta}, \tag{18}$$

where $\beta \in [0,1]$ is a hyperparameter, which regulates the degree of bias compensation. In particular, there is no compensation for non-uniform probabilities P^{φ}_{t} if $\beta = 0$; there ispartial compensation if $0 < \beta < 1$; and thre is full compensation if $\beta = 1$. As a result, The loss of a mini-batch φ after weight compensation is rewritten as

$$L = \mathop{\mathbb{E}}_{\left(\mathcal{S}_t, \mathcal{A}_t^p, \mathcal{S}_{t+1}, R_t, V_t^{\varphi}\right) \sim \mathcal{D}2} \left[\lambda_t^{\varphi} \left(V_t^{\varphi} \right)^2 \right]. \tag{19}$$

Algorithm 1 JPRL methodfor joint bandwidth and power allocation

Initialize:

Initialize the network parameters in MADDQN and P-MADDPG respectively, ω, ω^Q;
Initialize the replay buffer $\mathcal{D}1$ and the prioritized replay buffer $\mathcal{D}2$, $|\mathcal{D}_1|$, $|\mathcal{D}_2|$;
Initialize a sum tree for $\mathcal{D}2$, α, β.

Excute:

for episode $i = 1, \cdots, I$ **do**
  Receive initial observation state of all agents K, and input $s_k(t)$ to agent k.
  Initialize the actions of all agents.
  for step $t = 1, \cdots, T_i$ **do**
    for agent $k = 1, \cdots, K$ **do**
      if random number $\zeta < \epsilon_t$ **then**
        Randomly choose $a_k^b(t)$ from bandwidth allocation action space.
      else
        $a_k^b(t) = argmax_{a_k^b(t)} Q_k^b\left(s_k(t), a_k^b(t) \mid \omega\right)$.
      end if
      Choose power allocation $a_k^p(t) = [\pi(s_k(t) + \sigma_t)]_{P_{\min}}^{P_{max}}$.
      Execute actions $a_k^b(t)$, $a_k^p(t)$ and observe next state $s_k(t+1)$.
    end for
    Calculate reward with all agents' actions by Equation (8).
    Store transition with bandwidth allocation $\left(\mathcal{S}_t, \mathcal{A}_t^b, \mathcal{S}_{t+1}, R_t\right)$ in $\mathcal{D}1$.
    Store transition φ with power allocation $\left(\mathcal{S}_t, \mathcal{A}_t^p, \mathcal{S}_{t+1}, R_t, V_t^{\varphi}\right)$ in $\mathcal{D}2$.
    if Both $\mathcal{D}1$ and $\mathcal{D}2$ are full **then**
      Sample a mini-batch of transition from $\mathcal{D}1$.
      Sample a mini-batch of transition from $\mathcal{D}2$ according to sample importance.
      Compute the action-value function of MADDQN and P-MADDPQ according to Equations (9) and (14), respectively.
      Update evaluated Q network of MADDQN by Equation (10).
      Update actor online network by Equation (15).
      Update critic online network by Equation (19).
      Update the MADDQN and P-MADDPG networks by soft updating.
    end if
  end for
end for

For a mini-batch with D samples, directly traversing the experience buffer $\mathcal{D}2$ for sampling requires D times, and the complexity is intolerable. To tackle this matter, a sum-tree frame is designed for $\mathcal{D}2$, where the sample φ is stored with the sampling probability P_t^{φ}. As shown in Figure 2, the structure is a binary tree with a root node at the top, and there are only two child nodes for each node of the upper level. For the leaf nodes at the bottom, the tuple $\left(\mathcal{S}_t, \mathcal{A}_t^p, R_t, \mathcal{S}_{t+1}, V_t^{\varphi}\right)$ of transition φ is stored with its probability according to Equation (16). Note that the value of each node is the sum of its child nodes' value. We divided the value of the root node (the sum of the probabilities of all samples) into D segments, which have an equal interval. In each interval, a random value, which is no more than the range of the interval generated to backtrack the leaf node from top to bottom. The specific backtracking rules are listed as follows, until the leaf node is selected, if the random value is less than or equal to the value of the left child node, and we continue backtracking from left child node; otherwise, we continue backtracking from the right child node and calculates the difference between this value and the value of the left child node as the basis for the next backtracking. Then, the critic and actor networks are updated by the selected transition samples. The proposed JPRL method is summarized in Algorithm 1.

3.3. Time Analysis of the Proposed JPRL Method

We analyzed the time complexity of our proposed JPRL method. In Algorithm 1, let I be the total number of training episodes and T_i be the training steps in the episode i. Therefore, the total amount of training iterations implies the time complexity, that is $\mathcal{O}\left(\sum_{i=1}^{I} iT_i\right)$. For each iteration, the computational efficiency is subjected to the the size of the neural network, i.e., the number of parameters. According to [44], the time complexity for a fully connected layer is $\mathcal{O}\left(\sum_{l=1}^{L} c_l c_{l-1}\right)$, where l is the fully connected layer and c_l denotes the number of neural units in layer l. In the JPRL method, each agent utilizes two algorithms to output the bandwidth and power actions. Note that the two algorithms are run simultaneously. Thus, the time complexity is $c = \mathcal{O}\left(\max\left(\sum_{\{\mathrm{MADDQN,P-MADDPG}\}}{}_{l=1}^{L} c_l c_{l-1}\right)\right)$. The total time complexity of the JRPL method is $\mathcal{O}\left(c\sum_{i=1}^{I} iT_i\right)$.

4. Simulations

In this section, we evaluate the performance of the proposed JPRL method. First of all, the simulation setup is portrayed. Then, the experience results are discussed in terms of the convergence, learning rate analysis, and performance comparison. Lastly, the performance of our proposed method compared to different models is exhibited.

4.1. Setup

Parameter Setting of Environment: We set the location of seven base stations in the cell center, and four users wererandomly distributed in each cell. The uplink user power was limited to $P_{\min} = -40$ dBm and $P_{\max} = 23$ dBm [40]. The total bandwidth of the system was $W = 20$ MHz. The minimum throughput requirement of all the users was $TH^{\mathrm{th}} = 0.15$ bit/s, and the power spectral density n_f was -174 dBm/Hz.

The size of the cells and channel model change according to different scenarios [39], which are referenced from the test scenior in the 3GPP protocol, such as UMa, RMa, RMi. By default, the outsider scenario of the non-line-of-sight of the RMa was selected to evaluate the proposed method. The RMa stipulates the radius of cell r, and the pathloss is defined as

$$PL_{l_{m,n}} = \max\left(PL_{l_{m,n},1}, PL_{l_{m,n},2}\right), \tag{20}$$

where $PL_{l_{m,n},1}$ and $PL_{l_{m,n},2}$ denote the line-of-sight and non-line-of-sight pathloss, respectively, which are written as

$$PL_{l_{m,n},1} = \begin{cases} PL_{l_{m,n},11}, & 10\ \mathrm{m} < d_h < d_{\mathrm{BP}}, \\ PL_{l_{m,n},12}, & d_{\mathrm{BP}} < d_h < 5\ \mathrm{km}, \end{cases} \tag{21}$$

where

$$\begin{aligned} PL_{l_{m,n},11} = {} & \min(0.03h_{\mathrm{b}}^{\varepsilon}, 10)\lg(d_{\mathrm{s}}) - \min(0.044h_{\mathrm{b}}^{\varepsilon}, 14.77) \\ & + 0.002\lg(h_{\mathrm{b}})d_{\mathrm{s}} + 20\lg(40\pi d_{\mathrm{s}} f_{\mathrm{c}}), \end{aligned} \tag{22}$$

$$PL_{l_{m,n},12} = PL_{l_{m,n},11} + 40\lg\left(\frac{d_{\mathrm{s}}}{2\pi h_{\mathrm{a1}} h_{\mathrm{a2}} f_{\mathrm{c}}/v}\right), \tag{23}$$

and

$$\begin{aligned} PL_{l_{m,n},2} = {} & 161.4 - 7.11\lg(l_{\mathrm{w}}) + 7.5\lg(h_{\mathrm{b}}) - \left(24.37 - 3.7\frac{h_{\mathrm{b}}^2}{h_{\mathrm{a1}}^2}\right)\lg(h_{\mathrm{a1}}) \\ & + (43.42 - 3.1\lg(h_{\mathrm{a1}}))(\lg d_{\mathrm{s}} - 3) + 20\lg(f_{\mathrm{c}}) - 10.24(\lg(11.75h_{\mathrm{a2}}))^2 + 4.97. \end{aligned} \tag{24}$$

Here, $d_s = \sqrt{d_h^2 + (h_{a1} - h_{a2})^2}$ and $d_h = d_{l_{m,n},m}$ denote the straight distance and horizontal distance between BS and user respectively, where h_{a1} and h_{a2} are the heights of the antenna in the BS and user, respectively. h_b is the building height, l_w is the average width of the road, and ε is the excitation factor. For the long distance line-of-light pathloss $PL_{l_{m,n},12}$, f_c is the central frequency, and v denotes the propagation velocity. These parameter settings are listed in Table 1. In this paper, the five benchmarks were considered:

(1) DDQN and DDPG: The existing DDQN for bandwidth assignment and the DDPG for allocating the power. The architecture with a one-layer fully connected network was used in the DDQN, and the DDPG deployed two-layer fully connected networks in the actor and critic networks. Both of them adopted the uniform sampling-based experience replay.

(2) DDQN andP-DDPG: The settings were the same as (1), except that the DDPG used the prioritized experience replay.

(3) MADDQN and MADDPG(ct): We treated each user as an agent and deployed the DDQN and DDPG on each agent. Centralized training was adopted.

(4) MADDQN and MADDPG(dt): The MADDQN and MADDPG with distributed training. Note that each agents had the exclusive reward and loss.

(5) Genetic algorithm (GA): The GA framework in the DEAP was used to realize this benchmark [45]. The bandwidth and power allocation schemes were encoded into the chromosome of each individual, which is the action sequence about the bandwidth and power allocation of all the users. We set the population size to 200. The crossover rate and mutation rate were set as 0.8 and 0.05, respectively.

Note that the GA depends on the fitness rather than the learning-based reward; thus it is appropriate to compare the results after final convergence instead of comparing the entire optimization process with the learning-based approach.

Table 1. Environmental parameters.

Parameters	Values	Description
M	7	The number of cells
N	4	The number of users per cell
P_{min}	−40 dBm	The minimum transmitting power
P_{max}	23 dBm	The maximum transmitting power
W	2 GHz	The total bandwidth
TH^{th}	0.15 bit/s	The minimum throughput constraint
n_f	−174 dBm/Hz	The power spectral density of noise
r	866 m	The radius of cells
h_b	10 m	The average height of building
h_{a1}	10 m	The antenna height of BS
h_{a2}	1.5 m	The antenna height of user
ε	1.72	Excitation factor
f_c	1 GHz	The center frequency
v	3×10^8 m/s	The propagation velocity
l_w	20 m	The average road width

Hyperparameter Setting of JPRL: The JPRL method contains an MADDQN algorithm and a P-MADDPG algorithm. There are the same size of the experience buffer for the two algorithms, which are set to $|\mathcal{D}1| = |\mathcal{D}2| = 10000$. The learning rate, including the MADDQN, actor network, and critic network of the P-MADDPG was set as 0.0001. Furthermore, we set the hyperparameters of the prioritized replay buffer in the P-MADDPG $\mathcal{D}2$ as $\alpha = 1$ and $\beta = 0.1$. In the training phase, both the MADDQN and P-MADDPG used the Adam optimizer to optimize the loss function. The sampling batch size was $D = 128$, and the reward discount factor was $\gamma_1 = \gamma_2 = 0.89$. The system began to train the neural network when the memory buffer was full, and it updated the neural parameters at one-step frequency after training. Besides, we set the number of episodes to $I = 500$. Note that every episode did not have fixed steps. To determine whether an episode was completed, a done flag was designed, where the done flag was true if the reward R increased by 200 steps; otherwise, it was false (the learning of this episode was not finished). the other parameters of each neural network are listed in Table 2.

All experiences were operated by a computer with the 12-th Gen Intel(R) Core(TM) i7-12700F @2.10 GHz, 16-GB RAM. The simulation results were presented using *Numpy 1.21.5* and *Tensorflow 2.3.0* on the *Python 3.6 platform*.

Table 2. Network parameters.

Network	Neural Units	Activation	Optimizer
MADDQN	64	sigmoid	Adam Optimizer
Actor Network of P-MADDPG	32, 16	tanh	Adam Optimizer
Critic Network of P-MADDPG	32, 16	ReLu	Adam Optimizer

4.2. Results

The setting of the learning rate has a profound impact on the learning of the distribution scheme of the proposed method, which determines the ability to explore action space. Specifically, higher learning rates are detrimental to the exploration of the action space, as well as to the updating of network parameters in large systems with large action spaces. Moreover, in large systems with large action spaces, a lower learning rate implies finer-grained exploration, which does not mean that better actions can be explored, since having more actions in a large action space degrades performance. Thus, it is necessary to study the setting of the learning rate in a multi-cell multi-user system. Firstly, Figure 5 compares the loss values of the multiple networks under different learning rates. In order to view the variation and performance clearly, the loss values within the first 3000 steps after training are given. Figure 5a–c imply an interaction between the MADDQN and P-MADDPG in the proposed method. It is worth noting that the curve values in Figure 5b show a clear loss reduction in the P-MADDPG with a lower learning rate. It reveals the fact that the MADDQN exploring bandwidth influenced the P-MADDPG training. However, as shown in Figure 5c, a decrease in loss value did not signify an increase in throughput, and it may have also been trapped in sub-optimality. As a result, we set the learning rate of the MADDQN to 0.0001 to achieve a high throughput and fast convergence speed of the P-MADDPG.

Figure 6 illustrates the loss values and throughput of the actor and critic networks in the P-MADDPG at different learning rates. For the learning rates of the actor network, the proposed JPRL achieved the best in terms of the loss value and throughput whenthe learning rate was 0.0001. The loss curves of the MADDQN in Figure 6a show a slight increase after 1000 steps, and a similar trend appears in Figure 6d. The reason is that the power actions selected from the P-MADDPG affected the training process of the MADDQN. As shown in Figure 6b,c, it is noticed that, the smaller the learning rate, the better the performance, since the larger learning rate may skip various actions within the infinite action space. Finally, from Figure 6d–f, in the large action spaces, the critic network with a higher learning rate converged faster but converged to a worse value. The reason is that a larger learning rate of the critic network implies a more coarse-grained exploration, which is prone to learning sub-optimality. As a result, when the learning rate of the actor and critic networks were set to 0.0001, our method could jump out of the local optimal.

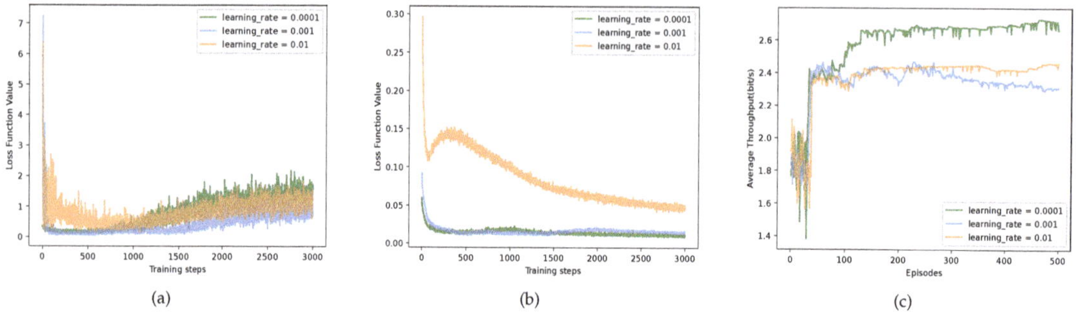

Figure 5. The loss value and throughput under different learning rates of MADDQN. The learning rates of both actor networks and critic networks of P-MADDPG are set to 0.0001, and the loss value was extracted at 3000 steps after the beginning of network training. (**a**) MADDQN loss function value. (**b**) P-MADDPG loss function value. (**c**) Average throughput.

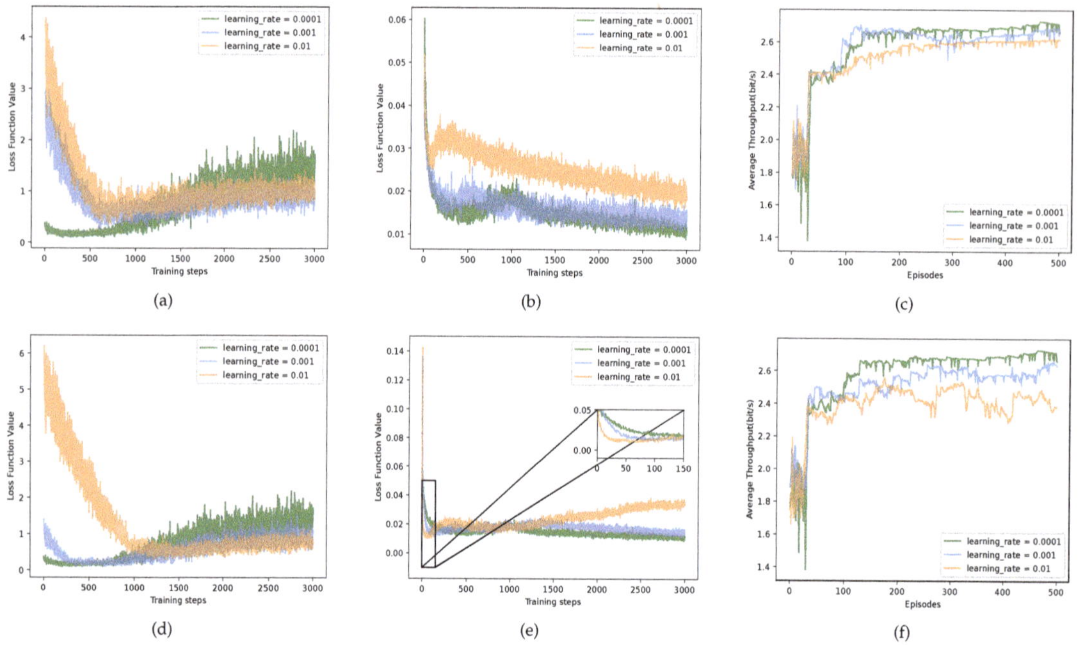

Figure 6. The loss function value and throughput of the two networks of P-MADDPG under different learning rates. (**a**–**c**) show the effect of variable learning rates on actor network, and (**d**–**f**) are the exhibitions within the changing learning rates of critic network. Note that the learning rates of other networks were set as default when a network varied in learning rate.

With respect to recording the reward for every 200 steps, Figure 7 plots the reward values of the proposed method and benchmarks; the benchmarks included the DDQN and DDPG, DDQN and P-DDPG, MADDQN and MADDPG based on the centralized training (ct) and decentralized training (dt). In the process of early random exploration (before the buffers are full), rewards decrease to negative values. The reason is that there are users whose throughput does not meet the QoS requirement. As the system begins to train, all five curves have a sharp augment. After a period of training, the moving average of the average throughput $\widetilde{TH}_{t,\tau}$ will be close to the average throughput $\overline{TH}_t$, e.g., the reward is close to 0, which indicates that the methods fall into a local optimal or converge to an optimal. It is seen that the curve of the MADDQN and MADDPG(dt) swung more than that of the MADDQN and MADDPG(ct). As a result, Figure 7 indicates that the JPRL method has an excellent ability to jump out of sub-optimal conditions and obtain good feedback.

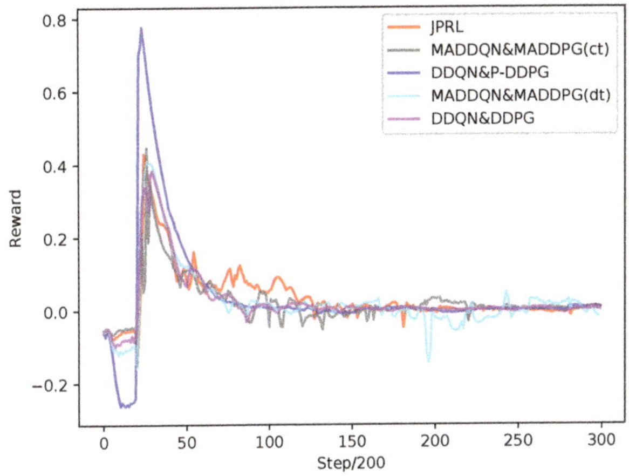

Figure 7. Reward comparison of different methods as the number of steps increased.

Figure 8 illustrates the average throughput of the different methods after 500 episodes. In the random exploration stage, the throughput is unstable and relatively small because of the impacts of Gaussian noise and the randomly selected actions. All methods are prone to get stuck in the local optimum during the learning process, and there is a small fluctuation for the average throughput because of the existence of the Gaussian noise. Since a small change in power of any user may cause a large variation for co-channel interference, the benchmarks fall into the local optimum easily and are difficult to jump out of it. We can also see that the joint method MADDQN and MADDPG(dt) was extremely unstable, since the distributed training favored the individual performance of the agent at the expense of the overall performance. In other words, an agent, which follows its own wishes while neglecting the other characteristics for increasing power, will increase interference and decrease throughput. It was observed that the proposed JPRL outperformed the other methods in terms of throughput, since it explored the action spaces fully.

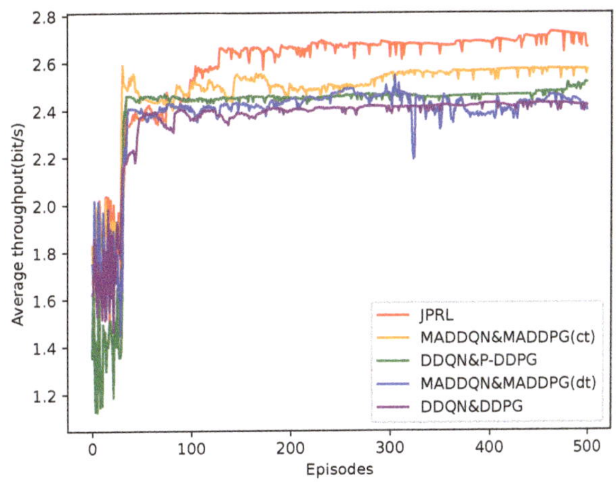

Figure 8. Average throughput comparison of different methods as the number of episodes increased.

Figure 9 depicts a comparison of the average throughput for the six methods versus the cell number M. It should be observed that the average throughput decreased as the number of cells M increased. This is because fewer cells mean less interference from users $l_{m,n}$, which leads to a lower amount of co-channel interference. Obviously, it can be seen that the RL-based approach was far superior to the GA, which is because the GA fell into the local optimum easily. We also see that the proposed JPRL had a steeper curve than the others, since it had better exploration in the small action spaces as cells decreased. Therefore, the JPRL method could achieve the high throughput.

As shown in Figure 10, we further tested the average throughput of the proposed JPRL under some different channel models, including the RMa, RMi, and UMa. The average throughput of the users for the urban environment (UMa model) is generally less than that of the users in rural scenarios (RMa and RMi models). This is because severe interference is caused by a lot of users in a small range. It can be seen that the JPRL method is universally applicable to different environments.

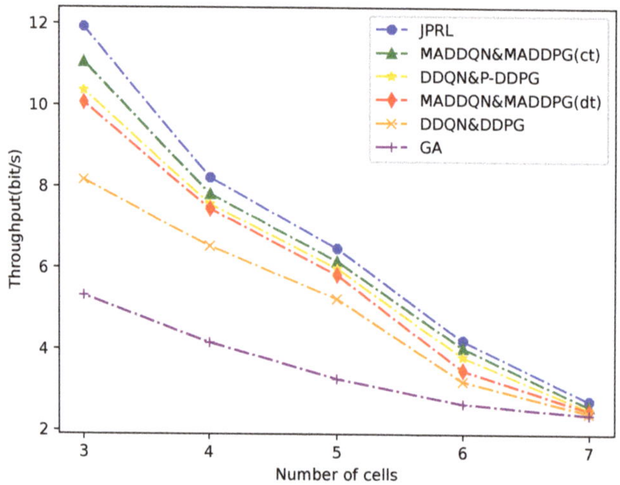

Figure 9. Comparison of average throughput for the different methods versus the number of cells.

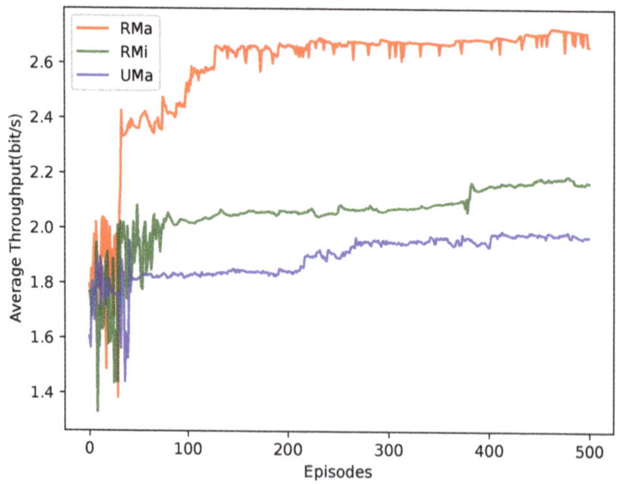

Figure 10. Comparison of average throughput for the different channel models versus the number of episodes.

5. Conclusions

This paper mainly studied the resource allocation to maximize the throughput by jointly optimizing the bandwidth assignment and power allocation subject to the QoS constraint for the multi-cell multi-user uplink system. According to the variable attributes of the joint resource allocation problem, we proposed a JPRL method to decouple the optimization problem into two sub-problems, where the MADDQN was used to allocate bandwidth, and the P-MADDPG assigned uplink power with the given importance of transition. In order to compare the loss value and learning performance of the different networks with various learning rates, we set the appropriate parameters for the proposed JPRL method and analyzed the impact of the different learning rates. Furthermore, we evaluated the reward value and throughput of the proposed JPRL method against other existing methods. the simulation results showed that our approach can (1) obtain a better performance and be more applicable to the complex environments than other alternative methods (e.g., the average throughput was approximately 10.4–15.5% better than the average throughput of the benchmarks.) and (2) be universally applicable to other large-scale scenarios.

It is worth noting that, for simplicity, the single antenna system was used in this work. As for multi-antenna systems such as MIMO, the impact of more complex channel matrices caused by multiple antennas on user interference needs to be considered. In future work, the multiple antennas, the users' trajectory, and cloud computing will be taken into consideration in multi-cell systems to facilitate communication–computing integration. By considering the interference corresponding to the complex channel matrix, the optimization is relevant to the compromised performance of the computing delay and energy consumption, which is based on the resource allocation and task offloading under various constraints, such as QoS constraints and offloading decisions. Moreover, multi-dimensional and deep analysis will be researched to validate the system tradeoff.

Author Contributions: Conceptualization, C.Z. and T.L.; methodology, P.H.; software, Z.L.; validation, C.Z., T.L., and P.H.; formal analysis, J.Z.; investigation, Y.R.; resources, T.L.; data curation, C.Z.; writing—original draft preparation, C.Z.; writing—review and editing, T.L.; visualization, C.Z.; supervision, T.L.; project administration, Z.L.; funding acquisition, Z.L. All authors have read and agreed to the published version of the manuscript.

Funding: This research was funded in part by the National Natural Science Foundation of China under Grant number 62271068 and 61827801, the Beijing Natural Science Foundation under Grant number L222046, and the Basic Scientific Research Project under Grant NS2022046.

Institutional Review Board Statement: Not applicable.

Informed Consent Statement: Not applicable.

Data Availability Statement: Not applicable.

Conflicts of Interest: The authors declare no conflict of interest.

References

1. Liu, G.; Cai, B.; Xie, W. Research on 5G Wireless Networks and Evolution. In Proceedings of the IEEE International Symposium on Broadband Multimedia Systems and Broadcasting (BMSB), Chengdu, China, 4–6 August 2021; pp. 1–5.
2. Shah, A.F.M.S.; Qasim, A.N.; Karabulut, M.A.; Ilhan, H.; Islam, M.B. Survey and Performance Evaluation of Multiple Access Schemes for Next-Generation Wireless Communication Systems. *IEEE Access* **2021**, *9*, 113428–113442.
3. Song, H.J.; Lee, N. Terahertz Communications: Challenges in the Next Decade. *IEEE Trans. Terahertz. Sci. Technol.* **2022**, *12*, 105–117.
4. Vaezi, M.; Azari, A.; Khosravirad, S.R.; Shirvanimoghaddam, M.; Azari, M.M.; Chasaki, D.; Popovski, P. Cellular, Wide-Area, and Non-Terrestrial IoT: A Survey on 5G Advances and the Road Toward 6G. *IEEE Commun. Surv. Tutor.* **2022**, *24*, 1117–1174.
5. Sharma, S.K.; Wang, X. Toward Massive Machine Type Communications in Ultra-Dense Cellular IoT Networks: Current Issues and Machine Learning-Assisted Solutions. *IEEE Commun. Surv. Tutor.* **2020**, *22*, 426–471.
6. Mughees, A.; Tahir, M.; Sheikh, M.A.; Ahad, A. Energy-Efficient Ultra-Dense 5G Networks: Recent Advances, Taxonomy and Future Research Directions. *IEEE Access* **2021**, *9*, 147692–147716.

7. Mardian, R.D.; Suryanegara, M.; Ramli, K. Measuring Quality of Service (QoS) and Quality of Experience (QoE) on 5G Technology: A Review. In Proceedings of the IEEE International Conference on Innovative Research and Development (ICIRD 2019), Jakarta, Indonesia, 28–30 June 2019; pp. 1–6.
8. Xu, Y.; Gui, G.; Gacanin, H.; Adachi, F. A Survey on Resource Allocation for 5G Heterogeneous Networks: Current Research, Future Trends, and Challenges. *IEEE Commun. Surv. Tutor.* **2021**, *23*, 668–695.
9. Lin, M.; Zhao, Y. Artificial intelligence-empowered resource management for future wireless communications: A survey. *China Commun.* **2020**, *17*, 58–77.
10. Bossy, B.; Kryszkiewicz, P.; Bogucka, H. Energy-Efficient OFDM Radio Resource Allocation Optimization with Computational Awareness: A Survey. *IEEE Access* **2022**, *10*, 94100–94132.
11. Xu, J.; Niu, H.; Zhao, T.; Gao, X.; Ye, J.; Liang, C.; Liu, Z.; Liang, J. Robust Optimal Power Control and Subcarrier Allocation in Uplink OFDMA Network With Assistance of Mobile Relay. *IEEE Access* **2021**, *9*, 57475–57485.
12. Liu, R.; Yu, G.; Yuan, J.; Li, G.Y. Resource Management for Millimeter-Wave Ultra-Reliable and Low-Latency Communications. *IEEE Trans. Commun.* **2021**, *69*, 1094–1108.
13. Sun, Q.; Wu, H.; Petrosian, O. Optimal Power Allocation Based on Metaheuristic Algorithms in Wireless Network. *Mathematics* **2022**, *10*, 3336.
14. Cao, L.; Wang, Z.; Wang, Z.; Wang, X.; Yue, Y. An Energy-Saving and Efficient Deployment Strategy for Heterogeneous Wireless Sensor Networks Based on Improved Seagull Optimization Algorithm. *Biomimetics* **2023**, *8*, 231.
15. Zeng, M.; Nguyen, N.P.; Dobre, O.A.; Ding, Z.; Poor, H.V. Spectral-and energy-efficient resource allocation for multi-carrier uplink NOMA systems. *IEEE Trans. Veh. Technol.* **2019**, *68*, 9293–9296.
16. Sharif, Z.; Jung, L.T.; Razzak, I.; Alazab, M. Adaptive and Priority-Based Resource Allocation for Efficient Resources Utilization in Mobile-Edge Computing. *IEEE Internet Things J.* **2023**, *10*, 3079–3093.
17. Xue, J.; An, Y. Joint Task Offloading and Resource Allocation for Multi-Task Multi-Server NOMA-MEC Networks. *IEEE Access* **2021**, *9*, 16152–16163.
18. Brahmi, I.; Koubaa, H.; Zarai, F. Genetic Algorithm based Resource Allocation for V2X Communications. In Proceedings of the International Conference on Communications and Networking, ComNet, Hammamet, Tunisia, 27–30 October 2020; pp. 1–5.
19. Dun, H.; Ye, F.; Jiao, S.; Li, Y.; Jiang, T. The Distributed Resource Allocation for D2D Communication with Game Theory. In Proceedings of the 2019 IEEE APS Topical Conference on Antennas and Propagation in Wireless Communications (APWC), Granada, Spain, 9–13 September 2019; pp. 104–108.
20. Li, M.; Peng, T.; Wu, H. Power Allocation to Achieve Maximum Throughput in Multi-radio Multi-channel Mesh Network. In Proceedings of the IEEE 6th International Conference on Computer and Communications (ICCC), Chengdu, China, 11–14 December 2020; pp. 293–297.
21. Wang, C.; Yan, F. Graph Theory based Resource Allocation Algorithm in Terahertz Communication Networks. In Proceedings of the 2021 IEEE International Conference on Information Networking, Jeju Island, Republic of Korea, 13–16 January 2021; pp. 304–308.
22. Xiong, Z.; Zhang, Y.; Niyato, D.; Deng, R.; Wang, P.; Wang, L.C. Deep Reinforcement Learning for Mobile 5G and Beyond: Fundamentals, Applications, and Challenges. *IEEE Trans. Veh. Technol.* **2019**, *14*, 44–52.
23. Du, Z.; Deng, Y.; Guo, W.; Nallanathan, A.; Wu, Q. Green Deep Reinforcement Learning for Radio Resource Management: Architecture, Algorithm Compression, and Challenges. *IEEE Trans. Veh. Technol.* **2021**, *16*, 29–39.
24. Han, K.; Ye, C. Power Control Research for Device-to-Device Wireless Network Underlying Reinforcement Learning. In Proceedings of the Global Conference on Robotics, Artificial Intelligence and Information Technology (GCRAIT), Chicago, IL, USA, 30–31 July 2022; pp. 351–354.
25. Mnih, V.; Kavukcuoglu, K.; Silver, D.; Rusu, A.A.; Veness, J.; Bellemare, M.G.; Graves, A.; Riedmiller, M.; Fidjeland, A.K.; Ostrovski, G.; et al. Human-level control through deep reinforcement learning. *Nature* **2015**, *518*, 529–533.
26. Liu, J.; Ma, X.; Han, W.; Wang, L. Resource Allocation in OFDMA Networks with Deep Reinforcement Learning. In Proceedings of the IEEE 8th International Conference on Information, communication and networks (ICICN), Xi'an, China, 17–20 August 2020; pp. 111–117.
27. Guan, X.; Lv, T.; Lin, Z.; Huang, P.; Zeng, J. D2D-Assisted Multi-User Cooperative Partial Offloading in MEC Based on Deep Reinforcement Learning. *Sensors* **2022**, *22*, 7004.
28. Rahman, G.M.S.; Dang, T.; Ahmed, M. Deep reinforcement learning based computation offloading and resource allocation for low-latency fog radio access networks. *Intell. Converg. Netw.* **2020**, *1*, 243–257.
29. Iqbal, A.; Tham, M.L.; Chang, Y.C. Double Deep Q-Network for Power Allocation in Cloud Radio Access Network. In Proceedings of the 2020 IEEE 3rd International Conference on Computer and Communication Engineering Technology (CCET), Beijing, China, 14–16 August 2020; pp. 272–277.
30. Lillicrap, T.P.; Hunt, J.J.; Pritzel, A.; Heess, N.; Erez, T.; Tassa, Y.; Silver, D.; Wierstra, D. Continuous control with deep reinforcement learning. *arXiv* **2015**, arXiv:1509.02971.
31. Meng, F.; Chen, P.; Wu, L.; Cheng, J. Power Allocation in Multi-User Cellular Networks: Deep Reinforcement Learning Approaches. *IEEE Trans. Wirel. Commun.* **2020**, *19*, 6255–6267.
32. Zheng, K.; Jia, X.; Chi, K.; Liu, X. DDPG-Based Joint Time and Energy Management in Ambient Backscatter-Assisted Hybrid Underlay CRNs. *IEEE Trans. Comm.* **2023**, *71*, 441–456.

33. Yue, Y.; Cao, L.; Lu, D.; Hu, Z.; Xu, M.; Wang, S.; Li, B.; Ding, H. Review and empirical analysis of sparrow search algorithm. *Artif. Intell. Rev.* **2023**, 1–53. [CrossRef]
34. Zhang, K.; Yang, Z.; Başar, T. Multi-agent reinforcement learning: A selective overview of theories and algorithms. In *Handbook of Reinforcement Learning and Control*; Springer: Cham, Switzerland, 2021; pp. 321–384.
35. Rosenberger, J.; Urlaub, M.; Schramm, D. Multi-agent reinforcement learning for intelligent resource allocation in IIoT networks. In Proceedings of the 2021 IEEE Global Conference Artificial Intelligence and Internet of Things (GCAIoT), Dubai, United Arab Emirates, 12–16 December 2021; pp. 118–119.
36. Hu, J.; Wang, X.; Li, D.; Xu, Y. Multi-agent DRL-Based Resource Allocation in Downlink Multi-cell OFDMA System. In Proceedings of the 2020 International Conference on Wireless Communications and Signal Processing (IWCSP), Nanjing, China, 21–23 October 2020; pp. 257–262.
37. Jiang, M.; Hai, T.; Pan, Z.; Wang, H.; Jia, Y.; Deng, C. Multi-Agent Deep Reinforcement Learning for Multi-Object Tracker. *IEEE Access* **2019**, *7*, 32400–32407.
38. Tian, J.; Liu, Q.; Zhang, H.; Wu, D. Multiagent Deep-Reinforcement-Learning-Based Resource Allocation for Heterogeneous QoS Guarantees for Vehicular Networks. *IEEE Internet Things J.* **2022**, *9*, 1683–1695.
39. Zhu, Q.; Wang, C.X.; Hua, B.; Mao, K.; Jiang, S.; Yao, M. 3GPP TR 38.901 channel model. In *The Wiley 5G Ref: The Essential 5G Reference Online*; Wiley Press: Hoboken, NJ, USA, 2021; pp. 1–35.
40. Morais, D.H.; Morais, D.H. 5G NR Overview and Physical Layer. In *Key 5G Physical Layer Technologies: Enabling Mobile and Fixed Wireless Access*; Springer: Cham, Switzerland, 2022; pp. 233–297.
41. Modak, K.; Rahman, S. Multi-cell Interference Management in In-band D2D Communication under LTE-A Network. In Proceedings of the 2021 International Conference on Computing, Electronics & Communications Engineering (ICCECE), Virtual, 16–17 August 2021; pp. 13–18.
42. Jia, R.; Liu, L.; Zheng, X.; Yang, Y.; Wang, S.; Huang, P.; Lv, T. Multi-Agent Deep Reinforcement Learning for Uplink Power Control in Multi-Cell Systems. In Proceedings of the 2022 IEEE International Conference on Communications Workshops (ICC Workshops), Seoul, Republic of Korea, 16–20 May 2022; pp. 324–330.
43. Haarnoja, T.; Zhou, A.; Hartikainen, K.; Tucker, G.; Ha, S.; Tan, J.; Kumar, V.; Zhu, H.; Gupta, A.; Abbeel, P.; et al. Soft actor-critic algorithms and applications. *arXiv* **2018**, arXiv:1812.05905.
44. Liu, Y.; Wang, H.; Peng, M.; Guan, J.; Wang, Y. An Incentive Mechanism for Privacy-Preserving Crowdsensing via Deep Reinforcement Learning. *IEEE Internet Things J.* **2021**, *8*, 8616–8631.
45. Fortin, F.A.; De Rainville, F.M.; Gardner, M.A.G.; Parizeau, M.; Gagné, C. DEAP: Evolutionary algorithms made easy. *J. Mach. Learn. Res.* **2012**, *13*, 2171–2175.

electronics

MDPI

Article

Federated Deep Reinforcement Learning for Joint AeBSs Deployment and Computation Offloading in Aerial Edge Computing Network

Lei Liu [1,*], Yikun Zhao [2], Fei Qi [1], Fanqin Zhou [2], Weiliang Xie [1], Haoran He [2] and Hao Zheng [2]

[1] Beijing Research Institute, China Telecom Corporation Limited, Beijing 102209, China
[2] State Key Laboratory of Networking and Switching Technology, Beijing University of Posts and Telecommunications, Beijing 100876, China
* Correspondence: liulei6@chinatelecom.cn

Abstract: In the 6G aerial network, all aerial communication nodes have computing and storage functions and can perform real-time wireless signal processing and resource management. In order to make full use of the computing resources of aerial nodes, this paper studies the mobile edge computing (MEC) system based on aerial base stations (AeBSs), proposes the joint optimization problem of computation the offloading and deployment control of AeBSs for the goals of the lowest task processing delay and energy consumption, and designs a deployment and computation offloading scheme based on federated deep reinforcement learning. Specifically, each low-altitude AeBS agent simultaneously trains two neural networks to handle the generation of the deployment and offloading strategies, respectively, and a high-altitude global node aggregates the local model parameters uploaded by each low-altitude platform. The agents can be trained offline and updated quickly online according to changes in the environment and can quickly generate the optimal deployment and offloading strategies. The simulation results show that our method can achieve good performance in a very short time.

Keywords: aerial network; mobile edge computing; 6G; computation offloading

Citation: Liu, L.; Zhao, Y.; Qi, F.; Zhou, F.; Xie, W.; He, H.; Zheng, H. Federated Deep Reinforcement Learning for Joint AeBSs Deployment and Computation Offloading in Aerial Edge Computing Network. *Electronics* **2022**, *11*, 3641. https://doi.org/10.3390/electronics11213641

Academic Editors: Yichuang Sun, Haeyoung Lee and Oluyomi Simpson

Received: 26 September 2022
Accepted: 31 October 2022
Published: 7 November 2022

1. Introduction

Due to the explosive development of various intelligent applications in the 6G era, user's demand for computing and communication is expected to increase dramatically. To meet this challenge, mobile edge computing (MEC) is considered as an efficient paradigm, which can improve network computing capability and user experience [1]. At the same time, the aerial network based on new mobile communication systems such as high-altitude platforms (HAPs) is considered as a potential architecture for 6G and has aroused much attention recently. The aerial network acts as a supplement and extension of the terrestrial mobile communication network to provide collaborative and efficient information services for various network applications in a wide spatial area [2]. The aerial network primarily consists of low-altitude platforms and high-altitude platforms and plays a key role in enhancing coverage, enabling edge services, and enabling flexible network reconfiguration. Communication, computing, caching, sensing, and navigation services will be possible on a global scale through the fusion of aerial networks and edge computing [3].

Nevertheless, there are numerous obstacles to integrating aerial networks with MEC systems. A classic MEC system typically deploys edge servers using the ground infrastructure. The aerial base station (AeBS) can be formed by carrying the MEC server and other communication equipment on the aerial platform. An aerial edge computing node differs from its terrestrial counterpart, since it is capable of flexibly adjusting its deployment position to achieve better communication conditions [4]. The traditional offloading strategy for terrestrial MEC networks may not be appropriate for aerial edge computing. In addition, in order to extend the coverage or increase the number of users served, a single AeBS is not feasible and multiple AeBSs can be cooperatively deployed to complete the task [5]. How to coordinate the deployment of multiple AeBSs is also a critical issue. On the other hand, although the flexibility of AeBSs brings a higher degree of freedom to the deployment design of MEC nodes, they often suffer from resource and energy constraints. Therefore, how to make the AeBSs better provide communication and computing services under the condition of limited resources is also a problem that needs to be solved [1].

Edge computing and deep learning can naturally be combined with each other. On the one hand, they complement each other technically [6]. Some recent work has focused on the generalization problem of deep neural network (DNN) models [7] or designing resource-friendly models [8,9] to help the DNN model be better applied in actual edge deployment scenarios. On the other hand, their application and popularization are mutually beneficial. The combination of the two technologies has promoted a wide range of intelligent applications, from face recognition [10] and drone navigation [11] to the Industrial Internet of Things [12]. Deep learning has strong perception and expression ability, while reinforcement learning has decision-making ability. Deep reinforcement learning (DRL), which introduces deep neural networks in deep learning into reinforcement learning, holds promise for generating network optimization decisions in complex and dynamic environments.

Traditional single-agent DRL approaches usually follow a centralized paradigm, performing poorly in scalability and flexibility due to their large state space and action space [13]. Recently, researchers have discovered that multi-agent deep reinforcement learning (MADRL), which handles modeling and computation in a distributed manner, can obtain better performance in solving multi-AeBS cooperation tasks [14]. However, in MADRL, agents need to interact with each other to exchange state and action information in order to maintain the stability of the environment. In practical situations, frequent communication between AeBSs will consume the communication resources of AeBSs and increase the complexity of the optimization problem. As a distributed training framework, federated learning can simplify the model and improve the convergence speed in large-scale MADRL models. The benefits of applying federated learning to MADRL can be mainly summarized as follows: (1) federated learning avoids direct data leakage and, thus, can protect data privacy [15,16], (2) federated learning can make DRL models converge quickly, so it performs well in some scenarios sensitive to model training time [17], (3) incorporating federated learning into DRL can improve system scalability [18], (4) federated learning can also address the data island problem [16,18].

In this paper, we focus on the joint optimization of a mobile device (MD) offloading scheme and AeBS deployment to fully utilize the resources of AeBSs. Considering the dynamic nature of communication networks and computational tasks, we propose a federated DRL-based scheme that simultaneously addresses AeBS deployment and MD computational offloading. The contributions of our work can be concluded as:

1. A federated DRL algorithm was designed to jointly optimize the AeBSs' deployment and computation offloading to achieve lower energy consumption and task processing delay.
2. A new training mechanism is presented in the aerial edge computing network where low-altitude AeBSs are controlled by their own agents and cooperate in a distributed manner, and an HAP acts as a global node for model aggregation to improve the training efficiency.
3. Two neural networks trained together were set up for each agent to deploy the AeBSs and generate the computation offloading policies, respectively.

The content of this paper is organized as follows. Section 2 introduces some related work. Section 3 shows the system model of the aerial edge computing network and analyzes the joint deployment and offloading optimization problem that needs to be addressed. Section 4 describes the detailed flow and architecture of our proposed federated deployment and computational offloading (FedDCO) algorithm. Section 5 presents the results and analysis of our simulation for FedDCO. Section 6 is a summary of the full paper.

2. Related Work

Several works on integrating edge computing into aerial networks have been conducted. In [3], the survey introduced several desirable attributes and enabling technologies of aerial computing. In [19], Jia et al. studied the offloading problem in a hierarchical aerial computing framework composed of HAPs and AeBSs, and the flexible mobility of AeBSs was ignored. Reference [20] adopted an AeBS to provide computation offloading services for mobile users, but they did not study the dynamic computation offloading strategy. In [21], Truong et al. investigated an aerial edge computing network where an HAP plays the role of MEC and an offloading optimization problem is formulated, aiming to minimize the cost for task completion.

Researchers have adopted some heuristic algorithms to solve the offloading decision problem. To increase the system's weighted computing efficiency, Reference [4] proposed a heuristic algorithm for maximizing computational efficiency. In order to reduce the energy used by the AeBSs, Reference [22] jointly optimized the offloading of computation bits and the trajectory of the AeBS in an AeBS-enabled MEC system. In Reference [19], a matching game-theory-based algorithm and a heuristic approach for offloading optimization were presented. However, considering the dynamic nature of the multi-AeBS scenario, network optimization decisions are expected to be real-time. These aforementioned algorithms usually take many iterations to reach a local optimum, which makes them unsuitable for practical computation offloading situations. Besides, their computational complexity tends to rise significantly with the expansion of MEC network scale.

Recently, deep learning has made a series of achievements in the field of wireless communication, and researchers have also investigated some advanced models to help the deep neural network be better applied in practice. To address the generalization issue of the deep neural network, a two-stage training method was devised to optimize the feature boundary of the convolution neural network (CNN) to reduce the over-fitting problem in [7]. Several works were devoted to designing a resource-friendly edge artificial intelligence model. Reference [8] designed a graphics processing unit (GPU)-accelerated faster mean-shift algorithm, which is valuable for accelerating the speed of the training of the DNN model and saving computing resources. Reference [9] implemented a classification system based on the multi-channel CNN, which can work in a hardware environment with limited computing resources. Some researches also discussed the application of deep learning in MEC networks. Reference [23] utilized distributed deep learning to make offloading decisions for MEC networks in parallel. Reference [24] developed a hierarchical deep learning task distribution framework to deal with the tradeoff between latency and energy consumption, where the unmanned Aerial Vehicles are embedded with lower layers of the pretrained CNN model, while the MEC server handles the higher layers. These studies demonstrate the potential of combining deep learning with edge computing and also reveal the importance of generalization and resource issues in practical applications.

DRL, a combination of the DNN and reinforcement learning, aims to create an intelligent agent that can execute effective strategies and maximize the return of long-term tasks through controllable actions. Reference [25] designed a fast deep-Q-network (DQN)-based offloading approach to boost computation performance. Reference [26] provided a DQN-based online computation offloading policy with random task arrivals in a similar network setup. The authors in [27] adopted a centralized DRL algorithm to settle the offloading issue and a differential-evolution-based approach to address the deployment issue. The optimization issue of maximizing the migration throughput of user workloads in aerial MEC systems was solved with a DRL method in [28]. Reference [21] utilized the deep deterministic policy gradient (DDPG) to reduce the overall cost of performing the tasks. DRL was also used to jointly tackle the optimization problem of user association and resource allocation in aerial edge computing systems [29].

MADRL has numerous advantages over single-agent DRL. It enables agents to work cooperatively to handle high-complexity tasks in a distributed manner. Different MADRL algorithms have different agent-to-agent interaction forms and communication costs. A multi-agent imitation learning technique was presented in [30] to reduce the average task completion time in edge computing networks. To reduce overall energy usage, Reference [31] employed a multi-agent path planning strategy for energy consumption minimization. Reference [32] devised an MADRL-based trajectory control approach, which plans the trajectory of each AeBS individually. To reduce the overall computation and communication cost, Reference [33] developed a decentralized value-iteration-based reinforcement learning approach to make joint computation offloading and resource allocation decisions. The above research discovered that the multi-agent algorithm performs effectively in the multi-AeBS control scenario. This is because the MADRL framework considers the system as a whole and can jump out of the local optimal solution, which maximizes the benefit of each agent.

Some researchers have introduced federated learning into the DRL algorithm. Federated learning can accelerates the convergence speed of the model and enhances the generalization ability of the model by aggregating parameters. Reference [34] jointly optimized resource allocation, user association, and power control in a multi-AeBS MEC system via a federated DQN approach. In a multi-AeBS MEC system, massive amounts of data have to be transmitted from UEs to the parameter center. The practical deployment and operation of the algorithm are challenging because of the corresponding communication delay. To solve this problem, the authors fused federated learning (FL) with the MADRL framework and proposed a semi-distributed multi-agent federated reinforcement learning algorithm with the integration of FL and DRL. The proposed algorithm enables the UEs to quickly learn models by keeping their data training locally. In [35], an edge federated multi-agent actor–critic approach for resource management, collaborative trajectory planning, and data scheduling was provided. For cooperation in MEC systems, a federated heterogeneous multi-agent actor–critic algorithm was designed in [36]. Reference [37] designed a federated DRL-based cooperative edge caching architecture, which enables base stations to cooperatively learn a shared model and addresses the complicated and dynamic control concerns. A hierarchical federated DRL approach was described in [38] in a content replacement scenario.

The aforementioned works inspired us to design a federated deep reinforcement learning algorithm in which each AeBS is managed by a separate agent and cooperates in a distributed way, aiming to reduce the overall task processing time and energy consumption. Table 1 lists a comparison of our work to the relevant research.

Table 1. Comparison between our work and the existing literature.

Reference	Optimization Goal	Offloading	Deployment	Method
[4]	Maximize the weighted computational efficiency of the system	Proportional	✓	Alternative computational efficiency maximization
[19]	Maximize the total IoT data computed by the aerial MEC platforms	Binary	/	Matching game-theory-based algorithm and a heuristic algorithm
[20]	Minimize the total energy consumption	/	✓	Successive convex approximation (SCA)
[21]	Minimize the total cost function of the system	Proportional	/	Deep deterministic policy gradient (DDPG)
[22]	Minimize the energy consumed at the AeBS	Binary	✓	Alternative optimization

Table 1. *Cont.*

Reference	Optimization Goal	Offloading	Deployment	Method
[23]	Minimize overall system utility including both the total energy consumption and the delay in finishing the task	Binary	✓	DNN
[24]	Minimize the delay and energy consumption, while considering the data quality input into the DNN and inference error	Binary and proportional	/	CNN
[25]	Optimal offloading policy	Proportional	/	Fast deep-Q-network (DQN)
[26]	Maximize the long-term utility performance	Binary	/	Double DQN
[27]	Average slowdown for offloaded tasks	One-to-one correspondence	✓	DQN
[28]	Maximize the migration throughput of user tasks	Binary	/	DQN
[29]	Maximize the average throughput of user tasks	Binary	✓	Q-learning
[30]	Minimize average task completion time	Binary	/	Multi-agent imitation learning
[31]	Minimize the total energy consumption of AeBSs	Binary	✓	Multi-agent deep deterministic policy gradient (MADDPG)
[32]	Maximize the fairness among all the user equipment (UE) and the fairness of the UE load of each AeBS	Binary	✓	MADDPG
[33]	Minimize the total computation and communication overhead of the joint computation offloading and resource allocation strategies	Binary	/	Multi-agent double-deep Q-learning
[34]	Minimize the overall consumed power	Binary	/	Federated DQN
[35]	Minimize the average source age (elapsed time)	Binary	✓	Federated multi-agent actor–critic
[36]	Minimize the average age of all data sources	Binary	✓	Federated multi-agent actor–critic
[37]	Maximize the expected long-term reward	Three-way	✓	Federated DQN
[38]	Improve the hit rate	Binary	/	Federated DQN
Our work	Jointly minimize overall task latency and energy consumption	Binary	✓	Federated DQN

Notes: Binary: tasks can be offloaded to the AeBSs or not; proportional: tasks have the offloading rate, and tasks can be offloaded partially; one-to-one correspondence: tasks must be offloaded to an associated AeBS; three-way: tasks can be offloaded to the AeBSs, processed locally, or processed by their neighbors.

3. System Model

Figure 1 depicts our scenario in an aerial edge computing network. There are M MDs, N low-altitude AeBSs, and an HAP in the system. For simplicity, we abbreviate low-altitude AeBSs as AeBSs and consider all AeBSs and MDs to be computationally capable, and additionally, we used a binary offloading strategy for the computational tasks, i.e., offloading to AeBSs or executing them locally [39]. AeBSs perform local training, and the model parameters are uploaded to an HAP for federated model aggregation. We chose an HAP as a global node because of its powerful computing power and lower latency compared to satellites. In addition, there are almost no other obstacles in the air except the aircraft itself, which makes the communication link better and more reliable than the ground communication. AeBS n trains its own two networks Q_1^n and Q_2^n together, in which Q_1^n is responsible for deployment and Q_2^n is responsible for the offloading policy.

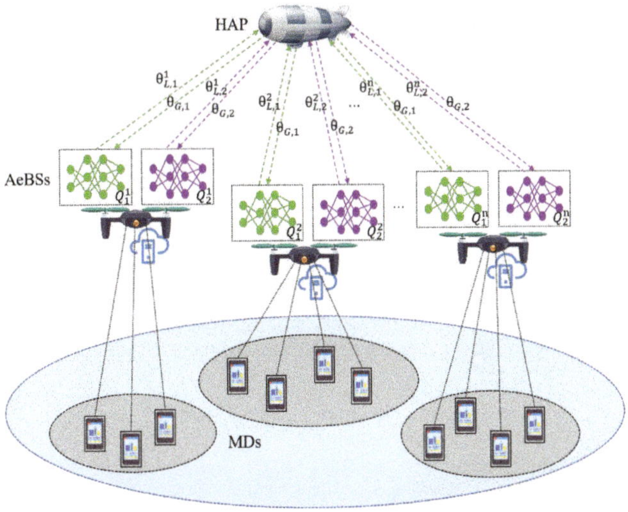

Figure 1. Federated deep-reinforcement-learning-based joint computation offloading and deployment control.

3.1. Communication Model

Since the line-of-sight (LoS) channel is dominant in air-to-ground links, we only considered the LoS propagation characteristics between AeBSs and MDs. Therefore, the channel gain between MD i and AeBS j can be obtained as:

$$h_{i,j} = g_0 d_{i,j}^{-2}, \tag{1}$$

where $d_{i,j}$ is the distance between MD i and AeBS j, which can be calculated by the locations of MD i and AeBS j. g_0 represents the channel gain at the reference distance of 1 m.

The data transmission rate between MD i and AeBS j is

$$R_{i,j} = B\log\left(1 + \frac{P_i h_{i,j}}{\sigma^2}\right), \tag{2}$$

where P_i denotes the transmit power of MD i and σ^2 represents the noise power.

3.2. Computation Model

For MD i with a computation task, D_i represents the size of the input data and S_i represents the CPU cycles required to process 1 bit of data. Let $A = \{a_{i,j}\}_{M\times(N+1)}$ represent the offloading decision of MDs. If the computation task in MD i is offloaded to AeBS j for computing, $a_{i,j} = 1$. If the computation task in MD i is computed locally, $a_{i,0} = 1$. Therefore, there are two kinds of computation modes that each MD can choose: local execution and offloading execution.

3.2.1. Local Execution

According to the definition of a task, the total CPU cycles required to execute the task is $S_i \times D_i$. Then, the time required to process a task in MD i is

$$T_i^{local} = \frac{S_i \times D_i}{f_i^{MD}}, \tag{3}$$

where f_i^{MD} is the computational capacity of MD i.

3.2.2. Offloading Execution

In this case, processing the task requires the following three steps. Firstly, MD i transmits the task data that need to be processed to AeBS j, which takes time $T_{i,j}^{tran}$. Secondly, AeBS j computes the data, which takes time T_j^{AeBS}. Thirdly, AeBS j transmits the result back to MD i, which takes a tiny amount of time $T_{j,i}^{back}$, and we usually neglect it. Thus, we have

$$T_{i,j}^{tran} = \frac{D_i}{R_{i,j}}, \tag{4}$$

$$T_j^{\mathrm{AeBS}} = \frac{S_i \times D_i}{f_j^{\mathrm{AeBS}}}, \tag{5}$$

where f_j^{AeBS} denotes the computational capacity of AeBS j. Therefore, the total time cost of offloading execution can be obtained as:

$$T_{i,j}^{off} = \frac{D_i}{R_{i,j}} + \frac{S_i \times D_i}{f_j^{\mathrm{AeBS}}}. \tag{6}$$

3.3. *Energy Model*

For local execution, only the computation energy consumption needs to be considered as there is no data transmission. The energy consumption can be calculated as [40]:

$$E_i^{local} = \delta f_i^{MD^2} S_i D_i, \tag{7}$$

where δ is an energy efficiency parameter, which is related to the chip architecture.

For offloading execution, the energy consumption includes transmission consumption and computation consumption, which can be written as:

$$E_{i,j}^{off} = P_i T_{i,j}^{tran} + \delta f_j^{\mathrm{AeBS}^2} S_i D_i. \tag{8}$$

3.4. *Problem Formulation*

In order to provide better service while taking into account the limited energy of AeBSs, it is necessary to minimize the system task processing delay and energy consumption. In this paper, we jointly optimized the AeBSs deployment and offloading strategy, aiming to minimize the weighted sum of task processing time and AeBS energy consumption, which can be written as:

$$\begin{aligned}
\min_{X,Y,A} \; & w_1 \sum_{j=1}^{N} \sum_{i=1}^{M} \left(a_{i,0} T_i^{local} + a_{i,j} T_{i,j}^{off} \right) \\
& + w_2 \sum_{j=1}^{N} \sum_{i=1}^{M} \left(a_{i,0} E_i^{local} + a_{i,j} E_{i,j}^{off} \right), \\
& C1 : x_{min} < x_j < x_{max}, \\
& C2 : y_{min} < y_j < y_{max}, \\
& C3 : \sum_{j=0}^{N} a_{i,j} = 1, \\
& C4 : a_{i,j} \in \{0,1\}, \forall i, j,
\end{aligned} \tag{9}$$

where $\forall i \in \{1,2,\ldots,M\}$ and $\forall j \in \{1,2,\ldots,N\}$. w_1 and w_2 denote the weights of the task processing time and energy consumption, respectively. Constraints C1 and C2 limit the range of movement of the AeBSs, and C3 and C4 indicate that each MD can either offload its task to an AeBS or execute the task locally and cannot partially offload its task. It is worth mentioning that, for such mixed-integer programming problems, with multi-objective optimization, it is difficult for traditional optimization algorithms to find the optimal solution in a short time, which is unacceptable for user computing tasks that change in real-time. To meet the real-time and complexity requirements, we propose FedDCO based on deep reinforcement learning, which can complete the deployment of AeBSs and obtain the offloading solution in a short time.

4. Federated Deep-Reinforcement-Learning-Based AeBS Deployment and Computation Offloading

The optimization problem in Equation (9) is defined as a mixed-integer programming problem, which is often difficult to find solutions to quickly. We designed an algorithm based on federated deep reinforcement learning called FedDCO, where each AeBS is equipped with a Q_1^n network responsible for generating deployment schemes and a Q_2^n network responsible for generating offloading policies. For an AeBS, it is unnecessary to focus on the MDs with weak channel gain, which will lead to information redundancy. Each AeBS only needs to pay attention to the information of the MDs that are within a certain distance. Therefore, we used the K-nearest-neighbor algorithm to divide the association between AeBSs and MDs. Then, each AeBS first moves to the optimal location using the Q_1^n network, followed by using the Q_2^n network to generate the offloading policies for nearby MDs. The basic components of FedDCO are as follows.

State: For Q_1^n of AeBS n, the state $s_1^n(t)$ includes the computational capacity of AeBS n and its associated MDs, the amount of computational tasks for its associated MDs, and the channel gain between AeBS n and MDs. The state $s_1(t)$ is the collection of each AeBS's state $s_1^n(t)$.

For Q_2^n of AeBS n, the state $s_2^n(t)$ also consists of the computational capacity of AeBS n and its associated MDs, the amount of computational tasks for its associated MDs, the channel gain between AeBS n and MDs, the number of MDs, which still have not generated a computation offloading decision, and a vector indicating the computation offloading policy for each MD. The state $s_2(t)$ is the collection of each AeBS state $s_2^n(t)$.

Action: For Q_1^n, the action $a_1^n(t)$ is the movement of AeBS n, including moving backward or forward, left or right, or remaining stationary. For Q_2^n, the action is the offloading policy of one MD, i.e., local execution or offloading execution.

Reward: We combine the impact of task latency and energy consumption in the overall task and define the reward function as $r_c - w_1 \sum_{j=1}^{N} \sum_{i=1}^{M} \left(a_{i,0} T_i^{local} + a_{i,j} T_{i,j}^{off} \right)$ $- w_2 \sum_{j=1}^{N} \sum_{i=1}^{M} \left(a_{i,0} E_i^{local} + a_{i,j} E_{i,j}^{off} \right)$, where r_c is a nonnegative constant.

We let $\theta_{L,1}^n$ represent the weight of Q_1^n, which is responsible for generating deployment decisions. $\theta_{L,2}^n$ represents the weight of Q_2^n, which generates the offloading policies. The updating processes of the two networks follow the strategy and principle of the classical DQN. The optimal Q function for AeBS n can be defined as:

$$Q^*(s_n, a_n) = \max_{\pi} {}_n \mathbb{E}\left[r_t + \gamma r_{t+1} + \gamma^2 r_{t+2} + \ldots \mid s_t = s_n, a_t = a_n, \pi_n \right], \tag{10}$$

which denotes the maximum expectation of the sum of the future reward at each time step t that can be obtained by the strategy $\pi_n = P(a_n \mid s_n)$ after observing state s_n and taking action a_n and γ is the discount factor. The DQN algorithm adopts the neural network to parameterize the approximation function $Q(s_n, a_n; \theta_n)$, where $\theta_{L,\zeta}^n$ is the weight parameter of the local neural network of AeBS n. The DQN utilizes the experience pool and stores the experiences of the agent at each time step t all in the dataset. During algorithm learning, the minibatches of the experiences are randomly sampled from the pool of stored samples to update the network. The loss function is calculated to measure the error between the prediction and the target value, which is:

$$\mathcal{L}_n\left(\theta_{L,\zeta}^n\right) = \mathbb{E}\left\{\left[y_n - Q\left(s_n, a_n; \theta_{L,\zeta}^n\right)\right]^2\right\}, \zeta \in \{1,2\}, \tag{11}$$

where $Q\left(s_n, a_n; \theta_{L,\zeta}^n\right)$ is the Q value in the Q-network with the current state s_n as the input; y_n is the target value and can be calculated as $y_n = r + \gamma \cdot \max_{a_n'} Q\left(s_n, a_n'; \theta_{L,\zeta}^n{}'\right)$. The Q-network is optimized in each iteration to minimize the loss function. We adopted the stochastic gradient descent method to optimize the loss function and update the weight parameters of the Q-networks. The DQN algorithm introduces two neural networks, the Q-network and the target Q-network, which have the same structure, but different parameters, and the parameters of the target Q-network are periodically updated according to the parameters in the Q-network.

After each AeBS trains its two networks based on its state, it uploads the parameters of Q_1 and Q_2 to the HAP for model aggregation at regular intervals f_a. The parameter f_a is set because it is unnecessary for AeBSs to upload model parameters to the HAP for each episode of training to save costs. The weight of the global model can be obtained as follows:

$$\theta_{G,\zeta} = \frac{1}{N} \sum_{n=1}^{N} \theta_{L,\zeta}^n, \zeta \in \{1,2\}, \tag{12}$$

where $\theta_{G,\zeta}$ is the weight of the global network and N is the number of AeBSs. The global network parameters are sent back to the AeBSs for updating their own local networks.

The detailed steps of the FedDCO algorithm are shown in Algorithm 1. We also depict the whole training process of the proposed FedDCO algorithm in Figure 2.

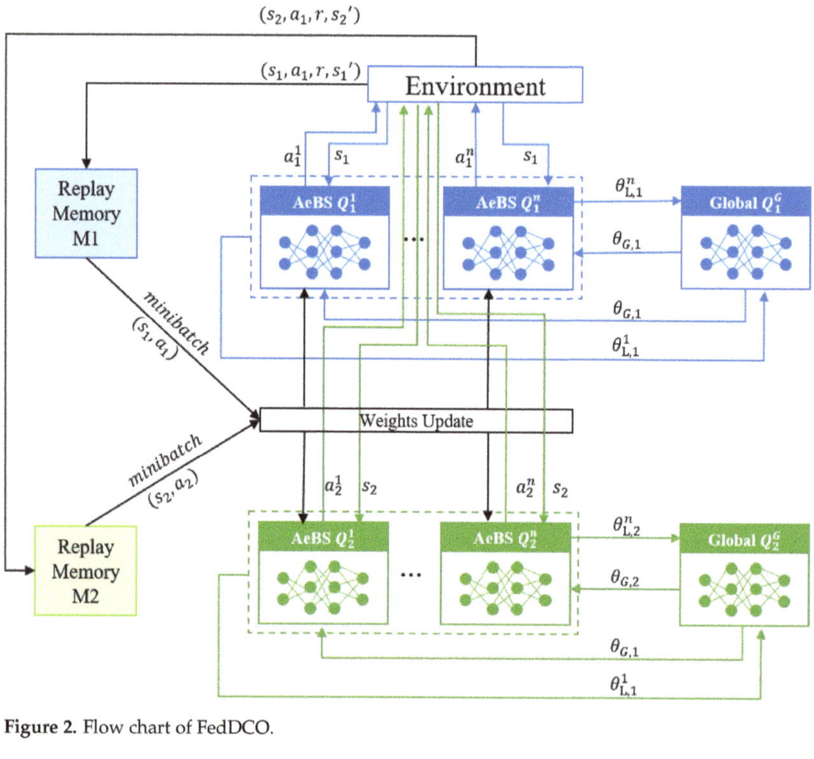

Figure 2. Flow chart of FedDCO.

Algorithm 1: Proposed FedDCO algorithm

Input: maximum number of episodes, states of AeBSs, locations of AeBSs and MDs
Output: deployment and offloading policy for each AeBS and MD
Initialize: replay memory size M_1, M_2, online network Q_1^n, Q_2^n and target network $Q_1^{n'}, Q_2^{n'}$ for each AeBS n, the distribution of MDs, and the aggregation frequency of federated learning f_a
Initialize:
for *episode* $\leftarrow 1 : max\_episode$ **do**
    Initialize the locations of AeBSs, and obtain the initial state s_1(t), s_2(t)
    for *step* $\leftarrow 1 : max\_step$ **do**
        Obtain the AeBS-MD association via the K-nearest-neighbor algorithm
        for *AeBS* $\leftarrow 1 : N$ **do**
            Choose an action $a_1^n(t)$ according to the ε-greedy policy: with probability $1 - \varepsilon$; AeBS n selects action $a_1^n(t) = argmaxQ_1^n(s_1^n(t), a_1^n(t); \theta_{L,1}^n)$, otherwise it selects a random action
            Execute action $a_1^n(t)$
        for *MD* $\leftarrow 1 : M$ **do**
            Choose an action $a_2^n(t)$ according to the ε-greedy policy: with probability $1 - \varepsilon$; its associated AeBS selects action $a_2^n(t) = argmaxQ_2^n(s_2^n(t), a_2^n(t); \theta_{L,2}^n)$, otherwise it selects a random action
            Execute action $a_2^n(t)$
            Observe reward r, and obtain the new state
            Store $(s_2(t), a_2(t), r(t), s_2(t+1))$ in replay memory M_2
        Store $(s_1(t), a_1(t), r^*(t), s_1(t+1))$ in replay memory M_1 (r^* is the best reward in the current episode)
        for *AeBS* $\leftarrow 1 : N$ **do**
            Perform a gradient descent step on the loss function according to Equation (11)
            Every C steps, update the target network for each AeBS n
    if *episode mod* f_a*=0* **then**
        Each AeBS uploads its $\theta_{L,1}^n$ and $\theta_{L,2}^n$ weights to the global node for model aggregation according to Equation (12), respectively.
        The global node sends the aggregated global model weight of $\theta_{G,1}$ and $\theta_{G,2}$ back to the AeBSs, and each AeBS updates its own model.

5. Simulation Results and Discussions

We selected a 1 km × 1 km area for simulation, and the performance metrics of the simulation were the total task processing time and energy consumption. The main simulation parameters are given in Table 2, which refer to [41]. The Q1- and Q2-networks of each AeBS were three-layer fully connected neural networks and used ReLU as the activation function. The simulation environment was Python 3.7 and Pytorch 1.9.1 and was run on a computer with i5-12500H processor from Intel and an RTX3060 GPU from Nvidia.

Table 2. Main simulation parameters.

Simulation Parameters	Values
AeBS altitude H	100 m
Transmit power P_i	0.5 W
Channel bandwidth B	1 MHz
Reference channel gain g_0	10,096
Energy efficiency parameter δ	2
Noise σ^2	5×10^{-5} W
ϵ in ϵ-greedy	0.1
Memory size	10,000
Batch size	512
Discount factor	0.97

We compared our FedDCO scheme with three other approaches named MADCO, K-means-based, and throughput-first, to observe the performance gain brought by federated learning, simultaneous training of two deep Q-networks, and location adjustment of the AeBSs, respectively [41]. The description of these schemes is listed as follows:

1. FedDCO: Each AeBS has two deep Q-networks, Q_1 and Q_2, and they are trained simultaneously to generate deployment and offloading policies. During the training process, AeBSs upload the Q_1 and Q_2 weights instead of the raw state and action data to the global node for model aggregation, and the global node sends the aggregated global model weight of Q_1 and Q_2 back to the AeBSs, then each AeBS updates its own model.
2. MADCO: MADCO optimizes the AeBS deployment schemes and offloading strategies by training two neural networks together to minimize the latency and energy consumption of computational task processing. Each AeBS exchanges action and state information with each other when making decisions. Its settings for the input/output, parameters, and DNN structure are consistent with FedDCO.
3. K-means: AeBSs are deployed based on the MD distribution through the K-means algorithm. The number of clusters of K-means was set as the number of AeBSs, and then, each AeBS is deployed directly above each cluster center of MDs. Specifically, the maximum number of iterations of K-means was 300, and if the sum of squares within all clusters between two iterations is less than 1×10^{-4}, the iteration is terminated. After the location of AeBSs is fixed, the offloading policy is generated through the Q_2-network, whose input/output settings, parameter settings, and network structure are the same as Q_2 in FedDCO.
4. Throughput-first: AeBSs are first deployed based on the Q_1-network with the goal of maximizing throughput, and the offloading policy is later generated through the Q_2-network. The settings of the input/output, parameters. and DNN structure in throughput-first are also consistent with FedDCO.

Figure 3 describes the convergence process of the above four algorithms, in the case that the number of AeBSs is 8 and the number of MDs is 60. We normalized the reward to reduce the magnitude difference between the task processing delay and energy consumption. The average reward of the proposed FedDCO increased rapidly in the first 100 episodes. After 150 episodes, the algorithm converged, and the reward became stable. It can be found that the convergence speed of FedDCO was faster than that of MADCO, which indicates that federated learning can improve the convergence speed of the model. K-means uses a clustering algorithm to deploy AeBS positions; although its training time was fast, its performance was the lowest. This is because it does not take advantage of the mobility and self-adjustment of the deployment location of AeBSs to obtain better performance.

In Figure 4, we mainly evaluate the effect of different algorithms with different numbers of MDs in the following three typical scenarios:

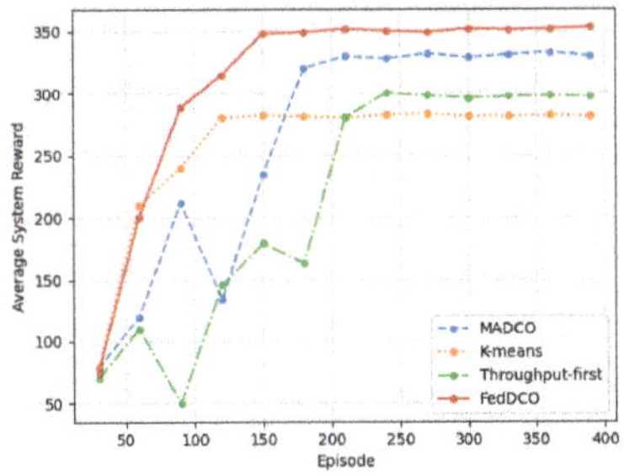

Figure 3. The reward curve of the four algorithms in the training process.

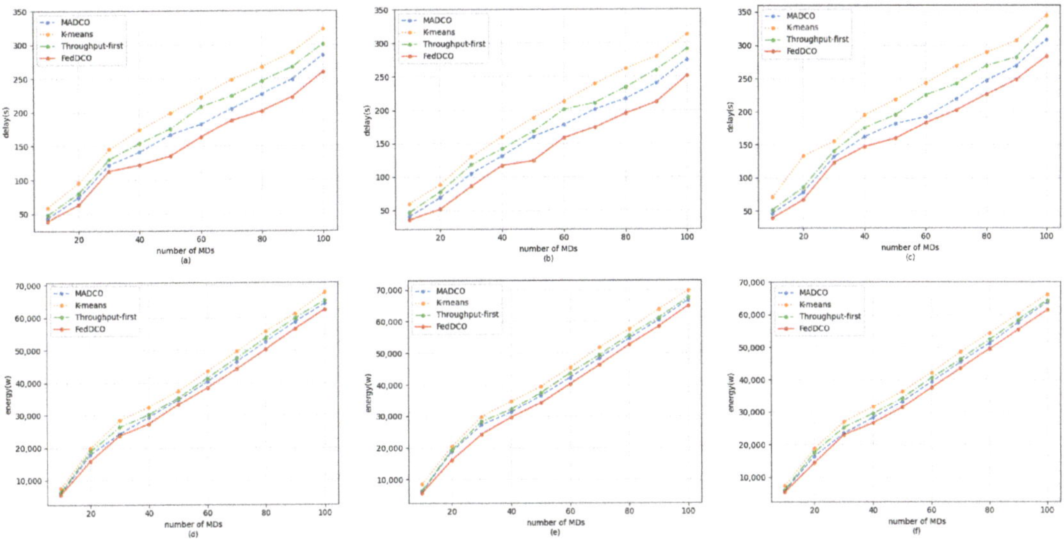

Figure 4. Comparison of network performance metrics at different MD quantities in different communication scenarios. (**a–c**) are the task processing delay in the general communication scenario, delay-sensitive scenario, and energy-sensitive scenario, respectively. (**d–f**) are the energy consumption in the general communication scenario, delay-sensitive scenario, and energy-sensitive scenario, respectively.

1. General communication scenario: In this scenario, we regarded delay and energy consumption as equally important indicators. Thus, we set $(w_1, w_2) = (1, 1)$ in the simulation.
2. Delay-sensitive scenario: There may be some real-time services in the network, which makes the MDs sensitive to delay. For this scenario, we set $(w_1, w_2) = (2, 1)$ in the simulation.
3. Energy-sensitive scenario: For some aerial platforms with limited payload capacity, such as small multi-rotor unmanned aerial vehicles, the battery capacity is limited, so it is necessary to reduce the energy consumption as much as possible to ensure the completion of the mission. For this scenario, we set $(w_1, w_2) = (1, 2)$ in the simulation.

Figures 4a–c depict the task processing delay for the three different scenarios, with different numbers of MDs ranging from 10 to 100 [42]. By comparing different algorithms, it can be observed that our proposed FedDCO obtained the lowest task processing latency in all of three cases. It can be seen that K-means had a longer task processing time in most cases, which may be due to the fact that it does not use a neural network to obtain a better deployment position. Figure 4d–f show the energy consumption under each scenario. FedDCO still achieved the best results compared to the other three algorithms. K-means and throughput-first performed similarly, probably because Q_2 is effective in minimizing energy consumption, leading them to make similar offloading strategies. By comparing the network performance metrics in different scenarios, it can be discovered that the delay of each scheme in the delay-sensitive scenario was basically smaller than that in the general communication scenario, but at the cost of more energy consumption. In the energy-constrained scenario, the agent's strategy prefers a lower energy consumption, and the task processing latency in this scenario increases slightly. To sum up the above figures, with the increase of the MDs' number, the key indicators, task processing delay, and energy consumption of the system were all on the rise. Taking the three indicators into consideration, the proposed FedDCO achieved the optimal performance in various communication scenarios compared to the other algorithms. The K-means-based scheme performed worst among these algorithms since it does not take advantage of adjusting the position of the AeBSs. This indicates that the deployment design of the AeBSs is a very important factor to be considered in the aerial edge computing network.

Figure 5 shows the performance of the four algorithms described in this article at different data sizes of the tasks. It can be noticed that, with the increasing data size of computing tasks, task processing delay and energy consumption both showed an upward trend. This was attributed to the fact that there were more data to be transmitted and processed, which brought a burden to the system and led to the degradation of the system performance. In the case of different data sizes of the tasks, our proposed FedDCO had a lower delay and energy consumption than the other three algorithms.

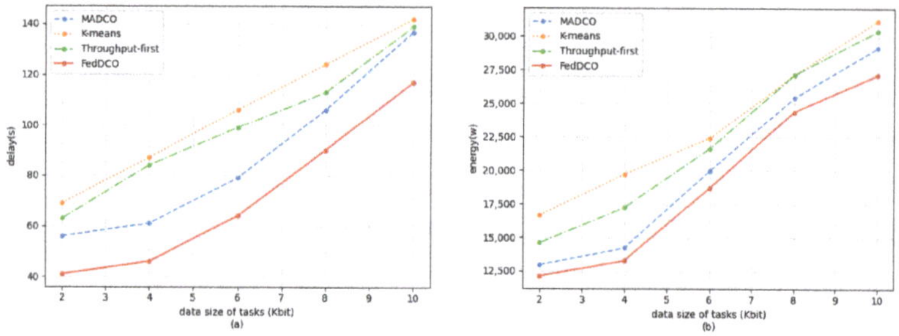

Figure 5. Comparison of network performance metrics at different data sizes of the tasks. (**a**) Task processing delay. (**b**) Energy consumption.

Figure 6 illustrates the performance of the four algorithms described in this article at different computational capacities of the AeBSs. With the enhancement of the computing capability of the AeBSs, the delay of the system decreased, but the system consumed more energy. This is because, when the computational power of the AeBSs increased, the tasks on the MDs tended to be offloaded to the AeBSs, which resulted in an increase in both the signal transmission energy consumption and AeBS computation energy consumption. By comparing the four algorithms, it can be found that our proposed algorithm FedDCO had the best performance, while the traditional machine learning algorithm K-means had a relatively poor performance.

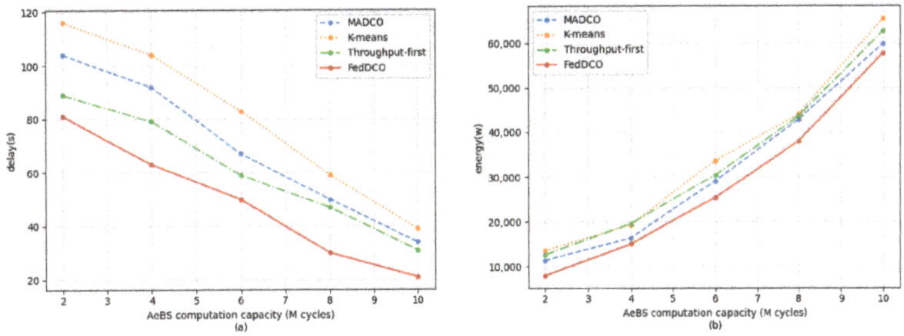

Figure 6. Comparison of network performance metrics at different computational capacities of the AeBSs. (**a**) Task processing delay. (**b**) Energy consumption.

Figure 7 shows the training time of FedDCO, MADCO, K-means, and throughput-first with different numbers of AeBSs. As the number of AeBSs in the system increased, the problem complexity increased, so the training time of FedDCO and MADCO also increased. The training time of FedDCO was smaller than that of MADCO and throughput-first, which indicates that federated learning can accelerate convergence and improve algorithm efficiency. This is owed to the fact that federated learning can solve the problem of the difficult convergence of agents by exchanging experience through parameter aggregation. In the K-means-based scheme, the K-means algorithm was adopted to decide the deployment position of the AeBSs instead of training a DRL model, which made its training speed relatively fast. Especially when the number of AeBSs increased, this gap became larger. However, the performance of the K-means-based scheme was not as good as FedDCO.

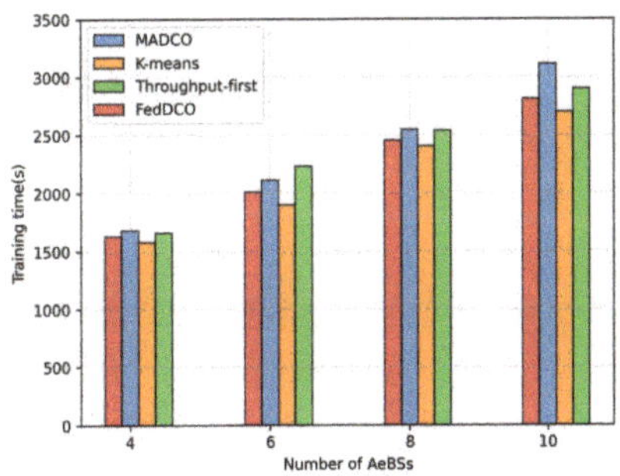

Figure 7. Training time of different algorithms.

Generally speaking, the simulation results above proved that the proposed FedDCO outperformed the other algorithms in different communication scenarios. Moreover, FedDCO also had the advantages of fast convergence and a short training time, which is very suitable for the dynamic network environment in aerial edge computing networks.

6. Conclusions

In this paper, we proposed an approach called FedDCO to address the joint optimization problem of AeBSs' deployment and computation offloading in an aerial edge computing network with the goal of minimizing the task processing time and energy consumption. We designed a training mechanism based on federated deep reinforcement learning, where low-altitude AeBSs train their local neural networks individually and an HAP plays the role of a global node for model aggregation. The simulation results showed that our proposed approach can achieve better performance in various communication scenarios compared with other benchmark schemes.

Author Contributions: Conceptualization, L.L. and F.Q.; funding acquisition, L.L. and W.X.; investigation, Y.Z. and W.X.; methodology, F.Q. and F.Z.; supervision, L.L.; validation, H.H. and H.Z.; visualization, H.H. and H.Z.; writing—original draft, Y.Z. and H.H.; writing—review and editing, F.Q. and F.Z. All authors have read and agreed to the published version of the manuscript.

Funding: This research was funded by the National Key R&D Program of China (No. 2020YFB1806700).

Conflicts of Interest: The authors declare no conflict of interest.

Abbreviations

The following abbreviations are used in this manuscript:

MEC	Mobile edge computing
HAP	High altitude platform
AeBS	Aerial base station
MD	Mobile device
FL	Federated learning
RL	Reinforcement learning
DRL	Deep reinforcement learning
DQN	Deep Q-network
DDPG	Deep deterministic policy gradient
MADRL	Multi-agent deep reinforcement learning
LOS	Line-of-sight

References

1. Ji, B.; Wang, Y.; Song, K.; Li, C.; Wen, H.; Menon, V.G.; Mumtaz, S. A Survey of Computational Intelligence for 6G: Key Technologies, Applications and Trends. *IEEE Trans. Ind. Informatics* **2021**, *17*, 7145–7154. [CrossRef]
2. Cui, H.; Zhang, J.; Geng, Y.; Xiao, Z.; Sun, T.; Zhang, N.; Liu, J.; Wu, Q.; Cao, X. Space-air-ground integrated network (SAGIN) for 6G: Requirements, architecture and challenges. *China Commun.* **2022**, *19*, 90–108. [CrossRef]
3. Pham, Q.V.; Ruby, R.; Fang, F.; Nguyen, D.C.; Yang, Z.; Le, M.; Ding, Z.; Hwang, W.J. Aerial Computing: A New Computing Paradigm, Applications, and Challenges. *IEEE Internet Things J.* **2022**, *9*, 8339–8363. [CrossRef]
4. Xu, Y.; Zhang, T.; Liu, Y.; Yang, D.; Xiao, L.; Tao, M. UAV-Assisted MEC Networks With Aerial and Ground Cooperation. *IEEE Trans. Wirel. Commun.* **2021**, *20*, 7712–7727. [CrossRef]
5. Shakeri, R.; Al-Garadi, M.A.; Badawy, A.; Mohamed, A.; Khattab, T.; Al-Ali, A.K.; Harras, K.A.; Guizani, M. Design Challenges of Multi-UAV Systems in Cyber-Physical Applications: A Comprehensive Survey and Future Directions. *IEEE Commun. Surv. Tutorials* **2019**, *21*, 3340–3385. [CrossRef]
6. Zhou, Z.; Chen, X.; Li, E.; Zeng, L.; Luo, K.; Zhang, J. Edge Intelligence: Paving the Last Mile of Artificial Intelligence With Edge Computing. *Proc. IEEE* **2019**, *107*, 1738–1762. [CrossRef]
7. Zheng, Q.; Yang, M.; Yang, J.; Zhang, Q.; Zhang, X. Improvement of Generalization Ability of Deep CNN via Implicit Regularization in Two-Stage Training Process. *IEEE Access* **2018**, *6*, 15844–15869. [CrossRef]
8. You, L.; Jiang, H.; Hu, J.; Chang, C.H.; Chen, L.; Cui, X.; Zhao, M. GPU-accelerated Faster Mean Shift with euclidean distance metrics. In Proceedings of the 2022 IEEE 46th Annual Computers, Software, and Applications Conference (COMPSAC), Alamitos, CA, USA, 27 June–1 July 2022; pp. 211–216. [CrossRef]
9. Zhao, M.; Chang, C.H.; Xie, W.; Xie, Z.; Hu, J. Cloud Shape Classification System Based on Multi-Channel CNN and Improved FDM. *IEEE Access* **2020**, *8*, 44111–44124. [CrossRef]
10. Jin, B.; Cruz, L.; Gonçalves, N. Pseudo RGB-D Face Recognition. *IEEE Sens. J.* **2022** . [CrossRef]
11. Arshad, M.A.; Khan, S.H.; Qamar, S.; Khan, M.W.; Murtza, I.; Gwak, J.; Khan, A. Drone Navigation Using Region and Edge Exploitation-Based Deep CNN. *IEEE Access* **2022**, *10*, 95441–95450. [CrossRef]
12. Liang, F.; Yu, W.; Liu, X.; Griffith, D.; Golmie, N. Toward Edge-Based Deep Learning in Industrial Internet of Things. *IEEE Internet Things J.* **2020**, *7*, 4329–4341. [CrossRef]
13. Zhao, Y.; Zhou, F.; Li, W.; Gao, Y.; Feng, L.; Yu, P. 3D Deployment and User Association of CoMP-assisted Multiple Aerial Base Stations for Wireless Network Capacity Enhancement. In Proceedings of the 2021 17th International Conference on Network and Service Management (CNSM), Izmir, Turkey, 25–29 October 2021; pp. 56–62. [CrossRef]
14. Zhang, Y.; Mou, Z.; Gao, F.; Jiang, J.; Ding, R.; Han, Z. UAV-Enabled Secure Communications by Multi-Agent Deep Reinforcement Learning. *IEEE Trans. Veh. Technol.* **2020**, *69*, 11599–11611. [CrossRef]
15. Xu, Y.; Bhuiyan, M.Z.A.; Wang, T.; Zhou, X.; Singh, A. C-fDRL: Context-Aware Privacy-Preserving Offloading through Federated Deep Reinforcement Learning in Cloud-Enabled IoT. *IEEE Trans. Ind. Inform.* **2022**. [CrossRef]
16. Yang, W.; Xiang, W.; Yang, Y.; Cheng, P. Optimizing Federated Learning With Deep Reinforcement Learning for Digital Twin Empowered Industrial IoT. *IEEE Trans. Ind. Inform.* **2022**. [CrossRef]

17. Tehrani, P.; Restuccia, F.; Levorato, M. Federated Deep Reinforcement Learning for the Distributed Control of NextG Wireless Networks. In Proceedings of the 2021 IEEE International Symposium on Dynamic Spectrum Access Networks (DySPAN), Angeles, CA, USA, 13–15 December 2021; pp. 248–253. [CrossRef]
18. Zhang, X.; Peng, M.; Yan, S.; Sun, Y. Deep-Reinforcement-Learning-Based Mode Selection and Resource Allocation for Cellular V2X Communications. *IEEE Internet Things J.* **2020**, *7*, 6380–6391. [CrossRef]
19. Jia, Z.; Wu, Q.; Dong, C.; Yuen, C.; Han, Z. Hierarchical Aerial Computing for Internet of Things via Cooperation of HAPs and UAVs. *IEEE Internet Things J.* **2022**. [CrossRef]
20. Jeong, S.; Simeone, O.; Kang, J. Mobile Edge Computing via a UAV-Mounted Cloudlet: Optimization of Bit Allocation and Path Planning. *IEEE Trans. Veh. Technol.* **2018**, *67*, 2049–2063. [CrossRef]
21. Truong, T.P.; Tran, A.T.; Nguyen, T.M.T.; Nguyen, T.V.; Masood, A.; Cho, S. MEC-Enhanced Aerial Serving Networks via HAP: A Deep Reinforcement Learning Approach. In Proceedings of the 2022 International Conference on Information Networking (ICOIN), Jeju-si, Korea, 12–15 January 2022; pp. 319–323. [CrossRef]
22. Zhou, F.; Wu, Y.; Sun, H.; Chu, Z. UAV-Enabled Mobile Edge Computing: Offloading Optimization and Trajectory Design. In Proceedings of the 2018 IEEE International Conference on Communications (ICC), Kansas City, MO, USA, 20–24 May 2018; pp. 1–6. [CrossRef]
23. Huang, L.; Feng, X.; Feng, A.; Huang, Y.; Qian, L.P. Distributed Deep Learning-based Offloading for Mobile Edge Computing Networks. *Mob. Networks Appl.* **2022**, *27*, 1123–1130. [CrossRef]
24. Yang, B.; Cao, X.; Yuen, C.; Qian, L. Offloading Optimization in Edge Computing for Deep-Learning-Enabled Target Tracking by Internet of UAVs. *IEEE Internet Things J.* **2021**, *8*, 9878–9893. [CrossRef]
25. Min, M.; Xiao, L.; Chen, Y.; Cheng, P.; Wu, D.; Zhuang, W. Learning-Based Computation Offloading for IoT Devices With Energy Harvesting. *IEEE Trans. Veh. Technol.* **2019**, *68*, 1930–1941. [CrossRef]
26. Chen, X.; Zhang, H.; Wu, C.; Mao, S.; Ji, Y.; Bennis, M. Optimized Computation Offloading Performance in Virtual Edge Computing Systems Via Deep Reinforcement Learning. *IEEE Internet Things J.* **2019**, *6*, 4005–4018. [CrossRef]
27. Yang, L.; Yao, H.; Wang, J.; Jiang, C.; Benslimane, A.; Liu, Y. Multi-UAV-Enabled Load-Balance Mobile-Edge Computing for IoT Networks. *IEEE Internet Things J.* **2020**, *7*, 6898–6908. [CrossRef]
28. Li, J.; Liu, Q.; Wu, P.; Shu, F.; Jin, S. Task Offloading for UAV-based Mobile Edge Computing via Deep Reinforcement Learning. In Proceedings of the 2018 IEEE/CIC International Conference on Communications in China (ICCC), Beijing, China, 16–18 August 2018; pp. 798–802. [CrossRef]
29. Wang, L.; Huang, P.; Wang, K.; Zhang, G.; Zhang, L.; Aslam, N.; Yang, K. RL-Based User Association and Resource Allocation for Multi-UAV enabled MEC. In Proceedings of the 2019 15th International Wireless Communications & Mobile Computing Conference (IWCMC), Tangier, Morocco, 24–28 June 2019; pp. 741–746. [CrossRef]
30. Wang, X.; Ning, Z.; Guo, S. Multi-Agent Imitation Learning for Pervasive Edge Computing: A Decentralized Computation Offloading Algorithm. *IEEE Trans. Parallel Distrib. Syst.* **2021**, *32*, 411–425. [CrossRef]
31. Wang, Z.; Rong, H.; Jiang, H.; Xiao, Z.; Zeng, F. A Load-Balanced and Energy-Efficient Navigation Scheme for UAV-Mounted Mobile Edge Computing. *IEEE Trans. Netw. Sci. Eng.* **2022**, *9*, 3659–3674. [CrossRef]
32. Wang, L.; Wang, K.; Pan, C.; Xu, W.; Aslam, N.; Hanzo, L. Multi-Agent Deep Reinforcement Learning-Based Trajectory Planning for Multi-UAV Assisted Mobile Edge Computing. *IEEE Trans. Cogn. Commun. Netw.* **2021**, *7*, 73–84. [CrossRef]
33. Waqar, N.; Hassan, S.A.; Mahmood, A.; Dev, K.; Do, D.T.; Gidlund, M. Computation Offloading and Resource Allocation in MEC-Enabled Integrated Aerial-Terrestrial Vehicular Networks: A Reinforcement Learning Approach. *IEEE Trans. Intell. Transp. Syst.* **2022**. [CrossRef]
34. Nie, Y.; Zhao, J.; Gao, F.; Yu, F.R. Semi-Distributed Resource Management in UAV-Aided MEC Systems: A Multi-Agent Federated Reinforcement Learning Approach. *IEEE Trans. Veh. Technol.* **2021**, *70*, 13162–13173. [CrossRef]
35. Zhu, Z.; Wan, S.; Fan, P.; Letaief, K.B. An Edge Federated MARL Approach for Timeliness Maintenance in MEC Collaboration. In Proceedings of the 2021 IEEE International Conference on Communications Workshops (ICC Workshops), Montreal, QC, Canada, 14–23 June 2021; pp. 1–6. [CrossRef]
36. Zhu, Z.; Wan, S.; Fan, P.; Letaief, K.B. Federated Multiagent Actor–Critic Learning for Age Sensitive Mobile-Edge Computing. *IEEE Internet Things J.* **2022**, *9*, 1053–1067. [CrossRef]
37. Wang, X.; Wang, C.; Li, X.; Leung, V.; Taleb, T. Federated Deep Reinforcement Learning for Internet of Things with Decentralized Cooperative Edge Caching. *IEEE Internet Things J.* **2020**, *7*, 9441–9455. [CrossRef]
38. Majidi, F.; Khayyambashi, M.R.; Barekatain, B. HFDRL: An Intelligent Dynamic Cooperate Cashing Method Based on Hierarchical Federated Deep Reinforcement Learning in Edge-Enabled IoT. *IEEE Internet Things J.* **2022**, *9*, 1402–1413. [CrossRef]
39. Mao, Y.; You, C.; Zhang, J.; Huang, K.; Letaief, K.B. A Survey on Mobile Edge Computing: The Communication Perspective. *IEEE Commun. Surv. Tutorials* **2017**, *19*, 2322–2358. [CrossRef]
40. Huang, L.; Bi, S.; Zhang, Y.J.A. Deep Reinforcement Learning for Online Computation Offloading in Wireless Powered Mobile-Edge Computing Networks. *IEEE Trans. Mob. Comput.* **2020**, *19*, 2581–2593. [CrossRef]

41. Xu, Z.; Zhou, F.; Feng, L.; Wu, Z. MARL-Based Joint Computation Offloading and Aerial Base Stations Deployment in MEC Systems. In Proceedings of the 2022 IEEE International Symposium on Broadband Multimedia Systems and Broadcasting (BMSB), Bilbao, Spain, 15–17 June 2022; pp. 1–6. [CrossRef]
42. Dai, B.; Niu, J.; Ren, T.; Hu, Z.; Atiquzzaman, M. Towards Energy-Efficient Scheduling of UAV and Base Station Hybrid Enabled Mobile Edge Computing. *IEEE Trans. Veh. Technol.* **2022**, *71*, 915–930. [CrossRef]

electronics

MDPI

Article

Beamforming Optimization with the Assistance of Deep Learning in a Rate-Splitting Multiple-Access Simultaneous Wireless Information and Power Transfer System with a Power Beacon

Mario R. Camana [1,2], Carla E. Garcia [2] and Insoo Koo [1,*]

1 Department of Electrical, Electronic and Computer Engineering, University of Ulsan, Ulsan 680-749, Republic of Korea; mario_camana@hotmail.com

2 Interdisciplinary Centre for Security, Reliability and Trust, University of Luxembourg, 4365 Luxembourg City, Luxembourg; carli.garcia27@hotmail.com

* Correspondence: iskoo@ulsan.ac.kr

Citation: Camana, M.R.; Garcia, C.E.; Koo, I. Beamforming Optimization with the Assistance of Deep Learning in a Rate-Splitting Multiple-Access Simultaneous Wireless Information and Power Transfer System with a Power Beacon. *Electronics* **2024**, *13*, 872. https://doi.org/10.3390/electronics13050872

Academic Editors: Juan-Carlos Cano, Yichuang Sun, Haeyoung Lee and Oluyomi Simpson

Received: 18 December 2023
Revised: 10 February 2024
Accepted: 21 February 2024
Published: 23 February 2024

Abstract: This study examined the implementation of rate-splitting multiple access (RSMA) in a multiple-input single-output system using simultaneous wireless information and power transfer (SWIPT) technology. The coexistence of a base station and a power beacon was considered, aiming to transmit information and energy to two sets of users. One set comprises users who solely harvest energy, whereas the other can decode information and energy using a power-splitting (PS) structure. The main objective of this optimization was to minimize the total transmit power of the system while satisfying the rate requirements for PS users and ensuring minimum energy harvesting (EH) for both PS and EH users. The non-convex problem was addressed by dividing it into two subproblems. The first subproblem was solved using a deep learning-based scheme, combining principal component analysis and a deep neural network. The semidefinite relaxation method was used to solve the second subproblem. The proposed method offers lower computational complexity compared to traditional iterative-based approaches. The simulation results demonstrate the superior performance of the proposed scheme compared to traditional methods such as non-orthogonal multiple access and space-division multiple access. Furthermore, the ability of the proposed method to generalize was validated by assessing its effectiveness across several challenging scenarios.

Keywords: simultaneous wireless information and power transfer; SWIPT; power beacon; rate-splitting multiple access; RSMA; deep neural network; DNN; semidefinite relaxation; SDR

1. Introduction

Rate-splitting multiple access (RSMA) has become a promising multiple-access framework that is being considered for implementation in future 6G networks and beyond [1]. Recent studies have shown that RSMA outperforms traditional multiple-access techniques, such as non-orthogonal multiple access (NOMA) and space-division multiple access (SDMA), in terms of energy and spectral efficiency for multi-antenna systems [1–4]. The concept behind RSMA involves dividing user messages into multiple parts, which are transmitted using superposition coding at the transmitter and decoded using successive interference cancellation (SIC) at the receivers. The most common approach is based on single-layer rate splitting (RS), involving two parts: a common part and a private part. The common parts from all users are transmitted simultaneously using a shared codebook and must be decoded by all users, whereas the private parts are transmitted using private codebooks. This approach allows RSMA to transmit all the messages simultaneously and in the same frequency band using the power and spatial domains. Consequently, RSMA offers flexibility, where interference can be treated as noise or fully decoded, making it a versatile framework encompassing SDMA and power-division NOMA as special cases [3,5].

Modern wireless systems consider communication trends, such as the Internet of Things (IoT) and machine-type communications (MTC). The key priorities in wireless communications involve advancing green technology and reducing device power consumption [6]. Various studies have been undertaken to establish self-sustainable communication systems using energy-harvesting (EH) techniques. One of the EH technologies contributing to these objectives is simultaneous wireless information and power transfer (SWIPT) [7]. SWIPT is an effective technology that enables the base station (BS) to transmit energy and information simultaneously to wireless users. Within SWIPT, the most well-known architectures are the power-splitting (PS) architecture and time-switching (TS) architecture, where both schemes consider an EH module and an information decoding (ID) module at the receiver. In the TS architecture, the receiver periodically alternates between the ID and EH modules based on a TS sequence. In the PS architecture, the incoming radio frequency (RF) signal is divided into two streams based on a PS ratio, which are sent to the ID and EH modules. The authors of [7] focused on the PS architecture because it has been established in the literature as achieving the best balance between energy harvesting and information decoding.

A cost-effective solution was proposed in [8] using low-cost stations called power beacons (PBs) for wirelessly recharging devices using RF energy. PBs primarily serve as a source of wireless energy, extending the operational lifespan of battery-powered devices through wireless recharging. PBs do not necessitate complex computations and have low backhaul link requirements, enabling cost-effective and adaptable placement, making them a valuable addition to the system.

SWIPT has been widely studied, ranging from single-antenna systems [9] to multiple-antenna systems [10,11]. Shi et al. [10] optimized the precoding vectors and PS ratios in a multiuser multiple-input single-output (MU MISO) system to minimize the transmission power while considering both the minimum signal-to-interference-plus-noise ratio (SINR) and minimum EH at the user side. An extension to a multiuser multi-input multi-output (MIMO) system was explored in [11], considering the EH constraints and a maximum tolerable mean square error (MSE) for received information. SWIPT has also been implemented along with the NOMA method in multi-antenna systems to maximize data rates [12], minimize transmit power [13], and enhance energy efficiency [14]. On the other hand, there is limited research on the performance of networks implementing SWIPT in conjunction with the RSMA method.

In the context of RSMA, a pioneering study introduced the concept of RS [2]. The authors elucidated the principal limitations of conventional methods while emphasizing the potential advantages of RS in terms of spectral and energy efficiencies compared to traditional techniques. The initial investigations of RS in a multiuser MISO system [15,16] aimed to maximize the minimum rate and maximize the sum rate. The solutions to these optimization problems were based on the weighted minimum mean square error (WMMSE) method and the alternate optimization (AO) algorithm for optimizing the precoding vectors and common rate variables. These studies reported that the RS framework consistently outperformed conventional methods, as evidenced by a comprehensive analysis of data rates and complexity in scenarios involving imperfect channel state information at the transmitter (CSIT). In [17], an extension analysis was conducted on massive MIMO systems to maximize the minimum achievable rate of the common message. A hierarchical rate-splitting approach was proposed to address the challenges posed by the extensive array of antennas at the transmitter. The simulation results revealed the superior performance of RS over conventional broadcasting methods, considering perfect and imperfect CSIT. The RSMA method was initially introduced for downlink multiuser MISO systems [3]. The authors highlighted the generality of RSMA compared to SDMA and NOMA. They also addressed the maximization of the weighted sum rate while considering the minimum rate constraints and power limitations. The validity of the authors' assertions was confirmed by the simulation results, emphasizing the superior performance of RSMA across various scenarios with different network loads and numbers of users. An exhaustive analytical analysis was conducted for a two-user case in [5], demonstrating the generality of RSMA by showcasing how it encompasses SDMA, NOMA, and orthogonal multiple access (OMA) as particular cases. Furthermore, the energy efficiency maximization and sum-rate maximization in multiuser MISO systems examined using RS emphasized the outstanding performance of RSMA, demonstrating its superior spectral and energy efficiency compared to SDMA and NOMA [18].

An initial study of RSMA and SWIPT was conducted to maximize the sum rate of users [19], considering two types of users: one for decoding information only and the other for harvesting energy only. A previous study [20] investigated the performance of RSMA with SWIPT in multiuser MISO systems, considering that users can decode information and harvest energy based on a PS factor. Furthermore, the scenario of RSMA with SWIPT when an IRS is deployed in the system was investigated in [21]. These studies reported significant improvements in performance provided by RSMA compared to traditional SDMA and NOMA methods. On the other hand, the aforementioned studies primarily focused on users close to the BS, driven by the EH requirements imposed by the optimization problem. This limitation becomes evident in real-world deployments, where users may be located far from the BS, posing challenges in meeting the EH requirements. Therefore, this paper investigated the deployment of a PB with RSMA as an efficient approach to address the previously mentioned challenge.

Huang et al. [22] introduced a system involving PBs for wireless power transfer (WPT). These PBs can wirelessly charge receivers and be strategically placed alongside femtocell base stations (BSs) to provide short-range SWIPT to wireless devices. A PB-assisted wireless-powered communication network (WPCN) was proposed in [23], consisting of a single-antenna PB and several single-antenna access points (APs). The PB provides RF energy to the APs, which transmit their information using the energy harvested from the PB. An extension of the PB-assisted WPCN, considering multi-antenna PBs, was reported in [24] to optimize the energy beamformer vector and maximize the spectrum efficiency. A previous study [25] considered the coexistence of a multi-antenna PB with a multi-antenna BS, assuming one ID user and several EH users, where the authors addressed the maximization of the total harvested energy. On the other hand, the aforementioned works did not consider SWIPT users equipped with a PS architecture and did not incorporate the RSMA framework.

Regarding state-of-the-art SWIPT systems assisted by a PB, the minimization of transmit power in a single-antenna system was investigated in [10]. Extensions to a multiuser MISO system incorporating SDMA and NOMA were introduced in [26,27], respectively. Vu et al. [28] considered a scenario in which a multi-antenna PB transmits RF energy to a single-antenna transmitter, serving two users with several relays. The transmitter employed the NOMA method and harvested energy from the PB to transmit the two messages. The relay applied the PS architecture to harvest energy and decode the message for the distant user, which was then forwarded. A part of this investigation was presented in a conference article [29], where the SWIPT system with RSMA was introduced with the aid of a PB, where the optimal scenario of decoding the whole interference from the PB at the user side was assumed. On the other hand, this paper considered the general case of treating the interference from the PB as noise. In addition, a low-complexity and efficient scheme was proposed based on deep learning (DL).

DL is gaining popularity as a technique for resource optimization in wireless networks because of its ability to significantly reduce computational complexity compared to traditional optimization methods. A review of the most common schemes based on machine learning and reinforcement learning methods for network optimization problems was published in [30,31]. A deep neural network (DNN) was introduced as an approximation method for solving the sum-rate maximization problem using the WMMSE algorithm in a single-antenna system [32]. The results of the simulations demonstrated the effectiveness of the DNN, which closely approximated the solution of the WMMSE algorithm while reducing the computational time. Xia et al. [33] addressed three popular optimization problems in a multiuser MISO system with SDMA. Their proposed solution leveraged a neural network module to predict key features based on the channel vectors and a beamforming module to construct the beamforming vector from the predicted vital features. Building upon the previous system model, the sum-rate maximization problem was investigated in [34], and two types of schemes based on a DNN were proposed: one where the beamformers are directly generated by the DNN output and the other utilizing a beamformer recovery module after the DNN module. The simulation results demonstrated performance close to traditional optimization methods, with the model incorporating the beamformer recovery module yielding the best results. On the other hand, the aforementioned work did not consider SWIPT technology and the RSMA method. The solutions are problem-dependent and rely on expert knowledge. Furthermore, state-of-the-art methods consider the channel vectors as direct inputs to the deep learning module. This approach involves using hundreds of features as the number of antennas and users increases, increasing the complexity of the deep learning model. A previous work [35] considered a MISO SWIPT system with RSMA, where a deep learning-based solution was proposed. The scheme comprised three modules: an autoencoder for dimension reduction, a DNN to predict a set of target variables, and a precoding module to obtain the precoding vectors. On the other hand, the authors did not account for EH users or the deployment of a PB, which significantly increased the number of input features. Furthermore, the autoencoder module required problem-dependent adjustments of its hyperparameters, such as the number of hidden layers, hidden nodes, activation functions, and learning rate. These factors made it difficult to generalize the scheme to other scenarios.

The RSMA framework has showcased significant enhancements in spectral and energy efficiencies compared to conventional SDMA and NOMA techniques. This establishes RSMA as a promising candidate for future 6G networks, prompting a thorough examination of its performance across diverse multi-antenna systems. Moreover, the deployment of PBs holds vital significance in extending the lifespan of wireless devices, especially in IoT scenarios where battery replacement poses challenges in hard-to-access areas. However, the interference generated by PBs during the information decoding process at the user side necessitates a comprehensive investigation into an efficient beamforming design at both the BS and PB. Consequently, this study aims to develop a high-performance, low-complexity solution to jointly optimize the beamforming vectors and PS ratios in a multi-antenna system. The proposed system model integrates a multi-antenna transmitter implementing RSMA with SWIPT, along with a multi-antenna PB. The primary objective of this study is to minimize the total transmission power while fulfilling data rates and EH requirements for EH and PS users. The main contributions of this paper are as follows:

- In the considered system model, the objective is to minimize the total transmission power of the BS and PB while meeting the minimum EH requirements for EH and PS users and ensuring a minimum data rate for PS users.
- A two-step approach is adopted to address the non-convex problem presented in this study. In the first step, we optimize the common rate variables through a DL-based scheme that combines the principal component analysis (PCA) technique for dimensionality reduction with a DNN. The second step focuses on optimizing the beamforming vectors of the BS and PB, along with the PS factors. The SDR technique is used to accomplish this.
- As a comparative scheme, the proposed minimization problem is addressed using a PSO-SDR approach, which is an iterative-based method that provides near-optimal solutions to the proposed problem. This scheme serves as a reference to analyze the performance of the proposed DL-based solution.
- Simulation results show that the proposed DL-based method can perform similarly to traditional iterative-based schemes while significantly reducing computational complexity. Furthermore, the proposed RSMA solution is compared with NOMA and SDMA, showing that the RSMA approach achieves the lowest transmit power. Moreover, the generalization performance of the proposed DL-based method is validated

by testing its performance across several challenging scenarios not included in the DL model training.

The remainder of this paper is structured as follows. Section 2 introduces the system model and formulates the problem. Section 3 outlines the proposed solution, and presents the comparative schemes. Section 4 provides the simulation results, and Section 5 reports the conclusions.

2. System Model and Problem Formulation

A MU MISO RSMA system assisted by a PB with SWIPT is considered, as shown in Figure 1. The number of antennas at the BS is $M \geq 2$, the number of antennas at the PB is $N \geq 2$, and there are K single-antenna PS users and G single-antenna EH users. EH users are exclusively dedicated to harvesting RF energy, whereas PS users are equipped with a PS mechanism, allowing them to partition the incoming RF signal into two components, serving both information decoding and EH based on a designated PS ratio, θ_k. The BS transmits information signals to both harvest energy at the EH module and decode information at the ID module for PS users. The PB, on the other hand, transmits energy-carrying signals used for energy harvesting by EH users and the EH module of PS users. However, these energy signals are considered interference at the ID module of PS users.

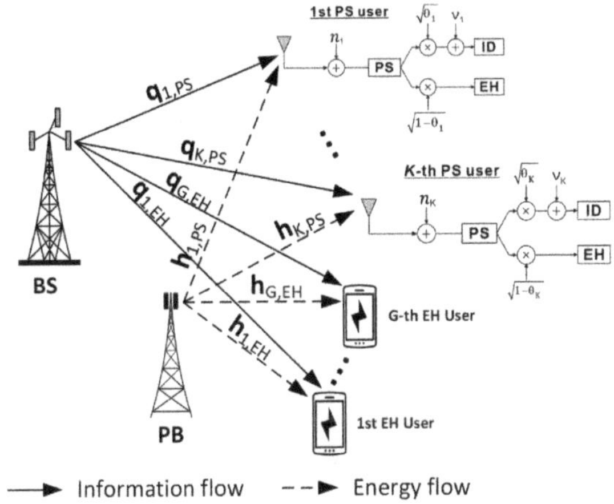

Figure 1. MISO SWIPT system aided by a PB.

In one-layer RSMA, the original message intended for the kth PS user, W_k, is separated into a common message, $W_{k,c}$, and a private message, $W_{k,p}$. A super-common message is produced by combining all the K common messages and encoding them into the common stream z_c^{PS}, which needs to be decoded by all PS users. The K private messages are independently encoded into K private streams $\{z_k^{PS}\}$ to be decoded by their respective user. At the PB, the energy-carrying signals are represented by $\left\{z_g^{EH}\right\}, g = 1, \ldots, G$. Therefore, the signals transmitted at the BS and PB after precoding are $\mathbf{x}_{BS} = \mathbf{p}_0 z_c^{PS} + \sum_{k=1}^{K} \mathbf{p}_k z_k^{PS}$ and $\mathbf{x}_{PB} = \sum_{g=1}^{G} \mathbf{e}_g z_g^{EH}$, respectively, where $\mathbf{p}_0 \in \mathbb{C}^{M\times 1}$ and $\mathbf{p}_k \in \mathbb{C}^{M\times 1}$ represent the information beamforming vectors, and $\mathbf{e}_g \in \mathbb{C}^{N\times 1}$ corresponds to the energy beamforming vector for z_c^{PS}, z_k^{PS}, and z_g^{EH}. Figure 2 shows the RSMA transmission scheme in the considered system model.

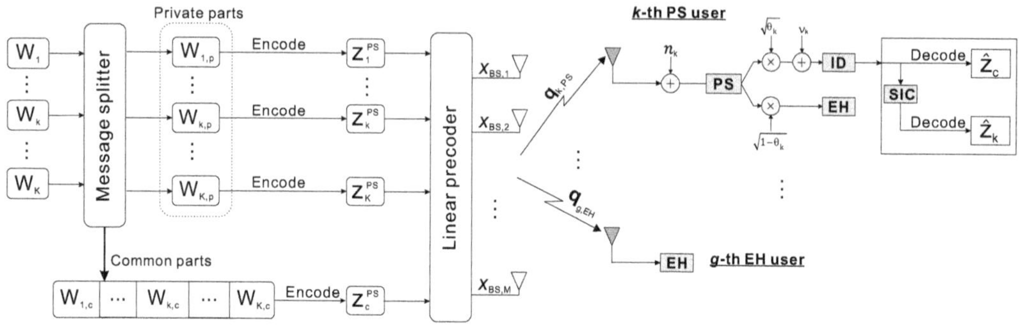

Figure 2. Proposed RSMA transmission scheme.

At the user side, the kth PS user first decodes the common stream z_c^{PS} with an achievable rate given by

$$R_{k,0} = \log_2\left(1 + \frac{\left|\mathbf{q}_{k,PS}^H \mathbf{p}_0\right|^2}{\sum\limits_{i=1}^{K}\left|\mathbf{q}_{k,PS}^H \mathbf{p}_i\right|^2 + \sum\limits_{g=1}^{G}\left|\mathbf{h}_{k,PS}^H \mathbf{e}_g\right|^2 + \sigma_k^2 + \frac{\gamma_k^2}{\theta_k}}\right), \forall k, \quad (1)$$

where $\mathbf{q}_{k,PS} \in \mathbb{C}^{M\times 1}$ represents the channel vector from the BS to the kth PS user; $\mathbf{h}_{k,PS} \in \mathbb{C}^{N\times 1}$ is the channel vector from the PB to the kth PS user; $\theta_k \in (0,1)$ is the PS factor; $n_k \sim \mathcal{CN}(0,\sigma_k^2)$ represents the Gaussian noise at the antenna of the kth PS user; and $v_k \sim \mathcal{CN}(0,\delta_k^2)$ is the data processing noise at the ID module of the kth PS user.

After decoding z_c^{PS}, RSMA employs the SIC procedure to eliminate interference due to the common stream z_c^{PS}. The kth PS can decode its private stream z_k^{PS} with an achievable rate given by

$$\mathrm{R}_k = \log_2\left(1 + \frac{\left|\mathbf{q}_{k,PS}^H \mathbf{p}_k\right|^2}{\sum\limits_{i=1,i\neq k}^{K}\left|\mathbf{q}_{k,PS}^H \mathbf{p}_i\right|^2 + \sum\limits_{g=1}^{G}\left|\mathbf{h}_{k,PS}^H \mathbf{e}_g\right|^2 + \sigma_k^2 + \frac{\gamma_k^2}{\theta_k}}\right), \forall k. \quad (2)$$

The rate at which z_c^{PS} is transmitted, expressed as R_0, must satisfy $R_0 \leq \min\{R_{1,0},\ldots,R_{K,0}\}$ because the common stream z_c^{PS} needs to be decoded by all users. Moreover, the rate R_0 is composed of the rates to transmit each $W_{k,c}$ and can be expressed as $R_0 = \sum\limits_{k=1}^{K}\alpha_k$, where α_k is the rate to transmit the common part of the kth message, $W_{k,c}$.

The energy harvested at the EH module of the kth PS user can be expressed as follows:

$$\mathrm{EH}_k^{PS} = \varsigma_k^{PS}(1-\theta_k)\left(\sum\limits_{i=0}^{K}\left|\mathbf{q}_{k,PS}^H \mathbf{p}_i\right|^2 + \sum\limits_{g=1}^{G}\left|\mathbf{h}_{k,PS}^H \mathbf{e}_g\right|^2 + \sigma_k^2\right), \forall k, \quad (3)$$

where ς_k^{PS} is the EH efficiency at the kth PS user. Moreover, at the gth EH user, the harvested energy can be expressed as

$$\mathrm{EH}_g^{EH} = \varsigma_g^{EH}\left(\sum\limits_{i=0}^{K}\left|\mathbf{q}_{g,EH}^H \mathbf{p}_i\right|^2 + \sum\limits_{j=1}^{G}\left|\mathbf{h}_{g,EH}^H \mathbf{e}_j\right|^2\right), \forall g, \quad (4)$$

where ς_g^{EH} represents the EH efficiency at the gth EH user, and $\mathbf{h}_{g,EH} \in \mathbb{C}^{N\times 1}$ and $\mathbf{q}_{g,EH} \in \mathbb{C}^{M\times 1}$ are the channel vectors from the PB and BS to the gth EH user, respectively.

We seek to optimize the beamforming vectors, $\mathbf{p}_0, \{\mathbf{p}_k, \mathbf{e}_g\}$; the common rate variables, α_k; and the PS ratios, θ_k, which together achieve a minimum rate requirement, and at the same time, they can harvest a required minimum EH for future use. The minimization of the sum transmission power of the PB and BS can be formulated as follows:

$$\min_{\mathbf{p}_0, \{\mathbf{p}_k, \alpha_k, \theta_k, \mathbf{e}_g\}} \sum_{i=0}^{K} \|\mathbf{p}_i\|^2 + \sum_{g=1}^{G} \|\mathbf{e}_g\|^2 \tag{5a}$$

$$\text{s.t. } \alpha_k + R_k \geq \chi_k, \forall k \tag{5b}$$

$$\sum_{i=1}^{K} \alpha_i \leq R_{k,0}, \forall k \tag{5c}$$

$$\text{EH}_k^{PS} \geq \varepsilon_k^{PS}, \forall k \tag{5d}$$

$$\text{EH}_g^{EH} \geq \varepsilon_g^{EH}, \forall g \tag{5e}$$

$$0 < \theta_k < 1, \forall k \tag{5f}$$

$$\alpha_k \geq 0, \forall k, \tag{5g}$$

Constraint (5b) ensures that the kth PS user achieves a minimum rate requirement, denoted as χ_k. Constraint (5c) is set to guarantee that the common stream can be decoded by all PS users. Constraints (5d) and (5e) ensure that each kth PS user and gth EH user can harvest a minimum EH requirement, denoted as ε_k^{PS} and ε_g^{EH}, respectively. Solving the power minimization problem (5) is challenging because of its non-convex constraints (5b)–(5e). In the following, a near-optimal solution is proposed based on DL techniques and the SDR method.

3. Proposed Approach for Addressing Problem (5)

A DL-based scheme and SDR method are developed to address the non-convex problem (5). First, problem (5) is reformulated into two subproblems, denoted as follows:

$$\min_{\{\alpha_k\}} (\Upsilon(\alpha_k)) \tag{6a}$$

$$\Upsilon(\alpha_k) = \min_{\mathbf{p}_0, \{\mathbf{p}_k, \theta_k, \mathbf{e}_g\}} \sum_{i=0}^{K} \|\mathbf{p}_i\|^2 + \sum_{g=1}^{G} \|\mathbf{e}_g\|^2 \quad \text{s.t. (5b)–(5f)}, \tag{6b}$$

where the first subproblem, represented in (6a), optimizes the common rate variables, α_k, for the given beamforming vectors, $\mathbf{p}_0, \{\mathbf{p}_k, \mathbf{e}_g\}$, and PS ratios, θ_k. Section 3.1 presents a DNN-based scheme with PCA to solve this first subproblem. The second subproblem, represented by $\Upsilon(\alpha_k)$ in (6b), optimizes the beamforming vectors, $\mathbf{p}_0, \{\mathbf{p}_k, \mathbf{e}_g\}$, and the PS ratios, θ_k, for the given common rate variables, α_k. The proposed solution for the second subproblem is based on the SDR technique with the penalty function method and is described in Section 3.2.

Figure 3 illustrates the overall procedure of the proposed scheme during both the training and online stages. In the training phase, the initial step involves generating the dataset. The dataset is generated by solving a minimization problem using conventional optimization methods, such as combining the PSO algorithm and SDR method. The features of the dataset consist of the rate and EH requirements, along with a reduced representation of the channel vectors obtained through PCA. The target values in this dataset correspond to the common rate, α_k. Subsequently, the K-fold cross-validation method is used to partition the dataset into training and validation subsets. These subsets are used to determine the optimal hyperparameters for the DNN module, including the number of hidden layers, hidden nodes, learning rates, batch size, and others. Once the DNN is trained, it is deployed at the BS for the online stage. During the online stage, the current channel vectors undergo dimensionality reduction via PCA. The reduced representations, rate, and EH requirements serve as inputs for the trained DNN. The result of the DNN is the prediction of the common rate variables, α_k, which are used in the SDR module to optimize the PS factors, θ_k, and the beamforming vectors, $\mathbf{p}_0, \{\mathbf{p}_k, \mathbf{e}_g\}$.

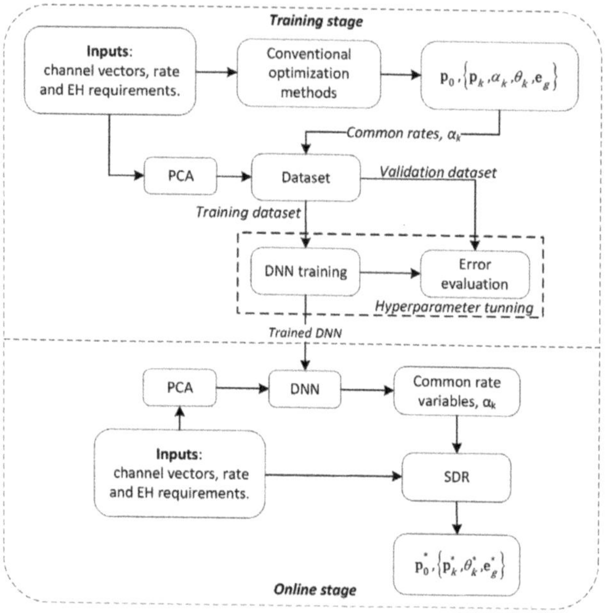

Figure 3. Overall procedure of the proposed approach.

Moreover, by utilizing the predicted common rate variables, α_k, the minimization problem (5) can be reformulated as follows:

$$\min_{\mathbf{p}_0, \{\mathbf{p}_k, \theta_k, \mathbf{e}_g\}} \quad \sum_{i=0}^{K} \|\mathbf{p}_i\|^2 + \sum_{g=1}^{G} \|\mathbf{e}_g\|^2 \tag{7a}$$

subject to

$$\frac{\left|\mathbf{q}_{k,PS}^H \mathbf{p}_k\right|^2}{\sum\limits_{i=1, i\neq k}^{K} \left|\mathbf{q}_{k,PS}^H \mathbf{p}_i\right|^2 + \sum\limits_{g=1}^{G} \left|\mathbf{h}_{k,PS}^H \mathbf{e}_g\right|^2 + \sigma_k^2 + \frac{\gamma_k^2}{\theta_k}} \geq \omega_k, \ \forall k \tag{7b}$$

$$\frac{\left|\mathbf{q}_{k,PS}^H \mathbf{p}_0\right|^2}{\sum\limits_{i=1}^{K} \left|\mathbf{q}_{k,PS}^H \mathbf{p}_i\right|^2 + \sum\limits_{g=1}^{G} \left|\mathbf{h}_{k,PS}^H \mathbf{e}_g\right|^2 + \sigma_k^2 + \frac{\gamma_k^2}{\theta_k}} \geq \varphi, \ \forall k \tag{7c}$$

along with (5d), (5e) and (5g),

where $\varphi = 2^{\sum_{i=1}^{K} \alpha_i} - 1$, and $\omega_k = \max\{0, 2^{\chi_k - \alpha_k} - 1\}$.

3.1. Deep Learning-Based Scheme for Optimizing α_k

This subsection describes the PCA module for dimensionality reduction and the DNN-based method to predict the common rate variables, α_k.

Figure 4 presents the PCA module used to derive a reduced representation of each channel vector. The PCA technique [36] projects the input data onto a lower-dimensional subspace to maximize the variance of the projected data. The training dataset is defined as $\mathbf{X} = \{\mathbf{x}_1, \ldots, \mathbf{x}_n, \ldots, \mathbf{x}_{N_D}\}$, where N_D is the total number of channel vectors used as the training dataset, and $\mathbf{x}_n$ is a W-dimensional vector composed of the real and imaginary components of the nth channel. The PCA module maps a channel vector onto an L-dimensional subspace that satisfies $L < W$. The projection of $\mathbf{x}_n$ can be represented by $\mathbf{y}_n = \mathbf{U}_L^T \mathbf{x}_n$, where $\mathbf{U}_L = [\mathbf{u}_1^T, \ldots, \mathbf{u}_L^T]$ with $\mathbf{u}_l^T \mathbf{u}_l = 1$ for $l = 1, \ldots, L$. A $\mathbf{U}_L$ that maximizes the trace of the covariance matrix of $\{\mathbf{y}_n\}$ is needed to maximize the variance of $\{\mathbf{y}_n\}$, which can be expressed as:

$$\mathbf{U}_L^* = \arg\max_{\mathbf{U}_L} (\mathrm{tr}(\mathbf{C}_\mathbf{y})), \tag{8}$$

where $\mathbf{C}_\mathbf{y} = \frac{1}{N_D} \sum_{n=1}^{N_D} (\mathbf{y}_n - \bar{\mathbf{y}})(\mathbf{y}_n - \bar{\mathbf{y}})^T$, and $\bar{\mathbf{y}} = \frac{1}{N_D} \sum_{n=1}^{N_D} \mathbf{x}_n$. Let $\mathbf{C}_\mathbf{x} = \frac{1}{N_D} \sum_{n=1}^{N_D} (\mathbf{x}_n - \bar{\mathbf{x}})(\mathbf{x}_n - \bar{\mathbf{x}})^T$ represent the covariance matrix of $\{\mathbf{x}_n\}$, and $\bar{\mathbf{x}} = \frac{1}{N_D} \sum_{n=1}^{N_D} \mathbf{x}_n$. The solution of (8), described in [36,37], is obtained when

$$\mathbf{C}_\mathbf{x} \mathbf{u}_l = \lambda_l \mathbf{u}_l, \tag{9}$$

which means that the variance of the projected data is maximized when $\mathbf{u}_l$ is an eigenvector of $\mathbf{C}_\mathbf{x}$. Therefore, the optimal $\mathbf{U}_L^*$ is a matrix composed of the first L eigenvectors of $\mathbf{C}_\mathbf{x}$ as columns. Algorithm 1 summarizes the steps for performing the PCA technique.

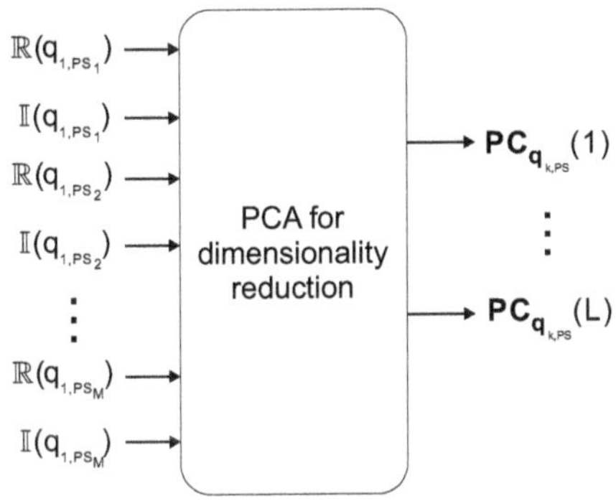

Figure 4. PCA module representation for the channel $\mathbf{q}_{1,PS}$.

Algorithm 1 PCA for dimensionality reduction

1: **inputs:** Training dataset $\mathbf{X} = \{\mathbf{x}_1, \ldots, \mathbf{x}_n, \ldots, \mathbf{x}_{N_D}\}$ containing the wireless channels and the number of components to keep, L.
2: Evaluate the mean $\bar{\mathbf{x}}$.
3: Calculate the covariance matrix $\mathbf{C}_\mathbf{x}$.
4: Decompose $\mathbf{C}_\mathbf{x}$ to obtain the eigenvectors and eigenvalues.
5: Select L eigenvectors corresponding to the L largest eigenvalues and create $\mathbf{U}_L$.
6: Map the W-dimensional channel vector into an L-dimensional representation by $\mathbf{y}_n = \mathbf{U}_L^T \mathbf{x}_n$.
7: **output:** Reduced representation of the channel vectors $\mathbf{q}_{k,PS}, \mathbf{q}_{g,EH}, \mathbf{h}_{k,PS}, \mathbf{h}_{g,EH}$ by performing step 6 for each channel vector.

Figure 5 shows the DNN model for the proposed approach, composed of an input layer, H_D hidden layers, and an output layer. The number of nodes in the input layer is determined by the user's rate requirements, EH requirements, and the reduced representation of the channel vectors after the PCA module, $\left\{\mathbf{PC}_{\mathbf{q}_{k,PS}}, \mathbf{PC}_{\mathbf{q}_{g,EH}}, \mathbf{PC}_{\mathbf{h}_{k,PS}}, \mathbf{PC}_{\mathbf{h}_{g,EH}}\right\}$. Each hidden layer consists of N_H hidden nodes, where the number of hidden layers and the number of hidden nodes are determined by fine-tuning the hyperparameters, as discussed in Section 4. The common rate variables specify the number of nodes in the output layer, $\{\alpha_k\}$. During the training stage, the weights of the DNN are optimized using the backpropagation algorithm [38], given a training dataset with $\{\alpha_k\}$ as the real-valued targets. In conjunction with the grid search method, the K-fold cross-validation method is employed to select the best hyperparameters, including the number of hidden layers, hidden nodes, learning rates, and activation functions. Subsequently, during the online stage, after successfully training the DNN, the predicted common rate variables for the current channel vectors are constrained to feasible ranges, defined by the constraints (5b) and (5g), and are expressed as follows:

$$\alpha_k = \begin{cases} 0, & \text{if } \alpha_k < 0, \ \forall k, & \text{(10a)} \\ \alpha_k, & \text{if } 0 \leq \alpha_k \leq \chi_k, \ \forall k, & \text{(10b)} \\ \chi_k, & otherwise, \ \forall k & \text{(10c)} \end{cases}$$

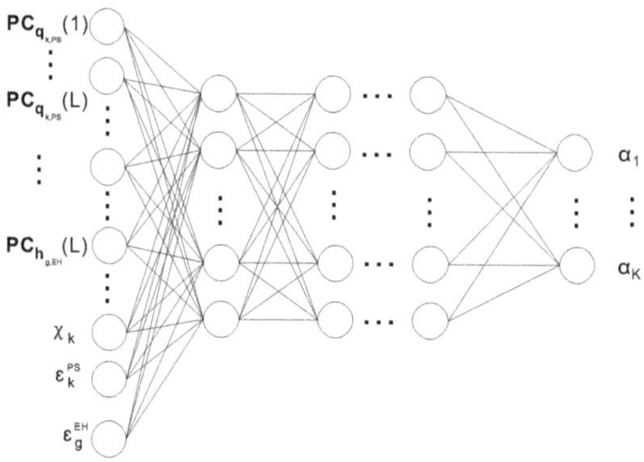

Figure 5. DNN module scheme for the proposed approach.

The modified common rate variables are passed to the SDR-based scheme to optimize the PS ratios, θ_k, and the beamforming vectors, $\mathbf{p}_0, \{\mathbf{p}_k, \mathbf{e}_g\}$. Within the DNN, we analyze the computational complexity in the online stage, as the training process takes place offline. This complexity is closely tied to the number of nodes and layers within the DNN. In particular, there are a total of $L(2K+2G)+2K+G$ nodes in the input layer, K nodes in the output layer, and each ith hidden layer has $N_{H,i}$ nodes, where $i = 1, \ldots, H_D$. Consequently, the computational complexity of the DNN module can be expressed as $\mathcal{O}((L(2K+2G)+2K+G)N_{H,1} + N_{H,1}N_{H,2} + \ldots + N_{H,H_D}K)$, even though this can be approximated to $\mathcal{O}(L(2K+2G))$ when the DNN parameters are considered fixed.

3.2. SDR-Based Method for Optimizing $\mathbf{p}_0, \{\mathbf{p}_k, \mathbf{e}_g, \theta_k\}$

We introduce an approach based on the SDR technique to jointly optimize the PS factors and beamforming vectors, with a fixed value for $\{\alpha_k\}$. The matrix variables are denoted as $\mathbf{P}_i = \mathbf{p}_i\mathbf{p}_i^H$, $\mathbf{E}_g = \mathbf{e}_g\mathbf{e}_g^H$, $\mathbf{Q}_{k,PS} = \mathbf{q}_{k,PS}\mathbf{q}_{k,PS}^H$, $\mathbf{H}_{k,PS} = \mathbf{h}_{k,PS}\mathbf{h}_{k,PS}^H$, $\mathbf{Q}_{g,EH} = \mathbf{q}_{g,EH}\mathbf{q}_{g,EH}^H$, and $\mathbf{H}_{g,EH} = \mathbf{h}_{g,EH}\mathbf{h}_{g,EH}^H$. In addition, the matrix variable $\mathbf{P}_i$ enforces the conditions $\mathbf{P}_i \succeq 0$, $\text{rank}(\mathbf{P}_i) = 1$, while matrix $\mathbf{E}_g$ enforces $\mathbf{E}_g \succeq 0$ and $\text{rank}(\mathbf{E}_g) = 1$. Problem (7) can be transformed into a convex problem by removing the rank-one constraints and given $\{\alpha_k\}$, as follows:

$$\min_{\mathbf{P}_0, \{\mathbf{P}_k, \theta_k, \mathbf{E}_g\}} \sum_{i=0}^{K} \text{Tr}(\mathbf{P}_i) + \sum_{g=1}^{G} \text{Tr}(\mathbf{E}_g) \tag{11a}$$

subject to

$$\left(\sum_{i=1, i\neq k}^{K} \text{Tr}(\mathbf{Q}_{k,PS}\mathbf{P}_i) + \sum_{g=1}^{G} \text{Tr}(\mathbf{H}_{k,PS}\mathbf{E}_g) + \sigma_k^2 + \frac{\gamma_k^2}{\theta_k} \right)\omega_k - \text{Tr}(\mathbf{Q}_{k,PS}\mathbf{P}_k) \leq 0, \ \forall k \tag{11b}$$

$$\left(\sum_{i=1}^{K}\mathrm{Tr}(\mathbf{Q}_{k,PS}\mathbf{P}_i)+\sum_{g=1}^{G}\mathrm{Tr}\big(\mathbf{H}_{k,PS}\mathbf{E}_g\big)+\sigma_k^2+\frac{\gamma_k^2}{\theta_k}\right)\varphi-\mathrm{Tr}(\mathbf{Q}_{k,PS}\mathbf{P}_0)\le 0,\ \forall k \tag{11c}$$

$$\frac{\varepsilon_k^{PS}}{\varsigma_k^{PS}(1-\theta_k)}-\sum_{i=0}^{K}\mathrm{Tr}(\mathbf{Q}_{k,PS}\mathbf{P}_i)-\sum_{g=1}^{G}\mathrm{Tr}\big(\mathbf{H}_{k,PS}\mathbf{E}_g\big)-\sigma_k^2\le 0,\ \forall k \tag{11d}$$

$$-\varsigma_g^{EH}\sum_{i=0}^{K}\mathrm{Tr}\big(\mathbf{Q}_{g,EH}\mathbf{P}_i\big)-\varsigma_g^{EH}\sum_{j=1}^{G}\mathrm{Tr}\big(\mathbf{H}_{g,EH}\mathbf{E}_j\big)+\varepsilon_g^{EH}\le 0,\ \forall g \tag{11e}$$

$$\mathbf{P}_0,\mathbf{P}_k,\mathbf{E}_g\succeq 0,\ \forall k,\ \forall g \tag{11f}$$

$$0<\theta_k<1,\ \forall k. \tag{11g}$$

Problem (11) is convex and can be solved efficiently using the CVX toolbox in MATLAB [39]. Problem (11) involves $K+1$ matrices of size $M\times M$ and G matrices of size $N\times N$, along with the $L_C = 3K+G$ linear constraint variables. Therefore, the computational complexity of addressing problem (11) amounts to $\mathcal{O}\Big(\sqrt{(K+1)M+GN}\Big((K+1)^3M^6 + G^3N^6 + L_c\big((K+1)M^2+GN^2\big)\Big)\log(1/\zeta)\Big)$ while maintaining a solution accuracy of $\zeta > 0$ [40]. The optimal solutions are represented in problem (11) as $\left\{\mathbf{P}_k^*,\mathbf{E}_g^*\right\}$. When the matrix solutions have a rank of one, the optimal beamforming vectors, $\left\{\mathbf{p}_i^*,\mathbf{e}_g^*\right\}$, are given by

$$\mathbf{p}_i=\sqrt{\lambda_H(\mathbf{P}_i)}\mathbf{v}_{H,\mathbf{P}_i},\ i=0,\ldots,K \tag{12a}$$

$$\mathbf{e}_g=\sqrt{\lambda_H(\mathbf{E}_g)}\mathbf{v}_{H,\mathbf{E}_g},\ g=1,\ldots,G, \tag{12b}$$

where $\lambda_H(\mathbf{A})$ represents the largest eigenvalue of the matrix $\mathbf{A}$, and $\mathbf{v}_{H,\mathbf{A}}$ denotes its corresponding eigenvector. If $\left\{\mathbf{P}_i^*,\mathbf{E}_g^*\right\}$ are rank-one, alternative approaches, such as the penalty function method [41] or the Gaussian randomization technique [20], can be employed to approximate the beamforming vectors.

The following details the penalty function method for the scenario where the matrix variables $\left\{\mathbf{P}_i^*,\mathbf{E}_g^*\right\}$ are not rank-one. The proposed penalty-based method is based on the definition of $\left\{\mathbf{P}_i^*,\mathbf{E}_g^*\right\}$ being positive semidefinite matrices, which satisfy the conditions $\mathrm{Tr}(\mathbf{P}_i)\ge\lambda_H(\mathbf{P}_i)$ and $\mathrm{Tr}(\mathbf{E}_g)\ge\lambda_H(\mathbf{E}_g)$, where the matrices $\left\{\mathbf{P}_i^*,\mathbf{E}_g^*\right\}$ are rank-one if $\mathrm{Tr}(\mathbf{P}_i)=\lambda_H(\mathbf{P}_i)$ and $\mathrm{Tr}(\mathbf{E}_g)=\lambda_H(\mathbf{E}_g)$. Hence, the proposed penalty-based approach aims to minimize $\mathrm{Tr}(\mathbf{P}_i)-\lambda_H(\mathbf{P}_i)$ and $\mathrm{Tr}(\mathbf{E}_g)-\lambda_H(\mathbf{E}_g)$ by introducing a penalty factor, κ, and incorporating these terms into the objective function as follows:

$$\begin{aligned}\min_{\mathbf{P}_0,\{\mathbf{P}_k,\theta_k,\mathbf{E}_g\}}\ &\sum_{i=0}^{K}\mathrm{Tr}(\mathbf{P}_i)+\sum_{g=1}^{G}\mathrm{Tr}(\mathbf{E}_g)+\kappa\sum_{i=0}^{K}\left(\mathrm{Tr}(\mathbf{P}_i)-\lambda_H(\mathbf{P}_i)\right)+\kappa\sum_{g=1}^{G}\left(\mathrm{Tr}(\mathbf{E}_g)-\lambda_H(\mathbf{E}_g)\right)\\ &\text{subject to (11b)},\ldots,\text{(11g)}.\end{aligned} \tag{13}$$

Problem (13) is non-convex because of the terms $-\lambda_H(\mathbf{P}_i)$ and $-\lambda_H(\mathbf{E}_g)$. Therefore, the following inequality that holds for any $\mathbf{Z}_k\ge 0$ is used to handle the aforementioned non-convex terms:

$$\lambda_H(\mathbf{P}_i)\ge\lambda_H(\mathbf{Z}_i)+\mathbf{v}_{H,\mathbf{Z}_i}^H(\mathbf{P}_i-\mathbf{Z}_i)\mathbf{v}_{H,\mathbf{Z}_i}, \tag{14}$$

where $\mathbf{v}_{H,\mathbf{Z}_i}$ represents the unit-norm eigenvector associated with the largest eigenvalue of $\mathbf{Z}_i$. Hence, (14) is used to approximate $\lambda_H(\mathbf{P}_i)$ based on a feasible matrix $\mathbf{P}_i^{(j)}$ at the jth iteration as follows:

$$\lambda_H(\mathbf{P}_i)\ge\lambda_H\Big(\mathbf{P}_i^{(j)}\Big)+\mathbf{v}_{H,\mathbf{P}_i^{(j)}}^H\Big(\mathbf{P}_i-\mathbf{P}_i^{(j)}\Big)\mathbf{v}_{H,\mathbf{P}_i^{(j)}},\ \forall k. \tag{15}$$

By applying a similar procedure as in (15), the term $\lambda_H(\mathbf{E}_g)$ can be equivalently transformed. As a result, problem (13) can be reformulated as follows:

$$\begin{aligned} &\min_{\mathbf{P}_0,\{\mathbf{P}_k,\theta_k,\mathbf{E}_g\}} \sum_{i=0}^{K} \mathrm{Tr}(\mathbf{P}_i)+\sum_{g=1}^{G} \mathrm{Tr}(\mathbf{E}_g) \\ &+\kappa \sum_{i=0}^{K} \left(\mathrm{Tr}(\mathbf{P}_i) - \lambda_H\left(\mathbf{P}_i^{(j)}\right) + \mathbf{v}^H_{H,\mathbf{P}_i^{(j)}} \left(\mathbf{P}_i - \mathbf{P}_i^{(j)}\right) \mathbf{v}_{H,\mathbf{P}_i^{(j)}} \right) \\ &+\kappa \sum_{g=1}^{G} \left(\mathrm{Tr}(\mathbf{E}_g) - \lambda_H\left(\mathbf{E}_g^{(j)}\right) + \mathbf{v}^H_{H,\mathbf{E}_g^{(j)}} \left(\mathbf{E}_g - \mathbf{E}_g^{(j)}\right) \mathbf{v}_{H,\mathbf{E}_g^{(j)}} \right) \\ &\text{subject to (11b)}, \ldots, \text{(11g)}. \end{aligned} \tag{16}$$

Note that problem (16) is convex, and the solution can be obtained using the CVX toolbox in MATLAB. Algorithm 2 lists the proposed iterative scheme based on the penalty method, which is applied when the solutions to problem (11), $\{\mathbf{P}_i^*, \mathbf{E}_g^*\}$, are not rank-one.

Algorithm 2 Penalty-based method for solving problem (11)

1: **inputs:** Matrix solutions of problem (11), $\{\mathbf{P}_i^*, \mathbf{E}_g^*\}$, tolerance value, ϕ, penalty factor, κ, channel vectors, rate, and EH requirements.
2: Set iteration counter $j = 0$.
3: Set initial feasible matrices, $\mathbf{P}_i^{(j)} = \mathbf{P}_i^*, \mathbf{E}_g^{(j)} = \mathbf{E}_g^*$.
4: **repeat**
5: Solve problem (16) given the feasible matrices $\left\{\mathbf{P}_i^{(j)}, \mathbf{E}_g^{(j)}\right\}$ and denote the solutions as $\{\mathbf{P}_i^*, \mathbf{E}_g^*\}$.
6: Increase counter $j = j + 1$.
7: Update the feasible matrices for the next iteration: $\mathbf{P}_i^{(j)} = \mathbf{P}_i^*, \mathbf{E}_g^{(j)} = \mathbf{E}_g^*$.
8: **until** $\sum_{i=0}^{K} (\mathrm{Tr}(\mathbf{P}_i) - \lambda_H(\mathbf{P}_i)) + \sum_{g=1}^{G} (\mathrm{Tr}(\mathbf{E}_g) - \lambda_H(\mathbf{E}_g)) \leq \phi$
9: Obtain the beamforming vectors with (12a) and (12b).
10: **Output:** $\{\mathbf{p}_i^*, \mathbf{e}_g^*\}$.

3.3. PSO-Based Approach for Optimizing $\{\alpha_k\}$ with a Given $\mathbf{p}_0, \{\mathbf{p}_k, \mathbf{e}_g, \theta_k\}$

This subsection presents a comparative scheme for optimizing the common rate variables using a PSO-based approach [20] with given PS factors and beamforming vectors. PSO is a potent metaheuristic algorithm inspired by the social behavior of flocking birds, where the collective knowledge of the swarm guides each particle through the search space to discover the optimal solution. In the proposed PSO-based scheme, there is a population of S particles whose position represents the common ratio variables to be optimized, i.e., the position of the sth particle is given by $\mathbf{x}_s = [\alpha_1^s, \ldots, \alpha_K^s], s = 1, \ldots, S$. The position of each particle is initialized randomly within the range $[0, \chi_k]$. The local best position for the sth particle, denoted as $\mathbf{x}_s^l$, represents the best location of the sth particle. Moreover, the global best position, denoted as $\mathbf{x}_g$, represents the best location of all the swarm particles. The location of the sth particle is based on its velocity and is expressed as

$$\mathbf{v}_s \leftarrow w\mathbf{v}_s + a_c u_1^n \left(\mathbf{x}_s^l - \mathbf{x}_s\right) + b_c u_2^n (\mathbf{x}_g - \mathbf{x}_s), \tag{17}$$

where w is the inertia weight; a_c, b_c are the acceleration parameters; and $u_1^n, u_2^n \sim U(0,1)$. Subsequently, the position of the sth particle is modified according to the following:

$$\mathbf{x}_s \leftarrow \mathbf{x}_s + \mathbf{v}_s. \tag{18}$$

The objective function, $f(\mathbf{x}_s)$, is defined by the sum transmission power (7a), obtained by solving problem (11), with $\{\alpha_k\}$ being determined by the position of the sth particle, $\mathbf{x}_s$. The computational complexity of the PSO-based method is given by $\mathcal{O}(S \cdot T_{\max} \cdot \mathcal{O}_{SDR})$, where $\mathcal{O}_{SDR}$ represents the complexity of solving problem (7) using the SDR method detailed in Section 3.2. Algorithm 3 presents the algorithm based on PSO designed to optimize $\{\alpha_k\}$.

Algorithm 3 Comparative scheme based on PSO for optimizing $\{\alpha_k\}$

1: **inputs:** Number of particles, S, maximum number of iterations, $T_{\max}$, rate, and EH requirements.
2: Set iteration counter $t = 0$ and initialize the position of each particle and its velocity, $\mathbf{v}_{s,t} = 0$.
3: Set the initial $\mathbf{x}^l_{s,t} = \mathbf{x}_{s,t}$ and $\mathbf{x}_{g,t} = \arg\min_{1 \le s \le S} f\left(\mathbf{x}^l_{s,t}\right)$.
4: **While** $t \le T_{\max}$ **do**
5:   **for** $s = 1, \ldots, S$, **do**
6:     Update the velocity and the position based on (17) and (18) to obtain $\mathbf{v}_{s,t+1}$ and $\mathbf{x}_{s,t+1}$.
7:     Restrict the value of the position based on (10).
8:     Solve problem (11) with the common rates given by $\mathbf{x}_{s,t+1}$ to get $f(\mathbf{x}_{s,t+1})$.
9:     **if** $f(\mathbf{x}_{s,t+1}) < f\left(\mathbf{x}^l_{s,t}\right)$ **then**
      $\mathbf{x}^l_{s,t} = \mathbf{x}_{s,t+1}$
    end if
10:   **end for**
11:   $\mathbf{x}_{g,t} = \arg\min_{1 \le s \le S} f\left(\mathbf{x}^l_{s,t}\right)$
12:   Increase counter $t = t + 1$.
13: **end while**
14: **Output:** Best common rates, $\{\alpha^*_k\}$, defined by $\mathbf{x}_g$.

4. Simulation Results

Numerical simulations were conducted to assess the performance of the proposed DL-based solution in minimizing the sum transmission power of the BS and PB in the considered MISO SWIPT system with RSMA. The simulation parameters were set to $K = 3$ PS users, $G = 2$ EH users, $M = 8$ antennas at the BS, $N = 8$ antennas at the PB, $\sigma^2_k = \gamma^2_k = -60$ dBm, $\varsigma^{PS}_k = \varsigma^{EH}_g = 1$, $\chi_k = \chi$, and $\varepsilon^{PS}_k = \varepsilon^{EH}_g = \varepsilon$. The channel vector between the kth PS user and the BS is given by

$$\mathbf{q}_{k,PS} = \sqrt{\beta d^{-\nu}_{BS-PS,k}}\tilde{\mathbf{q}}_{k,PS}, \forall k, \tag{19}$$

where $\nu = 2.2$ defines the path-loss exponent; $\beta = 10^{-3}$; $d^{-\nu}_{BS-PS,k}$ denotes the distance between the BS and the kth PS user; and $\tilde{\mathbf{q}}_{k,PS}$ is subject to independent Rician fading. $\mathbf{q}_{g,EH}, \mathbf{h}_{k,PS}, \mathbf{h}_{g,EH}$ are channels established according to (13). The BS is positioned at coordinates (8 m, 20 m), whereas the PB is located at (13 m, 7 m). PS users are randomly distributed within a designated region defined by $x_{PS} \in [13\text{ m}, 18\text{ m}]$ and $y_{PS} \in [13\text{ m}, 25\text{ m}]$. Similarly, EH users are randomly distributed within an area bounded by $x_{EH} \in [17\text{ m}, 22\text{ m}]$ and $y_{EH} \in [2\text{ m}, 12\text{ m}]$.

The effectiveness of the proposed DL-based method was evaluated by comparing it with a PSO-SDR method and conventional techniques, such as SDMA and NOMA. The proposed scheme is denoted as DNN-RSMA, and the comparative scheme based on PSO and SDR is denoted as PSO-RSMA. In SDMA [26], the kth PS user's message, W_k, is encoded directly into the data stream, z^{PS}_k, without common rate variables, and interference originating from other users is considered noise. In addition, a comparative scenario was incorporated by assuming the absence of the power beacon (PB) deployment in the network.

In NOMA [27], interference originating from other users is mitigated by employing multiple layers in the SIC process. The message intended for the kth PS user, W_k, is encoded into the private stream, $z_{k,N}^{PS}$. The transmitted signal at the BS is given by $\mathbf{x}_{BS,NOMA} = \sum_{k=1}^{K} \mathbf{p}_{k,N} z_{k,N}^{PS}$, where $\mathbf{p}_{k,N} \in \mathbb{C}^{M\times 1}$ is the information beamforming vector for $z_{k,N}^{PS}$. The signal transmitted at the PB is the same as in the proposed RSMA-based framework. On the user side, the decoding order for PS users is determined by their respective channel strengths, denoted as $\|\mathbf{q}_{1,PS}\| \geq \ldots \geq \|\mathbf{q}_{K,PS}\|$. In particular, the kth PS user begins by decoding messages intended for the Kth, $(K-1)$th, …, $(k+1)$th PS users. Subsequently, the kth PS user proceeds to decode its intended message while treating the interference from the remaining messages as noise. The achievable rate to decode the lth message at the kth PS user is given by

$$\mathrm{R}_{k,NOMA}^{l} = \log_2\left(1 + \frac{\left|\mathbf{q}_{k,PS}^{H}\mathbf{p}_{l,N}\right|^2}{\sum_{k'=1}^{l-1}\left|\mathbf{q}_{k,PS}^{H}\mathbf{p}_{k',N}\right|^2 + \sum_{g=1}^{G}\left|\mathbf{h}_{k,PS}^{H}\mathbf{e}_g\right|^2 + \sigma_k^2 + \frac{\gamma_k^2}{\theta_k}}\right), \forall k, \forall l \geq k. \quad (20)$$

Two datasets were considered in the simulation analysis: a training/validation dataset and a testing dataset. As described in Section 3, the PSO-SDR method was used to process all the datasets considered. A fivefold cross-validation method was used to divide the training/validation dataset into separate training and validation datasets. These datasets were used to determine the optimal hyperparameters for the DNN module, including the number of hidden layers, hidden nodes, learning rates, activation functions, and batch sizes. Subsequently, the testing dataset was used to evaluate the ability of the model to generalize. The testing dataset encompassed scenarios not encountered during the training and validation phases to assess the generalization performance of the model. These scenarios included different user distances and varying numbers of antennas at the BS. The training/validation dataset consisted of 14,000 samples, covering data rates ranging from $\chi = 1$ bits/s/Hz to $\chi = 10$ bits/s/Hz, EH requirements from $\varepsilon = -16$ dBm to $\varepsilon = -30$ dBm, eight antennas at the BS, and the previously mentioned range of distances. The specifics of the testing dataset for each scenario are outlined later in this paper. In addition, the exponential linear unit (ELU) was selected as the activation function for the hidden layers, and the Adam algorithm was used as the optimizer. The training data for the PCA module were composed of 6000 channel vectors. Regarding the PSO-SDR method, the simulation parameters considered for the PSO algorithm after fine-tuning the parameters were $S = 15$, $w = 0.7$, $a_c = b_c = 1.494$, and $T_{max} = 25$.

Firstly, this paper presents an analysis of the convergence behavior of the DNN module while considering different hyperparameters. Figure 6 shows the convergence performance of the DNN module under varying learning rates (lr), numbers of hidden layers, and numbers of principal components (PCs) in the PCA module. In the legend in Figure 6, the last term represents the number of units $\times$ the number of hidden layers, where all hidden layers have the same number of hidden units. Regarding the learning rate, a high learning rate leads to rapid convergence, but it can also result in overfitting, as observed in the case of lr = 0.005. On the other hand, low learning rate values require a significantly higher number of epochs to achieve convergence, and they may not guarantee the lowest error value, as observed in the case of lr = 0.0001. Regarding the number of PCs, it is important to note that the PCA module was applied to each channel vector, resulting in each channel vector being represented by a certain number of PCs. The lowest error was achieved when the number of PCs was 2 or 3. In particular, when PC = 2, it implies a lower number of required input nodes in the DNN module. For the remainder of the simulations, a learning rate of lr = 0.001 and PC = 2 were selected based on these observations.

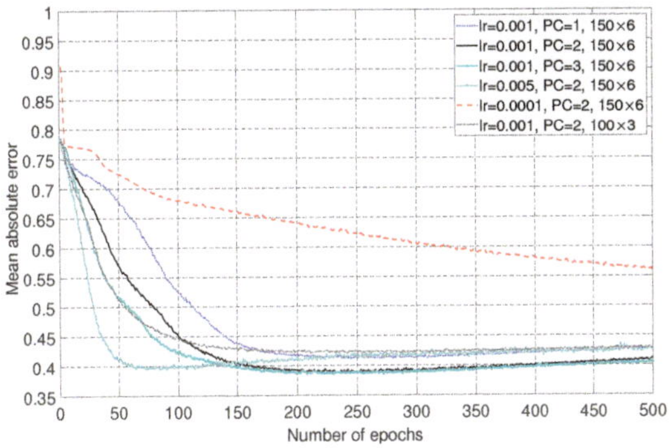

Figure 6. Convergence behavior of the proposed DNN module.

We utilized the recurrent neural network (RNN) model [38] in our simulation results for comparative analysis. RNN, a type of neural network, is specifically designed to analyze sequences of vectors, where values in successive vectors exhibit interrelationships. However, it is noteworthy that each sample in our dataset comprised a singular vector containing the user's rate requirements, EH requirements, and the reduced representation of the channel vectors post-PCA module. To incorporate the RNN into our simulations, we configured the number of vectors in the sequence to one and set the size of the hidden layer to 150.

Figure 7 shows the variation of the sum transmission power of the PB and BS with respect to the data rate requirements, χ, considering a minimum EH requirement of $\varepsilon = -24$ dBm. The data rate requirement serves as a constraint in the proposed optimization problem (5) and must be satisfied by the BS for all users in the system. As the rate requirement increases, the BS must allocate more transmission power to meet the specified rate, thereby increasing the total transmission power. Although the PB does not transmit information signals, their energy-carrying signals aid in meeting the EH requirements but are considered interference at the ID module of PS users. In particular, we observed that deploying the PB enabled a reduction of up to 3 dBm in the total transmission power for all the multiple-access methods compared. Furthermore, the RSMA scheme achieved significantly lower transmission power with and without the PB compared to their respective SDMA- and NOMA-based counterparts. Moreover, the proposed DNN-RSMA-based solution achieved a similar result to the near-optimal scheme of the PSO-SDR RSMA while significantly reducing the computational complexity. In particular, the complexity of the PSO-SDR method is denoted as $\mathcal{O}(S \cdot T_{\max} \cdot \mathcal{O}_{SDR})$, and the complexity of the DNN-based method is represented as $\mathcal{O}(\mathcal{O}_{DNN} + \mathcal{O}_{SDR})$, where $\mathcal{O}_{SDR}$ is the computational complexity of solving problem (7) using the SDR method, and $\mathcal{O}_{DNN}$ represents the computational complexity of predicting the common rate variables. Consequently, the proposed DNN-based method was approximately $S \times T_{\max}$ times faster than the PSO-SDR near-optimal scheme. In addition, we observed that the RNN model achieved performance comparable to that of the proposed DNN. This similarity arises because when the input is a single vector, the RNN exhibits a layer-wise architecture similar to that of a basic neural network. Furthermore, the proposed DNN demonstrated lower computational time compared to the RNN. Specifically, the RNN required a computational time of 191.232 s for training and 0.002 s for testing, whereas the proposed DNN required a computational time of 40.295 s for training and 0.001 s for testing. These simulations were conducted on a computer equipped with an Intel Core i7-6700 CPU and 16 GB of RAM. Therefore, the proposed DNN emerges as the most suitable neural network architecture for solving the proposed optimization problem.

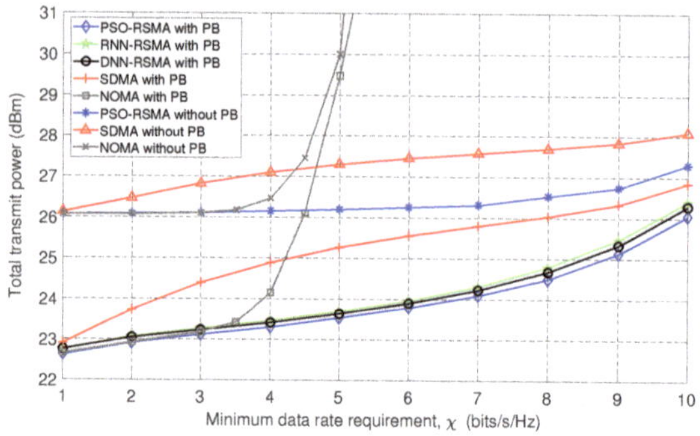

Figure 7. Total transmission power of the PB and BS versus the required data rate targets of PS users.

RSMA outperformed SDMA because of its capability to decode a portion of the interference through the SIC process on the common message (Figure 7), whereas SDMA treats all interference as noise. In the NOMA scheme, the difference arises from using the SIC process. In RSMA, the SIC process aims to cancel the interference from the common message. In contrast, in the NOMA scheme, the SIC process is employed to eliminate interference from the messages of users with weaker channel strengths. On the other hand, achieving excellent performance under the NOMA scheme in multi-antenna systems necessitates that users whose messages will be decoded by SIC have weaker channel strengths and sufficiently aligned channels. This alignment requirement is uncommon in real-world deployments and the channel model considered in Equation (13). Furthermore, as the rate requirement of the PS users increases, the transmission power needed for the beamforming vectors also increases, leading to performance degradation in the NOMA scheme after reaching $\chi = 4$ bits/s/Hz.

The variation of the sum transmission power of the PB and BS with respect to the EH requirements, ε, is presented in Figure 8, considering a minimum rate requirement of $\chi = 4$ bits/s/Hz. Similar to Figure 7, a significant reduction of approximately 3 dBm was observed in the total transmission power due to the deployment of the PB. Furthermore, the benefit of applying RSMA compared to conventional methods, such as NOMA and SDMA, in considerably reducing the transmission power was demonstrated. This is because RSMA employed the SIC procedure, which improved the data rate at the ID module of PS users while simultaneously reducing the PS factor to enhance harvested energy, thus reducing the transmission power. Furthermore, the proposed DNN-based approach performed similarly to the PSO-based method while significantly reducing computational complexity.

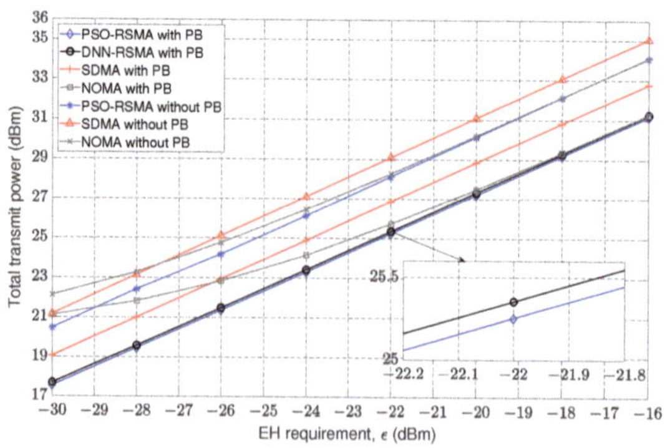

Figure 8. Total transmission power of the PB and BS versus the required EH of PS and EH users.

Next, we analyzed the generalization performance of the proposed DNN-based method across considerably different scenarios from those used in the training phase. In particular, we kept the trained DNN module fixed and tested its generalization performance by varying the number of antennas, resulting in a modification of the components of the channel vectors, and by varying the distance from the BS, resulting in a significant alteration of the channel strengths. The channel vectors, along with the requirements for data rates and energy harvesting, served as inputs for the PCA and DNN modules in the proposed scheme to generate the common rate variables, as detailed in Section 3.1.

Figure 9 presents the variation of the sum transmission power of the PB and BS with respect to the number of antennas equipped at the BS, considering a required minimum rate of $\chi = 4$ bits/s/Hz and a required minimum EH of $\varepsilon = -25$ dBm. As the number of antennas at the BS increased, there was a reduction in the transmission power due to the increased degrees of freedom. In the case of the DNN-based method, the training dataset solely consisted of samples representing a scenario with a BS equipped with eight antennas, as detailed at the beginning of Section 4. In contrast, the testing data in Figure 9 contained samples encompassing scenarios with varying numbers of antennas at the BS. The utilization of the same trained DNN module with different numbers of antennas can be attributed to the PCA module. In particular, independent PCA modules for each scenario were trained based on the number of antennas while maintaining the same number of principal components, L. Consequently, despite variations in the dimensionality of the channel vectors, the output of the PCA module remained consistent across all different numbers of antennas. Training a PCA module is a straightforward task because it only requires channel vector samples and does not necessitate labels or target values.

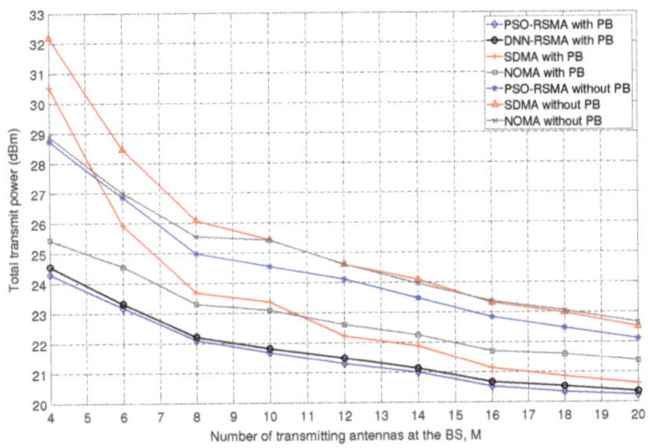

Figure 9. Total transmission power of the PB and BS versus the number of antennas at the BS.

The proposed DNN-based scheme demonstrated remarkable performance comparable to that of the PSO-based method, even when the number of antennas differed from the training scenarios (Figure 9). This indicates strong generalization capabilities and robustness to environmental changes. Furthermore, the RSMA-based methods consistently achieved lower total transmission power compared to the traditional NOMA and SDMA methods. Moreover, the deployment of the PB resulted in a significant decrease in the transmission power, even with an increasing number of antennas at the BS. This can be attributed to the ability of the PB to mitigate signal attenuation because of the distance from the transmitter, a critical factor in scenarios with EH requirements for users.

Figure 10 shows the variation of the sum transmission power of the PB and BS with respect to the position of the PS users, considering a minimum rate requirement of $\chi = 4$ bits/s/Hz and a required EH of $\varepsilon = -25$ dBm. In particular, the range of the position of the PS users on the x-axis, x_{PS}, was varied. This position was randomly selected within the region of $x_{PS} \in [d_x \text{ m}, (d_x + 5) \text{ m}]$, where d_x varied to analyze performance as the distance between the PS users and the BS increased. The transmission power increased as d_x increased because the average distance from the user to the BS also increased with higher values of d_x. Additionally, the difference in the transmission power between scenarios with and without the PB diminished as the value of d_x increased, owing to the fixed location of the PB across all d_x values. As d_x increased, users were positioned farther from the PB, resulting in increased attenuation of the energy signal. Although the energy signal at the PS users was utilized for energy harvesting at the EH module, it was considered interference for the ID module. Consequently, as the received energy signal power from the PB diminished due to increased distance, it contributed less to the EH at the PS users. Meanwhile, reliance on the received power from the information signal transmitted by the BS became more dominant, leading to a reduction in the impact of the PB deployment. Moreover, as shown in the previous figures, the RSMA method consistently achieved a considerable decrease in the total transmission power in comparison to the NOMA and SDMA methods.

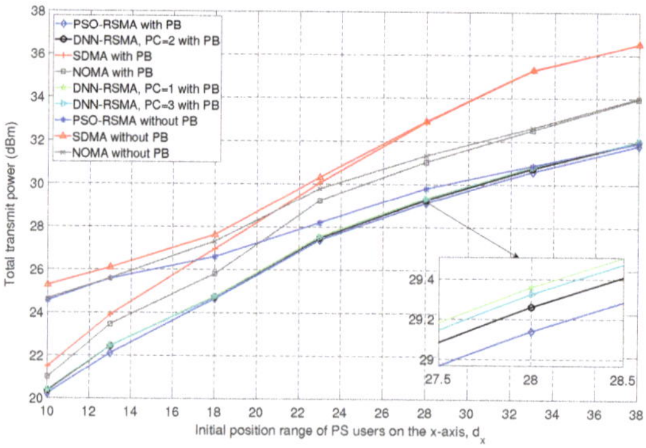

Figure 10. Total transmission power of the PB and BS versus the initial position range of PS users on the *x*-axis.

In the case of the DNN-based method, the training dataset consisted solely of samples representing a scenario with $d_x = 13$ m, as detailed at the beginning of Section 4. In contrast, the testing data in Figure 10 contained samples encompassing scenarios with varying values of d_x. The proposed DNN-based scheme exhibited high generalization performance, achieving comparable transmission power to the PSO-based method. Furthermore, this study analyzed the effect of slightly changing the number of PCs in the PCA module. Similar to the results shown in Figure 6, the cases of PC = 2 and PC = 3 achieved similar results, with PC = 2 having lower transmission power and demonstrating the best generalization performance.

5. Conclusions

A multiuser MISO SWIPT system using RSMA was evaluated with the assistance of a PB. The objective was to minimize the combined transmission power from the BS and PB while optimizing the beamforming vectors, common rate variables, and PS ratios. The proposed optimization problem was carried out under constraints that included EH requirements for both EH and PS users and data rate requirements for PS users. The proposed non-convex problem was divided into two parts. The first part was solved using the PCA method to reduce dimensionality and a DNN to predict the common rate variables. The second part used the SDR technique to optimize the PS factors and beamforming vectors. Comparative schemes were developed based on the PSO algorithm and SDR method for RSMA, along with a baseline scheme using NOMA.

Numerical simulations showed that RSMA significantly reduced the transmission power compared to conventional methods such as NOMA and SDMA. Moreover, the proposed DNN-based method achieved high performance, closely matching the results of the near-optimal PSO-based scheme while considerably reducing computational complexity.

Furthermore, this study tested the proposed DNN-based scheme across challenging scenarios that were significantly different from those used for training during data collection. Across these scenarios, the DNN-based scheme exhibited high generalization performance. Furthermore, strategically placing the PB close to EH users substantially decreased the overall transmit power, effectively meeting their EH requirements.

Author Contributions: Conceptualization, M.R.C. and C.E.G.; methodology, M.R.C.; software, M.R.C. and C.E.G.; validation, C.E.G. and I.K.; formal analysis, M.R.C.; investigation, M.R.C. and C.E.G.; resources, I.K.; data curation, M.R.C.; writing—original draft preparation, M.R.C.; writing—review and editing, C.E.G. and I.K.; visualization, M.R.C.; supervision, I.K.; project administration, I.K.; funding acquisition, I.K. All authors have read and agreed to the published version of the manuscript.

Funding: This work was supported in part by the National Research Foundation of Korea (NRF) through the Korean Government's Ministry of Science and ICT (MSIT) under Grant NRF-2021R1A2B5 B01001721, and in part by the Regional Innovation Strategy (RIS) through the NRF funded by the Ministry of Education (MOE) under Grant 2021RIS-003.

Data Availability Statement: Data are contained within the article.

Conflicts of Interest: The authors declare no conflicts of interest.

References

1. Mishra, A.; Mao, Y.; Dizdar, O.; Clerckx, B. Rate-Splitting Multiple Access for 6G—Part I: Principles, Applications and Future Works. *IEEE Commun. Lett.* **2022**, *26*, 2232–2236. [CrossRef]
2. Clerckx, B.; Joudeh, H.; Hao, C.; Dai, M.; Rassouli, B. Rate splitting for MIMO wireless networks: A promising PHY-layer strategy for LTE evolution. *IEEE Commun. Mag.* **2016**, *54*, 98–105. [CrossRef]
3. Mao, Y.; Clerckx, B.; Li, V.O. Rate-splitting multiple access for downlink communication systems: Bridging, generalizing, and outperforming SDMA and NOMA. *EURASIP J. Wirel. Commun. Netw.* **2018**, *2018* , 133. [CrossRef]
4. Mao, Y.; Dizdar, O.; Clerckx, B.; Schober, R.; Popovski, P.; Poor, H.V. Rate-Splitting Multiple Access: Fundamentals, Survey, and Future Research Trends. *IEEE Commun. Surv. Tutor.* **2022**, *24*, 2073–2126. [CrossRef]
5. Clerckx, B.; Mao, Y.; Schober, R.; Poor, H.V. Rate-splitting unifying SDMA, OMA, NOMA, and multicasting in MISO broadcast channel: A simple two-user rate analysis. *IEEE Wirel. Commun. Lett.* **2020**, *9*, 349–353. [CrossRef]
6. Ponnimbaduge Perera, T.D.; Jayakody, D.N.K.; Sharma, S.K.; Chatzinotas, S.; Li, J. Simultaneous Wireless Information and Power Transfer (SWIPT): Recent Advances and Future Challenges. *IEEE Commun. Surv. Tutor.* **2018**, *20*, 264–302. [CrossRef]
7. Bi, S.; Ho, C.K.; Zhang, R. Wireless powered communication: Opportunities and challenges. *IEEE Commun. Mag.* **2015**, *53*, 117–125. [CrossRef]
8. Huang, K.; Lau, V.K.N. Enabling Wireless Power Transfer in Cellular Networks: Architecture, Modeling and Deployment. *IEEE Trans. Wirel. Commun.* **2014**, *13*, 902–912. [CrossRef]
9. Pan, G.; Lei, H.; Yuan, Y.; Ding, Z. Performance Analysis and Optimization for SWIPT Wireless Sensor Networks. *IEEE Trans. Commun.* **2017**, *65*, 2291–2302. [CrossRef]
10. Shi, Q.; Liu, L.; Xu, W.; Zhang, R. Joint Transmit Beamforming and Receive Power Splitting for MISO SWIPT Systems. *IEEE Trans. Wirel. Commun.* **2014**, *13*, 3269–3280. [CrossRef]
11. Zhang, H.; Dong, A.; Jin, S.; Yuan, D. Joint Transceiver and Power Splitting Optimization for Multiuser MIMO SWIPT Under MSE QoS Constraints. *IEEE Trans. Veh. Technol.* **2017**, *66*, 7123–7135. [CrossRef]
12. Xu, Y.; Shen, C.; Ding, Z.; Sun, X.; Yan, S.; Zhu, G.; Zhong, Z. Joint Beamforming and Power-Splitting Control in Downlink Cooperative SWIPT NOMA Systems. *IEEE Trans. Signal Process.* **2017**, *65*, 4874–4886. [CrossRef]
13. Li, Z.; Chen, W.; Wu, Q.; Wang, K.; Li, J. Joint Beamforming Design and Power Splitting Optimization in IRS-Assisted SWIPT NOMA Networks. *IEEE Trans. Wirel. Commun.* **2022**, *21*, 2019–2033. [CrossRef]
14. Al-Obiedollah, H.M.; Cumanan, K.; Thiyagalingam, J.; Burr, A.G.; Ding, Z.; Dobre, O.A. Energy Efficient Beamforming Design for MISO Non-Orthogonal Multiple Access Systems. *IEEE Trans. Commun.* **2019**, *67*, 4117–4131. [CrossRef]
15. Joudeh, H.; Clerckx, B. Robust transmission in downlink multiuser MISO systems: A rate-splitting approach. *IEEE Trans. Signal Process.* **2016**, *64*, 6227–6242. [CrossRef]
16. Joudeh, H.; Clerckx, B. Sum-rate maximization for linearly precoded downlink multiuser MISO systems with partial CSIT: A rate-splitting approach. *IEEE Trans. Commun.* **2016**, *64*, 4847–4861. [CrossRef]
17. Dai, M.; Clerckx, B.; Gesbert, D.; Caire, G. A rate splitting strategy for massive MIMO with imperfect CSIT. *IEEE Trans. Wirel. Commun.* **2016**, *15*, 4611–4624. [CrossRef]
18. Mao, Y.; Clerckx, B.; Li, V.O.K. Rate-Splitting for Multi-Antenna Non-Orthogonal Unicast and Multicast Transmission: Spectral and Energy Efficiency Analysis. *IEEE Trans. Commun.* **2019**, *67*, 8754–8770. [CrossRef]

19. Mao, Y.; Clerckx, B.; Li, V.O.K. Rate-Splitting for Multiuser Multi-Antenna Wireless Information and Power Transfer. In Proceedings of the IEEE 20th International Workshop on Signal Processing Advances in Wireless Communications (SPAWC), Cannes, France, 2–5 July 2019.
20. Camana Acosta, M.R.; Moreta, C.E.G.; Koo, I. Joint Power Allocation and Power Splitting for MISO-RSMA Cognitive Radio Systems With SWIPT and Information Decoder Users. *IEEE Syst. J.* **2021**, *15*, 289–5300. [CrossRef]
21. Camana, M.R.; Garcia, C.E.; Koo, I. Rate-Splitting Multiple Access in a MISO SWIPT System Assisted by an Intelligent Reflecting Surface. *IEEE Trans. Green Commun. Netw.* **2022**, *6*, 2084–2099. [CrossRef]
22. Huang, K.; Zhong, C.; Zhu, G. Some new research trends in wirelessly powered communications. *IEEE Wirel. Commun.* **2016**, *23*, 19–27. [CrossRef]
23. Ma, Y.; Chen, H.; Lin, Z.; Li, Y.; Vucetic, B. Distributed and Optimal Resource Allocation for Power Beacon-Assisted Wireless-Powered Communications. *IEEE Trans. Commun.* **2015**, *63*, 3569–3583. [CrossRef]
24. Xu, W.; Chen, W.; Fan, Y.; Zhang, Z.; Shi, X. Spectrum efficiency maximization for cooperative power beacon-enabled wireless powered communication networks. *China Commun.* **2021**, *18*, 230–251. [CrossRef]
25. Park, J.-H.; Jeon, Y.-S.; Han, S. Energy Beamforming for Wireless Power Transfer in MISO Heterogeneous Network With Power Beacon. *IEEE Commun. Lett.* **2017**, *21*, 1163–1166. [CrossRef]
26. Camana, M.R.; Garcia, C.E.; Koo, I. Transmit Beamforming for a MISO SWIPT System with a Power Beacon. In Proceedings of the 2020 International Conference on Information and Communication Technology Convergence (ICTC), Jeju, Republic of Korea, 21–23 October 2020.
27. Camana, M.R.; Garcia, C.E.; Koo, I. Power Minimization in a MISO NOMA SWIPT System Assisted by a Power Beacon. In Proceedings of the 2022 Winter Comprehensive Conference of the Korea Information and Communications Society (KICS), Pyeongchang, Republic of Korea, 9–11 February 2022.
28. Vu, T.-H.; Nguyen, T.-V.; Kim, S. Cooperative NOMA-Enabled SWIPT IoT Networks with Imperfect SIC: Performance Analysis and Deep Learning Evaluation. *IEEE Internet Things J.* **2022**, *9*, 2253–2266. [CrossRef]
29. Camana, M.R.; Garcia, C.E.; Koo, I. Rate Splitting Multiple Access for a MISO SWIPT System Aided by a Power Beacon. In Proceedings of the 2022 Thirteenth International Conference on Ubiquitous and Future Networks (ICUFN), Barcelona, Spain, 5–8 July 2022 .
30. Zhang, W.; Zhang, Z.; Chao, H.-C.; Guizani, M. Toward intelligent network optimization in wireless networking: An auto-learning framework. *IEEE Wirel. Commun.* **2019**, *26*, 76–82. [CrossRef]
31. Liang, L.; Ye, H.; Yu, G.; Li, G.Y. Deep-learning-based wireless resource allocation with application to vehicular networks. *Proc. IEEE* **2020**, *108*, 341–356. [CrossRef]
32. Sun, H.; Chen, X.; Shi, Q.; Hong, M.; Fu, X.; Sidiropoulos, N.D. Learning to optimize: Training deep neural networks for interference management. *IEEE Trans. Signal Process.* **2018**, *66*, 5438–5453. [CrossRef]
33. Xia, W.; Zheng, G.; Zhu, Y.; Zhang, J.; Wang, J.; Petropulu, A.P. A deep learning framework for optimization of MISO downlink beamforming. *IEEE Trans. Commun.* **2020**, *68*, 1866–1880. [CrossRef]
34. Kim, J.; Lee, H.; Hong, S.-E.; Park, S.-H. Deep Learning Methods for Universal MISO Beamforming. *IEEE Wirel. Commun. Lett.* **2020**, *9*, 1894–1898. [CrossRef]
35. Camana, M.R.; Garcia, C.E.; Koo, I. Deep Learning-Assisted Power Minimization in Underlay MISO-SWIPT Systems Based on Rate-Splitting Multiple Access. *IEEE Access* **2022**, *10*, 62137–62156. [CrossRef]
36. Bishop, C.M. *Pattern Recognition and Machine Learning*, 1st ed.; Springer: Cambridge, UK, 2006.
37. Wang, Q. Kernel principal component analysis and its applications in face recognition and active shape models. *arXiv* **2014**, arXiv:1207.3538v3.
38. Aggarwal, C.C. *Neural Networks and Deep Learning: A Textbook*, 1st ed.; Springer: Cham, Switzerland, 2018.
39. Grant, M.C.; Boyd, S.P. CVX: Matlab Software for Disciplined Convex Programming, Version 2.2. 2020. Available online: http://cvxr.com/cvx/ (accessed on 1 November 2023).
40. Karipidis, E.; Sidiropoulos, N.D.; Luo, Z.-Q. Quality of service and max–min fair transmit beamforming to multiple cochannel multicast groups. *IEEE Trans. Signal Process.* **2008**, *56*, 1268–1279. [CrossRef]
41. Phan, A.H.; Tuan, H.D.; Kha, H.H.; Ngo, D.T. Nonsmooth Optimization for Efficient Beamforming in Cognitive Radio Multicast Transmission. *IEEE Trans. Signal Process.* **2012**, *60*, 2941–2951. [CrossRef]

sensors

MDPI

Article

Recent Advances in Machine Learning for Network Automation in the O-RAN

Mutasem Q. Hamdan [1], Haeyoung Lee [2,*], Dionysia Triantafyllopoulou [3], Rúben Borralho [4], Abdulkadir Kose [5], Esmaeil Amiri [4], David Mulvey [4], Wenjuan Yu [6], Rafik Zitouni [4], Riccardo Pozza [4], Bernie Hunt [4], Hamidreza Bagheri [7], Chuan Heng Foh [4], Fabien Heliot [4], Gaojie Chen [4], Pei Xiao [4], Ning Wang [4] and Rahim Tafazolli [4]

1. Samsung Electronics R&D Institute, Staines TW18 4QE, UK; m.hamdan@samsung.com
2. School of Physics, Engineering and Computer Science, University of Hertfordshire, Hatfield AL10 9AB, UK
3. Professorship of Communications Engineering, Chemnitz University of Technology, D-09111 Chemnitz, Germany; dionysia.triantafyllopoulou@etit.tu-chemnitz.de
4. 5GIC & 6GIC, Institute of Communication System, University of Surrey, Guildford GU2 7XH, UK; r.a.borralho@surrey.ac.uk (R.B.); e.amiri@surrey.ac.uk (E.A.); d.mulvey@surrey.ac.uk (D.M.); r.zitouni@surrey.ac.uk (R.Z.); r.pozza@surrey.ac.uk (R.P.); b.hunt@surrey.ac.uk (B.H.); c.foh@surrey.ac.uk (C.H.F.); f.heliot@surrey.ac.uk (F.H.); gaojie.chen@surrey.ac.uk (G.C.); p.xiao@surrey.ac.uk (P.X.); n.wang@surrey.ac.uk (N.W.); r.tafazolli@surrey.ac.uk (R.T.)
5. Department of Computer Engineering, Abdullah Gul University, Kayseri 38080, Turkey; abdulkadir.kose@agu.edu.tr
6. School of Computing and Communications, InfoLab21, Lancaster University, Lancaster LA1 4WA, UK; w.yu8@lancaster.ac.uk
7. School of Science, Technology and Health, York St John University, York YO31 7EX, UK; h.bagheri@yorksj.ac.uk

* Correspondence: h.lee@herts.ac.uk

Abstract: The evolution of network technologies has witnessed a paradigm shift toward open and intelligent networks, with the Open Radio Access Network (O-RAN) architecture emerging as a promising solution. O-RAN introduces disaggregation and virtualization, enabling network operators to deploy multi-vendor and interoperable solutions. However, managing and automating the complex O-RAN ecosystem presents numerous challenges. To address this, machine learning (ML) techniques have gained considerable attention in recent years, offering promising avenues for network automation in O-RAN. This paper presents a comprehensive survey of the current research efforts on network automation using ML in O-RAN. We begin by providing an overview of the O-RAN architecture and its key components, highlighting the need for automation. Subsequently, we delve into O-RAN support for ML techniques. The survey then explores challenges in network automation using ML within the O-RAN environment, followed by the existing research studies discussing application of ML algorithms and frameworks for network automation in O-RAN. The survey further discusses the research opportunities by identifying important aspects where ML techniques can benefit.

Keywords: open radio access networks; machine learning; artificial intelligence

Citation: Hamdan, M.Q.; Lee, H.; Triantafyllopoulou, D.; Borralho, R.; Kose, A.; Amiri, E.; Mulvey, D.; Yu, W.; Zitouni, R.; Pozza, R.; et al. Recent Advances in Machine Learning for Network Automation in O-RAN. *Sensors* **2023**, *23*, 8792. https://doi.org/10.3390/s23218792

Academic Editor: Jari Nurmi

Received: 7 September 2023
Revised: 14 October 2023
Accepted: 25 October 2023
Published: 28 October 2023

1. Introduction

Open Radio Access Network (O-RAN) is a revolutionary concept in the field of wireless telecommunications that aims to transform traditional, proprietary Radio Access Networks (RAN) into open, intelligent, and interoperable networks. The O-RAN concept involves separating the hardware and software components of RANs and enabling interoperability and integration of solutions coming from different vendors. This open architecture is made possible through the use of open Application Programming Interfaces (APIs), standardized interfaces, and virtualization technologies, which allow RAN components to be disaggregated and easily swapped out or upgraded. Thus, O-RAN is expected to bring greater flexibility, innovation, cost-efficiency, design flexibility, operational adaptability, system functionality, deployment scalability, and function expandability to RANs, while also supporting the demands of next-generation networks and services such as 5G/6G, Internet of Things (IoT), and edge computing.

Over the past couple of decades, wireless communications have gone through several transformations to support massive connectivity and to meet the demands of modern real-time and mission-critical applications, as the target 5G and 6G Key Performance Indicators (KPIs) include ultra-high reliability, low latency, and high-throughput. However, despite of all advancements made in wireless systems, particularly in 5G, achieving all these goals remains challenging. The key problems confronting stakeholders in 5G and 6G systems include efficiently supporting wireless access across diverse frequency bands, dealing with heterogeneous technologies, addressing a wide variety of application requirements leading to complex protocol stacks, and managing the rising capital and operational expenditure (CapEx and OpEx) needed for infrastructure upgrades and maintenance [1]. These challenges include the need to design an independent, service-focused network architecture due to the diversity of Quality of Service (QoS) requirements, achieving network agility while ensuring backward compatibility with existing equipment and support for future upgrades, and guaranteeing network efficiency to avoid increased computational complexity and a heavy load on the backhaul network [2].

The motivation behind the development of the O-RAN concept lies on the fact that traditional RAN systems are proprietary, i.e., closed systems, limiting mobile network operators (MNO) to obtaining all the radio, hardware, and software systems from a single supplier when deploying a network at each region. Aside from the considerable impact on RAN deployment CapEx and OpEx, this implies the lack of openness and interoperability, which can hinder innovation and agility. Traditional RANs are monolithic systems, which are designed to operate as integrated products, seen by the operators as black-boxes. Traditionally, at its most basic level, the RAN architecture consists of a radio unit (RU) or remote radio unit (RRU), a baseband unit (BBU), antennas, and various software-based interfaces. As described above, this results in difficulty meeting the very strict and diverse KPIs of modern networks. Consequently, the consensus was that the mobile network should be more software driven, virtualized, flexible, and intelligent to provide all the KPI goals and address the aforementioned challenges.

The RAN evolution towards O-RAN started with disaggregation, defined by 3GPP in Release 15 [3], where the 5G NR RAN (more specifically gNB) functionalities split into three logical nodes: the Central Unit (CU), the Distributed Unit (DU), and the Radio Unit (RU). The CU handles gNB functions like transfer of user data, radio access management, positioning, mobility, and session management. A DU function is dependent on the functional split option, but mainly manages baseband processing functions across cell sites. The CU operation is controlled by the CU. The RU component is located near or integrated into the antenna unit where the radio signals are transmitted, received, amplified, and digitized. In traditional RAN configuration, sometimes called distributed RAN (D-RAN), BBU and RRH are co-located in the same place in the cell site, in which they are directly connected via Common Public Radio Interface (CPRI). Disaggregation option provides new levels of flexibility and efficiency at RAN level by enabling the network operators to decide where to locate each function and maximize performance. In 2009, centralized or cloud RAN (C-RAN) has emerged as an efficient solution, exploiting disaggregation to move the BBU functionalities to a centralized location, called BBU pool, while leaving the RRU and antenna on cell site. The principle design idea of C-RAN is to move some of RAN functionalities to the cloud infrastructure, the BBU pool could be implemented on a cloud platform [4]. The path towards O-RAN making mobile networks "more software driven, virtualized, flexible, intelligent and energy efficient", as well as "cost-efficient and reliable" [2], is paved through the use of Network Function Virtualization (NFV) concepts. The O-RAN concept is supported by several standard bodies such as the O-RAN Alliance [5], the Third Generation Partnership Project (3GPP) [6], the European Telecommunications

Standards Institute (ETSI) [7], the Next Generation Mobile Networks (NGMN) [8], and the Optical Internetworking Forum (OIF) [9], to ensure interoperability and interconnection between O-RAN components from different vendors.

As O-RAN environments are inherently complex, characterized by heterogeneity and dynamism, with various hardware and software components from different vendors working together, Machine Learning (ML) is expected to be an invaluable tool. The effectiveness and efficiency of O-RAN architectures are intricately tied to the integration of ML capabilities. ML techniques offer a plethora of advantages within O-RAN architectures. ML algorithms excel at real-time analysis of extensive network data, including KPIs, end-user behaviors, and network traffic patterns, all in real-time. This analytical strength empowers the prompt identification of trends, anomalies, and performance issues, facilitating network optimization and predictive maintenance [10]. Moreover, ML plays a vital role in resource allocation by intelligently assigning radio resources, optimizing resource utilization, and enhancing QoS [11]. Additionally, ML algorithms facilitate data-driven decision-making in O-RAN management and orchestration functions, such as automated network configuration, dynamic spectrum allocation, and intelligent traffic steering. This streamlined approach reduces network management complexity and empowers MNOs to optimize network performance. Overall, ML is a very important tool for O-RAN as it allows for the provision of insights, efficient resource allocation, and automated management capabilities.

In this paper, we embark on a comprehensive exploration of the current landscape concerning ML applications in O-RAN. Our objective is to identify and analyze the prevailing challenges that remain unresolved, preventing the full harnessing of ML's potential for enhancing O-RANs. While there have been several O-RAN survey papers in the literature [1,4,12–19], our survey paper stands out as it concentrates on the applications and potential of ML within the O-RAN context, which has not been extensively addressed in prior surveys. Table 1 provides a summary of the topics covered in relevant surveys, along with their contributions, in order to provide a clear comparison with our work. It can be noted that most of the existing surveys aim to provide a detailed tutorial on RAN evolution, O-RAN architecture, and components and use cases. They all have different focuses compared to our paper. For example, ref. [16] specifically focuses on the security and privacy risks associated with Open RAN architecture, which complements our survey. Meanwhile, ref. [17] centers its attention on the Non-Terrestrial Network (NTN), offering an architectural solution for an O-RAN-based NTN system. In addition, although [18,19] both acknowledge the prevailing challenges and prospective research directions in this field, ref. [18] primarily examines the existing O-RAN specifications, while [19] focuses on how Explainable AI (XAI) can contribute to O-RAN networks. Among the existing survey papers, refs. [14,15] have a similar focus on ML in O-RAN. Ref. [14] mainly provides a tutorial on how intelligent applications can improve the efficiency of O-RAN and the future opportunities in O-RAN, while [15] looks into how deep learning solutions can be integrated to the O-RAN architecture, as well as case studies. However, ref. [15] primarily focuses on deep learning techniques rather than general ML techniques. Moreover, while [14,15] discuss open problems and future research directions, they primarily look at these from the perspective of O-RAN, rather than looking at the specific challenges and opportunities associated with ML-empowered O-RAN. Different from the existing surveys, in this article, we dive deep into the ML integration in O-RAN, covering recent research works that study the application of ML in O-RAN, discussing emerging issues and research opportunities that shall be addressed to fulfill its design commitment. We believe this survey marks a significant milestone as the first comprehensive endeavor aimed at summarizing recent studies and providing crucial technical guidance to researchers interested in ML-enabled O-RAN. Within this paper, we also present research opportunities across diverse areas encompassing data collection and analysis, as well as the development, deployment, maintenance, and operation of ML within the O-RAN domain.

Table 1. Summary of surveys relevant to ML-enabled O-RAN. H: High, M: Medium, and L: Low.

Year	Ref	O-RAN Architecture	ML Application in O-RAN	Research Opportunities	Remarks
2023	[1]	H	M	M	This paper provides a detailed tutorial on O-RAN, describing its architecture, design principles, and the O-RAN interfaces.
2021	[4]	H	L	L	This paper overviews the idea of O-RAN and presents ongoing O-RAN Alliance standardization activities in this context, followed by a study of traffic steering use case.
2022	[12]	H	L	L	This paper provides a comprehensive survey of O-RAN development, encompassing a summary of the RAN evolution history, an introduction to O-RAN technology, an overview of Open RAN-related projects and activities, a discussion of standardization efforts, challenges, and potential solutions.
2021	[13]	M	L	L	This paper provides an overview of the O-RAN Alliance RAN architecture, highlighting its core building blocks and, subsequently, it presents a practical use case that leverages AI/ML-based innovations.
2022	[14]	H	M	M	This paper presents an O-RAN architecture overview, delves into AI applications within Op-RAN, and discusses the challenges and opportunities in implementing intelligent solutions in 5G and B5G telecommunications.
2022	[15]	H	M	M	This paper focuses on mapping existing deep-learning-based studies to the O-RAN architecture, highlighting key technical challenges, open issues, and future AI-enabled O-RAN research directions.
2023	[16]	M	L	L	This paper examines security and privacy risks in O-RAN architecture, proposes possible solutions and presents relevant security standardization efforts.
2023	[17]	M	L	L	This paper focuses on Non-Terrestrial Networks exploring the possible implementation of an O-RAN-based NTN solution.
2022	[18]	H	L	L	This paper identifies critical limitations in current O-RAN specifications: security, latency, real-time control, and AI-based RAN control.
2023	[19]	H	M	L	This paper focuses on XAI methods and explores their deployment within the context of O-RAN.

The structure of the remaining sections in this paper is as follows. Section 2 provides an overview of the O-RAN architecture and its development, highlighting its design principles that support network automation. In Section 3, we survey the existing ML applications in the context of O-RAN. Section 4 focuses on the potential of O-RAN and discusses the research opportunities for applying ML techniques to enhance various aspects of mobile network operation within the O-RAN framework. Finally, Section 5 concludes the survey and discussion presented in this paper. We illustrate the organization of this paper in Figure 1.

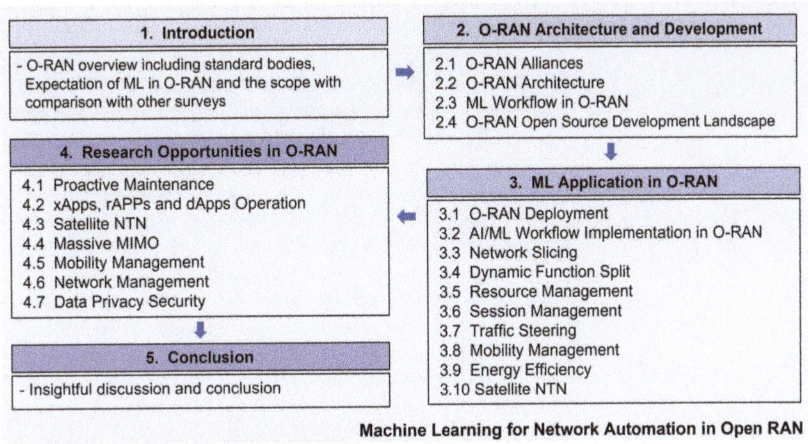

Figure 1. The overall structure of this paper.

2. O-RAN Architecture and Development

O-RAN architecture is a virtualized, software-driven, and open radio access network (RAN) architecture that enables the integration of hardware and software components from multiple vendors. It is designed to be modular, scalable, and flexible, with standardized interfaces that enable interoperability between different RAN components. The O-RAN architecture consists of multiple functional components that can be separated and managed independently. Some of the key components in O-RAN are Radio Unit (RU), Distributed Unit (DU), Central Unit (CU), RAN Intelligent Controller (RIC), and Service Management and Orchestration (SMO). In the following, we shall first describe the O-RAN alliance, architecture and softwarization developments, as well as their role in ML for O-RAN network automation.

2.1. O-RAN Alliance

The O-RAN Alliance is a global industry consortium that was founded in 2018 to drive the development and adoption of open and intelligent Radio Access Networks (RAN). The alliance is made up of more than 200 member organizations, including mobile network operators, network equipment vendors, and software companies.

The goal of the O-RAN Alliance is to create a more open, interoperable, and cost-effective RAN ecosystem that can accelerate the deployment of 5G networks and support new use cases and services. To achieve this goal, the O-RAN Alliance focuses on three main areas:

- Standardization: the O-RAN Alliance works to develop and promote open standards for RAN interfaces and APIs, enabling multi-vendor interoperability and reducing network deployment costs;
- Software: the O-RAN Alliance develops and promotes open software for RANs, including software-defined radio (SDR), virtualized RAN (vRAN), and open APIs for software integration;
- Testing and integration: the O-RAN Alliance provides specifications, conformance testing, and integration guidelines to ensure that O-RAN solutions can be easily integrated into existing network environments and can interoperate with other vendors' solutions.

Through its work in these areas, the O-RAN Alliance is helping to create a more open, flexible, and efficient RAN ecosystem that can meet the demands of 5G networks and support new use cases and services, such as industrial IoT, autonomous vehicles, and smart cities.

2.2. O-RAN Architecture

The O-RAN architecture is a set of open interfaces and protocols designed to enable multi-vendor interoperability and support a wide range of use cases and services in RAN. The architecture defines a modular and disaggregated approach to building RAN systems, where different functional components can be developed and deployed independently by a variety of vendors. Such an approach intends to increase innovation, reduce costs, and enable faster deployment of new services and features.

The O-RAN architecture specifies main components and interfaces connecting the components. The interfaces allow the Service Management and Orchestration (SMO) framework to connect with O-RAN network functions and O-Cloud. Figure 2 illustrates the high level O-RAN architecture, which can be viewed as Virtualized Network Functions (VNFs) placed above the O-Cloud and/or Physical Network Functions (PNFs). The A1 Interface between the Non-RT RIC in the SMO and the Near-RT RIC used for RAN Optimization, O1 Interface between the SMO and the O-RAN Network Functions used for Fault, Configuration, Accounting, Performance, Security (FCAPS) support and in the hybrid model. The Open Fronthaul M-plane interface between SMO and O-RU supports FCAPS too. While the O2 Interface between the SMO and the O-Cloud provides platform resources and workload management, the O-Cloud Notification interface allows event consumers such as an O-CU implemented on O-Cloud to subscribe to events or status. Moreover, the Y1 interface permits the Y1 consumers to subscribe or request the RAN analytics information delivered by Near-RT RIC. Where the Y1 consumer stands for an entity or more, within or outside of the public land mobile network (PLMN) trust domain that ingests analytics information services after mutual authentication and authorization by subscribing to or requesting the RAN analytics information via the Y1 service interface. There are three main control loops that run simultaneously in O-RAN, depending on the use cases, which are real-time (RT), which is limited to a maximum of 10 ms execution time t_{exe}; Near-RT, with $10 \leq t_{exe} \leq 1000$ ms; and None-RT, which can take $1000 \leq t_{exe}$ ms. Multi-vendor Slices use case targets enabling functions that belong to different vendors; there are many possible configurations to deploy the Multi-vendor slicing, all of which share that one O-RU is connected to one or more O-DUs. The advantages of such a use-case include a higher flexibility and rapid deployment of services to market by network operators, sharing RAN equipment among operators, optimizing CAPEX and OPEX among their existing assets, and future investments. In addition, reducing the supply chain risk; for example, if an existing vendor supplies a certain pair of vO-DU and vO-CU functions and if, for business reasons or even political situations, it has to withdraw from a certain market, then the operator can outsource and deploy alternative vO-DU and vO-CU that support multi-vendor slicing functions. The O-RAN specification work has been covered by eleven technical Work Groups (WG) to covers all the O-RAN Architecture parts, each WG has been supervised by the O-RAN alliance technical steering committee. Below is a brief overview of each WG:

- WG1: Use Cases and Overall Architecture. This WG is responsible for defining the overall architecture of Open RAN and the use cases that it will support;
- WG2: Non-Real-Time RAN Intelligent Controller and A1 Interface. This WG is responsible for defining the specifications for the Non-Real-Time RIC (Non-RT-RIC) and the A1 interface. The Non-RT-RIC is a centralized controller that manages the non-real-time aspects of the RAN. The A1 interface is the interface between the Non-RT-RIC and the radio units;
- WG3: Near-Real-Time RIC and E2 Interface. This WG is responsible for defining the specifications for the Near-Real-Time RIC (nRT-RIC) and the E2 interface. The nRT RIC is a centralized controller that manages the near-real-time aspects of the RAN. The E2 interface is the interface between the nRT RIC and the radio units;
- WG4: Open Fronthaul Interfaces. This WG is responsible for defining the specifications for the open fronthaul interfaces. The fronthaul is the interface between the baseband unit and the radio units;

- WG5: Open F1/W1/E1/X2/Xn Interfaces. This WG is responsible for defining the specifications for the open F1/W1/E1/X2/Xn interfaces. These interfaces are used to communicate between different parts of the RAN;
- WG6: Cloudification and Orchestration. This WG is responsible for defining the specifications for cloudification and orchestration of the RAN. Cloudification is the process of moving the RAN to the cloud. Orchestration is the process of managing the RAN;
- WG7: White-box Hardware. This WG is responsible for defining the specifications for white-box hardware for the RAN. White-box hardware is hardware that is not proprietary to a specific vendor;
- WG8: Stack Reference Design. This WG is responsible for defining the stack reference design for the RAN. The stack reference design is a model of the RAN that can be used to develop and test different RAN implementations;
- WG9: Open X-haul Transport. This WG is responsible for defining the specifications for open X-haul transport for the RAN. X-haul transport is the transport of data between the baseband units and the radio units;
- WG10: OAM. This WG is responsible for defining the specifications for operation, administration, and maintenance (OAM) for the RAN. OAM is the process of monitoring and managing the RAN;
- WG11: Security. This WG is responsible for defining the specifications for security for the RAN. Security is a critical part of the RAN, and this WG is responsible for ensuring that the RAN is secure.

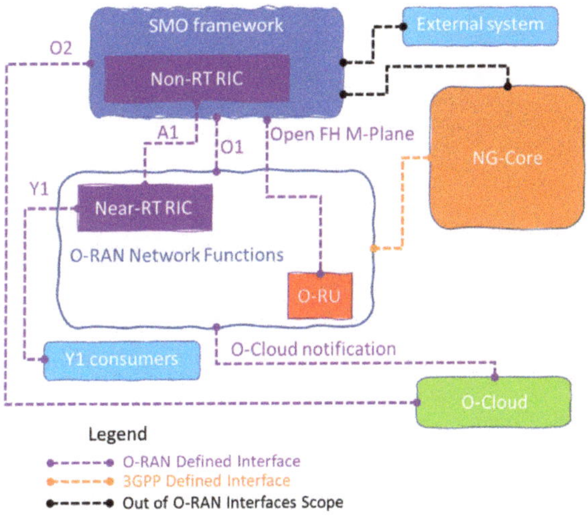

Figure 2. High-level architecture of O-RAN showing internal, 3GPP, and external system interfaces [20].

Figure 3 provides further detail of the O-RAN architecture. As can be seen, O-RAN consists of O-RU, O-DU, O-CU, Near-RT RIC, Non-RT RIC, and SMO. The Uu interface between UE and O-RAN components inside the green dashed area, as well as the UE and O-eNB, denote all the O-RAN functions required to support the Uu interface NR. On the other hand, the O-eNB terminates the Uu interface for LTE. The 3GPP defined and maintained interfaces and is considered part of the O-RAN architecture includes the E1, F1-c, F1-u, NG-c, NG-u, X2-c, X2-u, Xn-c, and Xn-u, as depicted in Figure 3 [20]. In the following, we shall elaborate these O-RAN components.

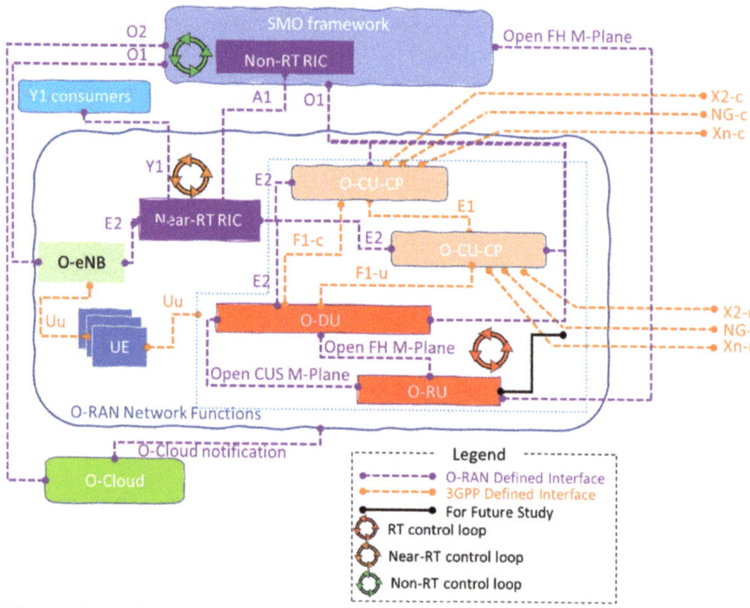

Figure 3. Logical architecture of O-RAN, its associated interfaces, and the three control loops.

2.2.1. O-RU

The major O-RU hardware and software components in Figure 4 highlight the internal and external interfaces that are required. The O-RU terminates the O-RAN Fronthaul (FH) interface, known as Lower Layer Split, as well as Low-PHY functions of the radio interface towards the UE. This is a physical node. The O-RU terminates the O-RAN Fronthaul M-Plane interface towards the O-DU and SMO. The O-RU termination of the O1 interface towards the SMO is under study under the O-RAN Operations and Maintenance Architecture. A single split point, known as "7−2x", but which allows a variation, with the precoding function located either "above" the interface in the O-DU or "below" the interface in the O-RU. For the most part, the interface is not affected by this decision, but there are some impacts, namely to provide the necessary information to the O-RU to execute the precoding operation. $\text{O-RU}_{(7-2)}$ within which the precoding is not done (therefore of lower complexity) are called "Category A" O-RUs, while $\text{O-RU}_{(7-2)}$ within which the precoding is done are called "Category B" O-RUs, as in Figure 5.

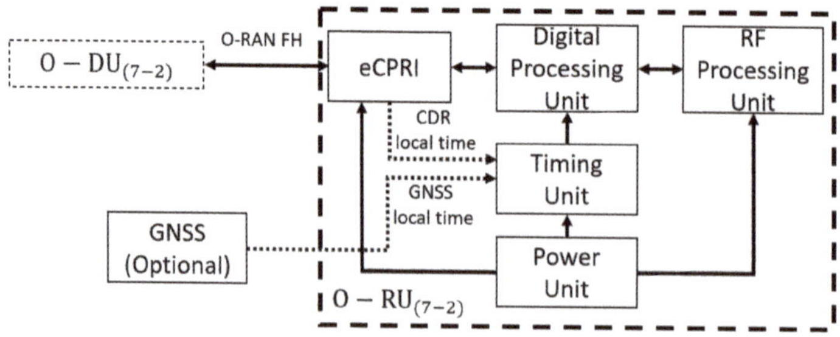

Figure 4. O-RU high-level architecture showing the main hardware components, and internal and external interfaces.

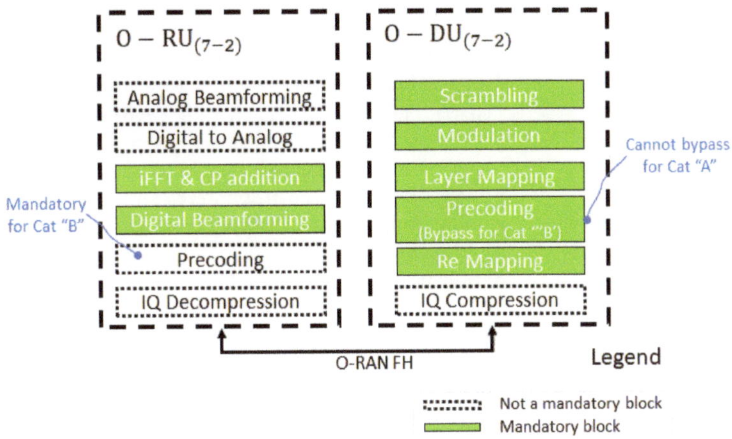

Figure 5. The O-DU$_{(7-2)}$/O-RU$_{(7-2)}$ split point option showing requirements for Category A and Category B for the O-RU.

2.2.2. O-DU

The O-DU is designed as a white box that performs the O-DU functions, such as upper L1 and lower L2 functions. The hardware includes a motherboard that contains a processing unit, memory, the internal I/O interfaces, and external connection ports. There are two split options for the O-DU, which are O-DU$_{(6)}$ and O-DU$_{(7-2)}$. WG4 considers the O-DU$_{(7-2)}$ functional split option due to the two competing interests. The first is to keep an O-RU as simple as possible, because size, weight, and power draw are the primary deciding considerations, and the more complex an O-RU, the larger, heavier, and more power-hungry the O-RU tends to be. The second is to have the interface at a higher level, which tends to reduce the interface throughput relative to a lower-level interface. However, the O-RU tends to be the more complex with higher levels of interface.

The fronthaul and backhaul interface are used to carry the traffic between O-RU$_{(7-2)}$, $FHM_{(7-2)}$, $FHGW_{(7-2)\rightarrow 8}$ and O-DU$_{(7-2)}$, as well as O-CU and O-DU$_{(7-2)}$. The O-DU$_{(7-2)}$ design may also provide an interface for hardware accelerator option design. The other hardware functional components include synchronization and timing, the storage for software, hardware and system debugging interfaces, and board management controller, just to name a few; the O-DU$_{(7-2)}$ designer will make decision based on the specific needs of the implementation. Note that the O-DU$_{(7-2)}$ hardware reference design is also feasible for O-CU and integrated O-CU/O-DU$_{(7-2)}$.

2.2.3. O-CU

The O-CU is another white box hardware that performs the O-CU function of upper L2 and L3. The O-CU hardware motherboard contains a processing unit, memory, the internal I/O interfaces, and external connection ports. The midhaul (MH) is used to carry the traffic between O-CU and O-DU$_{(7-2)}$, and the backhaul (BH) interface is for carrying the traffic between the O-CU and core network. Other hardware functional components, such as the storage for software, hardware and system debugging interfaces, board management controller, and more, are based on the specific needs of the implementation. The hardware of the O-CU is similar to the O-DU$_{(7-2)}$. However, the hardware accelerator is mandatory to offload computationally intensive functions and to optimize the performance under varying traffic and loading conditions.

2.2.4. Near-RT RIC

The Near-RT RIC is a logical function that can control and optimization RAN elements and resources in near real time by collecting detailed data from the O-RAN logical components and provide actions over the E2 interface. In addition, the A1 Interface enables Non-RT RIC to drive the policy guidance of the Near-RT RIC applications/functions and support AI/ML. Near-RT RIC hosts many functions in Figure 6, which include the following:

- Database and related Shared Data Layer (SDL) services: to exchange information between RAN and UE to support specific use cases;
- xApp subscription management: to manage subscriptions from different xApps and provides unified data distribution to xApps;
- Conflict mitigation: to resolve potentially overlapping or conflicting requests from multiple xApps;
- Messaging infrastructure: to allow message interaction within the Near-RT RIC functions;
- Security, which provides the security scheme for xApps;
- Management services: to manage fault, configuration and performance as a service producer to SMO;
- Logging service: to provide tracing and metrics collection which capture, monitor, and collect the status of Near-RT RIC internals and transfer to external systems for further evaluation if needed;
- Interface termination: to provide interfacing to other O-RAN components;
- Functions hosted by xApps: to allow services to be executed at Near-RT RIC;
- API-Enabled function: to support capabilities related to Near-RT RIC API operations such as API repository/registry, authentication, discovery, generic event subscription, etc.;
- AI/ML support: to feature data pipelining, training, and performance monitoring for xApps;
- xApp Repository function: to manage selection of xApps for A1 message routing, based on the A1 policy types and operator policies, and Access control of A1-EI types for xApps based on operator policies.

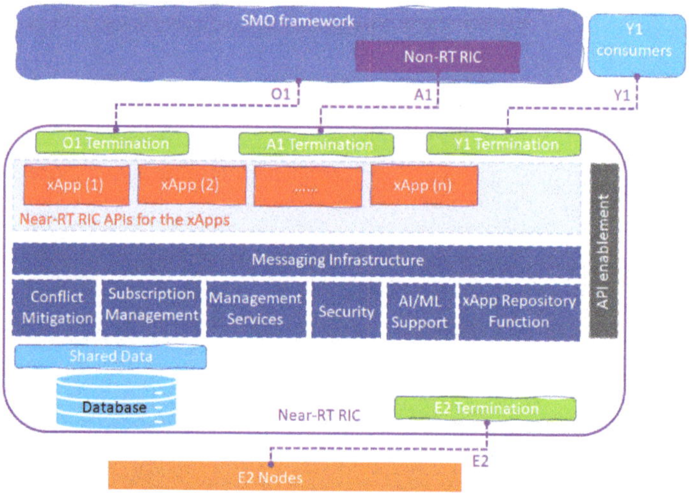

Figure 6. Near-RT RIC functional architecture including the Near-RT RIC API component for the xApps.

2.2.5. SMO and Non-RT RIC

The telecom industry widely considers the service-based architectural in network implementation to give flexibility and future-proof solutions. In addition, the choice of components that produce and/or consume certain services to the deployment allows multi-vendor interoperability through the definition of standardized services and service interfaces. It is important to have the perspective of two Non-RT RIC architecture views;

the first one is the "Functional" view in Figure 7, showing the internal SMO framework and the three categorical components: rApps, Non-RT RIC framework and the open APIs for the rApps, while the "Service-based" view allows wide range of flexibility for deployment and is future-proof, the main principles for this architecture illustrated in Figure 8 are modularity, extensibility, functional abstraction, discoverability, composability, reusability, and loose coupling.

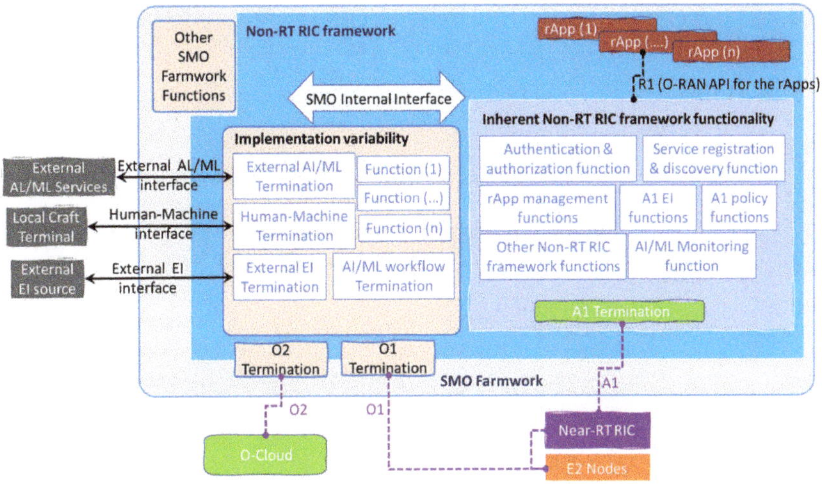

Figure 7. Non-RT RIC functional architecture, showing the R1 and external interfaces [21].

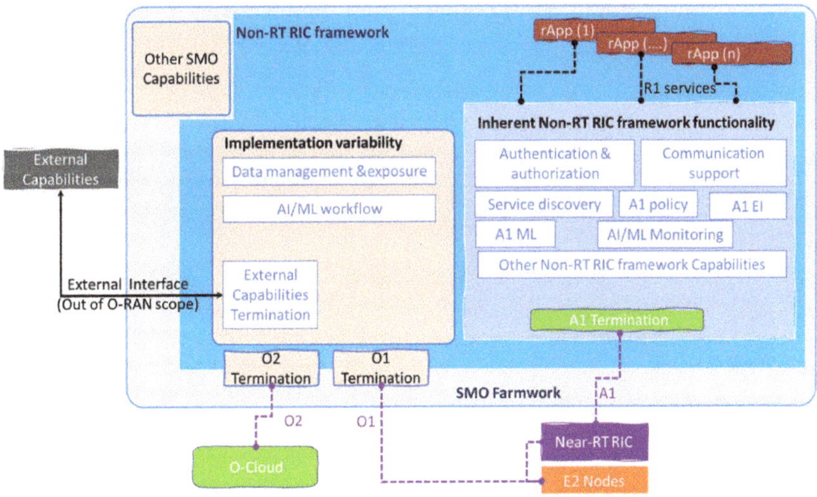

Figure 8. Non-RT RIC architecture service-based view, in which the services are exposed to rApps via the R1 [21].

2.2.6. O-RAN Interfaces

The O-RAN architecture is designed to promote interoperability, multi-vendor support, and innovation in the RAN. Here are the main interfaces specified in the O-RAN architecture:

- Open Front Haul (OFH) interface: this interface connects the O-RU to the O-DU and carries the digitized baseband signal and control information between the two units;

- Fronthaul interface: This interface connects the O-RU to the O-CU in the O-RAN Cloud RAN architecture. The interface carries the digitized baseband signal and control information between the O-RU and the O-CU;
- E1 interface: This interface connects the O-DU to the O-CU and carries control and management information between the two units;
- E2 interface: This interface connects the O-CU to the O-CU-CP and carries control and management information between the two units;
- A1 interface: This interface connects the O-RAN Controller to the O-RAN Element Management System (EMS) and provides management and monitoring capabilities for the O-RAN network;
- O1 interface: This interface connects the O-RAN SMO to the O-RAN Controller and provides service orchestration and management capabilities;
- O2 interface: This interface connects the O-RAN Controller to the O-RAN Radio Resource Management (RRM) and provides resource management and optimization capabilities for the O-RAN network;
- O3 interface: This interface connects the O-RAN Controller to the O-RAN Network Management (NM) and provides network management and monitoring capabilities.

2.2.7. Interface with 3GPP

3GPP interfaces are also used in O-RAN to provide message exchanges between O-RAN components following 3GPP signaling specifications. They are summarized as follows:

- X2 interface: The X2 interface is used to exchange control information and user data between different eNodeBs (eNBs) in a 4G/LTE network. It is also used to facilitate inter-cell handovers and load balancing. O-RAN uses the same X2 interface for communication between O-RAN radio units and between the O-RAN radio units and 3GPP core network;
- S1 interface: The S1 interface is used to exchange control and user plane information between the eNodeB and the 4G/LTE core network. This interface is responsible for mobility management, session management, and connection management. O-RAN uses the same S1 interface for communication between the O-RAN radio unit and the 3GPP core network;
- F1 interface: The F1 interface is a new interface introduced by O-RAN that connects the O-RAN radio unit to the O-RAN distributed unit. This interface carries the radio frequency (RF) signals and also supports the exchange of control and management information. The F1 interface is similar to the W1 interface used in 3GPP's split architecture;
- E2 interface: The E2 interface is used in the 5G RAN to exchange control and management information between different network functions, including the radio access network function (RANF), central unit (CU), and distributed unit (DU). The O-RAN Alliance has developed an E2 interface specification that is compatible with the 3GPP E2 interface.

2.3. *ML Workflow in O-RAN*

ML workflows in O-RAN involve a series of steps that enable the development, deployment, and optimization of ML models for network operations. The workflow consists of several stages of processing. We shall elaborate the processes involved in ML workflow in the following [22]:

- Data collection: The first step in the ML workflow is to collect and preprocess the data. This involves identifying the relevant data sources, collecting the data, and preparing it for analysis. This step is crucial as the quality of the ML model depends on the quality of the data used to train it;
- Data exploration and analysis: In this step, the collected data are explored to gain insights and identify patterns. This involves data visualization, statistical analysis, and

other data exploration techniques to understand the underlying structure of the data. This step is important for selecting appropriate ML algorithms and for identifying relevant features that can be used to train the models;

- Model development: In this step, ML algorithms are selected and trained using the data. This involves selecting the appropriate algorithms, feature engineering, and tuning the model hyper-parameters. Once the model is trained, it is evaluated and validated to ensure that it is accurate and reliable;
- Model deployment: In this step, the trained model is deployed into the O-RAN environment. This involves integrating the model into the network operations environment and deploying it in a way that allows it to access the relevant data and provide real-time predictions or recommendations. This step also involves monitoring the performance of the deployed model to ensure that it is performing as expected;
- Model optimization: Once the model is deployed, it needs to be optimized to improve its performance. This involves monitoring the performance of the model in real-time, identifying areas for improvement, and updating the model as necessary. This step is crucial for ensuring that the ML models continue to provide accurate and reliable predictions and recommendations over time;
- Model maintenance: The final step in the ML workflow is model maintenance. This involves maintaining the ML model, updating it as necessary, and ensuring that it remains aligned with the evolving needs of the O-RAN network operations environment.

Overall, the ML workflow in O-RAN involves a series of steps that enable the development, deployment, and optimization of ML models for network operations. By leveraging the power of ML, network operators can improve network performance, reduce energy consumption, and provide a better user experience.

2.4. O-RAN Open Source Development Landscape

During the past few years, open source platforms for cellular networks have been developed to move away from proprietary hardware and mitigate technological barriers [23–26]. The main open source projects implementing O-RAN specifications are OpenAirInterface (OAI) [24], srsRAN [25] and O-RAN Software Community (O-RAN SC) [23]. These three communities are sharing the source code of different modules with different states of progress. For example, the OAI has implemented the CU/DU split by supporting Software Defined Radio (SDR) USRP devices [27] for RU and on-the-shelf UEs. The testbed Colosseum [28] was built to provide remote access to OAI resources configured with 256 SDRs. It is a large-scale wireless testbed with a massive channel emulator, which enables the design, development, and testing of solutions at scale in various deployments and channel conditions. The testbed is open to the research community and can be used for experimental research with different applications. On the other hand, srsRAN has developed full stacks of UE and gNB with a simple setup compared to the OAI platform. The O-RAN SC has published partly their industrial solutions of the O-DU with Medium Access Control (MAC) and Radio Link Control (RLC) protocols. In parallel, the Software-Defined Radio Access Network (SD-RAN) paradigm enables RAN programmability and introduces new APIs for control extending platforms like OAI and srsRAN to support control/data plane separation [29]. The objective of SD-RAN is to focus on the L2 protocols: Radio Resource Control (RRC), RLC, MAC and Packet Data Convergence Protocol (PDCP) protocols. FlexRAN [30] is an example of SD-RAN platforms promising flexibility by supporting dynamic control functions and robustness by handling network applications with critical real-time requirements. Other new SD-RAN controllers, such as the 5G-EmPower [31], deal with other challenges like heterogeneity of mobile RANs or RAN slicing with NexRAN [32]. These open source platforms and APIs represent key first steps toward the availability of a fully open-source O-RAN solution.

These platforms allow the research community to experiment with new methods and replace the simulation hypothesis. The research results would be with significant impact. Furthermore, without the shift of the softwarization of the RAN functions as close as possible to the antenna, the ML algorithms cannot be applied efficiently for the reconfiguration of the network. In addition, the 3GPP specifications are upgraded frequently with new interfaces and protocols to improve the global network performances and handle new applications. For example, the 3GPP release 17 introduces new NTN and satellite communications and the new Multicast/Broadcast Session (MBS) in the definition of protocols' functions. If the RAN is closed with a full hardware implementation, the solution would not be maintained quickly and the research community would not have access to new challenges. However, the flexibility of the software radio would allow the designer and the researcher to quickly maintain the O-RAN solutions as well as design, prototype, demonstrate, and analyze the O-RAN functions in the real-world settings.

There remain challenges in the implementation and testing of open source O-RAN solutions. Limited use cases, missing functionalities in the current implementation, as well as the affordability of hardware devices are some obstacles to make further progress in O-RAN implementation. For example, at the time of writing, O-RAN SC [23] shares only the O-DU without the complete implementation of the CU. This O-DU handles the registration of one UE with one distributed unit and without the possibility to test it with an SDR. The srsRAN [25] and O-RAN SC [23] support the release 15 of the 3GPP specifications, and they do not support the side link V2X communications use case. The OAI [24] uses USRP X300, which is a capital outlay. Therefore, research and development efforts, such as developing a specific SD-RAN for existing RIC implementation (for example, FlexRIC [33]), or supporting new 3GPP specifications (for example, splitting of the CU into CU-UP and CU-CP [34]), are open for both research and industrial communities to contribute.

3. ML Application in O-RAN

One key advantage of the Open RAN architecture is its ability to separate intelligent controls from the core network. This architecture gives flexibility for RAN to incorporate intelligence toward network control. With this architecture, ML can be easily integrated into Open RAN, not only to automate and optimize network operations, but also to improve network efficiency, introducing new use cases and services that are traditionally challenging to implement in the RAN.

While being developed, the concept of Open RAN has already sparked many research works in investigating the potential of Open RAN and its performance benefits. Notably, substantial research works focusing on applying ML algorithms in Open RAN have appeared in the literature. Improvements, such as optimizing radio resource management and network slicing, automating component deployment for efficient use of computing and communication resources, or improving energy efficiency, are some research activities receiving attention. The structure of this section is summarized in Table 2, and the content in each subsection is summarized in Tables 3–12.

Table 2. The structure of Section 3 for ML applications in Open RAN.

Title	Ref	Title	Ref
Section 3.1 O-RAN Deployment	[15,35]	Section 3.2 AI/ML Implementation	[22,36]
Section 3.3 Network Slicing	[37–40]	Section 3.4 Dynamic Function Split	[41–43]
Section 3.5 Resource Management	[44–47]	Section 3.6 Session Management	[48–50]
Section 3.7 Traffic Steering	[4,51,52]	Section 3.8 Mobility Management	[53]
Section 3.9 Energy Efficiency	[54]	Section 3.10 Satellite NTN	[55]

3.1. O-RAN Deployment

The deployment of an Open RAN poses challenges in terms of computing and network resources. Integrating hardware and software components from multiple vendors can lead to issues with resource allocation, such as conflicts over computing power, storage, and bandwidth. Interoperability challenges can also lead to inefficiencies in resource utilization, as different components may have different requirements and capabilities.

In addition, security concerns related to access control, data protection, and privacy can require additional computing and network resources to address. Overall, the deployment of O-RAN requires careful planning and management of computing and network resources to ensure optimal performance and security.

In [15], the authors investigate how the deep-learning mechanisms could be deployed in an O-RAN architecture, via its hierarchical RIC modules. In particular, its O-RAN placement within the Near-RT RIC, O-CU, O-DU, and O-RU modules, as well as the associated functional blocks and O-RAN interfaces, is discussed. The authors describe the general procedure to implement automated DL models in O-RAN to achieve stable performance of these models by introducing ML system operations (MLOps) concept in O-RAN. They then go deeper and explore two case studies for DL deployment in O-RAN, which are classified as supervised and/or deep reinforcement learning (DRL).

For supervised learning, the authors identify two approaches to deployment: a centralized approach and a federated learning (FL) distributed architecture. The authors propose that for either approach, in order to enhance RAN performance and reduce operational cost, RICs would integrate embedded ML capability. To achieve this, local models (i.e., "xAPPs") would run in the Near-RT RIC and global model parameters would be generated by the Non-RT RIC. In the centralized case, the data would be held in the Non-RT RIC, but for FL, the xApps could be built in the Near-RT RIC using O-RU level data, and just the local parameters transmitted to the Non-RT RIC for aggregation, leaving the data to be held locally. In either approach, the Non-RT RIC would then send out the global parameters to the xApps to update their models, which then operate in real-time using data obtained from the O-RUs via the O1 interface.

In the case of reinforcement learning (RL), the authors focus the deployment strategy on the actions of the RL algorithm intelligent agent(s), which should be deployed near the Near-RT RIC to improve the performance of xApps. This agent then uses the E2 interface to communicate with the O-DU and O-CU-C/U modules to take periodic actions to update the policy of resource allocation and scheduling in the O-DU's MAC layer.

Additionally, the agent connects to the O-RU through the O1 interface to receive the obtained reward based on user experience quality and new state of the system expressed by the total number of allocated resource blocks and users' density. By using inputs and rewards, the ML model can be trained to make data-driven decision more accurately.

The authors then discuss options for control of the training and deployment process for ML systems. The first option is a fully manual process with review by network staff before deployment of an updated run-time system. The authors envisage that this will become inadequate, however, in cases where data profiles vary with time, requiring frequent retraining. They propose an automated pipeline process, triggered by various predefined criteria, to retrain, validate, and deploy the updated ML systems. Pipeline metadata are retained in case a roll-back to a previous model is required, and to assist in debugging. The performance of RAN, which is based on manually created and deployed ML models, may degrade due to the dynamics of the radio access environment, or even the data profiles of the environment; as a result, the authors propose a general procedure to implement automated DL models in O-RAN to achieve the stable performance of such O-RAN models, called MLOps. Furthermore, the Non-RT-RIC module can monitor the Near-RT-RIC performance for implementing the ML models using the A1 interface to pass the information to enable ML automation. In addition, the article shows that the O-RAN architecture supports the design of machine-learning-based schemes to provide optimization for the Automatic Neighbor Relation (ANR) function of a Self Organizing Network (SON), which allows gNodeB handovers process improvement and provides an example article [56], where both Acumos Framework and Open Network Automation Platform (ONAP) were used to create the ML models that the O-RAN RIC module can execute, monitor, and manage the model design workflow with.

Finally, the authors identify a number of open problems in the deployment of DL systems in O-RAN. A critical issue is security, where the large number of new interfaces and the potential lack of trust between the components results in a significant number of new threats. The authors also identify issues with the integration of network slicing, self-organizing network functions and edge computing entities, the use of online DL systems, scalability, and challenges with energy management.

In order to overcome the challenges with the integration of network slicing capabilities within the O-RAN environment, the authors of [35] tried to optimize the admission and placement of O-RAN slices using DRL. The authors emphasize that while previous works considered the placement of slices to optimize processing and bandwidth resources, they have not considered admission control or the long-term impact of admitting a slice. The authors propose an optimization model using a joint RL approach to intelligently admit and place network slices in the available resources of different load scenarios. The proposed solution is compared with two baseline methods, a greedy heuristic and a DRL-based solution (RMAX) [57], under two load scenarios, low and high load. The results show that the proposed solution outperforms the baseline methods in terms of revenue, cost, and total profit for both scenarios, and therefore maximizes the long-term profit of infrastructure providers by considering revenue factors of slices and the idle cost of servers to deploy them.

Table 3. Summary of current efforts on ML applications for O-RAN deployment.

Year	Ref	Contribution
2022	[15]	Proposed a centralized/federated learning approach for the deployment of deep learning mechanisms in the O-RAN architecture, while the integration of RL mechanisms is associated with the Near-RT RIC, O-CU, O-DU, and O-RU modules. The E2 and O1 interfaces are explored for RL-based resource allocation and scheduling.
2022	[35]	Proposed a joint DRL-based solution using PPO to solve an optimization problem aiming to intelligently admit and place network slices in the O-RAN environment considering the available resources and different network loads.

3.2. AI/ML Workflow Implementation in O-RAN

The O-RAN specification addresses the overall architecture and solution for AI/ML Workflow-related requirements in [22] for the O-RAN use cases. These requirements allow automating AI scaling, where Data, Model Training, and Model Evaluation pipelines are key points to make AI faster, easier to deploy, and able to scale to larger and more complex problems.

In [36], the authors adopt the importance of ML training pipeline automation to propose ML pipeline automation techniques to apply the MLOps level 1 (ML pipeline automation) to the RIC platform, where they use Kubeflow for supporting the end-to-end lifecycle of the model management and propose the training pipeline automation to the RIC Platform to conduct the online training process. However, the authors use the KFServing inference service to deploy Kubeflow's trained model. The new ML xApp type structure removed the RMR/gRPC adapter and replace it by using a Shared Data Layer (SDL) and RIC Message Router (RMR) libraries directly from the ML xApp. The authors show that the round trip time of the inference request between assisted xApp and ML xApp reduced significantly when the number of requests is more than 300. To find how an RL model application performs under the proposed RIC AI/ML workflow, the authors trained it to solve the resource allocation in the DU optimization problem using PPO to show the improvement in user throughput.

Table 4. Existing studies on ML applications for AI/ML implementation in O-RAN.

Year	Ref	Contribution
2021	[22]	Provides an initial O-RAN standard the terminology, workflow, and requirements, related to AI/ML model training, distribution and deployment.
2021	[36]	Proposed ML pipeline automation technique to manage ML training in O-RAN RIC.

3.3. *Network Slicing*

Network slicing in O-RAN refers to the ability to partition a physical network into multiple virtual networks, each tailored to specific use cases or applications. Each network slice has its own set of resources, including bandwidth, processing power, and storage, which can be dynamically allocated and managed according to the needs of the specific use case. O-RAN network slicing enables operators to offer customized services to their customers, such as low-latency communication for industrial applications or high-bandwidth streaming services for consumers. It also allows multiple services to be delivered over the same physical infrastructure, maximizing resource utilization and improving efficiency. The O-RAN architecture provides a framework for implementing network slicing, with the RAN and the CN working together to manage and allocate resources. The RAN is responsible for managing the radio access resources, while the CN is responsible for managing the core network resources. Together, they can allocate and manage resources across the network slices as required.

The synergies between O-RAN and network slicing, as well as SON and MEC technologies, were explored in [58], where the O-RAN platform was proposed as a common denominator for the integration of those technologies via proper modifications and extensions of its present architecture. It has been shown that an O-RAN-centric approach is beneficial, and such integration solves some of the issues not well-addressed by the current O-RAN implementation. Also, due to the integration, some components of the contributory technologies can be removed or reused.

ML algorithms can be used to automate the process of creating, modifying, and deleting network slices based on changing network conditions. By combining predictive analytics, real-time monitoring, optimization, and intelligent resource allocation, ML algorithms can create a closed-loop feedback system that can automatically adjust network slices to meet changing network conditions. The main point is to enable the third party to develop better ML algorithms. Particularly for network slicing, we can capture data for training purposes so as to improve the efficiency of network slicing. Based on real-time monitoring, ML algorithms can identify changes in network conditions. By monitoring network performance, ML algorithms can detect changes in network traffic, demand, and capacity, and can modify network slices in response. Then, ML algorithms can be trained to learn how to optimize network slicing based on changing network conditions. By considering factors such as network traffic, user demands, and network capacity, machine learning algorithms can adjust the size and characteristics of network slices to improve network performance and reduce energy consumption. With the network slicing, ML algorithms can then optimize the allocation of network resources to different network slices based on changing network conditions. By learning from historical data and network performance metrics, machine learning algorithms can adjust the allocation of network resources to different slices in real-time, and can optimize resource utilization to improve network performance and reduce energy consumption.

In particular, ML can be of help to network slicing by (i) traffic forecasting; (ii) admission control; and (iii) resource allocation [37]. They reflect three key network slicing building blocks that, together, aim at ensuring network slicing service level agreements (SLAs) are fulfilled. The traffic forecasting block allows us to predict the evolution of traffic load and resource usage for slices over future time instants. The outcome of the traffic forecasting solution can be fed into the slice admission control solution and slice resource allocation solution to enable better decisions (e.g., maximize system resource utilization).

The admission control decides upon the slices to be served over future time instants, based on different criteria, e.g., resource availability, resource efficiency, operator revenue, etc. It can also be built on the outcome of the traffic forecasting for refining admission decisions in an anticipatory manner. Once a slice/user is admitted, the resource allocation block assigns the resources to each slice/user by avoiding over-provisioning and under-provisioning of the resources and ensuring the SLAs are respected.

O-RAN is a promising RAN architecture that inherits all the necessary features, such as intelligence, open and standard interfaces, and closed control loops, to facilitate resource management in a network shared among verticals. In [38], AI techniques are used to perform predictions of future SLA violations and perform corrective actions in advance. Specifically, a recurrent neural network model is utilized to predict the amount of resources required over each slice, given the volume of traffic it carries. E2E O-RAN setup has been used for evaluation of the intelligent closed control loop and resource provisioning scheme, for network slicing and control of the radio and cloud resources of slices, respectively.

In O-RAN, distinct network slices must be dynamically controlled to avoid SLA variation caused by rapid changes in the environment. A novel framework is introduced in [39] to manage network slices through provisioned resources intelligently. Due to diverse heterogeneous environments, the intelligent machine learning approaches require sufficient exploration to handle the harsh situations in a wireless network and accelerate convergence. To tackle this issue, a solution based on evolutionary-based DRL (EDRL) is proposed to accelerate and optimize the slice management learning process in the RIC modules.

An elastic O-RAN slicing problem is addressed in [40] for industrial monitoring and control in the industrial internet of things (IIoT) networks. This work aims to reduce the age of information (AoI) penalty of fresh information updates from different IIoT devices while considering the energy consumption of the IIoT devices. A matching game for solving the IIoT association problem is introduced and an actor-critic-based DRL model applied for O-RAN slicing-based resource allocation.

Table 5. Summary of ML studies for network slicing in O-RAN.

Year	Ref	Contribution
2022	[37]	Introduced the application of ML to network slicing; discussed some open challenges and potential solutions.
2022	[38]	Provided an intelligent closed-loop SLA assurance scheme for O-RAN slicing. A real-world dataset of a large operator is used to train a learning solution for optimizing resource utilization in the proposed closed-loop service automation process.
2022	[39]	Developed a novel O-RAN slicing framework over an evolutionary-based DRL approach to manage network slices dynamically in the rapid changing environment.
2022	[40]	Addressed the elastic O-RAN slicing problem for industrial monitoring and control in IIoT and introduced a matching game for solving the IIoT association problem, and then applied an actor-critic-based deep reinforcement learning model for O-RAN slicing-based resource allocation.

3.4. *Dynamic Function Split*

Dynamic function splitting is an essential technique for enhancing the efficiency and flexibility of O-RAN. It involves breaking down a network node or application's functions into smaller and more modular components, which can be distributed across different computing resources. By leveraging ML, O-RAN can benefit from real-time intelligence and decision-making capabilities. ML algorithms can analyze network traffic patterns and resource usage to predict future demand and allocate computing resources accordingly, leading to better resource utilization, reduced latency, and improved overall network performance. Additionally, ML can optimize the selection of computing resources for specific functions based on factors like location, processing power, and energy efficiency, resulting in better dynamic function splitting and better service delivery to users.

In [41], the authors provide a RL-based approach to the problem of optimizing dynamic function splitting in O-RAN compliant disaggregated and virtualized RANs. Their paper addresses the specific scenario of function splitting between a CU and one or more DUs in a fully virtualized environment, as opposed to splitting between virtual and physical resources. They adopt a multi-agent RL approach using either Q-Learning or State Action Reward State Action (SARSA); a similar but slightly different alternative to Q-Learning. Optimization takes into account the traffic type (e.g., eMBB or URLCC) and also energy efficiency. It is assumed that the CU and each of the DUs run on separate physical environments, each with their own renewable energy source backed up by a battery-based energy storage facility, with a grid connection as a reserve input. Optimization is designed to minimize operational expenditure (Opex), including maximizing the usage of the renewable energy source, taking into account the energy remaining in the battery and also the variation in grid electricity prices during the day. Optimization takes place over a 48-h period and comparative Opex results are presented using solar radiation data from Stockholm, Cairo, Jakarta, and Istanbul, combined with broad traffic level assumptions.

In [42], a novel and efficient energy-efficient RAN disaggregation and virtualization method tailored for O-RAN is presented. This method effectively tackles challenges related to dynamic traffic conditions. By formulating the energy consumption as a multi-objective optimization problem, the authors integrate the Advantage Actor-Critic (A2C) algorithm with a sequence-to-sequence model to effectively address the sequential nature of RAN disaggregation and capture long-term dependencies. The results demonstrate the effectiveness of the proposed solution to reduce energy consumption for dynamic Virtual VNF splitting over traditional approaches like D-RAN and C-RAN.

In [43], the authors propose a novel DRL-based algorithm to jointly solve the optimal placement of network functions between the CU, DU, and user RU in an O-RAN architecture. In the meantime, the proposed algorithm aims to minimize the end-to-end delay and deployment cost, while considering constraints such as processing capacity and link bandwidth. The proposed method evaluates the impact of user mobility on the proposed DQN-based joint user association and CU-DU placement scheme (DJRCD) using the SUMO traffic simulation tool. The algorithm is tested using a mixed highway-urban region in the north of the Greater Toronto Area in Canada, and it is found to be superior to existing methods, achieving a reduction of up to 30% and 40% in end-to-end delay and deployment cost, respectively.

Table 6. Summary of current efforts on ML applications for dynamic function splitting in O-RAN.

Year	Ref	Contribution
2021	[41]	Applied RL to dynamically perform function split decisions for DUs and CU in a virtualized O-RAN architecture.
2023	[42]	Applied DRL to propose an energy-efficient RAN disaggregation and virtualization approach across edge sites, including DUs, and a cloud site, including CUs.
2022	[43]	Proposed a DRL-based algorithm to jointly solve the optimal placement of network functions between the CU, DU, and user RU in an O-RAN architecture.
2023	[59]	Proposed a DRL-based algorithm for a multi-objective optimization to minimize computational costs and the overhead associated with periodically reconfiguring dynamic VNFs splitting.

The dynamic nature of O-RAN environments often necessitates VNF reconfigurations during operation, resulting in additional overhead costs and traffic instability. To tackle this challenge, the work in [59] introduced a multi-objective optimization approach aimed at simultaneously minimizing VNF computational expenses and the periodic reconfiguration overhead. This solution relies on constrained combinatorial optimization with deep reinforcement learning, where an agent works to minimize a penalized cost function derived from the proposed optimization problem. The evaluation of this solution demonstrates substantial improvements, including a remarkable 76% reduction in VNF reconfiguration overhead, accompanied by a modest increase of up to 23% in computational costs. Furthermore, when compared to the most resilient O-RAN system, Centralized RAN (C-RAN), which does not necessitate VNF reconfigurations, this solution delivers savings of up to 76% in bandwidth while revealing a 27% overprovisioning of CPU resources.

3.5. Resource Management

In [44], the authors compare the performance of on-policy and off-policy DRL methods. The former is based on Proximal Policy Optimization (PPO) and the latter on a Sample Efficient Actor-Critic with Experience Replay (ACER). This process was conducted under an O-RAN setup. The O-RAN architecture is a suitable technology for DRL implementation, since it includes mechanisms that enable AI for more efficient network management and orchestration. In particular, both Near-RT and Non-RT RICs are designed for hosting AI workflows, namely DRL models. Since the objective is to optimize the resource allocation of a real-time surveillance video application, the types of services were classified as latency-sensitive and latency-tolerant. In this direction, two slices are established and managed by an O-RAN cross-slice resource orchestrator hosted by the SMO. One slice serves video surveillance cameras to transmit real-time videos via O-RAN to the 5G Core Network (CN) and then to a Control Center (CC) for real-time monitoring, and a slice serving the latency-tolerant users. The performance of the DRL on-policy mechanism was shown to provide better overall results namely in terms of implementation simplicity, performance stability, good trade-off between users latency and energy consumption, and faster convergence.

In [45], the authors consider the problem in which the eMBB and URLLC services compete for limited and insufficient computing resources, and the operator must balance the allocation of these resources to users of both services in multiple O-RUs/shared O-Cloud while maximizing fairness. The problem is initially modeled as an Integer Linear Programming (ILP) problem. However, given the high complexity of solving the NP-Hard ILP problem, a policy gradient-based RL algorithm aiming to solve a Markov Decision Process (MDP) is used instead. The latter is expected to perform similarly to the ILP solver, and both approaches are compared. Simulation results showed that the RL agent performed close to the optimal results of the ILP solver not deviating from the ILP by more than 6%, while being fairer at the same time.

The authors of [46] proposed an RL-based framework to manage traffic flows while taking advantage of the O-RAN ecosystem. The framework receives periodic reports from the O-RAN DU about the network status and dynamically adapts the per-flow resource allocation for which each traffic flow can compete, and identifies the corresponding modulation and coding scheme (MCS) that best fits the traffic flow KPIs and the channel quality. The RL-based dynamic resource controller solution that leverages a policy differential semi-gradient State-Action-Reward-State-Action (SARSA) targets the minimization of the maximum difference between desired and actual throughput, across all active traffic flows. The framework was integrated into an O-RAN platform, and is deployed as an xApp in the Near-RT RIC. The deployed framework is very flexible, and can adapt its architecture based on the number of traffic flows. Additionally, it is possible to create multiple policy instances, each independently and sequentially serving a subset of users, improving the framework's scalability.

The research conducted in [47] employs RL for adaptive resource allocation, demonstrating its utility in the context of Non-RT RIC. The ML agent deployed in the Non-RT RIC acquires knowledge and learns a radio resource allocation policy capable of adapting to dynamic environments, while simultaneously meeting diverse energy-driven criteria. Within the learning context, key information such as the mean and variance of the channel quality indicator (CQI) from the previous period and the bit number of new data are aggregated at the onset of each period. Subsequently, three transmit parameters, including transmit power, the highest MCS, and the maximum transmission airtime, are selected and transmitted to the Near-RT RIC so that the transmit parameters can be applied to

BSs. The chosen resource allocation policy's effectiveness is evaluated at the end of each period by the Near-RT RIC, which computes a reward indicative of the transmission rate. This reward is then sent to the Non-RT RIC. Through this iterative learning process, the ML agent gradually acquires knowledge and discerns the optimal resource allocation policy capable of adapting to dynamically changing environments.

Table 7. Summary of current efforts on ML applications for O-RAN resource management.

Year	Ref	Contribution
2022	[44]	Demonstrated the performance of on-policy and off-policy DRL methods in the form of PPO and ACER, respectively, and the performance was compared in an O-RAN setup for resource allocation optimization in a real-time surveillance video application.
2022	[45]	Employed a policy gradient-based RL algorithm as an alternative to the initially proposed ILP, to solve an MDP and address the challenge of fairly allocating limited resources to eMBB and URLLC users from multiple O-RUs while providing a significantly less complex solution.
2021	[46]	Proposed a RL-based dynamic resource controller leveraging policy differential semi-gradient SARSA to optimize traffic flow management by effectively and dynamically allocating per-flow resources within an O-RAN platform.
2021	[47]	Designed ML deployed in a Non-RT RIC to adapt the resource allocation policy to environment dynamics while satisfying various energy-driven criteria.

3.6. Session Management

Session management in O-RAN refers to the process of establishing, maintaining, and terminating communication sessions between network components. These sessions are used to transmit data, control signals, and other information between different components of the network. Session management is an essential function of O-RAN, as it enables the coordination and control of network operations and services. It involves the management of session parameters, such as session identifiers, session timeouts, and session initiation and termination procedures. Effective session management in O-RAN is critical for ensuring the efficient and reliable operation of the network, as well as for enabling the delivery of high-quality services to end-users.

The authors of [48] explore the efficiency of RL-based methods for intelligent session management when taking advantage of the intelligent gNB architecture of O-RAN. This architecture is again assumed to facilitate the inclusion of AI/ML algorithms mainly due to the introduction of the RIC, which is designed for sustainable deployment of these algorithms. The work is focused on the lack of effort to reduce the packet transmission latency in the Core and Data Networks, which can go up to hundreds of seconds compared to several milliseconds in the gNB fronthaul/backhaul links. By installing intelligent session management schemes based on policy evaluation RL methods such as Q-learning, double and State-Action-Reward-State-Action (SARSA) on the RIC of an O-RAN emulated gNB, it becomes able to effectively predict room to accommodate new PDU sessions with given service requirements. Therefore, the gNB can decide whether to grant a new PDU session or QoS flow, preventing existing and new sessions from violating the latency requirements.

In [49], the authors study connection management under the session management function (SMF), specifically for user-cells associating the user with a BS, considering sub-optimal and greedy solutions such as the received signal reference power (RSRP). Even though the greedy approach is simple and effective, it does not take into consideration the network's local and global status. This causes the lack of load balancing where a certain BS can be overloaded while the neighboring BSs are underutilized. So, the O-RAN architecture features support global RAN automation to balance the load over the network resources by deploying ML algorithms as rApps and xApps in the RIC. The authors proposed a Deep Q-learning algorithm to infer the weights of the graph neural networks (GNN) for optimal user-cell association.

While user access management in O-RAN refers to the process of controlling and securing access to network resources and services by users and applications, it includes the management of user identities, authentication, authorization, and accounting (AAA). The goal of user access management is to ensure that only authorized users and applications can access the network, and that they can do so in a secure and controlled manner. User access management in O-RAN typically involves the use of access control policies, authentication mechanisms, and auditing tools to manage user access and monitor network activity. It is a critical component of network security and privacy, as unauthorized access to network resources can lead to data breaches, network disruptions, and other security incidents. Effective user access management is essential for ensuring the secure and reliable operation of O-RAN with session management.

In [50], the authors address the anticipated handover-rate and load balancing issues if O-RAN is deployed under the conventional user access control schemes typically based on signal strength or capacity measurements. This is a result of the openness nature of O-RAN, where the BS functions are decomposed and virtualized into CUs, DUs, and RUs, where they are massively deployed throughout the network. Therefore, typical access control procedures would make this a practically intractable process due to the massive signaling overhead and system complexity. This can be considerably mitigated if each UE autonomously selects proper BSs (or CUs/DUs/RUs). In this direction, a federated DRL-based scheme to address user access control in the O-RAN is proposed to establish intelligent user-centric access control mechanism to optimize the overall throughput and avoid frequent handovers. This is achieved by enabling the UE to train two deep Q-networks (DQNs) using its own observations, and making access decisions based on its DQNs outputs. Then, the DQN parameters are forwarded to a global model server installed in the RIC. This server can select just a group of UEs in each instance to further mitigate the signaling overhead. Afterward, the global model is updated by aggregating DQN parameters obtained from the selected UEs. The DQN global parameters are then disseminated to each UE to further improve its access decision. The achieved results show that the proposed scheme can reduce the frequency of handover in a O-RAN environment up to 53% when compared to a conventional access control based on signal strength UE measurements.

Table 8. Summary of current efforts on ML applications for O-RAN session management.

Year	Ref	Contribution
2021	[48]	Applied an ML algorithm in O-RAN to maintain QoS satisfaction by controlling admission of PDU sessions.
2021	[49]	Applied DQN to optimize user-cell association in O-RAN by supporting the global RAN automation through load balancing over the network resources.
2021	[50]	Proposed a smart user-centric access control in O-RAN using Federated DRL-based learning to mitigate frequent handover.

3.7. Traffic Steering

O-RAN traffic steering research aims to optimize network traffic management by intelligently routing traffic based on real-time network conditions and user demand. This involves analyzing various network parameters, such as traffic load, network congestion, and user behavior, to determine the best path for traffic flow. Traffic steering techniques can also be used to dynamically split traffic across different network functions and resources in real-time, improving network performance and reducing latency. ML algorithms can play a critical role in traffic steering research, providing real-time intelligence and decision-making capabilities to optimize traffic flow and improve overall network performance. However, there are still challenges related to scalability, complexity, and limited resources that need to be addressed to achieve optimal traffic steering in O-RAN.

In O-RAN, more processing is done by placing virtual network functions (VNF) in the DU before transferring data over midhaul links. This means that placing VNFs in the CU needs more bandwidth compared to placing in the DU [60]. Therefore, the question is how many functions should be left in the DUs to prevent network congestion on midhaul links arises. This is especially the case when DUs have limited computing/storage capacities compared to CUs. The authors of [61] have proposed an optimization problem to select the split points in O-RAN. The objective of this study is to balance the load across CUs as well as midhaul links with respect to the required delay and bandwidth and processing capacity of the DUs and CUs. Going beyond the static optimization, it is also noted that in real-world scenarios under traffic demands dynamicity and uncertainty at RUs, methods like RL and DRL can be used to provide dynamic VNF splitting across CUs and DUs.

In [62], O-RAN decouples the Control Plane (CP) from the User Plane (UP) through the E1 interface, which is derived from the software defined network (SDN) architecture. This separation of CP and UP allows a network to be more flexible in programming. The CP is implemented in hierarchical RICs, and manages radio resource functions through A1 and E2 interfaces. The authors of [62] proposed using the hierarchical RICs to minimizes end-to-end delay of the data plane traffic by placing Containerized Network Functions (CNFs) effectively. In comparison to VNFs, CNFs are lighter and can be implemented through microservice architectures, enabling a dynamic, scalable, and flexible architecture towards 5G [63].

Additionally, paper [4] discusses the use of ML methods to achieve modular and flexible O-RAN implementations in 6G networks, with a focus on the traffic steering use case and O-RAN xApps. The authors describe several ML algorithms that can be used for traffic steering, including decision trees, k-nearest neighbor (KNN), and neural networks. They also discuss the use of RL to train an agent to make traffic steering decisions in real-time. In [51], a federated meta-learning approach for traffic steering in O-RAN systems is proposed. This approach allows multiple Radio Access Technologies (RATs) to learn from each other without sharing their private data. The authors present a neural network architecture that uses meta-learning to adapt to different RATs and learn to steer traffic in a decentralized manner.

Paper [52] proposes a traffic steering use-case in O-RAN systems that exploits the benefits of Non-Orthogonal Multiple Access (NOMA) to improve radio resource efficiency. The authors introduce a resource allocation algorithm that dynamically allocates radio resources based on the traffic demand and channel conditions of the users. The proposed algorithm leverages NOMA to allow multiple users to share the same radio resources, while ensuring a high-quality user experience. Paper [64] proposes a traffic steering approach that ensures the efficient coexistence of eMBB and uRLLC services in O-RAN systems. They introduce a multi-objective optimization problem and a traffic steering algorithm that dynamically steers traffic based on the network load and user demands while ensuring a minimum QoS requirement for both eMBB and uRLLC users.

Table 9. Summary of current efforts on ML applications for traffic steering in O-RAN.

Year	Ref	Contribution
2021	[4]	Proposed logistic regression for modular and flexible O-RAN in 6G networks, focusing on traffic steering and O-RAN xApps.
2022	[51]	Introduced federated meta-learning with DQN for privacy-preserving multi-RAT knowledge sharing.
2022	[52]	Proposed Q-learning-based algorithm for power and frequency allocation in O-RAN to minimize macro gNB interference and maximize device QoS.

3.8. Mobility Management

Mobility management is a key function for cellular communication to maintain service continuity and ensure a good level of service quality for users moving across a network.

This requires efficient coordination of radio resources to achieve predictive, timely, and successful handovers for preventing communication disruptions in highly dynamic mobile environments. To support this, the O-RAN architecture offers various capabilities, including the collection of, maintenance of, and access to historical traffic and radio data. Additionally, real-time monitoring of traffic and radio conditions is achievable through the support of the Near-RT RIC framework, which enables the deployment of AI/ML-based applications for detecting and predicting handover anomalies at the user level.

In [53], the authors proposed a new predictive handover method to predict target cells in advance, hence to reduce handover failures. Handover cases are simulated by random user movement within an environment with three eNBs and coverage holes. The handover prediction algorithm is implemented within a software developed by O-RAN Software Community (O-RAN SC), where an Anomaly Detection use case has been installed in the Near-RT RIC platform, which consists of three xApps: Anomaly Detection, Traffic Steering, and QoE Predictor. The process starts with the Anomaly Detection xApp, which analyzes UE data and sends notifications to the Traffic Steering xApp via the RMR protocol when anomalies are detected. The Traffic Steering xApp then requests a prediction of the target cell from the QoE Predictor xApp, which uses the Vector Autoregressive (VAR) algorithm to forecast time-series data based on past throughput data. As user mobility is not considered by the original QoE Predictor xApp, the paper has contributed by adding mobile users' RSRP measurements to the predict cells' throughput. The proposed intelligent prediction method achieves higher successful transmission rates than conventional handover algorithms and allows the traffic steering xApp to send commands to the RAN, such as a handover command using the REST API interface.

Table 10. Summary of ML studies for mobility management in O-RAN.

Year	Ref	Contribution
2022	[53]	Implemented a NN-based handover prediction method for the next target cell in Near-RT RIC using software developed by O-RAN Software Community (O-RAN SC).

3.9. Energy Efficiency

Energy efficiency is a significant challenge in O-RAN due to their highly flexible and scalable design, where several components and technologies must work together seamlessly. Achieving energy efficiency also requires balancing trade-offs with other performance metrics such as latency or throughput. Although dynamic function splitting and ML-based optimization techniques can be used to allocate computing resources efficiently, challenges related to limited resources, scalability, and the dynamic environment of O-RAN must be addressed to achieve optimal energy efficiency.

To address this, the authors in [54] propose an online learning-based energy-aware scheduling method for virtualized Base Stations (vBS) in O-RAN. The goal is to optimize scheduling policies that reduce energy consumption while maximizing vBS performance. The novelty of this work lies in the application of adversarial bandit learning to vBS operations. The authors introduce a Policy Decider application within Non-RT RIC to learn and implement optimal policies, which can be adjusted based on network conditions and user needs. The policy decision is shared with Near-RT RIC through A1 interface, and the Data Monitor calculates the reward based on achieved performance and energy cost, which is then sent to the Policy Decider via the O1 interface at the end of each decision time slot. Data-driven experiments based on real-world traffic traces and testbed measurements are conducted to evaluate the effectiveness of the proposed method and compare it to state-of-the-art approaches. The proposed approach outperforms state-of-the-art methods by achieving energy savings between 35.5% and 74.3%, as demonstrated through data-driven experiments based on real-world traffic traces and testbed measurements.

Table 11. Summary of ML studies for energy efficiency in O-RAN.

Year	Ref	Contribution
2022	[54]	Proposed online adversarial bandit learning for energy-aware scheduling policy optimization to maximize the performance of virtualized Base Stations (vBS) in O-RAN.

3.10. Satellite NTN

Interference management in satellite networks involves reducing interference between multiple satellites and traditional cellular communication systems, which can lead to degraded signal quality and decreased network performance. O-RAN can be used for interference management in satellite networks by leveraging its cutting-edge functionalities. One way to use O-RAN for interference management in satellite networks is by using a dynamic spectrum access techniques to dynamically allocate and manage spectrum resources based on real-time network conditions. This involves monitoring network parameters, such as channel quality and interference levels, and dynamically reconfiguring the network to avoid interference and optimize network performance. Another approach is to use machine learning algorithms to analyze network data and make intelligent decisions on interference management. For example, machine learning algorithms can be trained to identify patterns in network data that indicate interference and automatically adjust network parameters to mitigate interference. Finally, O-RAN can also be used to implement beamforming techniques to improve signal quality and reduce interference in satellite networks. Beamforming involves adjusting the phase and amplitude of transmitted signals to create directional beams focused on specific network areas. By optimizing beamforming parameters based on network conditions, interference can be minimized and network performance improved. Additionally, the radio spectrum is a finite and highly sought-after resource; therefore, spectrum sharing aims to help resolve this issue by creating regulatory frameworks and developing wireless technologies to share spectrum bands between heterogeneous users. The authors in [55] proposed O-RAN with machine learning for 5G/XG and closed-loop feedback via sensing can reduce harmful interference between heterogeneous 5G and Low Earth Orbit (LEO) satellite communication systems.

Table 12. Existing study of ML applications for Satellite NTN relevant to O-RAN.

Year	Ref	Contribution
2021	[55]	Investigated how O-RAN can be used for 5G/XG to mitigate interference between terrestrial and space communication systems.

4. Research Opportunities in O-RAN

Applying ML in O-RAN networks has the potential to significantly improve network performance, automate complex network operations, and enable new use cases and business models. ML algorithms can analyze large volumes of network data, enabling intelligent decision-making in real-time and optimizing network resources for enhanced user experiences. In the previous section, we provided overview of the state-of-the-art research works that demonstrate the use of ML to improve O-RAN and the underlying network elements. However, challenges remain to apply and integrate ML into O-RAN for network automation. These challenges include the need for high-quality and diverse training datasets, ensuring robustness and fairness of ML models, addressing privacy concerns, and developing efficient computational frameworks capable of handling the scale and complexity of O-RAN deployments. In this section, we shall describe the potential of ML applications in O-RAN and their challenges. Table 13 shows a summary of identified research opportunities.

Table 13. Research opportunities for ML applications in O-RAN.

	Category	Issues
Section 4.1	ProactiveMaintenance	- How to evolve the O-RAN framework in conjunction with ML system design; - To develop ML training approaches across the network irrespective of equipment vendor or site location (i.e., different configurations of multiple vendor equipment and unharmonized data across multiple sites).
Section 4.2	xApps,rApps,dAppsOperation	- Orchestration of xApps, rApps, and dApps in the O-RAN RIC when they are simultaneously operated for network automation; - Orchestration across the domain with SON functions in a core network and any newly added xApps, rApps, and dApps to the RAN.
Section 4.3	SatelliteNTN	- To optimize the resource allocation for capacity enhancement; - To mitigate network interference sources such as adjacent satellite interference and inter-system interference.
Section 4.4	MassiveMIMO	- To provide right interfaces to integrate multi-antenna processing in O-RAN to maximize the spectral efficiency; - To dynamically adjust the level of coordination/cooperation between DUs and to efficiently perform the RU clustering; - To effectively distribute the channel state information between the split baseband functions.
Section 4.5	MobilityManagement	- To jointly optimize the trajectory of UAV and the task offloading among diverse O-RAN elements.
Section 4.6	NetworkManagement	- To optimize the flexible functional split of RAN slices dynamically to respond to changing network environments.
Section 4.7	DataPrivacySecurity	- To make ML models to access and utilize the data without compromising user privacy; - To secure ML models against adversarial attacks and to develop the measure indicating protection against potential data breaches or cyberattacks.
Section 4.8	Big DataCollectionfor ML	- To improve data collection by using O-RAN, including collection and consolidation of hybrid empirical and synthetic data.

4.1. *Proactive Maintenance*

A significant fraction of the cost of maintaining a cellular radio network is due to the need for site visits. Virtualization of the RAN within the O-RAN framework may help considerably by extending the scope for remote or automatic intervention to mitigate hardware failures, but ultimately the only way to repair faulty on-site hardware is by making a site visit. Ideally, it would be possible to predict failures before they occur, allowing proactive scheduling of site visits. Deep learning systems can potentially assist here by trawling very large datasets to detect "precursor" events occurring before failures, which might be difficult to find by an engineer tasked with scanning the fault logs. Before this can happen, however, at least two open issues need to be addressed. The first is to evolve the O-RAN framework in conjunction with ML system design so that detailed fault-related data are provided at the O1 interface, including additional information required by the virtualization of the RAN. The second is related to the fact that in a reliable network fault data are relatively scarce, with the further issue that equipment from different vendors may operate in slightly different ways. Here, the challenge is to develop ML training approaches which can be applied in a standardized way across the network, irrespective of equipment vendor or site location.

In relation to the first issue, in [65], members of Chungwha Telecom Taiwan and their academic partner present a view of a proposed high level architecture for an O-RAN Network Management System from the network operator's perspective. The authors describe the architectures for 3GPP NG- and O-RAN, and provide a list outlining the key Operations and Maintenance (OAM) functions supported by the O-RAN O1 interface. They then provide a high-level layered architecture for an O-RAN-based network management system, together with a diagram indicating the relationships between the proposed NMS and other system entities. The paper references a TM Forum white paper published in 2003 on the New Generation Operations System and Software (NGOSS), but at this stage no attempt is made to evaluate this in relation to the current O-RAN context, where extensive use is made of virtualized network functions. The authors suggest that further work is needed to clarify the work distribution between the network management system and the entities managing the relevant clouds, as well as the approach to management of the virtualized O-RAN elements.

On the second issue, one opportunity to increase the amount of available fault data could be to make use of the emergence of self-healing networks, especially in the context of virtualization, in which faults being compensated for by the self-healing process may persist in dormant form for some considerable time and yield valuable information.

Another possibility is suggested by Mulvey et al. [66], in which they survey the literature on fault management in cellular networks and outline a number of suggestions for further work which are relevant to O-RAN. In particular, they consider the issue of multi-vendor equipment configurations, and suggest that transfer learning may be a possible approach, especially for distributed ML systems, which would allow ML subsystems to be trained on one vendor's equipment, and the learning approach transferred with appropriate adjustments to equipment provided by other vendors.

Recently, there has also been growing interest in federated learning [67], which can potentially harmonize data across multiple site locations, allowing a centralized ML system to utilize data from a large number of locations, when translated into a common format.

4.2. xApps, rApps and dApps Operation

The O-RAN architecture allows the RIC to host and run applications developed by third party for automation and intelligent orchestration to the network through ML and AI, which will leverage the enormous amount of data generated by the RAN and exposed through the O-RAN interfaces to analyze the current network conditions, forecast future traffic profiles and demand, and implement closed-loop network control strategies to optimize the RAN performance. The add-on xApps, rApps, and dApps will make the monolithic RAN "black-box" obsolete and provide open, programmable and virtualized solutions that expose status and offer control knobs through standardized interfaces [68]. The rApps are residence of the Non-RT RIC and control the optimization objectives such as policies, models and slicing that the time scale of the close control loop of the network requires more than one millisecond. The Near-RT RIC hosts the xApps that require the response of control loops time between 10 and 1000 ms for optimizing objectives such as the RRM and session management. However, the notion of dApps, custom and distributed applications can complement xApps/rApps by implementing RAN intelligence at the CUs/DUs for real-time use cases outside the timescales of the current RICs. The control loop response timescale in such use cases is $\leq$10 ms to optimize objects such as the beamforming and modulation management. The dApps receive real-time data and KPMs from the RUs (e.g., frequency-domain I/Q samples), DUs (e.g., buffer size, QoS levels), and CUs (e.g., mobility, radio link state), as well as enrichment information from the Near-RT RIC, and use it to execute real-time inference and control of lower-layer functionalities. Such dApps enable network intelligence at the edge of the O-RAN ecosystem [69].

It is clear that intelligent and dynamic xApps, rApps and dApps are key enablers for future network automation. However, it also introduces novel practical challenges concerning. One of the challenges is the orchestration of the existing xApps, rApps, and dApps in the O-RAN RIC. In addition, the question is how to maintain the orchestration cross domain with core network SON functions in 3GPP and any newly added xApps, rApps, and dApps to the RAN without creating conflicts between all these apps. As a result, the network intelligence orchestration for the different types such as the xApps, rApps, and dApps is an unprecedented problem that requires innovative, automated and scalable solutions. In [69], the authors formulate an orchestration problem where the orchestration policy variable

X is computed to maximize the total value of requests being accommodated by selecting (i) which requests can be accommodated; (ii) which AI/ML models should be instantiated; and (iii) at what location the AI/ML and requests should be executed to satisfy request performance and timescale requirements to avoid or mitigate the conflict between then and at the same time, complying with the requirements of each request [70].

4.3. Satellite NTN

One of the challenges of Satellite NTN low-latency communications is long latency due to the significant distance between the terrestrial UE and the satellite [71–73]. Using a distributed computing model, O-RAN can help a satellite solve its latency drawback. In this model, the O-RAN intelligent controller can be deployed closer to the end-user (for example, at the satellite gateway or user terminal) to reduce the round-trip time for control and management signals. The controller can also use advanced algorithms to intelligently allocate radio resources, reducing the need for frequent signaling between the user terminal and the satellite. Another way O-RAN can help reduce latency is through the use of edge computing. Edge computing involves moving tasks closer to the user or device, reducing the amount of data that must be sent back and forth between the user and the satellite. This approach can be used to run applications such as video streaming, gaming, or IoT applications, which require low-latency and high-bandwidth connections.

For massive devices involved in NTN, ML-empowered O-RAN architecture can play an essential role in optimizing the performance, e.g., throughput, coverage probability, latency, and energy efficiency. Firstly, for capacity optimization, ML can help predict the capacity requirements of the NTN and optimize the allocation of resources such as bandwidth, power, and antenna coverage. Secondly, ML can help identify and mitigate network interference sources, such as adjacent satellite interference, co-channel interference, and inter-system interference. Thirdly, ML can optimize the energy consumption of the O-RAN-assisted NTN, such as by reducing the power consumption of individual network elements or adjusting the power levels based on traffic demand. Finally, as latency remains one of the main challenges in Satellite NTN, reducing latency is important for Satellite NTN to support a wide range of applications. Some recent papers introduce the use of generative AI and digital twin in the communication system to deal with bandwidth limitation and, particularly, the latency [74]. As O-RAN offers convenient hosting of ML models, ML models can be pervasively deployed in O-RAN to encourage use of generative AI and digital twin in satellite communications.

4.4. Massive MIMO

In the last two decades, the idea of using multiple antennas for transmitting and receiving information over the air has evolved from the classic single-cell MIMO technology, to the distributed cooperative massive MIMO with no cell boundaries technology, also known as cell-free technology. With mMIMO and cell-free becoming ubiquitous in 5G and 6G, O-RAN will need to accommodate these technologies and provide the right interfacing for making the most of their large spectral efficiency (SE) potential. In this regard, the work of [75] looks at how mMIMO and cell-free can be integrated in O-RAN. They provide several options of integrating the multi-antenna processing (mMIMO precoding/beamforming) of cell-free MIMO in the O-RAN architecture, by adjusting the level of coordination/cooperation between the open-distributed units (ODUs) and open-radio units for performing this processing. It turns out that increasing the centralization (exchange of information between the various units) increases significantly the SE. Even though the current O-RAN architecture can, in their view, already support cell-free networks, there are opportunities for achieving higher SE in the future by specifying the inter-O-DU interface in O-RAN and performing the multi-antenna processing at the O-DU. Finally, they also point out the importance of RU clustering (which is, itself, tied to user grouping) for achieving better SE performance; this could be efficiently implemented at the Near-RT RIC in O-RAN. We can foresee that AI and ML will help to dynamically adjust the level of coordination/cooperation between the ODUs, as well as efficiently perform the RU clustering.

As pointed out by [75], the integration of the multi-antenna processing for the mMIMO/ cell-free technology is clearly an important issue in O-RAN, given the split of baseband functionalities, and the work [76] investigates this issue in a more practical manner. More specifically, it investigates how to effectively distribute the channel state information (CSI) between the split baseband functions to minimize the performance degradation it incurs when performing multi-antenna processing. This work also identifies further research opportunities on the same topic, as for instance, the optimization of the fronthaul bandwidth allocated to different users according to their mobility, priority, or channel conditions, and reduce the exchange of information over the fronthaul interface between O-DU and O-RU. This optimization process can obviously made more generic and efficient by using ML.

Another important technology that will deploy in 6G is the intelligent reflective surface (RIS) technology. The idea behind RIS is to increase the SE/EE of wireless communications systems by dynamically improving the propagation environment, without the need of deploying extra costly access points (APs). The work in [77] mentions the possible integration of RISs in the O-RAN architecture, where RISs are managed via a dedicated controller. This controller is then linked to O-RAN via a new interface at the Near-RT RIC. Accordingly, it is clear that the design of a practical and effective signaling interface will be the main challenge for integrating RISs into the O-RAN architecture. This work sees the deployment of RISs as an the opportunity to create fully inter-operable so-called 'smart-radio environments' which, in turn, can provides more openness and flexibility for the network operators. The management of such 'smart-radio environments' will require intelligence provided by AI/ML.

4.5. Mobility Management

In V2X communication, the meticulous design of an efficient handover strategy holds paramount importance in effectively managing challenges such as short stays, ping-pong effects, and remote cell scenarios. Furthermore, with the proliferation of UAVs across various applications, such as agricultural plant protection, police enforcement, and environmental monitoring, the reliability of connection, inherently influenced by mobility factors, emerges as a critical area of investigation.

Within the framework of the O-RAN architecture, ML emerges as a powerful tool for designing a proactive and data-driven strategy for mobility management. Specifically, leveraging Non-RT RIC, multi-dimensional data can be obtained, including metrics derived from vehicle-related measurements based on UE reports, trajectory information pertaining to vehicle paths, and spatial constraints. Subsequently, using this acquired data, the Non-RT RIC can construct machine learning models by leveraging historical information, enabling informed decisions to facilitate reliable connection support. These ML models, once constructed within the Non-RT RIC, can be effectively deployed and executed by the Near-RT RIC, enabling it to discern optimal radio resource configurations for establishing and maintaining dependable communication links [78].

As highlighted in [79], the integration of flying UAV BSs with O-RAN introduces notable challenges concerning agility, distributed computation, and dynamic mobility of UAVs. The efficient control of UAV-BSs can be significantly enhanced through the utilization of intelligent O-RAN functionalities, playing a pivotal role in addressing the requirements of unforeseen applications where terrestrial networks may prove inadequate. In this context, it would be imperative to explore innovative approaches rooted in ML for jointly optimizing the trajectory of UAVs acting as flying BSs and the task offloading among the diverse O-RAN elements, including O-RU, O-DU, and O-CU. The comprehensive evaluation of performance metrics, encompassing resource utilization, service acceptance rate, and utility values, alongside a multi-agent learning framework, becomes essential [79].

In mobility management, location information provides an additional dimension of inputs to improve its decision making. Localization techniques can also be enhanced using ML techniques [80]. With O-RAN, not only can ML techniques can be introduced into the network to enhance the localization, but the output of localization can, in turn, provide useful inputs to improve the mobility management in O-RAN.

4.6. Network Management

As mobile networks become increasingly complex, there is a growing need for advanced solutions to effectively manage network operations. Although research has explored the application of ML techniques for automating network management, further efforts are required to enhance various aspects of network management. O-RAN, with its open architecture, presents a convenient platform for leveraging ML techniques in network management. The flexibility and openness of O-RAN enable the seamless integration of ML-based approaches, providing opportunities to enhance and optimize various aspects of network management.

For supporting different network slices (e.g., eMBB, URLLC, and mMTC slices), the efficient placement of VNFs of slices onto the network infrastructure is crucial. The necessity of the optimization of the functional split of individual RAN slices between CU, DU, and RU entities, based on the functional split options defined by 3GPP, is studied in [81]. In addition, the optimal placement of RAN slices in a multi-tier 5G Open RAN architecture, including multi-tier aggregation sites, has been emphasized in [81] by demonstrating that a flexible functional split can lead to enhanced utilization of physical network resources. Considering the various types of data available, such as network traffic patterns, resource usage availability, and future resource demand, these can be leveraged to determine the VNF split for each network slice. By harnessing the power of ML, characterized by its remarkable ability to forecast future patterns and make data-driven decisions, the flexible function split can be dynamically adjusted in response to evolving network environments, thereby optimizing target objectives encompassing resource utilization, data rate, power efficiency, and cost-effectiveness.

4.7. Data Privacy and Security

As ML relies heavily on data, ensuring the privacy and security of sensitive network data are paramount. O-RAN handles vast amount of data, including user information and network configurations. It is crucial to manage ML algorithms to access and utilize these data without compromising user privacy. In addition, O-RAN often involves in collaborations between different operators and vendors. Then, secure data sharing protocols must be established to ensure that sensitive network information is shared only with authorized parties. ML models requiring data from multiple sources should adhere to strict data sharing policies. As O-RAN becomes more software-centric and dynamic, it can be vulnerable to cyberattacks. While ML can be used to detect and respond to threats, it is crucial to secure ML models themselves against adversarial attacks [16]. In this case, what measures can be implemented to protect against potential data breaches or cyberattacks targeting ML models is worth investigating. Striking the right balance between data accessibility for ML and maintaining robust data security would be critical for network operators and developers. Successfully addressing the data privacy and security challenge will be essential to foster trust in ML applications with O-RAN and ensure compliance with evolving data protection regulations.

4.8. Big Data Collection for Machine Learning

Given its openness, using O-RAN for big data collection in the context of ML offers substantial advantages. O-RAN promotes interoperability among different vendors' network components, allowing for diverse data sources, which is vital for ML model training and accuracy. It also fosters vendor neutrality, reducing lock-in and enabling network operators to select the best-suited hardware and software components, enhancing data collection capabilities. The scalability of O-RAN accommodates the growing volume of data generated by modern networks and encourages innovation by enabling custom software applications, including data collection and analytic tools tailored to specific ML use cases. O-RAN development can further advance with the more data available for ML, including use of synthetic data from network simulators [82]. Simulation tools such as OpenRAN Gym [83] and WiThRay [84] offer data generation for ML training when empirical data are insufficient or difficult to collect. However, data may be out-of-distribution due to different domains, and new learning techniques should be explored to tackle the out-of-distribution issue [85].

5. Conclusions

ML applications have sparked significant interest in O-RAN for their potential to revolutionize network automation. By exploiting ML to analyze vast amounts of network data in real-time, identifying performance issues and optimizing network parameters on the fly, ML is expected to facilitate predictive maintenance, intelligent resource allocation, and network optimization. Furthermore, predictive analytics can help anticipate and prevent network failures, reducing downtime and maintenance costs. Ultimately, ML holds the promise of making O-RAN networks more efficient, reliable, and responsive to the dynamic demands of modern communication environments.

In this paper, focusing on network automation in O-RAN using ML techniques, we first presented the design principles and architecture of O-RAN, highlighted the openness and disaggregation of RAN components, and its capability to extend network operation with native ML support. The current research landscape of ML applications in O-RAN was then presented. Several key aspects of network management were surveyed, including session and user access management, radio resource management, network slicing, mobility and traffic management, energy efficiency, O-RAN component deployment and function splits, ML workflow management, and support for NTN and satellite networks. For instance, ML can play a pivotal role in automating fault detection and recovery processes by meticulously analyzing data traffic and identifying anomalies. This reduces the necessity for manual intervention, subsequently bolstering network resilience. ML can also prove invaluable in the realms of capacity planning and elevating customer experiences. By forecasting network traffic growth, aiding in optimizing capacity, and analyzing user behavior and feedback, ML can contribute to a more efficient and customer-centric network management approach.

However, numerous challenges must be surmounted to unlock the full potential of ML in propelling O-RAN network forward. Upon identifying several pivotal research domains, these challenges manifest across diverse realms encompassing data collection and analysis, as well as the development, deployment, maintenance, and operation of ML models. A fundamental hurdle arises from the various types of data emanating from different vendor equipment, necessitating harmonization for effective data analysis. Utilization of synthetic data collectible from network simulators could be also considered when real-world empirical data are difficult to obtain. Once data acquisition hurdles are overcome, data-driven decision-making processes must be thoughtfully tailored for target optimizations, such as enhancing network capacity and mitigating interference. In complex scenarios (e.g., O-RAN-including UAVs), consideration of multiple objectives would be required simultaneously, such as optimizing the trajectory of UAVs and task offloading among heterogeneous O-RAN components, including UAVs. The multi-faceted nature of O-RAN components provide various options for deploying ML models, prompting careful deliberation on compatible components and interface for ML model and data exchange. When multiple components are selected for ML model deployment, the level of coordination and cooperation between components should be also investigated. Furthermore, the dependable operation of multiple ML modules such as xApps, rApps, and dApps within O-RAN RIC, mandates unwavering attention to reliability. Equally critical is the dynamic and autonomous responsiveness of deployed ML models to ever-changing network environments. Furthermore, we discussed the issue to find the balance between data accessibility for ML and maintaining robust data security.

These challenges, when conquered, will usher in an era where ML empowers O-RAN network automation, leading to heightened efficiency, reduced operational costs, fortified security, and an enriched user experience. They also position network operators to adapt to the evolving demands of the communication landscape, conferring a competitive edge in this rapidly transforming terrain.

Author Contributions: Writing—original draft, D.T., R.B., A.K., E.A., D.M., R.Z., R.P., B.H., H.B., F.H., G.C., P.X. and N.W.; Writing—original draft, review and editing, M.Q.H., H.L., W.Y. and C.H.F.: Supervision, R.T. All authors have read and agreed to the published version of the manuscript.

Funding: This research received no external funding.

Informed Consent Statement: Not applicable.

Data Availability Statement: No new data were created or analyzed in this study. Data sharing is not applicable to this article.

Acknowledgments: The authors would like to acknowledge the support of the 5GIC/6GIC members for this work.

Conflicts of Interest: The authors declare no conflict of interest.

References

1. Polese, M.; Bonati, L.; D'Oro, S.; Basagni, S.; Melodia, T. Understanding O-RAN: Architecture, Interfaces, Algorithms, Security, and Research Challenges. *IEEE Commun. Surv. Tutor.* **2023**, *25*, 1376–1411.
2. Singh, S.K.; Singh, R.; Kumbhani, B. The Evolution of Radio Access Network Towards Open-RAN: Challenges and Opportunities. In Proceedings of the 2020 IEEE Wireless Communications and Networking Conference Workshops (WCNCW), Virtual, 25–28 May 2020; pp. 1–6.
3. 3GPP. *Study on CU-DU Lower Layer Split for NR*; TR 38.816 V15.0.0; Release 15; 3GPP: Sophia Antipolis, France, 2017.
4. Dryjański, M.; Kułacz, Ł.; Kliks, A. Toward Modular and Flexible Open RAN Implementations in 6G Networks: Traffic Steering Use Case and O-RAN xApps. *Sensors* **2021**, *21*, 8173. [CrossRef] [PubMed]
5. O-RAN Alliance. Available online: https://www.o-ran.org/ (accessed on 6 September 2023).
6. 3rd Generation Partnership Project (3GPP). Available online: https://www.3gpp.org/ (accessed on 6 September 2023).
7. European Telecommunications Standards Institute (ETSI). Available online: https://www.etsi.org/ (accessed on 6 September 2023).
8. Next Generation Mobile Networks Alliance (NGMN). Available online: https://www.ngmn.org/ (accessed on 6 September 2023).
9. Optical Internetworking Forum (OIF). Available online: https://www.oiforum.com/ (accessed on 6 September 2023).
10. Noor-A-Rahim, M.; Liu, Z.; Lee, H.; Khyam, M.O.; He, J.; Pesch, D.; Moessner, K.; Saad, W.; Poor, H.V. 6G for Vehicle-to-Everything (V2X) Communications: Enabling Technologies, Challenges, and Opportunities. *Proc. IEEE* **2022**, *110*, 712–734.
11. Morocho-Cayamcela, M.E.; Lee, H.; Lim, W. Machine Learning for 5G/B5G Mobile and Wireless Communications: Potential, Limitations, and Future Directions. *IEEE Access* **2019**, *7*, 137184–137206.
12. Azariah, W.; Bimo, F.A.; Lin, C.W.; Cheng, R.G.; Jana, R.; Nikaein, N. A Survey on Open Radio Access Networks: Challenges, Research Directions, and Open Source Approaches. *arXiv* **2022**, arXiv:2208.09125.
13. Garcia-Saavedra, A.; Costa-Pérez, X. O-RAN: Disrupting the Virtualized RAN Ecosystem. *IEEE Commun. Stand. Mag.* **2021**, *5*, 96–103. [CrossRef]
14. Arnaz, A.; Lipman, J.; Abolhasan, M.; Hiltunen, M. Toward Integrating Intelligence and Programmability in Open Radio Access Networks: A Comprehensive Survey. *IEEE Access* **2022**, *10*, 67747–67770.
15. Brik, B.; Boutiba, K.; Ksentini, A. Deep Learning for B5G Open Radio Access Network: Evolution, Survey, Case Studies, and Challenges. *IEEE Open J. Commun. Soc.* **2022**, *3*, 228–250. [CrossRef]
16. Liyanage, M.; Braeken, A.; Shahabuddin, S.; Ranaweera, P. Open RAN security: Challenges and opportunities. *J. Netw. Comput. Appl.* **2023**, *214*, 103621.
17. Campana, R.; Amatetti, C.; Vanelli-Coralli, A. O-RAN based Non-Terrestrial Networks: Trends and Challenges. In Proceedings of the 2023 Joint European Conference on Networks and Communications & 6G Summit (EuCNC/6G Summit), Gothenburg, Sweden, 6–9 June 2023; pp. 264–269.
18. Abdalla, A.S.; Upadhyaya, P.S.; Shah, V.K.; Marojevic, V. Toward Next Generation Open Radio Access Networks–What O-RAN Can and Cannot Do! *IEEE Netw.* **2022**, *36*, 206–213. [CrossRef]
19. Brik, B.; Chergui, H.; Zanzi, L.; Devoti, F.; Ksentini, A.; Siddiqui, M.S.; Costa-Pérez, X.; Verikoukis, C. A Survey on Explainable AI for 6G O-RAN: Architecture, Use Cases, Challenges and Research Directions. *arXiv* **2023**, arXiv:2307.00319.
20. O-RAN ALLIANCE. *O-RAN-Architecture-Description*, v06.00; O-RAN: Alfter, Germany, 2021.

21. O-RAN ALLIANCE. *O-RAN Non-RT RIC: Functional Architecture 1.01-March 2021 (O-RAN. WG2. Non-RT-RIC-ARCH-TR-v01.01)*; v01.01; O-RAN: Alfter, Germany, 2021.
22. O-RAN. *AI/ML Workflow Description and Requirements 1.03*; v01.03; O-RAN.WG2.AIML-v01.03 Technical Specification; O-RAN: Alfter, Germany, 2021.
23. O-RAN Software Community. GitHub. Available online: https://github.com/o-ran-sc (accessed on 6 September 2023).
24. OpenAirInterface Software Alliance. GitHub. Available online: https://github.com/openairinterface (accessed on 6 September 2023).
25. GitHub—srsran/srsRAN: Open Source SDR 4G/5G Software Suite from Software Radio Systems (SRS). Available online: https://github.com/srsran/srsRAN_4G (accessed on 6 September 2023).
26. Upadhyaya, P.S.; Abdalla, A.S.; Marojevic, V.; Reed, J.H.; Shah, V.K. Prototyping Next-Generation O-RAN Research Testbeds with SDRs. 2022. Available online: https://doi.org/10.48550/arXiv.2205.1317 (accessed on 6 September 2023).
27. Ettus Research. High Performance Software Defined Radio (SDR). Ettus Research, a National Instruments Brand. The Leader in Software Defined Radio (SDR). Available online: https://www.ettus.com/product-categories/usrp-x-series/ (accessed on 6 September 2023).
28. Bonati, L.; Johari, P.; Polese, M.; D'Oro, S.; Mohanti, S.; Tehrani-Moayyed, M.; Villa, D.; Shrivastava, S.; Tassie, C.; Yoder, K.; et al. Colosseum: Large-Scale Wireless Experimentation Through Hardware-in-The-Loop Network Emulation. In Proceedings of the 2021 IEEE International Symposium on Dynamic Spectrum Access Networks, DySPAN 2021, Virtual, 13–15 December 2021; pp. 105–113.
29. Papa, A.; Durner, R.; Goshi, E.; Goratti, L.; Rasheed, T.; Blenk, A.; Kellerer, W. Marc: On modeling and analysis of software-defined radio access network controllers. *IEEE Trans. Netw. Serv. Manag.* **2021**, *18*, 4602–4615. [CrossRef]
30. Foukas, X.; Nikaein, N.; Kassem, M.M.; Marina, M.K.; Kontovasilis, K. FlexRAN: A flexible and programmable platform for software-defined radio access networks. In Proceedings of the 12th International Conference on Emerging Networking EXperiments and Technologies, CoNEXT 2016, Irvine, CA, USA, 12–15 December 2016; pp. 427–441.
31. Coronado, E.; Khan, S.N.; Riggio, R. 5G-EmPOWER: A software-defined networking platform for 5G radio access networks. *IEEE Trans. Netw. Serv. Manag.* **2019**, *16*, 715–728. [CrossRef]
32. Johnson, D.; Maas, D.; Van Der Merwe, J. NexRAN: Closed-loop RAN slicing in POWDER-A top-to-bottom open-source open-RAN use case. In Proceedings of the 15th ACM Workshop on Wireless Network Testbeds, Experimental Evaluation & CHaracterization, New Orleans, LA, USA, 31 January–4 February 2022; pp. 17–23.
33. Schmidt, R.; Irazabal, M.; Nikaein, N. FlexRIC: An SDK for next-generation SD-RANs. In Proceedings of the 17th International Conference on emerging Networking EXperiments and Technologies, CoNEXT 2021, Virtual, 7–10 December 2021; pp. 411–425.
34. NG-RAN. *E1 Application Protocol (E1AP)*; Release 17; 3GPP: Sophia Antipolis, France.
35. Sen, N.; A, A.F. Intelligent Admission and Placement of O-RAN Slices Using Deep Reinforcement Learning. In Proceedings of the 2022 IEEE 8th International Conference on Network Softwarization (NetSoft), Milan, Italy, 27 June–1 July 2022; pp. 307–311.
36. Lee, H.; Jang, Y.; Song, J.; Yeon, H. O-RAN AI/ML Workflow Implementation of Personalized Network Optimization via Reinforcement Learning. In Proceedings of the 2021 IEEE Globecom Workshops (GC Wkshps), Madrid, Spain, 7–11 December 2021; pp. 1–6.
37. Phyu, H.P.; Naboulsi, D.; Stanica, R. Machine Learning in Network Slicing—A Survey. *IEEE Access* **2022**, *11*, 39123–39153. [CrossRef]
38. Thaliath, J.; Niknam, S.; Singh, S.; Banerji, R.; Saxena, N.; Dhillon, H.S.; Reed, J.H.; Bashir, A.K.; Bhat, A.; Roy, A. Predictive Closed-Loop Service Automation in O-RAN Based Network Slicing. *IEEE Commun. Stand. Mag.* **2022**, *6*, 8–14. [CrossRef]
39. Lotfi, F.; Semiari, O.; Afghah, F. Evolutionary Deep Reinforcement Learning for Dynamic Slice Management in O-RAN. In Proceedings of the 2022 IEEE Globecom Workshops (GC Wkshps), Rio de Janeiro, Brazil, 4–8 December 2022; pp. 227–232.
40. Abedin, S.F.; Mahmood, A.; Tran, N.H.; Han, Z.; Gidlund, M. Elastic O-RAN Slicing for Industrial Monitoring and Control: A Distributed Matching Game and Deep Reinforcement Learning Approach. *IEEE Trans. Veh. Technol.* **2022**, *71*, 10808–10822. [CrossRef]
41. Pamuklu, T.; Erol-Kantarci, M.; Ersoy, C. Reinforcement Learning Based Dynamic Function Splitting in Disaggregated Green Open RANs. In Proceedings of the ICC 2021—IEEE International Conference on Communications, Virtual, 14–23 June 2021; pp. 1–6.
42. Amiri, E.; Wang, N.; Shojafar, M.; Tafazolli, R. Energy-Aware Dynamic VNF Splitting in O-RAN Using Deep Reinforcement Learning. *IEEE Wirel. Commun. Lett.* 2023, *early access*.
43. Joda, R.; Pamuklu, T.; Iturria-Rivera, P.E.; Erol-Kantarci, M. Deep Reinforcement Learning-based Joint User Association and CU-DU Placement in O-RAN. *IEEE Trans. Netw. Serv. Manag.* **2022**, *19*, 4097–4110. [CrossRef]
44. Hammami, N.; Nguyen, K.K. On-Policy vs. Off-Policy Deep Reinforcement Learning for Resource Allocation in Open Radio Access Network. In Proceedings of the 2022 IEEE Wireless Communications and Networking Conference (WCNC), Austin, TX, USA, 10–13 April 2022; pp. 1461–1466.
45. Sharara, M.; Pamuklu, T.; Hoteit, S.; Vèque, V.; Erol-Kantarci, M. Policy-Gradient-Based Reinforcement Learning for Computing Resources Allocation in O-RAN. In Proceedings of the 2022 IEEE 11th International Conference on Cloud Networking (CloudNet), Paris, France, 7–10 November 2022; pp. 229–236.

46. Mungari, F. An RL Approach for Radio Resource Management in the O-RAN Architecture. In Proceedings of the 2021 18th Annual IEEE International Conference on Sensing, Communication, and Networking (SECON), Rome, Italy, 6–9 July 2021; pp. 1–2.
47. Ayala-Romero, J.A.; Garcia-Saavedra, A.; Costa-Perez, X.; Iosifidis, G. Bayesian Online Learning for Energy-Aware Resource Orchestration in Virtualized RANs. In Proceedings of the IEEE INFOCOM 2021—IEEE Conference on Computer Communications, Virtual, 10–13 May 2021; pp. 1–10.
48. Lien, S.Y.; Deng, D.J.; Chang, B.C. Session Management for URLLC in 5G Open Radio Access Network: A Machine Learning Approach. In Proceedings of the 2021 International Wireless Communications and Mobile Computing (IWCMC), Harbin, China, 28 June–2 July 2021; pp. 2050–2055.
49. Orhan, O.; Swamy, V.N.; Tetzlaff, T.; Nassar, M.; Nikopour, H.; Talwar, S. Connection Management xAPP for O-RAN RIC: A Graph Neural Network and Reinforcement Learning Approach. In Proceedings of the 2021 20th IEEE International Conference on Machine Learning and Applications (ICMLA), Pasadena, CA, USA, 13–16 December 2021; pp. 936–941.
50. Cao, Y.; Lien, S.Y.; Liang, Y.C.; Chen, K.C.; Shen, X. User Access Control in Open Radio Access Networks: A Federated Deep Reinforcement Learning Approach. *IEEE Trans. Wirel. Commun.* **2022**, *21*, pp. 3721–3736. [CrossRef]
51. Erdol, H.; Wang, X.; Li, P.; Thomas, J.D.; Piechocki, R.; Oikonomou, G.; Inacio, R.; Ahmad, A.; Briggs, K.; Kapoor, S. Federated Meta-Learning for Traffic Steering in O-RAN. In Proceedings of the 2022 IEEE 96th Vehicular Technology Conference (VTC2022-Fall), London, UK, 26–29 September 2022; pp. 1–7.
52. Akhtar, M.W.; Mahmood, A.; Abedin, S.F.; Hassan, S.A.; Gidlund, M. Exploiting NOMA for Radio Resource Efficient Traffic Steering Use-case in O-RAN. In Proceedings of the GLOBECOM 2022—2022 IEEE Global Communications Conference, Rio de Janeiro, Brazil, 4–8 December 2022; pp. 5771–5776.
53. Prananto, B.H.; Iskandar; Kurniawan, A. O-RAN Intelligent Application for Cellular Mobility Management. In Proceedings of the 2022 International Conference on ICT for Smart Society (ICISS), Virtual, 10–11 August 2022; pp. 1–6.
54. Kalntis, M.; Iosifidis, G. Energy-Aware Scheduling of Virtualized Base Stations in O-RAN with Online Learning. In Proceedings of the GLOBECOM 2022—2022 IEEE Global Communications Conference, Rio de Janeiro, Brazil, 4–8 December 2022; pp. 6048–6054.
55. Smith, R.; Freeberg, C.; Machacek, T.; Ramaswamy, V. An O-RAN Approach to Spectrum Sharing Between Commercial 5G and Government Satellite Systems. In Proceedings of the MILCOM 2021—2021 IEEE Military Communications Conference (MILCOM), San Diego, CA, USA, 29 November–2 December 2021; pp. 739–744.
56. Lee, H.; Cha, J.; Kwon, D.; Jeong, M.; Park, I. Hosting AI/ML Workflows on O-RAN RIC Platform. In Proceedings of the 2020 IEEE Globecom Workshops (GC Wkshps), Taipei, Taiwan, 7–11 December 2020; pp. 1–6.
57. Bakri, S.; Brik, B.; Ksentini, A. On using reinforcement learning for network slice admission control in 5G: Offline vs. online. *Int. J. Commun. Syst.* **2021**, *34*, e4757. [CrossRef]
58. Kukliński, S.; Tomaszewski, L.; Kołakowski, R. On O-RAN, MEC, SON and Network Slicing integration. In Proceedings of the 2020 IEEE Globecom Workshops (GC Wkshps), Taipei, Taiwan, 7–11 December 2020; pp. 1–6.
59. Amiri, E.; Wang, N.; Shojafar, M.; Hamdan, M.Q.; Foh, C.H.; Tafazolli, R. Deep Reinforcement Learning for Robust VNF Reconfigurations in O-RAN. *IEEE Trans. Netw. Serv. Manag.* 2023, *early access*.
60. Larsen, L.M.; Checko, A.; Christiansen, H.L. A survey of the functional splits proposed for 5G mobile crosshaul networks. *IEEE Commun. Surv. Tutor.* **2018**, *21*, 146–172. [CrossRef]
61. Amiri, E.; Wang, N.; Shojafar, M.; Tafazolli, R. Optimizing Virtual Network Function Splitting in Open-RAN Environments. In Proceedings of the 2022 IEEE 47th Conference on Local Computer Networks (LCN), Edmonton, AB, Canada, 26–29 September 2022; pp. 422–429.
62. Kazemifard, N.; Shah-Mansouri, V. Minimum delay function placement and resource allocation for Open RAN (O-RAN) 5G networks. *Comput. Netw.* **2021**, *188*, 107809. [CrossRef]
63. Di Mauro, M.; Galatro, G.; Longo, M.; Postiglione, F.; Tambasco, M. Availability Analysis of IP Multimedia Subsystem in Cloud Environments. In Proceedings of the 2019 4th International Conference on System Reliability and Safety (ICSRS), Rome, Italy, 20–22 November 2019; pp. 111–115.
64. Kavehmadavani, F.; Nguyen, V.D.; Vu, T.X.; Chatzinotas, S. Traffic Steering for eMBB and uRLLC Coexistence in Open Radio Access Networks. In Proceedings of the 2022 IEEE International Conference on Communications Workshops (ICC Workshops), Seoul, Republic of Korea, 16–20 May 2022; pp. 242–247.
65. Wang, T.H.; Chen, Y.C.; Huang, S.J.; Hsu, K.S.; Hu, C.H. Design of a Network Management System for 5G Open RAN. In Proceedings of the 2021 22nd Asia-Pacific Network Operations and Management Symposium (APNOMS), Tainan, Taiwan, 8–10 September 2021; pp. 138–141.
66. Mulvey, D.; Foh, C.H.; Imran, M.A.; Tafazolli, R. Cell Fault Management using Machine Learning Techniques. *IEEE Access* **2019**, *7*, 124514–124539. [CrossRef]
67. Mulvey, D.; Foh, C.H.; Imran, M.A.; Tafazolli, R. Cellular Network Antenna Tilt Anomaly Detection Using Federated Unsupervised Learning. In Proceedings of the IEEE ICC May 2023, Rome, Italy, 28 May–1 June 2023; pp. 3048–3053.
68. Polese, M.; Bonati, L.; D'Oro, S.; Basagni, S.; Melodia, T. ColO-RAN: Developing Machine Learning-Based xApps for Open RAN Closed-Loop Control on Programmable Experimental Platforms. *IEEE Trans. Mob. Comput.* **2023**, *22*, 5787–5800. [CrossRef]
69. D'Oro, S.; Polese, M.; Bonati, L.; Cheng, H.; Melodia, T. dApps: Distributed Applications for Real-Time Inference and Control in O-RAN. *IEEE Commun. Mag.* **2022**, *60*, 52–58. [CrossRef]

70. D'Oro, S.; Bonati, L.; Polese, M.; Melodia, T. OrchestRAN: Network Automation through Orchestrated Intelligence in the Open RAN. In Proceedings of the IEEE INFOCOM 2022—IEEE Conference on Computer Communications, Virtual, 2–5 May 2022; pp. 270–279.
71. Araniti, G.; Iera, A.; Pizzi, S.; Rinaldi, F. Toward 6G Non-Terrestrial Networks. *IEEE Netw.* **2022**, *36*, 113–120. [CrossRef]
72. Lin, X.; Rommer, S.; Euler, S.; Yavuz, E.A.; Karlsson, R.S. 5G from Space: An Overview of 3GPP Non-Terrestrial Networks. *IEEE Commun. Stand. Mag.* **2021**, *5*, 147–153. [CrossRef]
73. Centenaro, M.; Costa, C.E.; Granelli, F.; Sacchi, C.; Vangelista, L. A Survey on Technologies, Standards and Open Challenges in Satellite IoT. *IEEE Commun. Surv. Tutor.* **2021**, *23*, 1693–1720. [CrossRef]
74. Akrout, M.; Mezghani, A.; Hossain, E.; Bellili, F.; Heath, R.W. From Multilayer Perceptron to GPT: A Reflection on Deep Learning Research for Wireless Physical Layer. *arXiv* **2023**, arXiv:2307.07359.
75. Ranjbar, V.; Girycki, A.; Rahman, M.A.; Pollin, S.; Moonen, M.; Vinogradov, E. Cell-Free mMIMO Support in the O-RAN Architecture: A PHY Layer Perspective for 5G and Beyond Networks. *IEEE Commun. Stand. Mag.* **2022**, *6*, 28–34. [CrossRef]
76. Hewavithana, T.; Chopra, A.; Mondal, B.; Wong, S.; Davydov, A.; Majmundar, M. Overcoming Channel Aging in Massive MIMO Basestations With Open RAN Fronthaul. In Proceedings of the 2022 IEEE Wireless Communications and Networking Conference (WCNC), Austin, TX, USA, 10–13 April 2022; pp. 2577–2582.
77. Strinati, E.C.; Alexandropoulos, G.C.; Wymeersch, H.; Denis, B.; Sciancalepore, V.; D'Errico, R.; Clemente, A.; Phan-Huy, D.T.; De Carvalho, E.; Popovski, P. Reconfigurable, Intelligent, and Sustainable Wireless Environments for 6G Smart Connectivity. *IEEE Commun. Mag.* **2021**, *59*, 99–105. [CrossRef]
78. O-RAN ALLIANCE. O-RAN Use Cases and Deployment Scenarios—Towards Open and Smart RAN. 2020. Available online: https://www.o-ran.org/resources (accessed on 6 September 2023).
79. Pham, C.; Fami, F.; Nguyen, K.K.; Cheriet, M. When RAN intelligent controller in O-RAN meets multi-UAV enable wireless network. *IEEE Trans. Cloud Comput.* **2022**, *11*, 2245–2259. [CrossRef]
80. Darbinyan, R.; Khachatrian, H.; Mkrtchyan, R.; Raptis, T.P. ML-based Approaches for Wireless NLOS Localization: Input Representations and Uncertainty Estimation. In Proceedings of the 2023 IEEE Ninth International Conference on Big Data Computing Service and Applications (BigDataService), Los Alamitos, CA, USA, 17–20 July 2023; pp. 87–94.
81. Sarikaya, E.; Onur, E. Placement of 5G RAN Slices in Multi-tier O-RAN 5G Networks with Flexible Functional Splits. In Proceedings of the 2021 17th International Conference on Network and Service Management (CNSM), Izmir, Turkey, 25–29 October 2021; pp. 274–282.
82. Wilhelmi, F.; Carrascosa, M.; Cano, C.; Jonsson, A.; Ram, V.; Bellalta, B. Usage of Network Simulators in Machine-Learning-Assisted 5G/6G Networks. *IEEE Wirel. Commun.* **2021**, *28*, 160–166. [CrossRef]
83. Bonati, L.; Polese, M.; D'Oro, S.; Basagni, S.; Melodia, T. OpenRAN Gym: An Open Toolbox for Data Collection and Experimentation with AI in O-RAN. *arXiv* **2022**, arXiv:2202.10318.
84. Choi, H.; Oh, J.; Chung, J.; Alexandropoulos, G.C.; Choi, J. WiThRay: A Versatile Ray-Tracing Simulator for Smart Wireless Environments. *IEEE Access* **2023**, *11*, 56822–56845. [CrossRef]
85. Akrout, M.; Feriani, A.; Bellili, F.; Mezghani, A.; Hossain, E. Domain Generalization in Machine Learning Models for Wireless Communications: Concepts, State-of-the-Art, and Open Issues. *arXiv* **2023**, arXiv:2303.08106.

electronics

MDPI

Article

A Multi-Task Classification Method for Application Traffic Classification Using Task Relationships

Ui-Jun Baek [1], Boseon Kim [2], Jee-Tae Park [1], Jeong-Woo Choi [1] and Myung-Sup Kim [1,*]

[1] Department of Computer and Information Science, Korea University, Sejong-si 30019, Republic of Korea; pb1069@korea.ac.kr (U.-J.B.); pjj5846@korea.ac.kr (J.-T.P.); choigoya97@korea.ac.kr (J.-W.C.)
[2] Korea Institute of Science and Technology Information, Daejeon 34141, Republic of Korea; boseon12@kisti.re.kr
* Correspondence: tmskim@korea.ac.kr

Abstract: As IT technology advances, the number and types of applications, such as SNS, content, and shopping, have increased across various fields, leading to the emergence of complex and diverse application traffic. As a result, the demand for effective network operation, management, and analysis has increased. In particular, service or application traffic classification research is an important area of study in network management. Web services are composed of a combination of multiple applications, and one or more application traffic can be mixed within service traffic. However, most existing research only classifies application traffic by service unit, resulting in high misclassification rates and making detailed management impossible. To address this issue, this paper proposes three multitask learning methods for application traffic classification using the relationships among tasks composed of browsers, protocols, services, and application units. The proposed methods aim to improve classification performance under the assumption that there are relationships between tasks. Experimental results demonstrate that by utilizing relationships between various tasks, the proposed method can classify applications with 4.4%p higher accuracy. Furthermore, the proposed methods can provide network administrators with information about multiple perspectives with high confidence, and the generalized multitask methods are freely portable to other backbone networks.

Keywords: application traffic classification; network management; multitask learning

Citation: Baek, U.-J.; Kim, B.; Park, J.-T.; Choi, J.-W.; Kim, M.-S. A Multi-Task Classification Method for Application Traffic Classification Using Task Relationships. *Electronics* **2023**, *12*, 3597. https://doi.org/10.3390/electronics12173597

Academic Editors: Yichuang Sun, Haeyoung Lee, Oluyomi Simpson and Martin Reisslein

Received: 20 July 2023
Revised: 6 August 2023
Accepted: 21 August 2023
Published: 25 August 2023

1. Introduction

With the recent advancement of IT technology, web services have become increasingly important in daily life, and due to the influence of COVID-19, the use of video streaming and online shopping has dramatically increased as indoor activity time has prolonged [1]. As a result, the demand for monitoring and analyzing network traffic, including application traffic classification and traffic prediction, has increased due to the emergence of complex and diverse application traffic resulting from the increase in the number and types of applications, such as SNS (Social Network Service), content streaming, and shopping. In particular, application traffic classification research is essential for effective network monitoring and analysis [2]. It can be widely used in areas such as cloud service pricing, resource planning, traffic control, and network security. For instance, in schools or public institutions, network resources can be restricted to limit non-work-related traffic, and companies subscribing to cloud services can classify the traffic of the services they use to subscribe to the appropriate services without unnecessary consumption.

Web services are software systems for application interaction between different types of computers on the network and can be composed of a combination of multiple applications. Therefore, the traffic generated by web services is also composed of a combination of traffic generated by various applications. However, most existing research only classifies network traffic by service unit and application unit, and this approach is similar to MCC (Multiclass Classification) shown in Figure 1. This can result in misclassification of mixed traffic in a service that includes multiple services or application traffic. For example, the traffic flow of *Googlefonts* in the *Naver* service and the traffic flow of *Googlefonts* in the *YouTube* service represent two different ground truths, even though they are both under the same sub-service. This can confuse the learning model when learning the characteristics of the traffic. Moreover, simply classifying multiple sub-service traffic or application traffic within a service only by service unit makes detailed management impossible. To address this issue, this paper proposes a method for classifying traffic using the relationships among four tasks, as shown in MTC (Multitask Classification) in Figure 1. Multitask learning (MTL) [3] is applied in a variety of fields, with the aim of simultaneously learning multiple related tasks so that the knowledge contained in one task is used for other tasks to improve the generalization performance of all tasks [4]. Ref. [5] performs multitask learning through the task lists provided by CICIDS 2017 [6], ISCX VPN-nonVPN 2016 [7], and ISCX Tor 2016 [8]. The task lists include normal/abnormal application categories, detailed applications, encryption, etc. Ref. [9] performed multitask learning through the task list provided by ISCX VPN-nonVPN 2016, which includes a total of three task lists. Ref. [10] performed multitask learning by creating a new task called Bandwidth and Duration from the QUIC and ISCX datasets. Unfortunately, there are not many multitask-based classification methods in the field of network traffic classification, and this is also often dependent on the task list provided by the dataset. Therefore, various tasks that can improve generalization performance in the field of network traffic classification need to be proposed. We set the goal of improving classification performance under the assumption that the tasks are not completely independent and have relationships among them. For instance, when using the *Edge* browser, a web browser released by *Microsoft*, *Microsoft* traffic mainly occurs when using the *Edge* browser to access web services. Similarly, when using the *Firefox* browser, *Mozilla* traffic mainly occurs when using the *Firefox* browser to access web services. In another case, the *YouTube* service communicates using the HTTP/3 protocol, and most of the traffic using the HTTP/3 protocol occurs within the *YouTube* service. The proposed method includes four tasks for traffic classification, and classifies accurate services and sub-services or applications by performing four tasks simultaneously using MTC. In addition, the proposed method provides detailed classification results for traffic, which can satisfy various requirements of network administrators. Our representative contributions include the following:

(i). Improved classification accuracy: improved classification performance considering relationships between multiple tasks (browsers, HTTP protocols, applications, services);
(ii). Generalizability and portability of the four multitask classification methods: the generalized classification model for multiple classifications improves classification performance across diverse backbone networks;
(iii). Possibility to monitor and analyze from multiple perspectives: network administrators can gain more detailed information and insights into the traffic occurring on the networks under their jurisdiction when monitoring and analyzing their networks.

This paper is structured as follows: Related research is described in Section 2 following the introduction. Section 3 provides a detailed explanation of the proposed method, and Sections 4 and 5 describe the experimental setup and results, respectively. Finally, in Section 6, the conclusion and future research are discussed, concluding this paper.

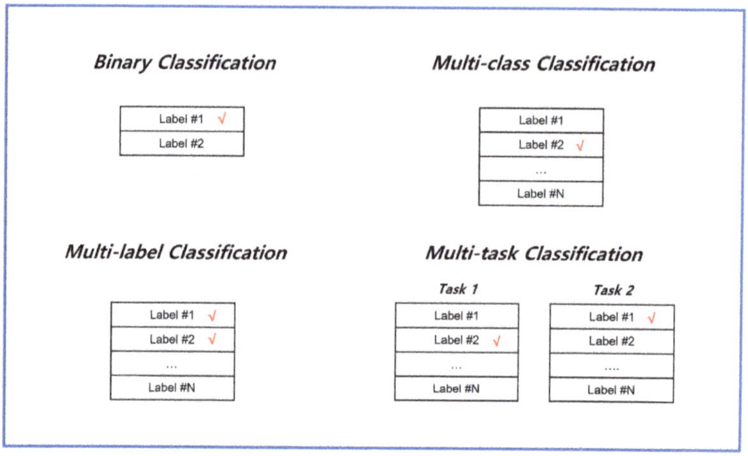

Figure 1. Overview of classification types.

2. Related Works

2.1. Task Description

This subsection describes four tasks of the proposed methodology. The first task is browser classification, where browsers are graphical user interface-based software applications that enable bidirectional communication between users and web servers, allowing the display of HTML documents or files. The labels included in the browser classification task are *Chrome*, *Edge*, and *Firefox*, whose combined usage accounts for the majority of global browser usage. Ref. [11] proposed real-time lightweight identification of HTTPS clients based on network monitoring and SSL/TLS fingerprinting and reported that 95.4% of HTTPS network traffic could be retrieved by the proposed method. Additionally, the study reported that the handshake fingerprints of SSL/TLS, including the cipher suite list of different clients, differ. This indicates that prior information about the browser or communication client can influence the classification results. The second task is protocol classification, where collected protocols include HTTP/1.1, HTTP/2 (HTTPS), and HTTP/3. HTTP/1.1 is one of the HTTP protocol versions released in 1999 and is still the most widely used. HTTPS refers to the second major version of the HTTP protocol, which was released in 2015. HTTP/3 is the third major version of the HTTP protocol, which was released in 2020, and uses the QUIC (Quick UDP Internet Connections) protocol instead of the TCP protocol used in previous versions, providing faster and more reliable data transfer. Web applications utilize various HTTP protocol versions, and the protocol version can serve as useful prior information for classifying specific applications. Ref. [12] proposes a method to improve the service classification performance by using the protocol classification results of the application traffic test dataset as prior information in the service classification process. The third task is service classification, where services are defined as software systems for interaction between different types of computers on the network, consisting of *Aladin*, *Amazon*, *Google*, *Nate*, *Naver*, and *YouTube*. For convenience, the services referred to in this paper denote the top-level service that includes multiple applications or sub-services. The fourth task is application classification, where applications or sub-services responsible for specific interactions within a service are identified. For example, *Google Fonts* and *Gstatic* are applications (sub-services) provided by *Google*, while *search.naver* and *pstatic* are applications provided by *Naver*. In other words, a service may consist of multiple applications or sub-services. The four tasks selected in this study are closely interconnected, and when performing each task, the other tasks can serve as valuable prior information.

2.2. Classification Type

This subsection describes the types and definitions of classification tasks. Classification tasks can be divided into BC (Binary Classification), MCC (Multiclass Classification), MLC (Multi-label Classification), and MTC (Multitask Classification), as shown in Figure 1.

BC is a classification task with two classes, where each sample can be labeled with only one class. For example, in an anomaly classification task to distinguish between anomaly and benign, the user can assign a label to each sample with only one of the two classes. There have been many studies on detecting the presence of malicious traffic in network traffic data [13–15].

MCC is a classification task with more than two classes, where each sample can be labeled with only one class. The majority of research in the field of application traffic classification is focused on MCC, where various application traffic types are labeled with a single label. For example, in the application classification task [7], to distinguish between Mail, File Transfer, P2P, VoIP, Streaming, and Chat, the user can assign a label to each sample with only one of the six classes.

MLC is a classification task in which multiple labels are assigned to each sample, equal to the number of possible classes when there are multiple classes. For example, in a weather classification task that includes seven classes, such as clear, cloudy, snow, rain, fog, thunder, and hail, the user can assign one or more labels from the seven classes to each sample.

MTC is a Multiclass–Multioutput Classification. MTC is used in the proposed methodology, where there are multiple tasks, and the user can assign only one label for each task.

2.3. Structured Inference Neural Network

The Structured Inference Neural Network (SINN) was inspired by a deep learning-based method that utilizes various label relationships to improve image classification performance by using a cumulative label prediction neural network [16]. In this neural network, structural graph formation is possible through relationships between labels, and different interpretations of various units are possible for representing images. For example, an image can be represented in terms of indoor or outdoor, specific location, and specific object units. As a result, SINN is a structural inference neural network that can model relationships between labels by considering dependencies between classification units through CNN and RNN. Figure 2a shows a baseball field image that can be represented as a structural graph, as shown in Figure 2b. The baseball field in Figure 2a belongs to the scene unit's artificial outdoor, the scene attribute unit's sports field or artificial element, the detailed scene unit's home plate, and the object unit's field, baseball bat, baseball, grass, and person classes, which are all represented by the structural graph in Figure 2b and are represented by red nodes. If the image belongs to the indoor class at the scene unit, the baseball bat object cannot be present, as the baseball bat object is dependent on artificial outdoor, which serves as evidence for using SINN. Inspired by these structured representations, we propose four units (browser, protocol, application, service) to represent traffic flows and use them to perform MTC.

(a)

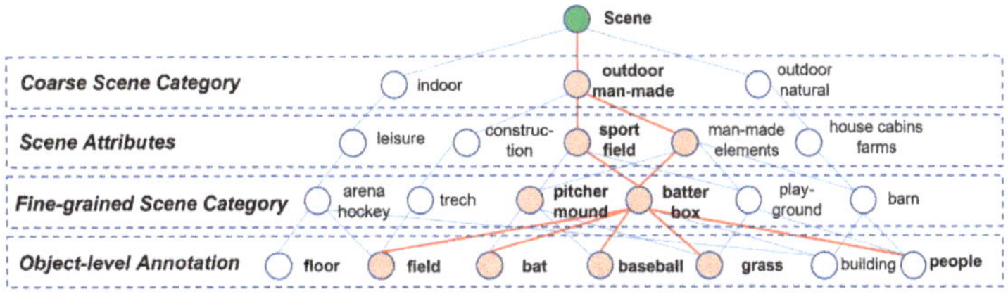

(b)

Figure 2. These represent a baseball field image (**a**) and its corresponding structural graph (**b**). (**a**) Baseball field image. (**b**) An image of a baseball field represented as a structural graph.

2.4. DL-Based Spatial-Temporal Feature Extraction

In this section, we describe the deep learning-based spatial and temporal feature extraction methods used in the multitask learning approach.

CNN is the most widely used method for extracting spatial features, especially for processing images or video data. CNNs are inspired by the structure of the visual cortex in animals, which has layers of neurons that are sensitive to specific visual features. Similarly, CNNs have layers that extract hierarchical representations of the input image or video, starting from low-level features, such as edges, and gradually moving towards high-level features, such as object parts and whole objects. The basic component of a CNN is the convolutional layer, which applies a set of filters or kernels to the input image, extracting local features that are then pooled and passed on to the next layer. We utilized two famous CNN-based backbone networks and two deep learning architectures proposed in previous studies to clearly demonstrate the contribution of applying the MTC method. The first one is Lenet [17], an initial model designed for handwritten digit recognition. Lenet consists of two convolutional layers, pooling layers and fully connected layers, and its model structure is shown in Figure 2. The second backbone network is Resnet [18] (Residual Network), a deep learning architecture proposed to solve the gradient vanishing problem that occurs in deep neural networks. Resnet solves the gradient vanishing problem by introducing skip connections, direct connections that skip several layers in the network, unlike traditional CNN architectures. Resnet still shows good performance in various fields. Ref. [19] proposes MISCNN (Multi-Input Shape Convolutional Neural Network) that utilizes various input forms that can be derived from fixed-length packet bytes. By observing packets from various angles through the different forms that can be derived from a single input, it shows a significant improvement in performance compared to previous research. Ref. [20] proposes HAST-IDS (Hierarchical Spatialtemporal Features-based Intrusion Detection System), an intrusion detection system that uses CNN to learn spatial features of packets and LSTM to learn temporal features between multiple packets. HAST-IDS performs a classification of multiple normal and abnormal traffic, and experiments show that HAST-IDS outperforms other approaches in terms of accuracy, detection rate, and FAR.

RNN (Recurrent Neural Network) is a type of deep learning that is used to handle sequential data [21]. RNN has the advantage of being able to solve the long-term dependency problem by using the output of the previous step as the input of the current step, thereby reflecting the previous information in the current processing. However, RNNs can suffer from vanishing gradient and exploding gradient problems. To solve this problem, a model based on RNN called GRU (Gated Recurrent Unit) was proposed [22], and this paper applies GRU to extract temporal features. GRU has the advantage of faster learning speed and the ability to handle longer sequences than RNN. GRU combines the hidden state and cell state used in RNN into one and updates it using two gates: the update gate and the reset gate. The update gate determines how much information to update using the current input and previous state, and the reset gate determines how much the previous state is forgotten. Through this, GRU can solve the long-term dependency problem while mitigating problems that arise during the learning process.

2.5. MTC-Based Traffic Classification

Ref. [5] proposes the use of multitask deep neural network in federated learning (MT-DNN–FL) to simultaneously perform network anomaly detection, VPN (Tor) traffic recognition, and traffic classification tasks. They report that the multitasking approach reduces training time overhead compared to multiple single-task models. Experimental results conducted on well-known datasets, such as CICIDS2017, ISCXVPN2016, and ISX-Tor2016, demonstrate that the proposed method achieves superior anomaly detection and classification performance compared to baseline models in a centralized training architecture. Ref. [10] proposes the use of multitask learning to predict the bandwidth and duration of network traffic flows while simultaneously classifying the traffic into different classes. Predicting bandwidth and duration does not require extensive labeling efforts or specific environments, allowing for the utilization of abundant training data. This approach significantly reduces the number of labeled samples required for traffic class prediction. Furthermore, the predicted bandwidth and duration can be applied in ISPs for resource allocation, routing, and QoS purposes. The experiments conducted on the QUIC and ISCX VPN-nonVPN datasets demonstrate that the multitask learning approach outperforms single-task learning and transfer learning methods. Ref. [9] proposed a novel multimodal multitask deep learning approach called DISTILLER. This approach is designed to address the challenges of encrypted traffic and diverse network visibility in traffic classification. DISTILLER leverages deep learning techniques to automatically extract complex patterns from various modalities of traffic and simultaneously solve multiple traffic categorization problems. The authors evaluate DISTILLER using public datasets and report superior performance compared to state-of-the-art deep learning architectures. Ref. [23] proposes a new multimodal deep learning framework called MIMETIC. MIMETIC overcomes performance limitations by leveraging the diversity of traffic data and achieves superior performance compared to existing single-modal deep learning-based traffic classification methods. It also highlights the effectiveness of multimodal deep learning in classifying traffic by capturing the characteristics of diverse traffic that carry information.

3. Proposed Method

In this chapter, we describe three multitasking learning methods that can learn relationships between tasks.

3.1. MTC-Based Traffic Classification

3.1.1. Single Task Single Inference

To compare with the proposed three multitask learning methods, we introduce the ST–SI (Single Task–Single Inference) learning method used in existing application traffic classification research. With ST–SI, a main classifier performs the main classification of one task through a single classifier. As shown in Figure 3, ST–SI uses four independent models for the four tasks of browser, protocol, service, and application. Figure 4a shows the structure of the model that performs the browser classification using the extracted features from the flow as the input to the backbone network. The output of the Figure 3a model is one of the three classes of the browser task. Similarly, Figure 3b–d are models responsible for each task, such as protocol, service, and application.

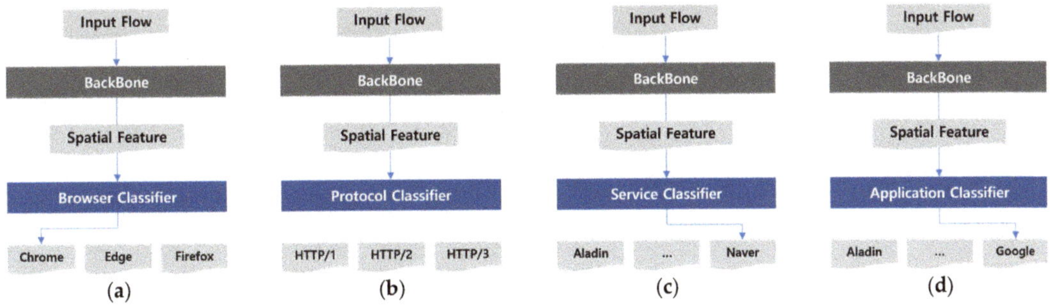

Figure 3. Overview of ST–SI (Single Task–Single Inference). (**a**) ST–SI-based browser classification; (**b**) ST–SI-based browser classification; (**c**) ST–SI-based browser classification; and (**d**) ST–SI-based browser classification.

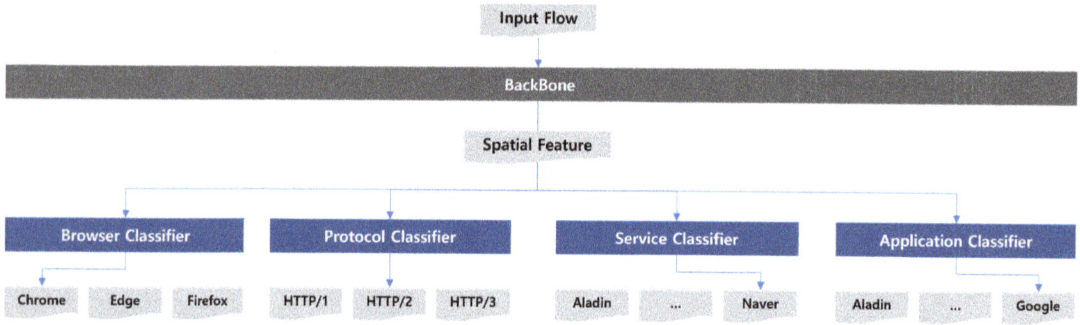

Figure 4. Overview of MT (Multitask).

3.1.2. Multitask

The first multitask learning method is MT (Multitask), which performs four main classifications simultaneously through four main classifiers. As shown in Figure 4, a single model is used to perform the four tasks of browser, protocol, service, and application. The features extracted from the backbone network are input to each main classifier responsible for the four tasks to predict a single class for each of the four tasks. In this case, each main classifier has an error weight equal to 0.25, meaning that the model learns with equal effort on all four tasks. There may be some common information between the four tasks, and the interactions between them can complement each other and improve performance. Also, by handling multiple tasks and learning common patterns, models that are more robust and flexible for new tasks can be created.

3.1.3. Multitask Single Inference

The second multitask learning method is MT–SI (Multitask–Single Inference), which performs the pre-classification of four tasks simultaneously through four pre-classifiers. Then, using the pre-classification results, a main classifier performs the main classification of one task. As shown in Figure 5, the MT–SI learning method uses four independent models for the four tasks of browser, protocol, service, and application. Figure 5a shows the structure of the model that performs the browser classification using the extracted features from the flow as input to the four pre-classifiers and one main classifier. Similarly, Figure 5b–d are models responsible for each task, such as protocol, service, and application. The error weight of the four pre-classifiers in the model is set to 0.1, and the main classifier is set to 0.6. The spatial features generated by the backbone network are input to the pre-classifiers to output classification results (probability of belonging to each class), which are merged with the previously generated spatial features. The main classifier performs the main task using the classification results and and spatial features of four pre-classifiers. The common information shared among tasks extracted during the training process of the pre-classifiers enhances the classification performance of the main classifier.

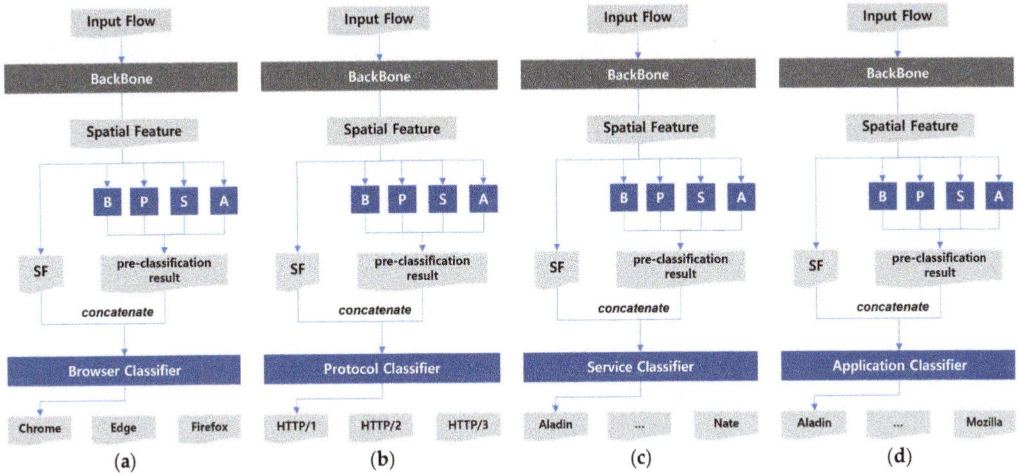

Figure 5. Overview of MT–SI (Multitask–Single Inference). (**a**) MT–SI-based browser classification; (**b**) MT–SI-based browser classification; (**c**) MT–SI-based browser classification; and (**d**) MT–SI-based browser classification.

3.1.4. Multitask Multi Inference

The third multitask learning method is MT–MI (Multitask–Multi Inference), which performs the pre-classification of four tasks simultaneously through four pre-classifiers. Then, using the pre-classification results, four main classifiers perform the main classification of four tasks simultaneously. As shown in Figure 6, the MT–MI learning method uses a single model for the four tasks of browser, protocol, service, and application. The error weight of each pre-classifier in the model is set to 0.05, and each main classifier is set to 0.2. The spatial features produced by the backbone network are input to the pre-classifiers, which output pre-classification results for each task. The pre-classification results are merged with the previously extracted spatial features and input to each main classifier to perform the corresponding task. The MT–MI method differs from MT in that it performs a brief pre-classification before performing each main classification and uses the results when performing the main classification.

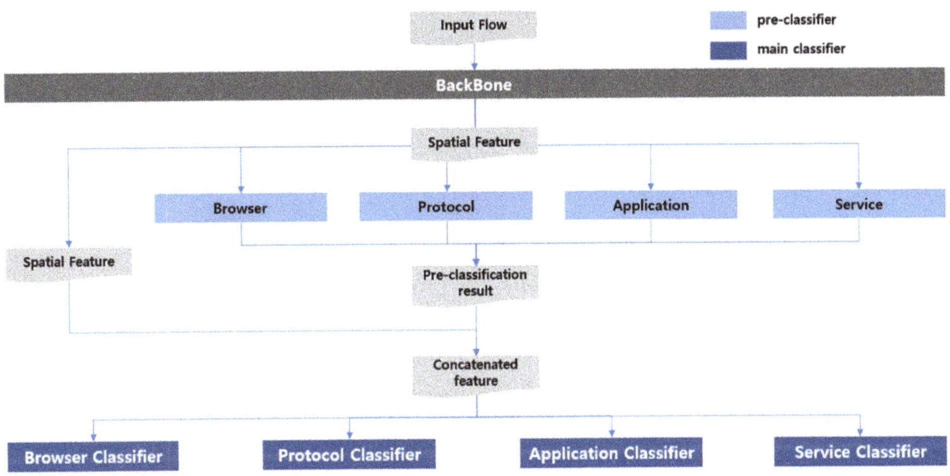

Figure 6. Overview of MT–MI (Multitask–Multi Inference).

3.2. Dataset Description

This section describes the dataset collected for validating the proposed methodology. When collecting data in a typical environment, traffic unrelated to the target web service can be collected due to the background services that are running. Therefore, a Docker platform that can be isolated from the host network is a good choice. The traffic dataset was collected using Selenium in containers on a Docker platform and consists of six types of web service traffic.

For the three additional tasks, apart from web service traffic, the labeling methods are as follows:

- Service: labeled at the time of collection;
- Browser: labeled at the time of collection;
- HTTP protocol: check the HTTP version of the GET or POST method response header when the protocol of the traffic flow is HTTP (perform the same process after decryption in the case of HTTPS);
- Application: check the Request URL for HTTP or the Service name indicator (SNI) in the Transport Layer Security (TLS) layer for HTTPS.

The collected dataset consists of 10,497 bidirectional flows, and the task-specific distribution is shown in Figure 7. In this figure, the value of each pie is in the form; the number of bi-flows, its percentages. The browser task consists of *Chrome*, *Edge*, and *Firefox*, with *Chrome* and *Firefox* accounting for a high percentage. The protocol task consists of HTTP/1.1, HTTP/2, and HTTP/3, with HTTP/2 accounting for a high percentage. The service task consists of *Aladin*, *Amazon*, *Google*, *Nate*, *Naver*, and *YouTube*, with all six accounting for an equal ratio. The application unit consists of *Aladin*, *Amazon*, *Google*, *Nate*, *Naver*, *YouTube*, *Microsoft*, *Mozilla*, and *Etc*, with *Google* and *Mozilla* accounting for a high percentage. To prevent excessive granularity, labeling was performed for traffic that belongs to major applications, and traffic that does not belong to the eight major applications was assigned to the *Etc* class.

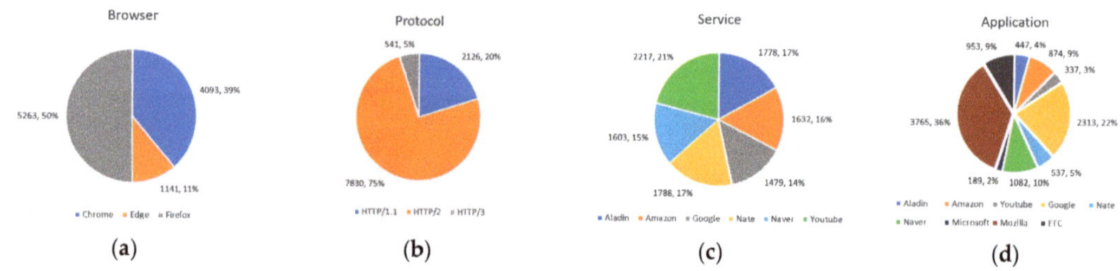

Figure 7. The task-specific distribution. (**a**) Browser distribution. (**b**) Protocol distribution. (**c**) Service distribution. (**d**) Application distribution.

Figure 8 shows the protocol ratio by service, with most services primarily using HTTP/2, but *Aladin* and *Nate* services have a relatively high percentage of using the HTTP/1.1 protocol. Also, since the *YouTube* service uses the QUIC protocol, the percentage of using the HTTP/3 protocol is high.

Figure 9 shows the application ratio by browser, and the ratio of applications used by the company that developed each browser is high.

Table 1 shows the application ratio by service, indicating that one or more applications are mixed within a single service traffic.

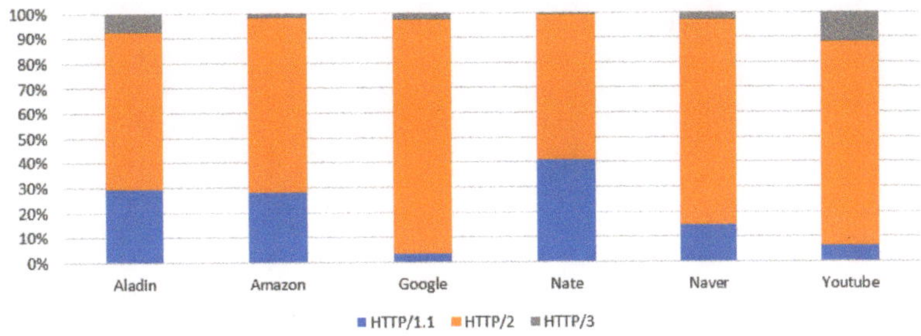

Figure 8. The protocol ratio by service.

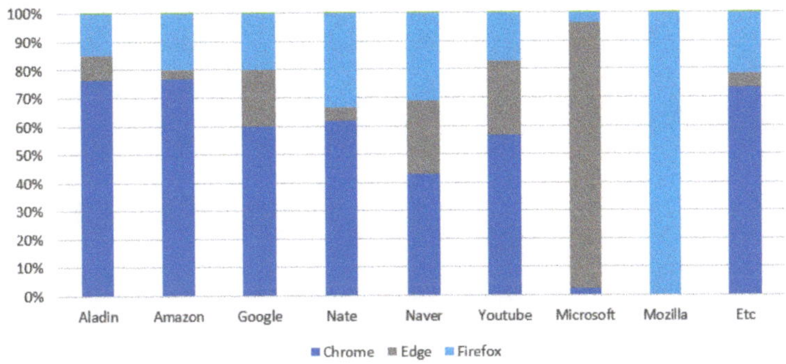

Figure 9. The browser ratio by service.

Table 1. The application ratio by service.

Service		Aladin	Amazon	Google	Nate	Naver	YouTube	Microsoft	Mozilla	Etc
Aladin	1778 100%	447 25.1%		364 20.5%		50 2.8%	5 0.3%	1 0.1%	660 37.1%	251 14.1%
Amazon	1632 100%		873 53.5%	27 1.7%			4 0.2%	2 0.1%	612 37.5%	114 7
Google	1479 100%			707 47.8%			102 6.9%	2 0.1%	668 45.2%	
Nate	1788 100%		1 0.1%	76 4.2%	537 30%	10 0.6%		2 0.1%	597 33.4%	565 31.6%
Naver	1603 100%			17 1.1%		1022 63.8%		101 6.3%	451 28.1%	12 0.7%
YouTube	2217 100%			1122 50.6%			226 10.2%	81 3.7%	777 35%	11 0.5%

4. Experiments

In this chapter, we describe the parameters used in the experiments. The experiments were performed with various parameters, and a total of 576 experiments were conducted by combining five types of parameters.

The first parameter is the four learning methods described in the methodology. The second parameter is the number of packets within a flow, which has three values of 4, 9, and 16. The third parameter is the packet size, which has four values of 324, 400, 484, and 576 (bytes). The fourth parameter is the backbone network, which has four values of LeNet, ResNet, HAST-IDS, and MISCNN. The four backbone networks are used as the backbone network within each learning method during the experiments. HAST-IDS [20] is an intrusion detection system that uses CNN to learn spatial features and LSTM to learn temporal features. MISCNN is a CNN-based service classification that utilizes various input forms that can be derived from fixed-length packet bytes [19]. The fifth parameter is the input form, which has three values: CP, MP, and MPG. CP (Concatenated Packet input) is an input form that collects and merges the first N bytes of packets within a flow, and extracts features by inputting them to the backbone network, as shown in Figure 10a. MP (Multiple Packet input) is an input form that inputs the first N bytes of packets within a flow to independent backbone networks, and merges the extracted features as shown in Figure 10b. MPG (Multiple Packet input with GRU) has the same form as MP but considers the temporal aspect of packets by inputting the extracted features to GRU, as shown in Figure 10c. A total of 576 experiments are conducted by combining the five parameters.

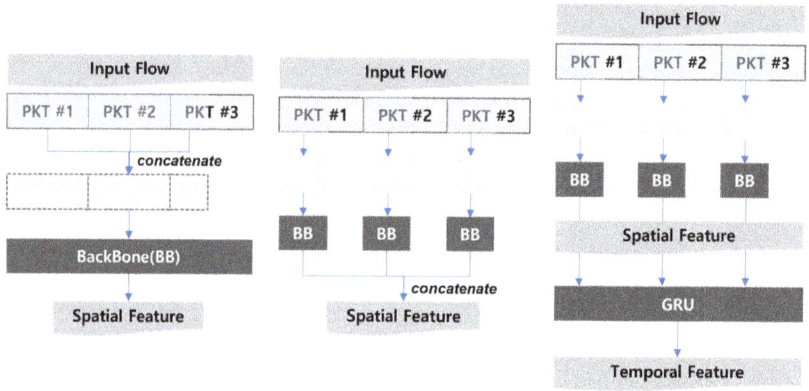

(**a**) CP (**b**) MP (**c**) MPG

Figure 10. Overview of flow input type. (**a**) Input type CP (Concatenated Packet Input); (**b**) input type MP (Multiple Packet Input); and (**c**) input type MPG (Multiple Packet Input w/GRU).

5. Experiment Results

In this chapter, we focus on comparing the classification performance based on the learning method, the number of packets within a flow, the packet size, the backbone network, and the input type.

5.1. Comparison of Task Performance According to Parameters

5.1.1. MT Method and Backbone Network

Figure 11 represents the browser classification accuracy based on the multitask approach and backbone network. Figure 11a shows the highest accuracy achieved when applying different experimental parameters to a fixed multitask approach and backbone network. From the perspective of the multitasking approach, methods that utilize the multitasking approach generally achieve higher accuracy compared to the single-task approach, except for when using the Resnet backbone. In terms of the backbone network, Resnet consistently demonstrates good results in browser classification, with the combination of Resnet and MT–SI showing the highest accuracy. Figure 11b represents the standard deviation of the accuracy for combinations of different experimental parameters applied to a fixed multitask approach and backbone network. Overall, experiments that apply the multitask approach tend to have lower standard deviations, indicating that providing prior information to the model is beneficial for generalization. Figure 12 represents the accuracy of HTTP protocol classification based on different multitask approaches and backbone networks. In terms of multitask approaches, except for the case with the Lenet backbone, the method with the application of MT-SI shows higher accuracy. In terms of backbone networks, MISCNN demonstrates overall better results in protocol classification.

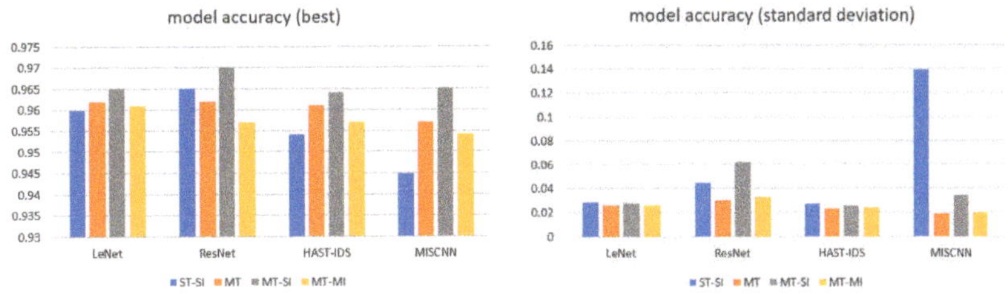

(a) (b)

Figure 11. Browser classification accuracy according to the MT method and backbone network. (**a**) The highest accuracy among the combinations of experimental parameters; (**b**) standard deviation of the results within the combination. (**a**) Best accuracy. (**b**) The standard deviation of accuracies.

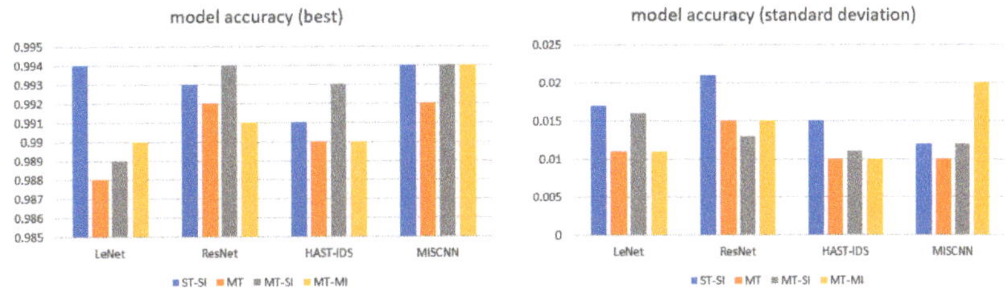

(a) (b)

Figure 12. Protocol classification accuracy according to the MT method and backbone network. (**a**) Best accuracy. (**b**) The standard deviation of accuracies.

Figure 13 represents the accuracy of service classification based on different multitask approaches and backbone networks. In terms of multitask approaches, except for the case with the Lenet backbone, the method with the application of multitask approaches shows higher accuracy. In terms of backbone networks, HAST-IDS demonstrates overall better results in service classification.

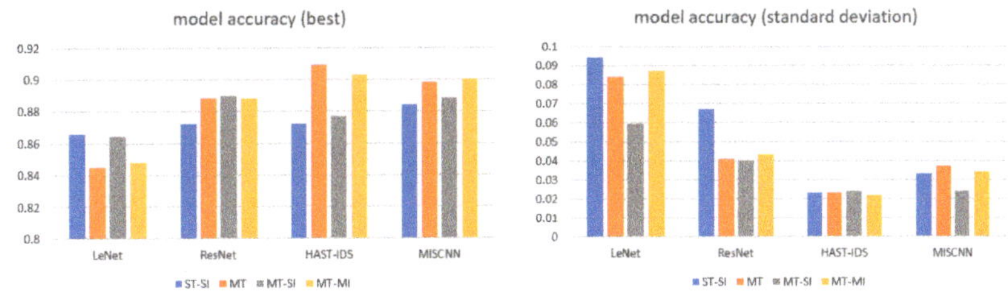

(a) (b)

Figure 13. Service classification accuracy according to the MT method and backbone network. (**a**) Best accuracy. (**b**) The standard deviation of accuracies.

Figure 14 illustrates the accuracy of application classification based on different multitask approaches and backbone networks. Except for the MISCNN backbone, single-task approaches show better performance. However, it can be observed that when applying multitask approaches, the model's variance is not significant.

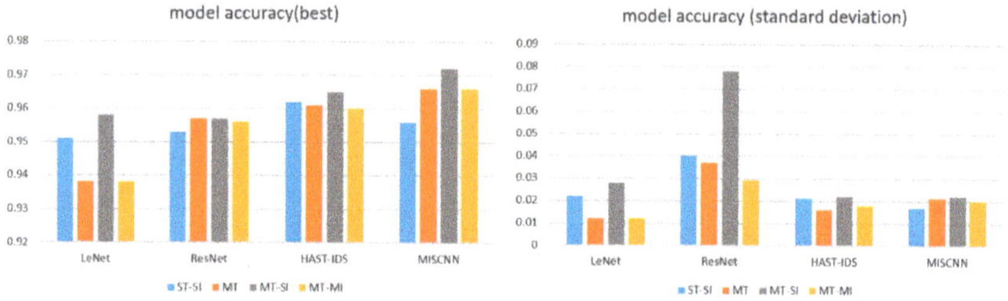

(a) (b)

Figure 14. Application classification accuracy according to the MT method and backbone network. (**a**) Best accuracy. (**b**) The standard deviation of accuracies.

5.1.2. Performance Comparison by the Number of Tasks

This section describes the performance comparison based on the number of tasks as an additional experiment. Assuming that tasks are not completely independent and have relationships with each other, we compare the performance based on the number of tasks to demonstrate more accurately the classification of other tasks by utilizing the relationships between tasks. We conduct experiments to see if the service classification accuracy improves when the service task is trained with other tasks. To compare the service classification accuracy, the service task is included in all experimental cases. The performance results based on the number of tasks are shown in Table 2. We can observe that the service classification accuracy improves when the service and application tasks are trained simultaneously and when the service, browser, protocol, and application tasks are trained simultaneously. In other words, when training with the application task included, the service classification accuracy improves. Furthermore, we can confirm that the service classification accuracy is the highest when all four tasks are trained simultaneously. Thus, we can see that the service classification task's accuracy improves when trained with other tasks.

Table 2. The average accuracy comparison by the number of tasks.

Task				Accuracy (Service)
Service	**Browser**	**Protocol**	**Application**	
√				86.124% ± 0.541%
√	√			88.698% ± 0.723%
√		√		89.028% ± 0.613%
√			√	90.357% ± 0.568%
√	√	√		89.52% ± 1.195%
√		√	√	89.81% ± 0.733%
√	√		√	90.286% ± 0.735%
√	√	√	√	90.512% ± 0.827%

5.2. Ablation Study

5.2.1. Number of Packets and Backbone Network

Figure 15 illustrates the difference in accuracy based on the change in packet count. In browser and service classification, the accuracy increases as the packet count increases. On the other hand, in protocol and application classification, the accuracy remains similar or decreases as the packet count increases. These results show a consistent trend regardless of the backbone network used.

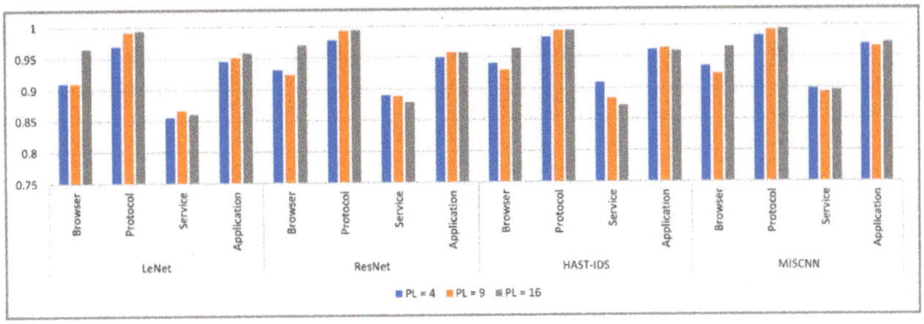

Figure 15. Task-specific classification accuracy based on the number of packets and backbone network.

5.2.2. Packet Length and Backbone Network

Figure 16 illustrates the difference in accuracy based on the change in packet length. Except for Lenet, there is not a significant variation in accuracy based on the packet length.

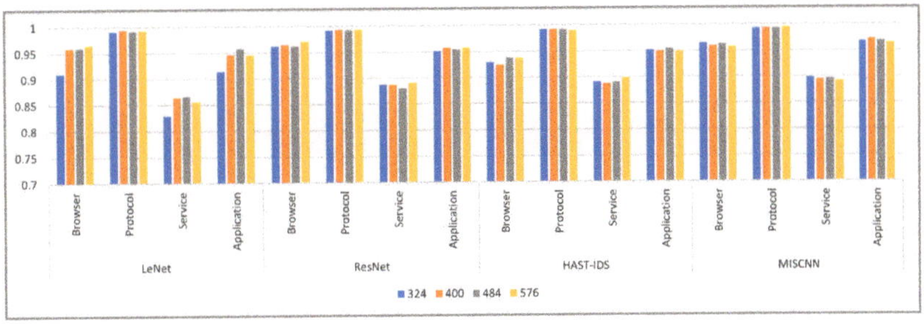

Figure 16. Task-specific classification accuracy based on packet length and backbone network.

5.2.3. Input Type and Backbone Network

Figure 17 represents the difference in accuracy based on the input type. In Lenet and Resnet, there is not a significant variation in accuracy based on the input type. However, in HAST-IDS and MISCNN, MPG generally exhibits higher accuracy in most tasks.

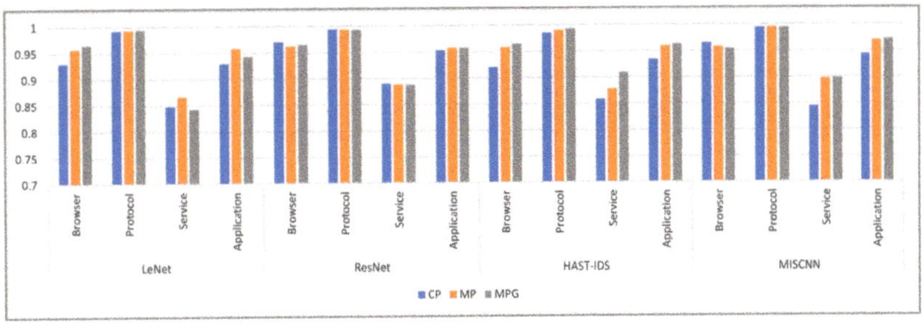

Figure 17. Task-specific classification accuracy based on input type and backbone.

5.2.4. Overall

Task-specific overall results based on parameters are shown in Table 3. The following table summarizes the parameters of the model that show the highest performance for each task and backbone network. The multitask learning-based model shows high classification accuracy in the browser, protocol, service, and application tasks, with high accuracy mainly observed between packet lengths of 9 and 16. High accuracy is also observed between packet sizes of 400 and 576, with MP or MPG input type showing high accuracy. When using the multitask learning method for all task classifications, higher classification accuracy is achieved than the conventional ST–SI learning method.

Table 3. Best task-specific classification accuracy based on parameters. LN: Lenet; RN: Resnet; HI: HAST-IDS; MC: MISCNN; NP: number of packets; PL: packet length; IT: input type; MT: MT method.

	Browser				Protocol				Service				Application			
	LN	RN	HI	MC	LN	RN	HI	MC	LN	RN	HI	MC	LN	RN	HI	MC
NP	16	16	16	16	16	16	16	16	9	4	4	4	16	9 or 16	9	16
PL	576	576	484	324	400	400	324	576	484	576	576	324	484	400 or 576	484	400
IT	MPG	CP	MPG	CP	MPG	CP	MPG	CP	MP	CP	MPG	MPG	MP	MP or MPG	MPG	MPG
MT	MTSI	MTSI	MTSI	MTSI	STSI	MTSI	MTSI	MTSI	STSI	MTSI	MT	MT	MTSI	MT or MTSI	MTSI	MTSI
Acc	0.965	0.97	0.964	0.965	0.994	0.994	0.993	0.994	0.866	0.89	0.909	0.898	0.958	0.957	0.965	0.972

5.2.5. Confusion Matrix for the Service Task

This section compares the confusion matrices of the service task for ST–SI and MT learning methods. Figure 18a shows the confusion matrix of the service task for the ST–SI learning method. The horizontal axis represents the actual labels, and the vertical axis represents the predicted labels. The result of predicting *YouTube* as *Aladin* in the service task classification using the ST–SI learning method is 8.5, which can be predicted to be a misclassification due to the *YouTube* streaming API call in the *Aladin* product description. Figure 18b shows the confusion matrix of the service task for the MT learning method. The result of predicting *YouTube* as *Aladin* in the service task classification using the MT learning method is 1.2, which has a lower probability of misclassification than the ST–SI learning method. In addition, the result of predicting *YouTube* as *Google* in the service task classification using the ST–SI learning method is 5.7, which can be predicted to be a misclassification due to similar traffic between *Google* and *YouTube* as they belong to the same company's platform. The result of predicting *YouTube* as *Google* in the service task classification using the MT learning method is 0.6, which has a lower probability of misclassification than the ST–SI learning method. The service task confusion matrix shows that the misclassification of *YouTube* into other classes has been improved. Figure 19a shows the confusion matrix of the application task for the ST–SI learning method. The result of predicting *Google* as *YouTube* in the application task classification using the ST–SI learning method is 54.6, which can be predicted to be a misclassification due to similar traffic between *Google* and *YouTube* as they belong to the same company's platform. Figure 19b shows the confusion matrix of the application task for the MT learning method. The result of predicting *Google* as *YouTube* in the application task classification using the MT learning method is 25.8, which has a lower probability of misclassification than the ST–SI learning method. The application task confusion matrix shows that the misclassification of *Google* into *YouTube* has been improved.

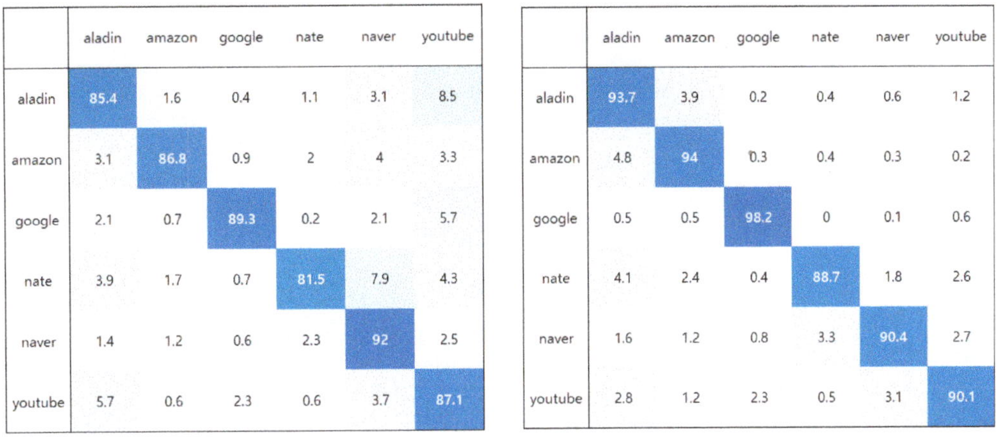

(**a**) (**b**)

Figure 18. Confusion matrix for the service task. (**a**) Confusion matrix for the service task of ST–SI; (**b**) confusion matrix for the service task of MT–SI.

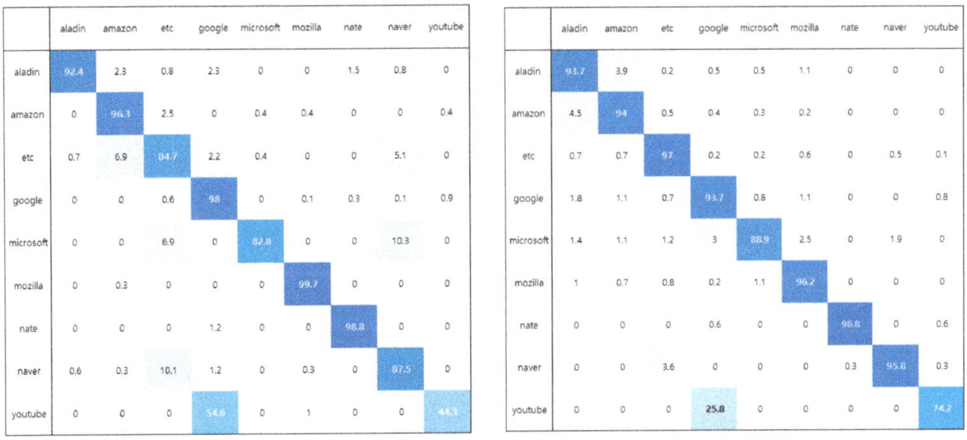

(**a**) (**b**)

Figure 19. Confusion matrix for the application task. (**a**) Confusion matrix for the application task of ST–SI; (**b**) confusion matrix for the application task of MT–SI.

6. Conclusions

This paper proposes a multitask learning method for application traffic classification using the relationships between browser, protocol, service, and application tasks. Three multitask learning methods, Multitask, Multitask Single Inference, and Multitask Multi-Inference, are proposed according to the pre-classification status and the number of main classifications. The experimental results show that more accurate application traffic classification can be achieved by utilizing the relationships between tasks. The practicality of the proposed learning methods has been demonstrated through experiments, and accuracy has been improved by 4.4%p on average while maintaining real-time performance. This means that simple model improvements allow network administrators to obtain more accurate classification results without further consideration of the resources of the hardware on which the existing classification model is installed. Specifically, experimental results suggest that if a service is called with another application that does not belong to it, it is likely to be misclassified in terms of service classification, and that the combination of service and application tasks solves this problem. In addition, applications provided in the form of APIs by *Google* appear to be confusing in classifying other services, but this is also mitigated by the application of multitask learning. Although the combination of service and application tasks has shown high classification accuracy improvements, there is also a small improvement in the combination with the proposed browser and protocol tasks. Also, the multitask learning method can provide accurate and detailed classification results for application traffic, making it widely applicable for various purposes of traffic analysis. Network administrators can receive classification results about the browser (96.5% accuracy), protocol (99.4% accuracy), service (90.9%), and application (97.2% accuracy) of web traffic. Moreover, the experimental results show that the proposed method can be applied to various existing research models. Furthermore, the proposed generalized method of multitasking learning can be combined with state-of-the-art high-performance classification models and has shown high performance in combination with the four backbone networks presented in this study. In future research, we plan to improve the multitask learning method to achieve higher performance by analyzing which classes are difficult to classify in the browser, protocol, service, and application tasks. Several limitations are set for future research. First, the consideration of the gradient conflict problem that may arise in multitask learning has not been addressed. The model's parameters need to be adjusted to satisfy the loss functions for various tasks simultaneously, but the loss functions of different tasks typically have gradients in different directions, which can degrade the performance of multitask learning or make the learning process more difficult. Therefore, there is a need to improve performance through appropriate techniques such as weight sharing or adjustment of loss functions [24]. Second, optimization of the backbone network is another challenge. The backbone network used in this study was only employed to validate multitask learning through newly proposed tasks, so there is a need for model structure adjustments and various parameter adjustments to enhance its capabilities.

Author Contributions: Conceptualization, M.-S.K.; methodology, U.-J.B.; software, B.K.; validation, B.K.; formal analysis, U.-J.B.; investigation, J.-T.P. and J.-W.C.; resources, B.K., J.-T.P. and J.-W.C.; data curation, U.-J.B.; writing—original draft preparation, U.-J.B. and B.K.; writing—review and editing, U.-J.B. and M.-S.K.; visualization, J.-W.C.; supervision, M.-S.K.; project administration, M.-S.K.; funding acquisition, M.-S.K. All authors have read and agreed to the published version of the manuscript.

Funding: This work was supported by the Technology Innovation Program grant funded By the Ministry of Trade, Industry and Energy (MOTIE, Republic of Korea) and the Korea Evaluation Institute of Industrial Technology (KEIT) (No. 20008902, Development of SaaS SW Management Platform based on 5Channel Discovery technology for IT Cost Saving) and was supported by "Regional Innovation Strategy (RIS)" through the National Research Foundation of Korea(NRF) funded by the Ministry of Education (MOE) (2021RIS-004).

Data Availability Statement: The dataset and codes presented in this study are available in request from the primary author https://github.com/pb1069/A-Multi-task-Classification-Method-for-Application-Traffic-Classification-Using-Task-Relationships (accessed on 20 August 2023).

Conflicts of Interest: The authors declare no conflict of interest.

References

1. GITNUX. Internet Traffic Statistics And Trends in 2023. Available online: https://blog.gitnux.com/internet-traffic-statistics/ (accessed on 7 August 2023).
2. Azab, A.; Khasawneh, M.; Alrabaee, S.; Choo, K.-K.R.; Sarsour, M. Network traffic classification: Techniques, datasets, and challenges. *Digit. Commun. Netw.* 2022, *in press*. [CrossRef]
3. Caruana, R. Multitask learning. *Mach. Learn.* **1997**, *28*, 41–75. [CrossRef]
4. Thung, K.-H.; Wee, C.-Y. A brief review on multi-task learning. *Multimed. Tools Appl.* **2018**, *77*, 29705–29725. [CrossRef]
5. Zhao, Y.; Chen, J.; Wu, D.; Teng, J.; Yu, S. Multi-Task Network Anomaly Detection Using Federated Learning. In Proceedings of the 10th International Symposium on Information and Communication Technology, Hanoi, Vietnam, 4–6 December 2019; Association for Computing Machinery: New York, NY, USA, 2019; pp. 273–279.

6. Sharafaldin, I.; Lashkari, A.H.; Ghorbani, A.A. Toward generating a new intrusion detection dataset and intrusion traffic characterization. In Proceedings of the International Conference on Information Systems Security and Privacy, Funchal, Portugal, 22–24 January 2018; Volume 1, pp. 108–116.
7. Draper-Gil, G.; Lashkari, A.H.; Mamun MS, I.; Ghorbani, A.A. Characterization of Encrypted and VPN Traffic Using Time-Related Features. In Proceedings of the 2nd International Conference on Information Systems Security and Privacy, Rome, Italy, 19–21 February 2016; SCITEPRESS—Science and and Technology Publications: Rome, Italy, 2016; pp. 407–414.
8. Lashkari, A.H.; Gil, G.D.; Mamun MS, I.; Ghorbani, A.A. Characterization of tor traffic using time based features. In Proceedings of the International Conference on Information Systems Security and Privacy, Porto, Portugal, 19–21 February 2017; SciTePress: Setúbal, Portugal, 2017; pp. 253–262.
9. Aceto, G.; Ciuonzo, D.; Montieri, A.; Pescapé, A. DISTILLER: Encrypted Traffic Classification via Multimodal Multitask Deep Learning. *J. Netw. Comput. Appl.* **2021**, *183–184*, 102985. [CrossRef]
10. Rezaei, S.; Liu, X. Multitask Learning for Network Traffic Classification. In Proceedings of the 2020 29th International Conference on Computer Communications and Networks (ICCCN), Honolulu, HI, USA, 3–6 August 2020; pp. 1–9.
11. Husák, M.; Čermák, M.; Jirsík, T.; Čeleda, P. HTTPS Traffic Analysis and Client Identification Using Passive SSL/TLS Fingerprinting. *Eurasip J. Info. Secur.* **2016**, *2016*, 6. [CrossRef]
12. Li, K.; Lang, B.; Liu, H.; Chen, S. SSL/TLS Encrypted Traffic Application Layer Protocol and Service Classification. In Proceedings of the Embedded Systems and Applications, Academy and Industry Research Collaboration Center (AIRCC), Vienna, Austria, 26 March 2022; pp. 237–252.
13. Hwang, R.-H.; Peng, M.-C.; Huang, C.-W.; Lin, P.-C.; Nguyen, V.-L. An Unsupervised Deep Learning Model for Early Network Traffic Anomaly Detection. *IEEE Access* **2020**, *8*, 30387–30399. [CrossRef]
14. Ullah, I.; Mahmoud, Q.H. Design and Development of a Deep Learning-Based Model for Anomaly Detection in IoT Networks. *IEEE Access* **2021**, *9*, 103906–103926. [CrossRef]
15. Su, T.; Sun, H.; Zhu, J.; Wang, S.; Li, Y. BAT: Deep Learning Methods on Network Intrusion Detection Using NSL-KDD Dataset. *IEEE Access* **2020**, *8*, 29575–29585. [CrossRef]
16. Hu, H.; Zhou, G.-T.; Deng, Z.; Liao, Z.; Mori, G. Learning Structured Inference Neural Networks with Label Relations. In Proceedings of the 2016 IEEE Conference on Computer Vision and Pattern Recognition (CVPR), Las Vegas, NV, USA, 27–30 June 2016; IEEE: Las Vegas, NV, USA, 2016; pp. 2960–2968.
17. Lecun, Y.; Bottou, L.; Bengio, Y.; Haffner, P. Gradient-Based Learning Applied to Document Recognition. *Proc. IEEE* **1998**, *86*, 2278–2324. [CrossRef]
18. He, K.; Zhang, X.; Ren, S.; Sun, J. Deep Residual Learning for Image Recognition. In Proceedings of the IEEE Computer Society Conference on Computer Vision and Pattern Recognition (CVPR), Las Vegas, NV, USA, 27–30 June 2016; pp. 770–778.
19. Baek, U.; Kim, B.; Park, J.; Choi, J.; Kim, M. MISCNN: A Novel Learning Scheme for CNN-Based Network Traffic Classification. In Proceedings of the 2022 23rd Asia-Pacific Network Operations and Management Symposium (APNOMS), Takamatsu, Japan, 28–30 September 2022; pp. 1–6.
20. Wang, W.; Sheng, Y.; Wang, J.; Zeng, X.; Ye, X.; Huang, Y.; Zhu, M. HAST-IDS: Learning Hierarchical Spatial-Temporal Features Using Deep Neural Networks to Improve Intrusion Detection. *IEEE Access* **2018**, *6*, 1792–1806. [CrossRef]
21. Learning Representations by Back-Propagating Errors | Nature. Available online: https://www.nature.com/articles/323533a0 (accessed on 16 July 2023).
22. Cho, K.; van Merrienboer, B.; Gulcehre, C.; Bahdanau, D.; Bougares, F.; Schwenk, H.; Bengio, Y. Learning Phrase Representations Using RNN Encoder-Decoder for Statistical Machine Translation. *arXiv* **2014**, arXiv:1406.1078.
23. Aceto, G.; Ciuonzo, D.; Montieri, A.; Pescapè, A. MIMETIC: Mobile Encrypted Traffic Classification Using Multimodal Deep Learning. *Comput. Netw.* **2019**, *165*, 106944. [CrossRef]
24. Liu, B.; Liu, X.; Jin, X.; Stone, P.; Liu, Q. Conflict-averse gradient descent for multi-task learning. *Adv. Neural Inf. Process. Syst.* **2021**, *34*, 18878–18890.

Journal of
Sensor and
Actuator Networks

MDPI

Article

Service-Aware Hierarchical Fog–Cloud Resource Mapping for e-Health with Enhanced-Kernel SVM

Alaa AlZailaa [1], Hao Ran Chi [2], Ayman Radwan [3,*] and Rui L. Aguiar [1]

[1] Instituto de Telecomunicações and DETI, Universidade de Aveiro, Campus Universitário de Santiago, 3810-193 Aveiro, Portugal; alz@ua.pt (A.A.); ruilaa@ua.pt (R.L.A.)
[2] Instituto de Telecomunicações and Universidade de Aveiro, Campus Universitário de Santiago, 3810-193 Aveiro, Portugal; ytchr@av.it.pt
[3] Department of Electrical and Computer Engineering and Instituto de Telecomunicações, Instituto Superior Técnico, University of Lisbon, 1049-001 Lisbon, Portugal
* Correspondence: aradwan@av.it.pt

Abstract: Fog–cloud-based hierarchical task-scheduling methods are embracing significant challenges to support e-Health applications due to the large number of users, high task diversity, and harsher service-level requirements. Addressing the challenges of fog–cloud integration, this paper proposes a new service/network-aware fog–cloud hierarchical resource-mapping scheme, which achieves optimized resource utilization efficiency and minimized latency for service-level critical tasks in e-Health applications. Concretely, we develop a service/network-aware task classification algorithm. We adopt support vector machine as a backbone with fast computational speed to support real-time task scheduling, and we develop a new kernel, fusing convolution, cross-correlation, and auto-correlation, to gain enhanced specificity and sensitivity. Based on task classification, we propose task priority assignment and resource-mapping algorithms, which aim to achieve minimized overall latency for critical tasks and improve resource utilization efficiency. Simulation results showcase that the proposed algorithm is able to achieve average execution times for critical/non-critical tasks of 0.23/0.50 ms in diverse networking setups, which surpass the benchmark scheme by 73.88%/52.01%, respectively.

Keywords: fog–cloud computing; task scheduling; service aware; support vector machine; network optimization; e-Health

Citation: AlZailaa, A.; Chi, H.R.; Radwan, A.; Aguiar, R.L. Service-Aware Hierarchical Fog–Cloud Resource Mapping for e-Health with Enhanced-Kernel SVM. *J. Sens. Actuator Netw.* **2024**, *13*, 10. https://doi.org/10.3390/jsan13010010

Academic Editors: Stefan Fischer, Yichuang Sun, Haeyoung Lee and Oluyomi Simpson

Received: 7 October 2023
Revised: 14 January 2024
Accepted: 18 January 2024
Published: 1 February 2024

1. Introduction

e-Health is becoming the backbone for upgrading the conventional healthcare system by validating remote services, which are specifically important under pandemic conditions (e.g., COVID-19) to release the pressure of lack of trained healthcare professionals [1,2].

Predicted to process 92% of the overall workload within 5 years, cloud computing has proved its high centralized computational and storage capacity in e-Health, e.g., developing health information technology (HIT) systems [3]. Additionally, to prevent inefficient massive data aggregation to centralized cloud computing in the conceived ultra-large-scale e-Health system with the rapid increase in the number of users, fog computing has emerged, since it is located physically closer to users [4]. To integrate the aforementioned strengths of cloud and fog computing, the fog–cloud hierarchical structure, as well as efficient fog–cloud resource management, has provided a glimmer of hope to support high portability and automatic provisioning for future e-Health development [5]. There are plenty of efforts for resource management among fog and cloud nodes, e.g., resource mapping [6] and task scheduling [4], where task classification is essential to identifying the features of both computing nodes (e.g., computational capacity, potential latency, etc.) [7] and tasks (e.g., payload, etc.) [8].

Despite the foreseen advantages, the fog–cloud hierarchical structure for e-Health is a significantly complicated system, as a cross-layer optimization problem, with new challenges. To start, the cross-layer task scheduling of the fog–cloud hierarchical structure for e-Health gains an irregular solution space due to the quantitatively large difference between features of fog and cloud nodes. Therefore, it requires higher complexity of modeling and solving, which might fail to support critical/latency-sensitive tasks due to tedious computation. To the best of the authors' knowledge, there is still a lack of research efforts for the real-time task scheduling of fog–cloud hierarchical structures for e-Health services. Additionally, task classification also encounters serious computational complexity, particularly for the foreseen ultra-large-scale e-Health systems. Improper task classification directly results in inefficient resource utilization regarding diverse demands for task scheduling, optimization of QoS, and latency. Moreover, due to the uniqueness of e-Health services, critical tasks are defined with very low latency tolerance margin, which is determined by services/patients' information, e.g., medical records/real-time symptoms [9]. Therefore, task scheduling in e-Health should fuse features at the service level and network level simultaneously, which directly complicates the computational process.

Targeting the aforementioned challenges, we propose an e-Health infrastructure for real-time fog–cloud hierarchical task scheduling by dynamically considering the real-time requirements of tasks with the modeling of both networking and computation simultaneously. The contributions in the paper can be summarized as follows:

1. We propose a task classification algorithm, fusing features at the network level and service level for e-Health, which is efficient in achieving user-centric QoS maximization, with latency minimized for critical tasks. Support vector machine (SVM)-based task classification which is efficient in handling the defined latency-sensitive critical tasks is proposed. It is necessary to note that although deep learning algorithms increasingly gain markets, shallow machine learning (e.g., SVM) with low computational costs still presents strengths for latency-sensitive e-Health applications [5].
2. A new kernel type is proposed for comprehensively classifying network-level and service-level features, fusing convolution, cross-correlation, and auto-correlation, which gains high overall classification accuracy for specificity and sensitivity enhancement.
3. We propose a task priority assignment algorithm and a resource-mapping algorithm, which achieve sufficient overall latency for the defined critical tasks while improving the overall resource utilization efficiency.

The rest of the paper is organized as follows: Section 2 presents a survey of related efforts. Section 3 introduces the system modeling with the formulation of communication and computational processes. Section 4 proposes optimization modeling and the SVM-based task-scheduling algorithm, with a newly developed kernel type to optimize resource utilization efficiency and communication latency. Section 5 explores three key algorithms. These algorithms collaboratively prioritize tasks, classify resources, and effectively allocate tasks to suitable fog and cloud nodes. Section 6 showcases the result analysis of the proposed scheme. Section 7 concludes the paper.

2. Related Work

Task scheduling for fog–cloudcomputing is highly demanded due to uneven distribution of workload and heterogeneous computing capacity. Conventionally, task scheduling is performed among cloud nodes to improve computational efficiency. For instance, tasks are prioritized according to their payload, which is further scheduled based on the cloud's MIPS, ensuring processing reliability for tasks with high priority [10]. Compared with cloud computing, the popularity of fog computing increases demand for task scheduling, resulting from the limited computational capacity of individual fog nodes (FNs) when dealing with heterogeneous data densification for e-Health. Depending on criteria such as user-preference-oriented features of FNs [11], energy efficiency maximization [12], and overall latency minimization [13], fog-based task scheduling is modeled and solved. As illustrated in Table 1, critical network- and service-level factors that comprehensively cover the typical demands and indicators to achieve optimal service-aware fog–cloud resource mapping for e-Health are selected and considered in this paper. Concretely, FN capacity relates to the number of fog nodes available, influencing performance and task offloading decisions. Task features, encompassing computational complexity, data size, and latency requirements, are the essential criteria to be considered when achieving service-level task scheduling in the network layer. Task priority is a critical indicator to guide scheduling based on service-level urgency. Latency, i.e., delay in task completion, impacts the quality of service, particularly in e-Health scenarios. In addition, execution time is specifically considered, besides overall latency, to further highlight the impact of task complexity and fog node resources simultaneously. Network and computation modeling are responsible for ensuring accurate latency and execution time predictions in service-aware task scheduling, respectively. Offloading decisions to the cloud, in contrast to fog-level offloading, should be considered a key criterion in fog–cloud resource mapping, with the cloud providing centralized computational/storage power to data-intensive tasks for a trade-off among network latency, balancing proximity, and computational power.

Many current research efforts focused on either fog-based or cloud-based task scheduling, where nodes share similar features. For example, FNs normally gain a similar computational capacity level (i.e., much lower than cloud nodes), leading to a relatively regularly shaped optimization solution space for task classification [4]. For instance, a deep learning-based fog-computing architecture was proposed in order to diagnose heart diseases, which offloaded tasks based on the CPU load of FNs as a metric [14]. Similarly, greedy algorithm-based resource allocation and classification for FNs based on the availability of CPU and bandwidth were proposed [15].

The fog–cloud-based hierarchical structure has drawn significant research attention in recent 5G development and research on next-generation communications thanks to its benefits of enhanced connectivity between sensors/devices/users and computing nodes while reducing transmission and computational latency with high QoS. In fact, fog–cloud hierarchical task scheduling has been foreseen to be dominant for future heterogeneous network (HetNet) deployment, requiring systemic optimization regarding service-centric and user-centric performance, e.g., energy efficiency, QoS, overall latency, etc. [16]. However, complicated features with multi-layer modeling cannot be tackled with the above-mentioned homogeneousfog–cloud-only task scheduling. There have been attempts to construct hierarchies among cloud nodes (e.g., cloudlet [1]) or FNs (e.g., multi-layer FNs [17]), aiming to increase utilization efficiency while reducing latency, as these are still not replaceable in fog–cloud hierarchical task scheduling with more diverse features. The most adopted strategies for fog–cloud task scheduling generally follow the principle that latency-tolerant and large-size tasks are assigned to cloud nodes, and latency-sensitive tasks, to FNs [9], based on which minimizing the overall makespan, maximizing resource utilization efficiency, or load balancing is targeted [16,18–21].

To facilitate task scheduling in fog–cloud hierarchical e-Health systems, task classification is essential, with tasks being normally categorized/prioritized (e.g., critical, moderate, normal [18]). Concretely, the priority of tasks is mostly determined with reference to the overall latency requirement [19], deadline and available resources in FNs [22], payload [23], makespan constraints with available resources [20], or task length [24]. For instance, a mobility-aware scheduling scheme to dynamically distribute tasks to the fog or the cloud was proposed for e-Health [25] by prioritizing tasks based on data size, response time, and complexity. A static scheduling method in a fog–cloud heterogeneous environment was also proposed to reduce CPU execution time and network usage [26]; it classifies tasks according to the required MIPS and the trade-off between CPU execution time and allocation memory of FNs. Task offloading to cloud nodes is still based on the aforementioned "latency-tolerant and large-size tasks to cloud nodes" principle.

Table 1. Literature review.

Ref.	FN Capacity	Task Features	Task Priority	Latency	Execution Time	Network Modeling	Computation Modeling	Offloading to Cloud
[10]	✘	✔	✔	✘	✔	✘	✘	✘
[11]	✘	✘	✘	✔	✔	✘	✘	✘
[12]	✔	✔	✘	✔	✔	✘	✘	✘
[16]	✔	✘	✘	✔	✔	✔	✔	✘
[18]	✔	✔	✔	✘	✔	✘	✘	✔
[19]	✔	✘	✔	✘	✔	✔	✘	✔
[20]	✔	✘	✔	✔	✔	✔	✘	✔
[21]	✔	✘	✘	✔	✔	✔	✔	✘
[23]	✘	✔	✔	✔	✘	✘	✘	✘
Proposed	✔	✔	✔	✔	✔	✔	✔	✔

However, due to the nature of e-Health, besides network-level requirements, tasks should be further classified based on service-level demands, e.g., patients' profile, etc., which should play a vital role in minimizing critical tasks (e.g., time-sensitive tasks, emergencies, etc.). For instance, detection of abnormal cardiovascular conditions might be much more serious for patients with related EHR than for normal users, which cannot be reflected by network-level features. Some attempts were made to reduce the computational process for mobile patients based on offloading high-priority tasks to the core with higher computational power based on the predefined priority of tasks in the application layer [27]. Comprehensive determination of task priority is still missing for e-Health regarding both network-level requirements and service-level demands. Therefore, we propose a task-scheduling scheme, comprehensively considering network-level and service-level features for task classification.

3. Fog–Cloud Hierarchical Infrastructure and Modeling for e-Health

This section outlines an innovative task-scheduling framework in a fog–cloud hierarchy specifically designed to enhance e-Health services by optimizing task allocation and reducing latency.

3.1. Fog–Cloud Hierarchical Infrastructure for e-Health

We propose an optimal cross-layer task-scheduling scheme based on a generic fog–cloud hierarchical infrastructure for e-Health, as shown in Figure 1. e-Health devices in the IoT layer are connected to base stations (BSs), which further relay the aggregated data to the task orchestrator (also known as fog-layer broker), centralized for task scheduling. Allocation between e-Health devices and BSs can be achieved using our previously published schemes, achieving optimal link quality with interference mitigation [6]. Conventionally, task-scheduling methods either schedule tasks among fog nodes (FNs) or forward latency-insensitive tasks to cloud nodes, which results in inefficient resource utilization of the fog–cloud hierarchical infrastructure while potentially failing to support critical tasks in e-Health according to service-level demands. Therefore, we focus on cross-layer task scheduling by embedding a support vector machine (SVM)-based task classification algorithm at the orchestrator, the output of which guides real-time task scheduling to fog–cloud nodes. Compared with the conventional methods, the proposed algorithm is capable of achieving real-time fog–cloud cross-layer task scheduling with minimized overall latency for critical tasks in e-Health while ensuring efficient computation for all users. The proposed task classification and task-scheduling algorithm will be discussed in Sections 4 and 5, respectively.

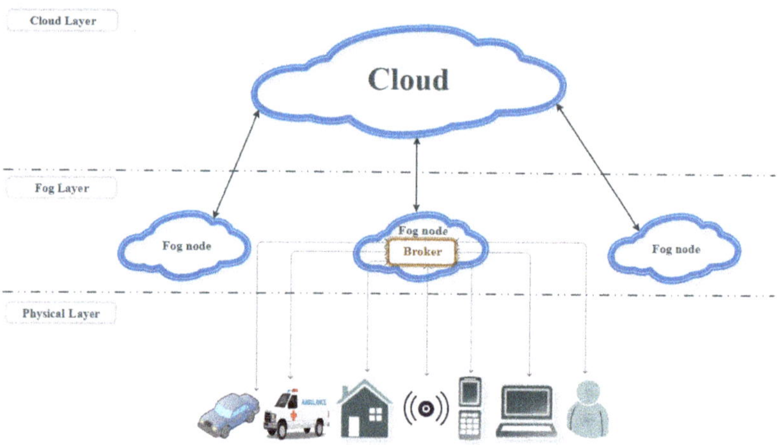

Figure 1. Generic cloud–fog hierarchical architecture for e-Health.

Suppose that there are M cloud nodes and N FNs in the infrastructure designed with heterogeneous computational capacities as

$$\begin{aligned} A_M^c &= \{a_1^c, a_2^c, \ldots, a_M^c\} \\ A_N^f &= \{a_1^f, a_2^f, \ldots, a_N^f\} \end{aligned} \tag{1}$$

where $M + N$ fog–cloud nodes support U_{IoT} e-Health devices, with each device u generating $x_1^u, x_2^u, \ldots, x_{K_u}^u$ tasks in unit time period t. To model this random generation, we have employed a Poisson distribution, which is commonly considered [28] and well proved to reflect the scenarios in our paper, i.e., events happening independently, at a steady rate, whose exact occurrence is random. Concretely, $D_{x_{k_u}^u}$ and $L_{x_{k_u}^u}$, with $k_u \leq K_u, u \leq U$, are the latency and the payload of the kth task generated by device u, respectively, which are well recognized for task classification, especially for emergent and critical e-Health applications [9]. However, specifically in e-Health, critical tasks should be defined according to service-level demands. For instance, it is intuitive that tasks related to acute diseases should gain higher critical level than chronic diseases, with much higher probability of leading to serious drawbacks.

To correlate tasks according to service-level demands, we consider the type of tasks according to their service-level functionality, $\boldsymbol{\theta}$, which holds four different values, defined as

$$\boldsymbol{\theta}_x^u = \{inquiry, backup, notification, alarm\} \tag{2}$$

where:

1. Inquiry: triggered for storing information in medical records;
2. Backup: generated periodically to update medical records;
3. Notification: set as reminders, e.g., pill time and therapy appointment for patients, medical status alert to medical workers, etc.;
4. Alarm: generated based on the diagnosis results, which are also affected/referenced by the proposed orchestrator, regarding task classification.

The final value of θ_x^u is set by the scheduler, after examining all indicated parameters in Algorithm 1.

Algorithm 1 Task priority determination algorithm.

/* 1.0 for each task $(x^u_{k_u})$ check $(\beta^u_{x_{k_u}})$, $(\mu^u_{x_{k_u}})$ fields */
if $(\beta^u_{x_{k_u}}) == (\mu_l^{\beta^u_x})$ **then**
  $SVM\_weight(x^u_{k_u}) == High$
else if $(\beta^u_{x_{k_u}})! = (\mu_l^{\beta^u_x})$ **then**
  $SVM\_weight(x^u_{k_u}) == Medium$
else
  $SVM\_weight(x^u_{k_u}) == Low$
end if
/* 2.0 assign the priority to the task */
$priority(x^u_{k_u}) = SVM_{weight}(x^u_k) + weight(PL(x^u_k))$
/* 2.1 label the task, assign the value (type of task) */
$\theta^u_{x_{k_u}} = \{inquiry, backup, notification, alarm\}$
/* 2.2 order tasks descending based on priority */
$P(x^u_k(p)) = \left\{x^u_k(p), x^u_{k-1}(p-1), \cdots, x^u_0(0)\right\}$
/* 3.0 send the prioritized tasks from the fog nodes */
$P(x^u_k(p)) \Rightarrow Orchestrator\{[X(p)^U_K][A^N_F]\}$

Additionally, patients' medical records, $\boldsymbol{\beta}^u_x$, are considered, in task scheduling, a key feature reflecting service-level demands and are generalized as acute and chronic diseases $(x \in \{x^u_1, \ldots, x^u_{K_u}\}, u \in \{1, \ldots, U_{IoT}\})$:

$$\boldsymbol{\beta}^u_x = \{Acute, Chronic\} \tag{3}$$

Similar to $\boldsymbol{\theta}$, tasks related to acute diseases should be considered critical tasks, compared with chronic diseases. To model $\boldsymbol{\beta}^u_x$ numerically, we define $\boldsymbol{\mu}$ as preliminary symptoms of diseases, according to patients' medical records:

$$\boldsymbol{\mu}^u = \begin{bmatrix} \mu_1^{\beta^u_1} & \cdots & \mu_L^{\beta^u_1} \\ \vdots & \ddots & \vdots \\ \mu_1^{\beta^u_{K_u}} & \cdots & \mu_L^{\beta^u_{K_u}} \end{bmatrix} \tag{4}$$

where $\forall u \in \{1, \ldots, U_{IoT}, \boldsymbol{\mu}^u \in \{0, 1\}$, with $\boldsymbol{\mu}^u = 1$ indicating that patient u has corresponding symptoms of acute or chronic diseases and $\boldsymbol{\mu}^u = 0$ indicating that no symptoms have been detected.

3.2. Network Modeling

Since $\boldsymbol{\theta}$ is considered a key parameter set for task scheduling, network modeling should be formulated to reflect the stringent latency and QoS requirements, which are modeled in a cross-layer fashion.

As illustrated in Figure 1, U_{IoT} devices are connected to FNs through a wireless network, with distance $\delta^f_{u,n}$ between device u and FN n. Similarly, with the conceived large throughput in the next generation of communications, the connection between fog and cloud layers is also assumed to be wireless, with distance being represented by $\delta^c_{n,m}$.

For each task $x^u_{k_u}, k_u \in \{1, \ldots, K_u\}, u \in \{1, \ldots, U_{IoT}\}$, the overall end-to-end transmission latency can be constrained as

$$D^u_{x_{k_u}} < \tau_{comm,x^u_{k_u}} + \tau_{wait,x^u_{k_u}} + \tau_{proc,x^u_{k_u}} \tag{5}$$

where $\tau_{comm,x^u_{k_u}}$, $\tau_{wait,x^u_{k_u}}$, and $\tau_{proc,x^u_{k_u}}$ refer to latency of propagation, waiting, and processing for task $x^u_{k_u}$, respectively.

$\tau_{comm,x^u_{k_u}}$ is affected by the transmission delay ($t_{trans,x^u_{k_u}}$), estimated as

$$t_{trans,x^u_{k_u}} = \frac{J^f_u}{R^{f,u}_n} + \frac{I^c_u}{R^{c,u}_m} \quad (6)$$

where $J^f_u = \sum x^u_{k_u}$ and $I^c_u = \sum x^u_{k_u}$ represent the number of bits of task $x^u_{k_u}$ from device u to FNs and cloud nodes, respectively. $R^{f,u}_n$ and $R^{c,u}_m$ refer to the data rates for transmitting the bits of $x^u_{k_u}$ to FN n and cloud nodes m, respectively, as [6]

$$R^{f,u}_n = (W^f_n) * \log(1 + SINR^{f,u}_n) \quad (7)$$

$$R^{c,u}_m = (W^c_m) * \log(1 + SINR^{c,u}_m) \quad (8)$$

where $W^z_i, i \in \{m,n\}, z \in \{c,f\}$, is the bandwidth assigned to node i and $SINR^{z,u}_i$ represents the signal-to-interference-plus-noise ratio for node i regarding device u. In the case of $i = n$, the $SINR$ is calculated as

$$SINR^{f,u}_n = \frac{Pw_u h_u (\delta^f_{u,n})^{-\alpha_{pl}}}{\sigma^2 + I^f_n} \quad (9)$$

where Pw_u refers to the transmission power of device u (uplink considered). h_u is the channel power coefficient. Path loss is defined according to δ_u with path loss coefficient α_{pl}. σ^2 and I^f_n are the noise variance and the residual interference power for FN n, respectively [29]. Similarly, for the connection between device u and cloud node m, the SINR is calculated as

$$SINR^{c,u}_m = \frac{Pw_u h_u (\delta^c_{u,m})^{-\alpha_{pl}}}{\sigma^2 + I^c_m} \quad (10)$$

Together with the propagation delay related to multi-hopping, $\phi^{z,u}_i, i \in \{m,n\}, z \in \{c,f\}$, has a proportional relationship with $\delta^f_{u,n}$ and $\delta^c_{n,m}$; $\tau_{comm,x^u_{k_u}}$ is derived as

$$\tau_{comm,x^u_{k_u}} = t_{trans,x^u_{k_u}} + \phi^{z,u}_i \quad i \in \{m,n\}, z \in \{c,f\} \quad (11)$$

$\tau_{wait,x^u_{k_u}}$ is derived as the summation of the delay for task $x^u_{k_u}$ requesting network access ($\gamma_{x^u_{k_u}}$) and the server response time ($\nu_{i,x^{z,u}_{k_u}}$) required for computing nodes to respond to the task:

$$\tau_{wait,x^u_{k_u}} = \gamma_{x^u_{k_u}} + \nu_{i,x^{z,u}_{k_u}} \quad i \in \{m,n\}, z \in \{c,f\} \quad (12)$$

$\tau_{proc,x^u_{k_u}}$ is affected by the payload of task $x^u_{k_u}$ and the computing capacity of the associated computing nodes, $i \in \{m,n\}, z \in \{c,f\}$, and is calculated as

$$\tau_{proc,x^u_{k_u}} = \frac{L^u_{x_{k_u}}}{a^z_i}, \quad i \in \{m,n\}, z \in \{c,f\} \quad (13)$$

3.3. Computation Modeling

Both cloud and fog nodes share basic features, e.g., computing capacity, including virtual central processing unit (vCPU) cores, MIPS, RAM, and storage, regarding A^f_n and A^c_m defined in (1):

$$A^z_i = \sum \{vCPU + MIPS + RAM + storage\}, i \in \{m,n\}, z \in \{c,f\} \quad (14)$$

We suppose that $T^z_{u,i}$, $i \in \{m,n\}, z \in \{c,f\}$, refers to the total number of time slots in the computing node, where

$$T^z_{u,i} = \sum_{t=1}^{T} \Delta\tau^z_{u,i}(t), \quad Z = c, f. \tag{15}$$

where T is the total processing time. $\Delta\tau^z_{u,i}(t)$ is the available time of computing node i in time slot t.

The required computing resource for task $x^u_{k_u}$ is defined as $rq^u_{x^u_{k_u}}(t)$ in a given time slot t. It is highlighted that $rq^u_{x^u_{k_u}}(t)$ is heterogeneous for diverse devices, which complicates task scheduling, with user/service-centric QoS and latency optimization. $rq^u_{x^u_{k_u}}(t)$ is constrained as

$$\forall u \in \{1, \ldots, U_{IoT}\} : \sum_{k=1}^{K_u} \sum_{t=1}^{T} rq^u_{x^u_{k_u}}(t) \leq \sum_{i=1}^{N} a^f_i + \sum_{j=1}^{M} a^c_j \tag{16}$$

4. Support Vector Machine-Based Multi-Layer Task Classification

This section addresses the proposed SVM-based multi-class algorithm for weighting each task. The result of this stage is used together with Algorithm 1 to determine the priority of the tasks. The proposed SVM-based algorithm determines the weight of the task into three categories, "high", "med", and "norm", according to two features, where the first one is the patient profile, which represents the medical history of the patient, and the second one is the symptom, which is related to the sensor's response to an action. Figure 2 shows the transitions between states of the weights for the three classes.

1. High: This state indicates that the notification from one of the sensors has one of the symptoms labeled risky, in addition to the fact that the patient has an illness history within their profile; additionally, the received symptom is directly connected to the patient's medical case.
2. Medium: This state includes two cases: The first one implies that the notification has one of the risky symptoms but the patient's medical profile is marked as healthy and has no illness history; this case is represented by (01). The second case is when the patient is labeled as having one of the chronic diseases which requires constant surveillance and the patient has no symptoms at the moment; this case is represented by (10).
3. Normal: This state refers to the situation where all incoming notifications are within safe limits, such as periodic readings, with a clear illness history for the patient.

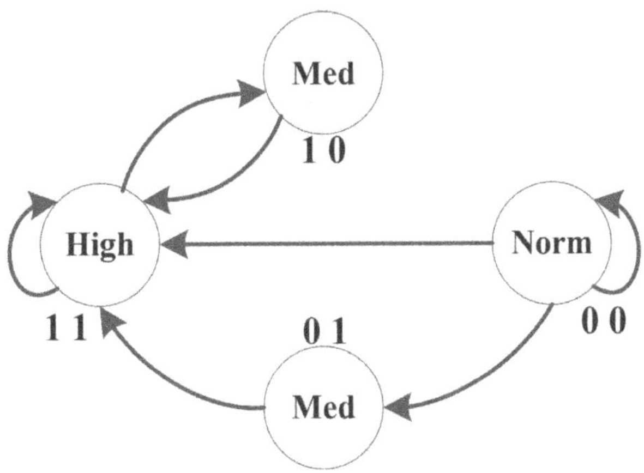

Figure 2. State diagram for task weight determination.

It is worth to highlight here that three classes are sufficient and efficient in targeting low latency for the high class and offering sufficient coverage and connectivity for the normal class. The medium class is classified as the transitional class; it is distinguished from the normal class and is potentially classified as high class once the extracted features tend to be more "critical", e.g., more severe symptoms are detected. To increase classification accuracy among the three classes, three classifiers are required for classifying high/med, med/norm, and high/norm [30].

Highly nonlinear and heterogeneous features lead the task priority determination in this paper to require machine learning-based algorithms [19]. Given the need for machine learning algorithms, SVM, well known for its capability of handling classification in nonlinear solution spaces with fast computation [30], is adopted as the backbone algorithm for task priority determination. Regarding the heterogeneous features considered in task priority determination, new kernel fusing cross-correlation, convolution, and auto-correlation are developed.

In the same manner as in Algorithm 1, the machine learning algorithm SVM is used in Algorithm 2 to classify FNs into three levels (high, medium, and low) based on three aspects: the first one is the physical characteristics of the FNs, which include MIPS, RAM, storage, and the number of CPUs; the second aspect reflects network communication features up/downlink between the FNs and the physical layer; and finally, the third aspect refers to the availability of the resources in each FN and their ability to receive and process new tasks. This procedure is periodically employed to scan the FNs in the service area and update the SVM value in Algorithm 2 to classify the FNs into three levels to meet the latency limits for critical tasks.

Task priority determination, together with fog/cloud computing ranking, which will be described in Section 5, contributes to Algorithm 3, i.e., resource mapping, for efficiently supporting critical tasks in e-Health with low latency and high QoS while ensuring effective connectivity and computation for normal tasks (e.g., monitoring). The resource-mapping algorithm is further illustrated in detail in Section 5.3.

Algorithm 2 Resource classification algorithm.

/* Scan for the available fog node A_N^f */
for $n = 0, 1, 2 \ldots, N$ **do**
  /* 1.0 SVM classify computational capacity (CC) */
  $$SVM\begin{Bmatrix} MIPS \\ RAM \\ CPUs \\ Storage \end{Bmatrix} \Rightarrow (CC(A_n^f))$$

  /* 1.1 SVM classify connection features (CF) */
  $$SVM\begin{Bmatrix} uplink \\ distance \end{Bmatrix} \Rightarrow (CF(A_n^f))$$

  /* 1.2 The Availability of the fog node (A) */
  $$Availabl\begin{Bmatrix} \text{Task spots} \\ \text{available resources} \end{Bmatrix} \Rightarrow (A(A_n^f))$$

end for
/* 2.0 Classify and order fog nodes using SVM */
$$SVM\begin{Bmatrix} (CC(A_n^f)) \\ (CF(A_n^f)) \\ (A(A_n^f)) \end{Bmatrix} \Rightarrow$$
$$\left\{(A_n^f)_{High},\ (A_n^f)_{Medium},\ \ldots,\ (A_n^f)_{Low}\right\}$$

Algorithm 3 Resource-mapping algorithm.

/* 1.0 Receiving prioritized fog nodes for Algorithm 2 */
$\left\{(A_n^f)_{High},\ (A_n^f)_{Medium},\ \ldots,\ (A_n^f)_{Low}\right\}$
/* 2.0 Receiving prioritized tasks for Algorithm 1 */
prioritized task= $\left\{(x_{high}^{f_n}),\ (x_{med}^{f_n}),\ (x_{low}^{f_n})\right\}$
/* checking task's parameters */
/* 2.1 checking the value of task's Payload */
if $PL(x_{k_u}^u) >= PL(high)$ **then**
    $weight(PL(x_{k_u}^u)) == high$
else if $PL(medium) =< PL(x_{k_u}^u) =< PL(high)$ **then**
    $weight(PL(x_{k_u}^u)) == medium$
else
    $weight(PL(x_{k_u}^u)) == low$
end if
/* 3.0 The Orchestrator maps & offloads tasks to FNs or CNs */

$$\begin{bmatrix} (A_1^f) \\ (A_2^f) \\ (A_n^f) \\ \vdots \\ (A_N^f) \\ \vdots \\ (A_m^c) \\ (A_M^c) \end{bmatrix} \Leftarrow \begin{bmatrix} (x_1^{f_1}) \\ (x_1^{f_1}) \\ (x_k^{f_n}) \\ \vdots \\ \vdots \\ (x_{K-k}^{f_{N-n}}) \\ \vdots \\ (x_K^{f_N}) \end{bmatrix}$$

/* 4.0 The connection capacity */
if $D(^u_{x_{k_u}}) < \tau_{(comm)}(x_{k_u}^u) + \tau_{(wait)}(x_{k_u}^u) + \tau_{(proc)}(x_{k_u}^u)$ **then**
    $(\theta^u_{x_{k_u}})$ = alarm
    Forward $x_{k_u}^u \Rightarrow$ Cloud node
end if

4.1. Feature and Database Determination

The data used for SVM classification algorithms are real open-source data for scientific research purposes; here, a dataset of heart attack data from [31,32] is used for the SVM algorithm as a critical application, taking into account many parameters, such as the type of chest pain, blood pressure, and cholesterol levels. Using the dataset, the parameters were identified as symptoms for each of the diseases, which helped determine the thresholds for SVM in Algorithm 1 to prioritize tasks.

On the other hand, the datasets used in Algorithm 2 are obtained through our experiments. The used parameters are extracted from the tests in the iFogSim simulator and input into SVM (Algorithm 2). The used parameters include MIPS, RAM, storage, the number of CPUs, connection features, and the available resources. These parameters are used to train the data to classify the FNs.

Both algorithms incorporate these carefully selected features to ensure accurate task classification. We employed five-fold cross-validation for robust model validation and to confirm the effectiveness of these features in practical scenarios.

4.2. New Kernel Design and Margin Maximization

It is commonly recognized that conventional kernels (e.g., linear, quadratic, etc.) cannot effectively be applied to ubiquitous applications [30]. Therefore, in this section, we develop a new type of kernel for enhancing classification accuracy. Conventional kernels and cross-correlation kernels are fused, reflecting symmetric features and anti-symmetric features, respectively, to enhance classification accuracy.

As mentioned previously, three classes, normal, medium, high, represented by C_0, C_1, C_2, respectively, are developed. For each class, q_{max} tasks are considered for the supervision of each class, derived as $X_{C_i} = \{x_{1,C_i}, \ldots, x_{q_{max},C_i}\}$, $i = \{0,1,2\}$. Suppose that L features are considered for each task; then, the similarity between tasks x_{q,C_i} and x_{q',C_j}, $i, j = 0, 1, 2$, $S_{x_{q,C_i}}^{x_{q',C_j}}$, is defined as a cross-correlation [30]:

$$S_{x_{q,C_i}}^{x_{q',C_j}}(l) = \sum_{n=-\infty}^{+\infty} x_{q,C_i}(n) \cdot x_{q',C_j}(n-l) \tag{17}$$

In general, tasks classified in the same class should gain higher value of $S_{x_{q,C_i}}^{x_{q',C_j}}$, with higher similarity compared with tasks in different classes.

Similarly, convolution and auto-correlation (auto-correlation only for tasks in the same class) [30], representing reversed similarity and self-similarity, respectively, are derived as

$$R_{x_{q,C_i}}^{x_{q',C_j}}(l) = \sum_{n=1}^{L} x_{q,C_i}(l) \cdot x_{q',C_j}(l-n) \tag{18}$$

$$SS_{x_{q,C_i}}^{x_{q',C_i}}(l) = \sum_{n=1}^{L} x_{q,C_i}(n) \cdot x_{q',C_i}(n-l) \tag{19}$$

By nature, cross-correlation reflects the similarity of two tasks, while convolution enhances the accuracy of similarity determination, reversely. Auto-correlation is also adopted, further improving classification performance in sensitivity and specificity.

In this paper, the information obtained by $S_{x_{q,C_i}}^{x_{q',C_j}}$, $R_{x_{q,C_i}}^{x_{q',C_j}}$, and $SS_{x_{q,C_i}}^{x_{q',C_i}}$ is used to model the kernels of task classification, which are derived based on the kernel matrix:

$$\mathbf{K}_{q,q'}^{S} = \begin{bmatrix} K_{1,1}^{S} & \cdots & K_{1,3q_{max}}^{S} \\ \vdots & \ddots & \vdots \\ K_{3q_{max},1}^{S} & \cdots & K_{3q_{max},3q_{max}}^{S} \end{bmatrix} \tag{20}$$

$$\mathbf{K}_{q,q'}^{R} = \begin{bmatrix} K_{1,1}^{R} & \cdots & K_{1,3q_{max}}^{R} \\ \vdots & \ddots & \vdots \\ K_{3q_{max},1}^{R} & \cdots & K_{3q_{max},3q_{max}}^{R} \end{bmatrix} \tag{21}$$

$$\mathbf{K}_{q,q'}^{i,SS} = \begin{bmatrix} K_{1,1}^{i,SS} & \cdots & K_{1,q_{max}}^{i,SS} \\ \vdots & \ddots & \vdots \\ K_{q_{max},1}^{i,SS} & \cdots & K_{q_{max},q_{max}}^{i,SS} \end{bmatrix}, i = \{0,1,2\} \tag{22}$$

where

$$K_{q,q'}^{S} = \sum_{n=1}^{L} \omega_S(n) \cdot S_{x_{q,C_i}}^{x_{q',C_j}}(n) \tag{23}$$

$$K_{q,q'}^{R} = \sum_{n=1}^{L} \omega_R(n) \cdot R_{x_{q,C_i}}^{x_{q',C_j}}(n) \tag{24}$$

$$K_{q,q'}^{i,SS} = \sum_{n=1}^{L} \omega_{SS}(n) \cdot SS_{x_{q,C_i}}^{x_{q',C_i}}(n), i = \{0,1,2\} \tag{25}$$

The weighting of features, i.e., $\{\omega_S(n), \omega_R(n), \omega_{SS}(n)\}$, should be determined according to the demand of applications. For instance, weighting for the feature related to "patients' medical record" should be designed as relatively higher, to potentially increase the specificity of defined critical tasks classified with high priority.

Mercer kernels (i.e., inner product kernels, $K^S_{q,q'}$, $K^R_{q,q'}$, $K^{SS}_{q,q'}$) obey the Mercer theorem, which satisfies the symmetric and positive semi-definite requirements [30]. Therefore, the kernel for task classification is defined as the sum of the inner product kernels:

$$K^o_{q,q'} = \mu_S \cdot K^S_{q,q'} + \mu_R \cdot K^R_{q,q'} + \sum_{i=0}^{2} v_{i,SS} \cdot K^{i,SS}_{q,q'} \tag{26}$$

where $\{\mu_S, \mu_R, \mu_{SS}, v_{i,SS}\}, i = \{0,1,2\}$, are weights for the defined inner product kernels. $K^o_{q,q'}$, as the sum of Mercer kernels, is also a Mercer kernel, which fuses the strengths of cross-correlation, convolution, and auto-correlation for enhancements in overall classification accuracy. The maximum margin function of $K^o_{q,q'}$ is derived as

$$\max \tilde{M}(\boldsymbol{\alpha}) = \sum_{q=1}^{q_{max}} \alpha_q + \frac{1}{2} \sum_{q=1}^{q_{max}} \sum_{q'=1}^{q_{max}} \alpha_q \alpha_{q'} y_q y_{q'} K^o_{q,q'} \tag{27}$$

subject to

$$\forall q \in [1, q_{max}], \alpha_q \geq 0 \tag{28a}$$

$$\sum_{q=1}^{q_{max}} \alpha_q y_q = 0 \tag{28b}$$

where α is the Lagrange multiplier and $y \in \{0,1\}$ is the output of classification, with 0 and 1 representing the binary side of each classifier. $\tilde{M}(\boldsymbol{\alpha})$ is developed for all the classifiers and its maximization optimizes overall accuracy in task classification.

Figure 3 clarifies the margins to detect one of the classified (high, medium, and low) cases, when a new notification is generated. The "high" case is highlighted in the middle of the figure with blue color; it represents the matching between high-risk symptoms which are connected to the patient's medical history (patient profile). On the other hand, the other two cases, "medium" and "low", highlighted in red color on either side of the blue zone, reflect either that the present symptoms are not dangerous under normal limits or that the following notifications are normal for those without a previous history of diseases.

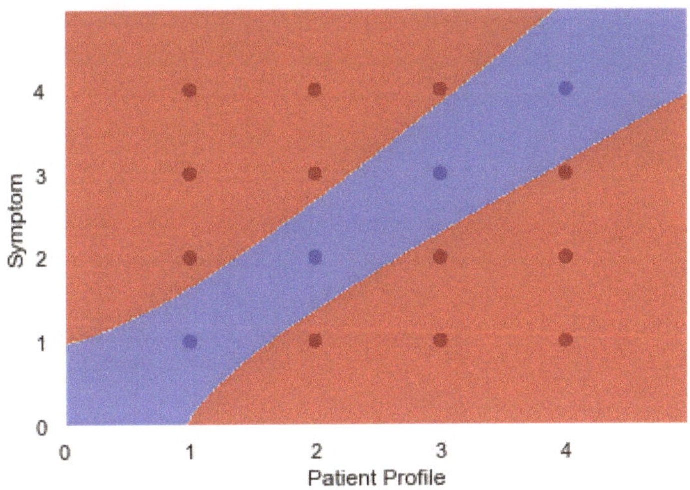

Figure 3. Training dataset and margins for the SVM-based task-scheduling algorithm.

5. Task Scheduling Based on Resource Mapping

In this section, we delve into the intricate process of task scheduling through resource mapping. We introduce three pivotal algorithms that collaborate to ensure efficient task prioritization, resource classification, and the effective mapping of tasks to the most suitable resources. Algorithm 1 focuses on determining the priority of tasks based on various factors, including patient profiles and symptoms. Algorithm 2 classifies fog nodes (FNs) based on their capacity and availability, setting the stage for optimal task allocation. Finally, Algorithm 3 oversees the mapping and offloading of tasks to the appropriate fog or cloud nodes, ensuring that each task is handled by the most fitting resources. Figure 4 shows the collaborative execution and compatibility of these algorithms to serve the proposed priority-based task-scheduling and resource allocation framework.

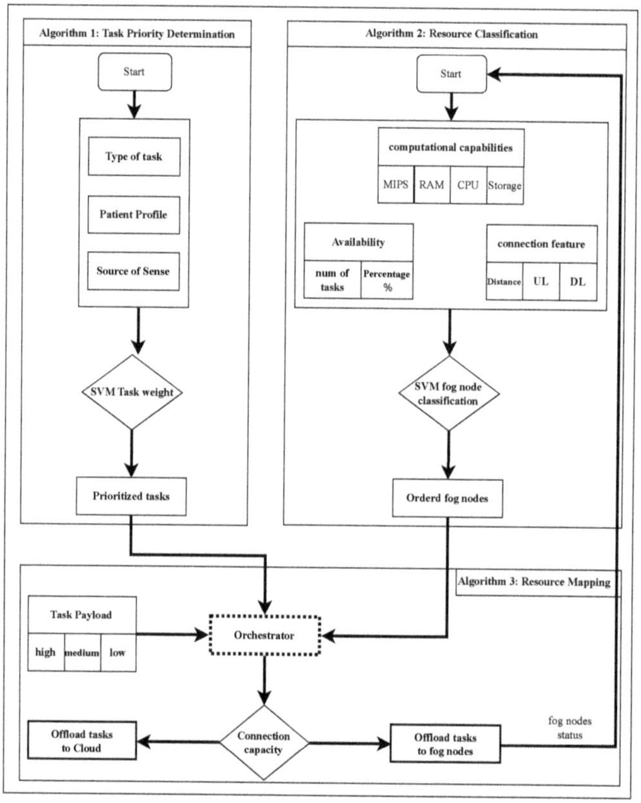

Figure 4. Flowchart showing collaborative execution of the three algorithms.

5.1. Task Priority Determination Algorithm

The task priority determination algorithm (Algorithm 1) contains the following steps, where the orchestrator is responsible for extracting the information from each incoming task:

1. Algorithm 1 checks the values of two fields (patient profile and symptom) in addition to other parameters, such as the payload of the task. If the two values of patient profile and symptom are equal, the SVM_weight of the task is high. On the contrary, if the values of patient profile and symptom are not equal, the SVM_weight of the task is medium. The case of SVM_weight equal to low is when the patient's health record is labeled healthy and the symptom field contains vital indicators in the normal limits.
2. The task's priority value is assigned based on the previously mentioned values.
3. The value of the field type of task is assigned based on the task's priority value.

4. The tasks are ordered in descending order based on their priority.
5. The prioritized-labeled tasks are sent to the orchestrator to be distributed to the proper FNs.

5.2. Resource Classification Algorithm

The resource classification algorithm (Algorithm 2) is a recursive procedure responsible for classifying FNs based on their capacity and availability and includes the following steps:

1. The FN computational capacity, which includes MIPS, RAM, storage, and the number of CPUs and their capacity, is extracted.
2. The topology of the service area is scanned to determine the characteristics of the connection, including the uplink/downlink bandwidth and the distance between the FNs and the devices that should connect to them; this distance is divided into three levels (near, medium, and far) based on the area where the device is located.
3. The FN sends the processing occupancy percentage, i.e., the volume of resources occupied in favor of processing tasks and the percentage of resources available to process new tasks.
4. According to the previous parameters, using the SVM algorithm, the FNs are classified and ordered in descending order into three levels: high, medium, and low.
5. The order of the classified FNs is sent to the orchestrator.

The output of Algorithm 2 is a set of FNs classified and ordered based on their capacity and availability. The capacity is considered a fixed attribute related to an FN's physical characteristics. In contrast, availability is treated as a dynamic attribute, reflecting the current resource usage within each FN. Algorithm 2 performs periodic assessments of the FNs, considering not only their computational capacity and connectivity features but also the processing occupancy percentage. The resulting order of these classified FNs is then communicated to the orchestrator for further task distribution and resource mapping processes.

5.3. Resource-Mapping Algorithm

Algorithm 3 involves the orchestrator mapping tasks to the appropriate FNs based on the classifications provided by Algorithms 1 and 2 through the following steps:

1. The orchestrator receives the classified fog nodes from Algorithm 2.
2. The orchestrator receives the prioritized tasks from Algorithm 1.
3. It checks the value of the payload field and assigns it the label high or medium based on the SVM threshold.
4. The orchestrator maps and offloads tasks to the FNs or CNs based on priority and classification.
5. The orchestrator checks if the network connection capacity is sufficient to serve the incoming requests to meet the latency requirement. If not, the type of task ($\theta^{u}_{x_{k_u}}$) field is labeled with an alarm and forwarded to the cloud node.

In summary, the integrated use of these algorithms forms an effective strategy for managing task scheduling and resource mapping in complex computing environments. They collectively enhance resource utilization, prioritize critical tasks, and ensure optimal task distribution, leading to improved system performance and efficiency.

5.4. Complexity Analysis

The computational complexity of the three algorithms is as follows:

- Algorithm 1 task priority determination algorithm: The complexity of this algorithm is primarily dependent on the number of tasks. If N represents the total number of tasks, then the complexity is $O(N)$, as each task requires a constant amount of time for processing.
- Algorithm 2 resource classification algorithm: The complexity is influenced by the number of fog nodes, denoted by M. Since each node is classified independently, the algorithm exhibits linear complexity, $O(M)$.

- Algorithm 3 resource-mapping algorithm: This algorithm combines aspects of both task prioritization and resource classification. With N tasks and M fog nodes, the worst-case complexity could be $O(N \times M)$, particularly in scenarios where each task must be considered for every node.

5.5. *Offloading Scheme*

This section explains the collaborative execution of Algorithms 1–3 in order to effectively offload the ordered tasks to the corresponding fog nodes and cloud nodes.

As mentioned in Section 5, the prioritized task list from Algorithm 1 and the classified FNs from Algorithm 2 are then used as input for Algorithm 3. In this algorithm, the orchestrator maps the highest-priority tasks to the available FNs with the highest capacity. Tasks with medium priority are assigned to FNs classified as having medium capacity, and low-priority tasks are offloaded to FNs with a low classification.

The task distribution process is a continuous and dynamic operation managed by Algorithm 3. The orchestrator periodically scans all existing FNs, assessing their current capacity and availability. After each assessment round, the FNs are reordered based on their updated capacity and availability status, ensuring the most efficient utilization of resources for processing the remaining tasks.

Figure 5 shows the change in the order of the FNs after each round of scanning, where the FNs that were originally classified as high are placed at the end of the ordered FN list; this is due to the lack of available resources in these FNs, as these FNs are busy handling other tasks. The other FNs that were originally labeled medium are moved to be classified as high among the available FNs. Figure 5 is plotted to clarify the reordering and transformation process of the FNs and the received tasks. In Figure 5, state = 1 shows that the high-priority tasks were sent to the FNs labeled high, e.g., tasks ID = 14, ID = 75, and ID = 1 were assigned to FNs ID = 14, ID = 2, ID = 10, respectively, and the tasks with medium priority were sent to the FNs labeled medium. Tasks and FNs with a low classification are handled in the same manner. In the next round, as shown in Figure 5 (state = state + 1), we can notice that FN ID-14 is placed with the low-class nodes based on its weak availability to receive and process new tasks. In the same figure, we notice that there is a new FN (ID-13) listed in the high class due to its capacity and availability and that FN ID-23, which was primarily classified as a medium-capacity FN, has been reallocated in a new spot in the high-capacity class.

Resource classifying 'state = 1'

Fog ID	14	2	10	23	7	9	18	5	11
Class	High			Med			Low		
Capacity	high	high	high	med	med	med	low	low	low
Availability	90%	80%	80%	92%	80%	75%	90%	70%	70%

Tasks classifying 'state = 1'

Task ID	55	75	22	1	85	36	16	49	8	29	77	48
Class	High				Med				Low			

Resource mapping & allocating 'state = 1'

Fog ID	14	2	10	23	7	9	18	5	11
Task ID	55	1	75	85	36	49	8	29	77

Resource classifying 'state = state + 1'

Fog ID	13	9	23	7	10	18	2	5	14
Class	High			Med			Low		
Capacity	high	med	med	med	high	low	high	low	high
Availability	80%	92%	80%	80%	20%	90%	40%	70%	5%

Tasks classifying 'state = state + 1'

Task ID	22	18	45	17	16	54	58	86	48	24	62	20
Class	High				Med				Low			

Resource mapping & allocating 'state = state + 1'

Fog ID	13	9	23	7	10	18	2	8	14
Task ID	22	18	45	16	54	58	48	24	62

Figure 5. Resource allocation recursive procedure.

6. Performance Analysis

The performance of the proposed algorithm was evaluated by simulating task scheduling and resource mapping and allocating in an iFogSim simulator. In our simulations, the results of our algorithm were compared to those of the widely used first-come, first-served (FCFS) algorithm. FCFS has been well acknowledged in recent works, and its framework represents a widely represented state-of-the-art method for network-level resource mapping and allocation [33–35]. The simulated scenarios considered modifying the number of clusters, FNs per cluster, and their connected IoT devices. The values of the parameters (CPU, RAM, bandwidth, etc.) of the fog devices and the tasks were selected randomly among certain groups of simulation settings, as tabulated in Table 2. A tree-based network topology was used in the simulated scenarios, where the number of fog devices and related sensors is equal in each cluster. As a starting stage, many bio-potential signals from patients were frequently captured and analyzed. Then, we compared the analyzed data with the database that contains the health records of the patients in order to detect any critical bio-potential.

Table 2. Summary of experimental configuration.

Element	Parameter	Units	Value
Cloud	CPU	MIPS	44,800
	RAM	MB	40,000
	Uplink	bytes/ms	20,000
	Downlink	bytes/ms	20,000
Fog device	CPU	MIPS	{2048, 1024, 768, 512, 256}
	RAM	MB	{2048, 1024, 768, 512, 256}
	Uplink	bytes/ms	{8000, 4000, 2000}
	Downlink	bytes/ms	{8000, 4000, 2000}
Task	CPU length	MIPS	{2000, 1000, 700, 500, 200}
	Network length	bytes	{4000, 2000, 1000}

6.1. Execution Time

Figure 6 presents the difference between the proposed task scheduling/resource mapping and allocation and the built-in scheduling algorithm. We can notice that in the case of Figure 6a, which has one cluster, the difference in execution time between the two algorithms remains close when the number of tasks is 40 or 80, respectively; when the number of tasks becomes large (120 tasks), the difference between the two execution times in the case of one cluster becomes clear in favor of our proposed algorithm.

In the four cases (a, b, c, and d) illustrated in Figure 6, we can notice that increasing the number of the tasks per cluster, in the case of the built-in algorithm, leads to an exponential increase in execution time. This increase in execution time is clearly visible when there is a large number of tasks, in contrast to the proposed algorithm, which keeps the increase in execution time in direct proportion to the increase in the number of tasks per cluster. The proposed algorithm is able to achieve an average execution time for critical tasks of 0.2393 ms, and for non-critical/normal task, it achieves an average execution time equal to 0.5001 ms. In the case of utilizing the FCFS algorithm with the same architecture, we can notice that the average execution times for the critical tasks and normal tasks are 0.9162 ms and 1.0419 ms, respectively. Hence, the proposed algorithm is able to achieve better execution time for all cases (critical/normal) compared with the FCFS algorithm, which has almost similar average execution time values.

6.2. Latency

Latency is one of the main KPIs (key performance indicators) to be considered when implementing a real-time healthcare system and has to be reduced to achieve the required high efficiency. In the fog–cloud architecture, using FNs reduces latency and enhances the overall execution time by processing tasks in the FNs locally, utilizing the available resources and decreasing the number of tasks that should be transmitted to and handled by the cloud. For the purpose of emphasizing the efficiency of the proposed algorithm, a comparison of latency is evaluated in both cloud-only and fog–cloud architectures, where the variable factor is the number of connected devices. In the cloud-only architecture, we can notice that the increase in the number of sensors directly leads to a steady increase in latency. Moreover, when the number of connected devices is above 45, the latency starts increasing in a more rapid way from 224.91 ms and reaches 331.81 ms when the connected devices are 60, as illustrated in Figure 7. Contrarily, in the fog–cloud architecture, increasing the number of devices has a limited effect on the achieved delay. When varying the number of connected devices from 20 to 60, the delay limit only varies between 11.4 ms and

23.06 ms, achieving an almost 90% reduction in delay compared with the cloud-only architecture, as highlighted in Figure 7.

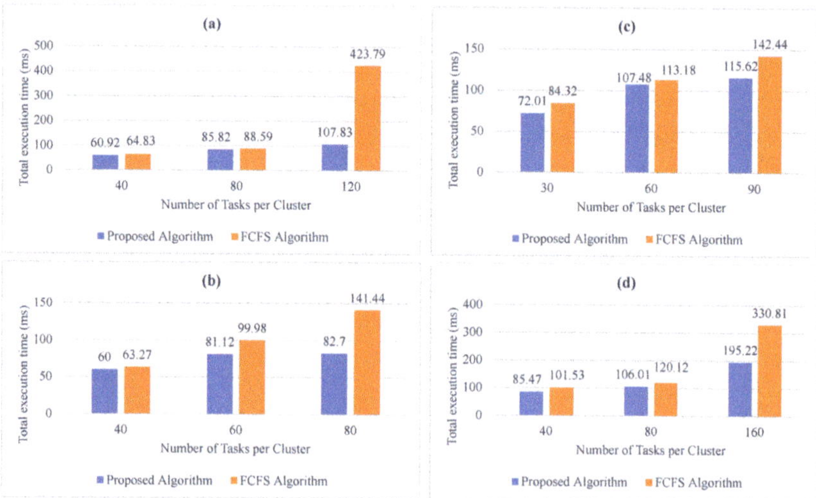

Figure 6. Total execution time in different clusters: (**a**) one cluster; (**b**) two clusters; (**c**) three clusters; (**d**) four clusters.

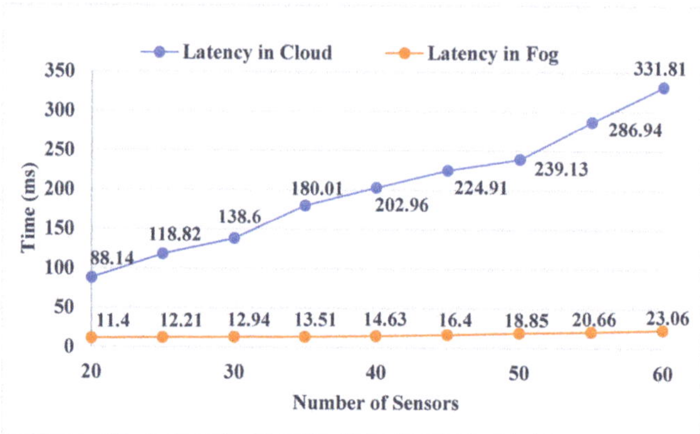

Figure 7. Comparison of latency.

6.3. Network Utilization

Regarding the network utilization efficiency aspect, which is measured in KByte per second, Figure 8 highlights the total network usage for four different cases (one cluster, two clusters, three clusters, four clusters). Each cluster is tested for three various groups of tasks (40, 80, 120), applied with FCFS-based algorithms (averaged based on [34,35], representing recent works related to machine learning-based FCFS and hierarchical FCFS, respectively) and the proposed algorithm. In the case of one, two, and three clusters, network usage in both scenarios is almost convergent, and our proposal attains lower network usage.

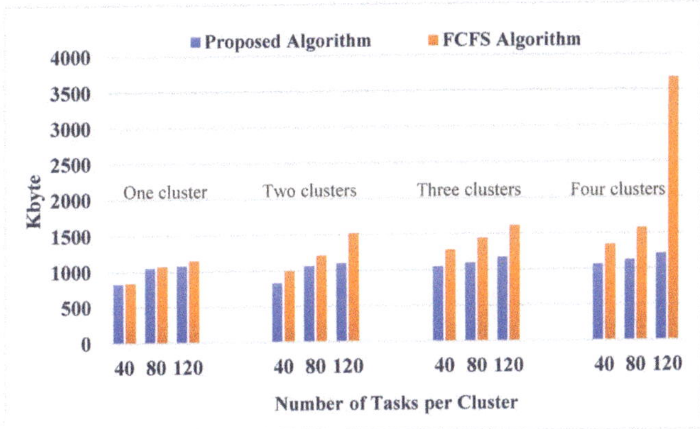

Figure 8. Network usage among different clusters.

Specifically, our proposed algorithm achieves better results when the number of tasks and clusters is high. In the last case, the FCFS algorithm results in about four times larger network usage compared with our work.

Consequently, the increase in network usage affects the total execution time, waiting time, and communication time. Furthermore, it also results in increased costs of network usage and power consumption. Concretely, comparing the results achieved in [34] for solving task scheduling with a time–cost-aware scheduling (TCaS) algorithm and the performance of our proposed solution, it can be seen that when the number of tasks is more than 100 in each cluster with five FNs, our proposed algorithm achieves better execution time. When the number of tasks is 120, 160, or 180, our solution achieves a delay of 80 ms, 83 ms, or 105 ms, whereas TCaS produces a delay of almost 150 ms, 195 ms, or 230 ms, respectively. Additionally, authors of the work in [35] proposed a method to reduce latency and network consumption in a remote pain-monitoring system. Comparing the achieved results, the delay obtained with our algorithm when the number of sensors is 20, 30, 40, or 50 in a cloud-only architecture is 88 ms, 138 ms, 202 ms, or 239 ms compared with 215 ms, 225 ms, 233 ms, or 238 ms in the cited paper, respectively. While in a fog–cloud architecture, our solutions produce slightly better values, especially when the number of sensors is high.

7. Conclusions

In this paper, we propose a fog–cloud hierarchical task-scheduling scheme for e-Health applications. Different from conventional task scheduling, we formulate features of tasks comprehensively, fusing network-level and service-level parameters simultaneously, which are further considered in the proposed support vector machine (SVM)-based task classification algorithm. Classified tasks are assigned priorities, giving guidance to be allocated to proper fog/cloud nodes, for network utilization efficiency maximization and overall latency reduction. In particular, the proposed task-scheduling scheme is capable of effectively achieving latency minimization for critical tasks as defined based on the demand of both networks and services. Simulation results show that the proposed algorithm was able to minimize the total execution time for all tasks and especially for the critical ones. The integration of SVM enhanced the latency and network usage in parallel with the increased number of tasks.

The current use of SVM algorithms limits the model's performance in complex datasets with higher dimensions to comprehensively serve ubiquitous healthcare applications. Future work will focus on integrating advanced algorithms, including deep learning and hybrid models combining SVM with other techniques, to discuss their feasibility with comprehensive consideration of computational complexity and algorithmic efficiency simultaneously.

Author Contributions: Conceptualization, A.R.; methodology, A.A. and H.R.C.; software, A.A.; validation, A.A. and H.R.C.; formal analysis, all authors; investigation, all authors; data curation, all authors; writing—original draft preparation, A.A. and H.R.C.; writing—review and editing, A.R. and R.L.A.; supervision, A.R. and R.L.A.; project administration, A.R. and R.L.A.; funding acquisition, A.R. and R.L.A. All authors have read and agreed to the published version of the manuscript.

Funding: This work is funded by FCT/MCTES through national funds and when applicable co-funded EU funds under the project UIDB/50008/2020-UIDP/50008/2020.

Data Availability Statement: The data presented in this study are available upon request from the corresponding author.

Conflicts of Interest: The authors declare no conflicts of interest.

Abbreviations

The following abbreviations are used in this manuscript:

A^c_M	Cloud node
A^f_N	Fog node
U_{IoT}	IoT device
$x^u_{K_u}$	Task
$C(x^u_{k_u})$	Cost
$P(x^u_{k_u})$	Priority of a task
$D(x^u_{k_u})$	Latency of a task
$L(x^u_{k_u})$	Payload of a task
$\theta(x^u_{k_u})$	Type of task
$\beta(x^u_{k_u})$	Patient profile
$\mu(x^u_{k_u})$	Patient preliminary symptoms
δ^f_u	Distance between an IoT device and a fog node
δ^c_u	Distance from a fog node to the cloud
$\tau_{(comm)}$	Communication time of a task
$\tau_{(wait)}$	Waiting time of a task
$\tau_{(proc)}$	Processing time of a task
$t_{(trans)}$	Transmission delay
R_{cloud}, R_{fog}	Data rate of a fog node and that of a cloud node
$W_{a^c_M}, W_{a^f_N}$	Link bandwidth of a fog node and that of a cloud node
$SINR$	Signal-to-interference-plus-noise ratio
$r^{UL}_{u_{(IoT)}}$	Uplink transmitting rate
$\psi(x^u_{k_u})$	Processing time of a task
$\Gamma(A^c_n), \Gamma(A^f_m)$	Computing capacity of a fog node and that of a cloud node
T^z_u	Total number of time slots in a processing node
$rq^u_x(t)$	Required resources for a task in a given slot time

References

1. Liu, J.; Ahmed, M.; Mirza, M.A.; Khan, W.U.; Xu, D.; Li, J.; Aziz, A.; Han, Z. RL/DRL Meets Vehicular Task Offloading Using Edge and Vehicular Cloudlet: A Survey. *IEEE Internet Things J.* **2022**, 9, 8315–8338. [CrossRef]
2. Chi, H.R.; Domingues, M.F.; Radwan, A. QoS-aware Small-Cell-Overlaid Heterogeneous Sensor Network Deployment for eHealth. In Proceedings of the 2020 IEEE SENSORS, Rotterdam, The Netherlands, 25–28 October 2020; pp. 1–4. [CrossRef]
3. Chi, H.R. Editorial: Edge Computing for the Internet of Things. *J. Sens. Actuator Netw.* **2023**, 12, 17. [CrossRef]
4. Kashani, M.H.; Mahdipour, E. Load Balancing Algorithms in Fog Computing. *IEEE Trans. Serv. Comput.* **2023**, 16, 1505–1521. [CrossRef]
5. Chi, H.R.; Domingues, M.d.F.; Zhu, H.; Li, C.; Kojima, K.; Radwan, A. Healthcare 5.0: In the Perspective of Consumer Internet-of-Things-Based Fog/Cloud Computing. *IEEE Trans. Consum. Electron.* **2023**, 1. [CrossRef]
6. Radwan, A.; Chi, H.R. Towards Cell-Free Networking: Analytical Study of Ultra-Dense On-Demand Small Cell Deployment for Internet of Things. In Proceedings of the 2023 International Wireless Communications and Mobile Computing (IWCMC), Marrakesh, Morocco, 19–23 June 2023; pp. 1202–1207. [CrossRef]
7. Strumberger, I.; Tuba, M.; Bacanin, N.; Tuba, E. Cloudlet Scheduling by Hybridized Monarch Butterfly Optimization Algorithm. *J. Sens. Actuator Netw.* **2019**, 8, 44. [CrossRef]

8. Mattia, G.P.; Beraldi, R. On real-time scheduling in Fog computing: A Reinforcement Learning algorithm with application to smart cities. In Proceedings of the 2022 IEEE International Conference on Pervasive Computing and Communications Workshops and other Affiliated Events (PerCom Workshops), Pisa, Italy, 21–25 March 2022; pp. 187–193. [CrossRef]
9. AlZailaa, A.; Chi, H.R.; Radwan, A.; Aguiar, R. Low-Latency Task Classification and Scheduling in Fog/Cloud based Critical e-Health Applications. In Proceedings of the ICC 2021-IEEE International Conference on Communications, Montreal, QC, Canada, 14–23 June 2021; pp. 1–6. [CrossRef]
10. Semmoud, A.; Hakem, M.; Benmammar, B.; Charr, J.C. Load balancing in cloud computing environments based on adaptive starvation threshold. *Concurr. Comput. Pract. Exp.* **2020**, *32*, e5652. [CrossRef]
11. Benblidia, M.A.; Brik, B.; Merghem-Boulahia, L.; Esseghir, M. Ranking Fog nodes for Tasks Scheduling in Fog-Cloud Environments: A Fuzzy Logic Approach. In Proceedings of the 2019 15th International Wireless Communications & Mobile Computing Conference (IWCMC), Tangier, Morocco, 24–28 June 2019; pp. 1451–1457. [CrossRef]
12. Abdel-Basset, M.; El-Shahat, D.; Elhoseny, M.; Song, H. Energy-Aware Metaheuristic Algorithm for Industrial-Internet-of-Things Task Scheduling Problems in Fog Computing Applications. *IEEE Internet Things J.* **2021**, *8*, 12638–12649. [CrossRef]
13. Hosseini, E.; Nickray, M.; Ghanbari, S. Optimized task scheduling for cost-latency trade-off in mobile fog computing using fuzzy analytical hierarchy process. *Comput. Netw.* **2022**, *206*, 108752. [CrossRef]
14. Tuli, S.; Basumatary, N.; Gill, S.S.; Kahani, M.; Arya, R.C.; Wander, G.S.; Buyya, R. HealthFog: An ensemble deep learning based Smart Healthcare System for Automatic Diagnosis of Heart Diseases in integrated IoT and fog computing environments. *Future Gener. Comput. Syst.* **2020**, *104*, 187–200. [CrossRef]
15. Azizi, S.; Shojafar, M.; Abawajy, J.; Buyya, R. Deadline-aware and energy-efficient IoT task scheduling in fog computing systems: A semi-greedy approach. *J. Netw. Comput. Appl.* **2022**, *201*, 103333. [CrossRef]
16. Kanbar, A.B.; Faraj, K. Region aware dynamic task scheduling and resource virtualization for load balancing in IoT–Fog multi-cloud environment. *Future Gener. Comput. Syst.* **2022**, *137*, 70–86. [CrossRef]
17. Okegbile, S.D.; Maharaj, B.T.; Alfa, A.S. A Multi-User Tasks Offloading Scheme for Integrated Edge-Fog-Cloud Computing Environments. *IEEE Trans. Veh. Technol.* **2022**, *71*, 7487–7502. [CrossRef]
18. Mutlag, A.A.; Khanapi Abd Ghani, M.; Mohammed, M.A.; Maashi, M.S.; Mohd, O.; Mostafa, S.A.; Abdulkareem, K.H.; Marques, G.; de la Torre Díez, I. MAFC: Multi-Agent Fog Computing Model for Healthcare Critical Tasks Management. *Sensors* **2020**, *20*, 1853. [CrossRef] [PubMed]
19. Chakraborty, C.; Mishra, K.; Majhi, S.K.; Bhuyan, H.K. Intelligent Latency-Aware Tasks Prioritization and Offloading Strategy in Distributed Fog-Cloud of Things. *IEEE Trans. Ind. Inform.* **2023**, *19*, 2099–2106. [CrossRef]
20. Gupta, S.; Iyer, S.; Agarwal, G.; Manoharan, P.; Algarni, A.D.; Aldehim, G.; Raahemifar, K. Efficient Prioritization and Processor Selection Schemes for HEFT Algorithm: A Makespan Optimizer for Task Scheduling in Cloud Environment. *Electronics* **2022**, *11*, 2557. [CrossRef]
21. Alatoun, K.; Matrouk, K.; Mohammed, M.A.; Nedoma, J.; Martinek, R.; Zmij, P. A Novel Low-Latency and Energy-Efficient Task Scheduling Framework for Internet of Medical Things in an Edge Fog Cloud System. *Sensors* **2022**, *22*, 5327. [CrossRef]
22. Khosroabadi, F.; Fotouhi-Ghazvini, F.; Fotouhi, H. SCATTER: Service Placement in Real-Time Fog-Assisted IoT Networks. *J. Sens. Actuator Netw.* **2021**, *10*, 26. [CrossRef]
23. Nagarajan, S.M.; Devarajan, G.G.; Mohammed, A.S.; Ramana, T.V.; Ghosh, U. Intelligent Task Scheduling Approach for IoT Integrated Healthcare Cyber Physical Systems. *IEEE Trans. Netw. Sci. Eng.* **2022**, *10*, 2429–2438. [CrossRef]
24. Chen, J.; He, Y.; Zhang, Y.; Han, P.; Du, C. Energy-aware scheduling for dependent tasks in heterogeneous multiprocessor systems. *J. Syst. Archit.* **2022**, *129*, 102598. [CrossRef]
25. Abdelmoneem, R.M.; Benslimane, A.; Shaaban, E. Mobility-aware task scheduling in cloud-Fog IoT-based healthcare architectures. *Comput. Netw.* **2020**, *179*, 107348. [CrossRef]
26. Ali, I.M.; Sallam, K.M.; Moustafa, N.; Chakraborty, R.; Ryan, M.; Choo, K.K.R. An Automated Task Scheduling Model Using Non-Dominated Sorting Genetic Algorithm II for Fog-Cloud Systems. *IEEE Trans. Cloud Comput.* **2022**, *10*, 2294–2308. [CrossRef]
27. Cheikhrouhou, O.; Mershad, K.; Jamil, F.; Mahmud, R.; Koubaa, A.; Moosavi, S.R. A lightweight blockchain and fog-enabled secure remote patient monitoring system. *Internet Things* **2023**, *22*, 100691. [CrossRef]
28. Tong, Z.; Deng, X.; Ye, F.; Basodi, S.; Xiao, X.; Pan, Y. Adaptive computation offloading and resource allocation strategy in a mobile edge computing environment. *Inf. Sci.* **2020**, *537*, 116–131. [CrossRef]
29. Balevi, E.; Gitlin, R.D. Optimizing the number of fog nodes for cloud-fog-thing networks. *IEEE Access* **2018**, *6*, 11173–11183. [CrossRef]
30. Chui, K.T.; Tsang, K.F.; Chi, H.R.; Ling, B.W.K.; Wu, C.K. An Accurate ECG-Based Transportation Safety Drowsiness Detection Scheme. *IEEE Trans. Ind. Inform.* **2016**, *12*, 1438–1452. [CrossRef]
31. Ramani, P.; Pradhan, N.; Sharma, A.K. Classification Algorithms to Predict Heart Diseases—A Survey. In Proceedings of the Computer Vision and Machine Intelligence in Medical Image Analysis, Accra, Ghana, 29–31 May 2019; Gupta, M., Konar, D., Bhattacharyya, S., Biswas, S., Eds.; WikiCFP: Singapore, 2020; pp. 65–71.
32. Kumar, P.; Chauhan, R.; Stephan, T.; Shankar, A.; Thakur, S. A Machine Learning Implementation for Mental Health Care. Application: Smart Watch for Depression Detection. In Proceedings of the 2021 11th International Conference on Cloud Computing, Data Science and Engineering (Confluence), Noida, India, 28–29 January 2021; pp. 568–574. [CrossRef]

33. Mahmud, R.; Pallewatta, S.; Goudarzi, M.; Buyya, R. iFogSim2: An extended iFogSim simulator for mobility, clustering, and microservice management in edge and fog computing environments. *J. Syst. Softw.* **2022**, *190*, 111351. [CrossRef]
34. Sing, R.; Bhoi, S.K.; Panigrahi, N.; Sahoo, K.S.; Bilal, M.; Shah, S.C. EMCS: An Energy-Efficient Makespan Cost-Aware Scheduling Algorithm Using Evolutionary Learning Approach for Cloud-Fog-Based IoT Applications. *Sustainability* **2022**, *14*, 15096. [CrossRef]
35. Hassan, S.R.; Ahmad, I.; Ahmad, S.; Alfaify, A.; Shafiq, M. Remote Pain Monitoring Using Fog Computing for e-Healthcare: An Efficient Architecture. *Sensors* **2020**, *20*, 6574. [CrossRef]

sensors

MDPI

Article

An Adaptive Bandwidth Management Algorithm for Next-Generation Vehicular Networks

Chenn-Jung Huang [1,*], Kai-Wen Hu [2] and Hao-Wen Cheng [1]

1 Department of Computer Science & Information Engineering, National Dong Hwa University, Shoufeng, Hualien County 974301, Taiwan; 2586611811v@gmail.com
2 Lookout, Inc., Taipei 110207, Taiwan; drive55555@gmail.com
* Correspondence: cjhuang@gms.ndhu.edu.tw

Abstract: The popularity of video services such as video call or video on-demand has made it impossible for people to live without them in their daily lives. It can be anticipated that the explosive growth of vehicular communication owing to the widespread use of in-vehicle video infotainment applications in the future will result in increasing fragmentation and congestion of the wireless transmission spectrum. Accordingly, effective bandwidth management algorithms are demanded to achieve efficient communication and stable scalability in next-generation vehicular networks. To the best of our current knowledge, a noticeable gap remains in the existing literature regarding the application of the latest advancements in network communication technologies. Specifically, this gap is evident in the lack of exploration regarding how cutting-edge technologies can be effectively employed to optimize bandwidth allocation, especially in the realm of video service applications within the forthcoming vehicular networks. In light of this void, this paper presents a seamless integration of cutting-edge 6G communication technologies, such as terahertz (THz) and visible light communication (VLC), with the existing 5G millimeter-wave and sub-6 GHz base stations. This integration facilitates the creation of a network environment characterized by high transmission rates and extensive coverage. Our primary aim is to ensure the uninterrupted playback of real-time video applications for vehicle users. These video applications encompass video conferencing, live video, and on-demand video services. The outcomes of our simulations convincingly indicate that the proposed strategy adeptly addresses the challenge of bandwidth competition among vehicle users. Moreover, it notably boosts the efficient utilization of bandwidth from less crowded base stations, optimizes the fulfillment of bandwidth prerequisites for various video applications, and elevates the overall video quality experienced by users. Consequently, our findings serve as a successful validation of the practicality and effectiveness of the proposed methodology.

Keywords: electric vehicle; bandwidth allocation; video service; 6G; data mining; machine learning; optimization

Citation: Huang, C.-J.; Hu, K.-W.; Cheng, H.-W. An Adaptive Bandwidth Management Algorithm for Next-Generation Vehicular Networks. *Sensors* **2023**, *23*, 7767. https://doi.org/10.3390/s23187767

Academic Editors: Yichuang Sun, Haeyoung Lee and Oluyomi Simpson

Received: 17 August 2023
Revised: 5 September 2023
Accepted: 6 September 2023
Published: 8 September 2023

1. Introduction

As the greenhouse effect worsens, environmental regulations for vehicles are now becoming more stringent in numerous countries. In order to meet the increasingly stringent regulations on greenhouse gas emissions and to solve the problems associated with traditional engine vehicles, such as air pollution, climate change, and high fuel prices, electric vehicles (EV), which use green energy to reduce carbon emissions, have been recognized as the future alternative to traditional automobiles.

Affected by the COVID-19 epidemic that started in late 2019, academics and organizations have been turning to video conferencing to match the COVID-19 precautionary measures, as well as to improve efficiency and convenience. Video conferencing applications account for an increasing share of Internet applications and bandwidth usage [1]. Nowadays, video conferencing plays an important role in people's lives. As video conferencing requires real-time transmission, the issue of resource allocation of uplinks is of concern. If a large amount of data is uploaded at the same time during rush hours, the smoothness and stability of the video conferencing may be affected. In addition, multimedia streaming plays a key role in various emerging in-vehicle infotainment applications [2]. The rise of video service providers such as TikTok and YouTube has emphasized the popularity of video services in everyday life. Video traffic has been growing in recent years [3], and these video service providers offer video on demand (VoD) streaming services to users through websites, mobile apps, or social networks [4]. Live video has also become one of the most popular consumer contents, such as online education, trade shows, sports events, concerts, and video games [5]. With the advancement of self-driving technology, autonomous electric vehicles are bound to become the mainstream of human transportation in the future, which will enable people to spend their time in the vehicle on entertainment and office work.

Encoding plays a vital role in modern digital media, ensuring efficient and reliable transmission and playback of videos in various situations. Currently, a significant portion of videos use Advanced Video Coding (AVC) [6] as their chosen encoding method, which was introduced in 2003 and is one of the most commonly used video coding standards. It is widely applied in various applications, such as video conferencing, mobile services, and high-definition video storage [7]. While a subsequent development, namely, High-Efficiency Video Coding (HEVC) [8], did lead to a notable reduction in the bitrate for 4K videos, it still proved insufficient in terms of efficiency for 8K applications [9]. The latest international video coding standard, Versatile Video Coding (VVC) [10], was introduced in 2020. Compared to HEVC, VVC offers higher compression efficiency [11]. VVC adopts various new coding technologies [12]. Multiple coding unit partitions and numerous coding tools improve compression performance, but also greatly increase coding complexity [13]. There are already studies comparing the performance of different codecs. Menasri and Skoudarli [14] proposed a performance comparison of throughput between context-based adaptive binary arithmetic decoding processes adopted in the AVC, HEVC, and VVC. Bonnineau et al. [9] evaluated 8K videos using HEVC and VVC, noting that VVC resulted in an average bitrate reduction of around 41%. Choi [15] conducted a comparison of the complexity between HEVC and VVC, revealing that VVC's encoding time was up to 27 times greater than HEVC in certain instances. Belda et al. [16] compared the encoding time of VVC, AVC, and HEVC. Their experimental results demonstrated that the encoding delay of VVC was significantly higher than that of AVC and HEVC. Accordingly, the above-mentioned experimental results indicate that VVC is not suitable for real-time transmission application demands.

Commencing in 2019, the official deployment of 5G witnessed the incorporation of sub-6 GHz and millimeter wave (mmWave) frequencies [17]. In response to the dramatic increase in video traffic, several technologies were employed to alleviate traffic burdens. However, if all video requests are served through wired backhaul links during rush hours, it would burden the wireless transmission channels heavily, leading to capacity bottlenecks for wireless video traffic [18]. Moreover, traditional data transmission within networks faces several limitations, including elevated latency, significant packet loss, and network congestion. These challenges must be effectively addressed by the next generation of wireless networks. Liu et al. [19] introduced a dual-layer algorithm to tackle the transmission challenge within a mobile wireless-powered communication network. The algorithm's primary objective is to optimize throughput by strategically pairing the energy consumption of a single transmission with the energy harvesting probability. Zheng et al. [20] conducted a study on the radio frequency-powered ambient backscatter-assisted hybrid underlay cognitive radio network. An adjusted deep deterministic policy gradient algorithm was proposed to establish an effective policy for time scheduling and energy management, with the ultimate aim of optimizing long-term secondary throughput. Furthermore, they integrated convex optimization into this algorithm to accelerate convergence and identify the optimal solution. Mei et al. [21] successfully adapted to the dynamic nature of the system and upheld stability in connection to task queues and battery levels through the application of Lyapunov optimization theory. They formulated an algorithm with the primary objective of optimizing system throughput to its maximum capacity. Wang et al. [22] placed their emphasis on effective throughput as the primary performance metric. They not only derived the mathematical expression for effective throughput, but also posed an optimization challenge with the objective of maximizing it.

Beyond tackling the issues of the next-generation network by enhancing network architecture and optimizing algorithms for resource allocation, the forthcoming generation of networks will predominantly revolve around 6G technology. Thus, 6G promises to usher in an era characterized by continuous connectivity, immersive experiences, robust support for a multitude of simultaneous users, ultra-low latency, universal accessibility, unmatched data capacity, unwavering reliability, and stringent security measures [23]. In recent literature, Pei et al. [24] introduced several pivotal technologies aimed at facilitating the implementation of 6G. They anticipated that 6G networks would not only rely on conventional spectrum usage, but also explore previously uncharted frequency bands within the context of cellular communication standards. These unexplored bands specifically included the terahertz (THz) frequency band and visible light communication (VLC). Expanding on these emerging concepts, the investigation of the THz frequency band has garnered significant attention in recent years. This untapped realm holds the potential to reshape the landscapes of wireless communication. The THz band occupies a frequency range situated between microwave and infrared within the electromagnetic spectrum. Its frequencies span from 300 GHz to 3 THz, accompanied by wavelengths ranging from 1 mm to 0.1 mm [25]. In order to support potential applications related to future vehicles, vehicle communication demands higher reliability and lower latency for transmitting large amounts of data. The THz band can provide ultra-high-speed data transmission, large bandwidth, and extremely low latency [26].

Scholars have already conducted research on the applications of THz in vehicle communications. Li et al. [27] analyzed vehicular communications in a 300 GHz urban scenario, providing a detailed characterization of path loss, shadow fading, and other properties. Lin et al. [28] addressed vehicle tracking and resource allocation for THz vehicle-to-infrastructure communication networks, proposing a solution using the Unscented Kalman Filter. Moltchanov et al. [29] established a mathematical framework for comparing multi-hop relaying systems with antennas installed at different positions on vehicles using IEEE 802.15.3d parameters and 300 GHz propagation measurements. They claimed that placing the antenna on the windshield effectively reduces the sensitivity of the technology penetration rate and increases transmission coverage. However, THz suffers from severe attenuation caused by molecular absorption due to its wavelength size being similar to atmospheric particles like raindrops and dust [30]. Additionally, THz waves encounter spreading loss caused by electromagnetic wave diffusion in the medium. As the transmission distance increases, the path loss from absorption and diffusion becomes more significant [31], limiting the long-range transmission capability of THz.

To address these limitations, the dense deployment of base stations is required to extend coverage. Moreover, the expenses associated with THz equipment are substantial [32]. To extensively implement THz base stations, telecommunication providers would be required to make substantial financial investments. Researchers have proposed deploying base stations in different frequency bands to overcome the transmission challenges of high-frequency base stations. Kouzayha et al. [33] deployed THz base stations for high data rates and radio frequency base stations for coverage. Wang and Chun [34] introduced a hybrid network architecture that includes both THz and mmWave frequencies. Moltchanov et al. [35] reviewed communication network deployment, propagation, antennas, blockage, micro mobility, beam searching, traffic, and service models. They explored the deployment and transmission challenges of mmWave and THz base stations in urban environments, considering different system and environmental conditions along with base stations operating at THz, mmWave, and sub-6 GHz frequencies. Yin et al. [36] deployed mmWave base stations to overcome the limited bandwidth of traditional microwave frequencies. They densely deployed mmWave base stations in existing base stations to form a heterogeneous mmWave network.

Alongside the THz frequency range, another set of frequencies currently unutilized in existing communication standards is VLC. This alternative spectrum brings forth specific advantages over other communication technologies. Firstly, VLC is relatively safe for human eyes [37] and can be easily integrated with existing LED lighting infrastructure, such as streetlights, traffic lights, and vehicle headlights, making it a convenient communication transmitter [38]. VLC not only integrates communication and lighting, but also presents benefits such as low power consumption, the utilization of a license-free spectrum, strong security, and resilience to electromagnetic interference [39]. With its limited field of view and line-of-sight transmission, VLC is well-suited for high-speed short-range wireless communication [40]. This makes VLC particularly suitable for vehicle-to-vehicle (V2V) communication, especially for applications requiring highly secure, reliable, and low-latency communication between vehicles, such as vehicle platoons [41]. Recently, Aghaei et al. [42] undertook a comprehensive analysis that contrasted the distinctions and benefits of VLC and mmWave within V2V communication. They presented the received signal strengths of both systems, delved into channel characteristics for inter-vehicle communication, and evaluated the impact of vehicle density on communication efficacy. Kamiya et al. [43] showcased the effective capturing of VLC signals during vehicular movement at a velocity of 40 km/h. The transmission occurred via an LED array, with the signals being received by an image sensor employing a rolling shutter mechanism. This investigation affirmatively established the feasibility of receiving VLC signals while in motion. Yang et al. [44] delineated the challenges linked with vehicular VLC and proposed an inventive architecture featuring tracking and environment sensing capabilities. The results of their simulations provided validation for the effectiveness of the proposed VLC system, showcasing its potential to achieve a bit error rate below 10^{-4}, even when confronted with substantial interference from external lighting sources. Additionally, a multiple input and multiple output (MIMO) VLC system with custom-designed pin arrays and headlights achieved data rates of 336 Mbps and 362 Mbps at distances of 100 m during the day and night, respectively [45]. The above-mentioned studies verify the feasibility of VLC in next-generation vehicular communication applications.

Beyond the realm of 6G technology, the significance of edge computing has risen substantially within contemporary communication networks. This strategy, involving the proximity of computation and resources to users, harmonizes seamlessly with the high-speed and low-latency benefits inherent in 6G. It is foreseeable that within the landscape of vehicular networks in the years to come, the fusion of 6G technology and edge computing will engender heightened efficacy in data processing and real-time applications. The advancement of mobile edge computing has also inspired the development of edge transcoding technology, where service providers can distribute transcoding tasks from central servers to edge servers closer to users for video transcoding. Edge transcoders can convert high-bitrate video into low-bitrate versions for user selection, improving the performance of video services by reducing transmission delay [46]. Furthermore, lightweight edge computing servers can be deployed at edge nodes like base stations. Video files can be stored on these servers for users to download, and if THz base stations are used, video transmission can become more efficient and faster, benefiting the provision of high-quality video services and meeting user demands [47]. Moreover, user equipment's computing and storage capabilities have significantly improved, enabling their participation in edge computing, storage, and communication as well [48].

As the demand for future multimedia traffic continues to rise, bandwidth shortages become unavoidable during peak periods. At present, multiple research endeavors have introduced techniques to allocate bandwidth and improve mobile network throughput, aiming to tackle the challenge of allocating network resources efficiently among users. Yuan et al. [49] introduced an algorithm supporting mobile video streaming applications in heterogeneous wireless networks. Their research allocated bandwidth among multiple users based on user experience quality and mobility-related information, enhancing user experience quality through a push strategy using the HTTP/3 protocol. Liu et al. [50] designed a federated deep reinforcement learning-based video streaming scheduling algorithm. Their algorithm predicted reasonable bandwidth allocation weights based on the current player's state and information provided by servers, subsequently allocating available bandwidth. Tung and Gündüz [51] utilized deep neural networks for end-to-end compression and channel coding of video frames, optimizing video frame bandwidth allocation through reinforcement learning.

The advent of 6G technology holds the promise of significantly enhancing vehicular mobility. Vehicles operating at high speeds can leverage the capabilities of 6G connectivity to establish cooperative communication, thereby ensuring safe following distances and smooth traffic flow [52]. Consequently, as we enter the era of 6G networks, the optimization of bandwidth allocation among vehicles is poised to become a pivotal concern. In recent years, a plethora of studies have surfaced, introducing inventive bandwidth allocation schemes customized for multimedia applications within vehicular networks. Xiong et al. [53] proposed an algorithm based on predictions of the vehicular communication and edge computing network states. It optimizes communication resource allocation, transmission paths, and power consumption for computation modes. Zhang et al. [54] employed radar communication rates, bandwidth allocation, and base station selection as parameters, utilizing reinforcement learning techniques to handle uncertainty in vehicle movement and fluctuations in multimedia data volume. Yun et al. [55] presented a video streaming strategy tailored for mobility-aware vehicular networks, harnessing the power of deep reinforcement learning. Within this scheme, millimeter-wave base stations were harnessed for the delivery of videos to users. Additionally, it crafted a dynamic video delivery approach that intelligently determined the content, quality, and quantity of video chunks. Cheng et al. [56] devised a cost-efficient task processing scheme for a dual-band cooperative vehicular network. This scheme allowed tasks to be processed locally or offloaded to either the macro-cell base station or road-side units (RSU). By optimizing task scheduling, computation, and communication resource allocation, while taking into account the vehicle's sojourn time, they aimed to minimize the total cost, considering both energy consumption and latency. Jiang et al. [57] conducted a comprehensive review of resource allocation strategies used for video streaming in vehicular ad hoc networks. They also explained in detail the widely adopted and practical optimization tools. Furthermore, they summarized the technologies that enable video streaming over vehicular ad hoc networks, with a particular focus on the integration of video communication, caching, and computing. Dai et al. [2] introduced a mobile edge computing-based framework for adaptive-bitrate multimedia streaming within the internet of vehicles. They employed deep Q-learning (DQN) to enhance solutions by revisiting past experiences and updating Q-functions using gradients. They also presented an adaptive-quality-based chunk selection (AQCS) algorithm that factored in service quality, available playback time, and freezing delay to determine both bandwidth allocation and video quality levels.

Based on the literature cited above, it is evident that most recent studies have primarily focused on RSUs and base stations for video downloading and transmission, while overlooking the potential of vehicular communication technology to enhance these processes and neglecting the prioritization of real-time applications. With the expanding user base of video applications, it is apparent that perceived video quality could be adversely affected by bandwidth constraints during peak traffic hours. In view of different characteristics and requirements of multimedia applications, future bandwidth allocation schemes should be planned according to the multimedia requirements so that the real-time multimedia applications can be prioritized to receive sufficient bandwidth.

To the best of our knowledge, there is a noticeable gap in the existing literature when it comes to exploring the utilization of the most recent 6G network communication technologies for distributing bandwidth in multimedia applications within upcoming vehicular networks. In light of this, our paper employs THz base stations, as well as mmWave and

In the realm of advancing chip technology, there is a prevailing anticipation that forthcoming next-generation vehicular networks will prominently integrate chips for a diverse range of computational functions. Addressing the burgeoning demand for decentralized computing, as introduced in this paper, dedicated servers are strategically deployed at both RSUs and base stations. Their primary purpose is to facilitate multimedia processing and computational transmissions.

Within the context of RSUs, these servers assume a pivotal role in several key aspects. They diligently record the bandwidth requirements of electric vehicle users within their respective managed segments. Concurrently, these servers store crucial information regarding the available bandwidth originating from base stations. Moreover, these servers harness their computational prowess to efficiently execute various RSU modules, optimizing their operational performance.

On the flip side, base stations also reap the benefits of these servers. Once an EV user initiates a video service application, the server situated at the base station undertakes the task of intelligently prioritizing the storage of specific video files. This strategic maneuver effectively mitigates potential download latency concerns, enhancing the overall user experience. Additionally, leveraging the robust computational capabilities of these servers, video transcoding emerges as a feasible capability. This translates to offering users an array of video resolutions to choose from, tailored to their preferences and requirements.

If the coverage area of the base station of the road section that the EV traverses cannot meet the bandwidth requirement for the EV user's video application, the managing RSU of the road section will activate the "Real-time Bandwidth Allocation" module to assist in allocating the bandwidth to meet the minimum requirement for the video application according to the characteristics of the video application and the pre-set video quality requirement of the EV user. In the case of video conferencing or live video, the RSU starts with the EV, adds other EVs arriving at the road section at the same time to form a fleet, and continues to expand the fleet until the last fleet member is on a road section where the base station's coverage can provide bandwidth to the requesting EV. For the scenario involving VoD, this paper proposes a strategy wherein a separate EV traversing the same road segment pre-downloads the necessary video segment when in a bandwidth-accessible area. Subsequently, when the two EVs cross paths, the pre-downloaded video segment is transmitted from the previously prepared EV to the requesting EV user via V2V communication. Notably, the EV using the video application will also track the EV users' satisfaction with the video playback quality at regular intervals during the video playback process, and if there is a need to adjust the video playback quality, it will notify the "Real-time Bandwidth Allocation" module of the managing RSU(s) for subsequent bandwidth adjustments of the video usage.

The modules shown in Figure 2 are described below.

2.1. Real-Time Route Planning for EVs

Before the start of the EV, the EV user starts the OBU to set the departure point, departure time, and destination, and then starts to run this module. The EV first downloads the global real-time road traffic information from the cloud, and uses Dijkstra's algorithm to estimate the shortest route from the departure point to the destination based on the average traveling time of each road section. Since the arrival time of EVs at each road segment is affected by the traffic conditions at the time of arrival at the road section, this module calculates the arrival time at each road section based on the latest traffic condition information of each road segment of the shortest driving route, and notifies the RSU managing the road section of the driving route and the estimated time of arrival at each road section.

To avoid discrepancies between the latest estimated arrival times of EVs and their expected arrival times at road sections due to ad hoc changes in itineraries by EV users or road congestion during peak hours, this module recalculates the arrival time at each road section at regular intervals based on the latest traffic condition information sent from the

RSU that manages each road section the EV travels through, and uses a machine learning technique to recalculate the arrival time at each road section. If the recalculated arrival times at the road sections are too different from the original predicted times, the module sends the updated times to the RSUs that manage the road sections of the traveled route. Given the effectiveness of machine learning in predicting road travel times in the literature [58,59], the support vector regression (SVR) technique [60] is used in this paper to predict the arrival time at each road section of a traveled route based on the relevant information.

The steps of this module are described below.

Step 1: Before the EV starts its journey, it first downloads the global real-time traffic condition information from the cloud, and after setting the starting location and time as well as the destination, it estimates the shortest path from the starting location to the destination using Dijkstra's algorithm based on the cost of the average traveling time of each road section.

Step 2: The estimated time of arrival at each road section of the traveled route is given by:

$$at_{p_{i+1}^{\sigma}} = at_{p_i^{\sigma}} + \mathrm{SVR}\left(sp_{p_i^{\sigma},p_{i+1}^{\sigma}}\left(at_{p_i^{\sigma}}\right), \rho_{p_i^{\sigma},p_{i+1}^{\sigma}}\left(at_{p_i^{\sigma}}\right), wd_{p_i^{\rho},p_{i+1}^{\rho}}\left(at_{p_i^{\sigma}}\right), wt_{p_i^{\rho},p_{i+1}^{\rho}}\left(at_{p_i^{\sigma}}\right)\right), 1 \leq i < h^{\sigma}, \quad (1)$$

where SVR$(\cdot)$ is the SVR library.

Step 3: The estimated time of arrival at each intersection is transmitted to the RSU that manages each road section, along with the EV's route.

Step 4: This module performs the background execution mode here. After the preset time interval, the latest traffic condition information is obtained from the RSU managing the road section, and the arrival time at each road section of the traveled route is recalculated using Equation (1).

Step 5: If the updated EV arrives at the traveled section in a time that exceeds the system's predefined thresholds due to a temporary trip change or peak hour congestion, the RSU governing the traveled section will be notified of the revised arrival time.

Step 6: Before the EV reaches its destination, go back to step 4 to continue.

2.2. Video Bandwidth Requirement Examination for EV Users

After an EV user activates the video application while the EV is in motion, this module obtains the requirements and specifications of the application from the server of the video application software provider, and adjusts the quality and bandwidth requirements of the video application in a timely manner according to the user video quality requirements pre-set by the EV user. Then, this module transmits the video bandwidth demand to the managing RSU of each road section that the application travels through during the video usage period, and each RSU carries out the bandwidth allocation according to the demand specification of the video application while the EV traverses on the managing road section.

The steps of this module are described below.

Step 1: After an EV user activates a video application, the requirements and specifications for the video application is retrieved from the video application software vendor's server.

Step 2: The RSU along the route informs the minimum bandwidth that can be provided by the base station coverage of the road section it manages.

Step 3: Using the driving durations of EVs across individual road sections and considering the resolution prerequisites for video applications by EV users, the calculation of the essential minimum bandwidth required for video application within each time slot or video segment throughout the driving duration takes place. This calculation is followed by an examination of the bandwidth provisioned by the base station within the coverage area of the corresponding road section to ascertain if it aligns with the stipulated requirements of the video application.

$$dub_t^{\sigma,v} = \sum_{\vartheta} ub_t^{\vartheta,\sigma} - \underline{\mathrm{UR}}(uvr_t^{\sigma,v}), \quad at_{p_i^\sigma} \leq t < at_{p_{i+1}^\sigma}, \\ 1 \leq i < h^\sigma, \text{ if } \kappa_1^{\sigma,v} = 1, \tag{2}$$

$$ddb_t^{\sigma,v} = \sum_{\vartheta} db_t^{\vartheta,\sigma} - \underline{\mathrm{DR}}(dvr_t^{\sigma,v}), \ at_{p_i^\sigma} \leq t < at_{p_{i+1}^\sigma}, \\ 1 \leq i < h^\sigma, \text{ if } \kappa_1^{\sigma,v} + \kappa_2^{\sigma,v} = 1, \tag{3}$$

$$at_{p_1^\sigma} \leq t_0^{\sigma,v} \leq t_e^{\sigma,v} \leq at_{p_{h^\sigma}^\sigma}, \text{ if } \kappa_1^{\sigma,v} + \kappa_2^{\sigma,v} = 1, \tag{4}$$

$$at_{p_1^\sigma} \leq pt_1^{\sigma,v} \leq pt_{S^{\sigma,v}}^{\sigma,v}, \text{ if } \kappa_3^{\sigma,v} = 1, \tag{5}$$

$$pt_s^{\sigma,v} \geq pt_{s-1}^{\sigma,v} + \mathrm{SS}_s^{\sigma,v}(br_s^{\sigma,v}) / \sum_{\vartheta} db_{\tau_s}^{\vartheta,\sigma}, \ pt_1^{\sigma,v} \leq \tau_s < pt_s^{\sigma,v}, 1 < s \\ \leq S^{\sigma,v}, \text{ if } \kappa_3^{\sigma,v} = 1, \tag{6}$$

$$buf_{\tau_s}^{\sigma} + \mathrm{SS}_s^{\sigma,v}(br_s^{\sigma,v}) \leq \overline{buf}^{\sigma}, \\ pt_1^{\sigma,v} \leq \tau_s < pt_s^{\sigma,v}, \quad 1 < s \leq S^{\sigma,v}, \quad \text{if } \kappa_3^{\sigma,v} = 1. \tag{7}$$

As shown in Equations (2) and (3), if the quantities of $dub_t^{\sigma,v}$ and $ddb_t^{\sigma,v}$ are greater than zero, it signifies that the base station's allocated bandwidth within the coverage zone of the route taken by the EV is sufficient to fulfill the requisite minimum upload and download bandwidth criteria for applications such as video conferencing or live video streaming, respectively. Equations (4) and (5) guarantee that the video conferencing or live video initiation time must be scheduled after the EV's departure time, and the application should be concluded prior to the EV reaching its destination.

Equation (6) clarifies that the pre-download of a video segment must be completed before the playback can begin. Meanwhile, Equation (7) guarantees that the buffer space available in the EV is sufficient to accommodate a video segment prior to its playback.

Step 4: If all of the base station coverage areas of the EV routes can meet the minimum bandwidth requirement, this module notifies each RSU of the road section of the bandwidth used in the road section and proceed to Step 6. On the other hand, this module notifies the RSUs along the road sections with insufficient bandwidth to assist in adjusting the bandwidth. The optimization objectives concerning the video quality requirements of EV users can be formulated by:

$$\begin{aligned} \mathrm{Max}\Bigg\{ & \kappa_1^{\sigma,v} \cdot \Bigg[\sum_{t_0^{\sigma,v} \leq t \leq t_e^{\sigma,v}} \left(\mathrm{QL}(ur_t^{\sigma,v}) - \phi^{\sigma,v} \cdot ult_t^{\sigma,v}\right) - \zeta^{\sigma,v} \\ & \cdot \sum_{t_0^{\sigma,v} < t \leq t_e^{\sigma,v}} \left|\mathrm{QL}(ur_t^{\sigma,v}) - \mathrm{QL}(ur_{t-1}^{\sigma,v})\right| \Bigg] + \left(\kappa_1^{\sigma,v} + \kappa_2^{\sigma,v}\right) \\ & \cdot \Bigg[\sum_{t_0^{\sigma,v} \leq t \leq t_e^{\sigma,v}} \left(\mathrm{QL}(dr_t^{\sigma,v}) - \phi^{\sigma,v} \cdot dlt_t^{\sigma,v}\right) - \zeta^{\sigma,v} \\ & \cdot \sum_{t_0^{\sigma,v} < t \leq t_e^{\sigma,v}} \left|\mathrm{QL}(dr_t^{\sigma,v}) - \mathrm{QL}(dr_{t-1}^{\sigma,v})\right| \Bigg] + \kappa_3^{\sigma,v} \\ & \cdot \Bigg[\sum_{1 \leq s \leq S^{\sigma,v}} \left(\mathrm{QD}(br_s^{\sigma,v}) - \psi^{\sigma,v} \cdot rbt_s^{\sigma,v}\right) - \beta^{\sigma,v} \\ & \cdot \sum_{1 < s \leq S^{\sigma,v}} \left|\mathrm{QD}(br_s^{\sigma,v}) - \mathrm{QD}(br_{s-1}^{\sigma,v})\right| - \omega^{\sigma,v} \cdot sd^{\sigma,v} \Bigg] \Bigg\}, \end{aligned} \tag{8}$$

subject to:

$$ult_t^{\sigma,v} = \frac{\mathrm{DS}_t^{\sigma,v}(uvr_t^{\sigma,v})}{ur_t^{\sigma,v}} \leq ud^{\sigma,v},\ t_0^{\sigma,v} \leq t < t_e^{\sigma,v},\ \text{if } \kappa_1^{\sigma,v} = 1, \tag{9}$$

$$dlt_t^{\sigma,v} = \frac{\mathrm{DS}_t^{\sigma,v}(dvr_t^{\sigma,v})}{dr_t^{\sigma,v}} \leq dd^{\sigma,v},\ t_0^{\sigma,v} \leq t < t_e^{\sigma,v},\ \text{if } \kappa_1^{\sigma,v} + \kappa_2^{\sigma,v} = 1, \tag{10}$$

$$\mathrm{QL}(ur_t^{\sigma,v}) = \log\left(\frac{ur_t^{\sigma,v}}{\underline{\mathrm{UR}}(uvr_t^{\sigma,v})}\right),\ t_0^{\sigma,v} \leq t < t_e^{\sigma,v},\ \text{if } \kappa_1^{\sigma,v} = 1, \tag{11}$$

$$\mathrm{QL}(dr_t^{\sigma,v}) = \log\left(\frac{dr_t^{\sigma,v}}{\underline{\mathrm{DR}}(dvr_t^{\sigma,v})}\right),\quad t_0^{\sigma,v} \leq t < t_e^{\sigma,v},\ \text{if } \kappa_1^{\sigma,v} + \kappa_2^{\sigma,v} = 1, \tag{12}$$

$$\mathrm{QD}(br_s^{\sigma,v}) = \log\left(\frac{dr_{\tau_s}^{\sigma,v}}{\underline{\mathrm{DB}}(br_s^{\sigma,v})}\right),\quad pt_1^{\sigma,v} \leq \tau_s < pt_s^{\sigma,v},\quad 1 < s \leq S^{\sigma,v},\ \text{if } \kappa_3^{\sigma,v} = 1, \tag{13}$$

$$uvr_t^{\sigma,v} \geq \underline{vr}^v,\ t_0^{\sigma,v} \leq t < t_e^{\sigma,v},\ \text{if } \kappa_1^{\sigma,v} = 1, \tag{14}$$

$$dvr_t^{\sigma,v} \geq \underline{vr}^v,\ t_0^{\sigma,v} \leq t < t_e^{\sigma,v},\ \text{if } \kappa_1^{\sigma,v} + \kappa_2^{\sigma,v} = 1, \tag{15}$$

$$br_s^{\sigma,v} \geq \underline{br}^v,\ 1 < s \leq S^{\sigma,v},\ \text{if } \kappa_3^{\sigma,v} = 1, \tag{16}$$

$$pt_s^{\sigma,v} \geq pt_{s-1}^{\sigma,v} + \frac{\mathrm{SS}_s^{\sigma,v}(br_s^{\sigma,v})}{dr_{\tau_s}^{\sigma,v}},\ pt_1^{\sigma,v} \leq \tau_s < pt_s^{\sigma,v},\quad 1 < s \leq S^{\sigma,v},\ \text{if } \kappa_3^{\sigma,v} = 1, \tag{17}$$

$$rbt_s^{\sigma,v} = \begin{cases} \frac{\mathrm{SS}_s^{\sigma,v}(br_s^{\sigma,v})}{dr_{\tau_s}^{\sigma,v}} - rbt_{s-1}^{\sigma,v}, & \frac{\mathrm{SS}_s^{\sigma,v}(br_s^{\sigma,v})}{dr_{\tau_s}^{\sigma,v}} > rbt_{s-1}^{\sigma,v},\ \text{if } \kappa_3^{\sigma,v} = 1 \\ 0 & \text{otherwise} \end{cases}, \tag{18}$$

$$rbt_1^{\sigma,v} = 0,\quad \text{if } \kappa_3^{\sigma,v} = 1, \tag{19}$$

$$\sum_{\vartheta} ub_t^{\vartheta,\sigma} \geq ur_t^{\sigma,v} \geq \underline{\mathrm{UR}}(uvr_t^{\sigma,v}),\ t_0^{\sigma,v} \leq t < t_e^{\sigma,v},\ \text{if } \kappa_1^{\sigma,v} = 1, \tag{20}$$

$$\sum_{\vartheta} db_t^{\vartheta,\sigma} \geq dr_t^{\sigma,v} \geq \underline{\mathrm{DR}}(dvr_t^{\sigma,v}),\ t_0^{\sigma,v} \leq t < t_e^{\sigma,v},\ \text{if } \kappa_1^{\sigma,v} + \kappa_2^{\sigma,v} = 1, \tag{21}$$

$$\sum_{\vartheta} db_{\tau_s}^{\vartheta,\sigma} \geq dr_{\tau_s}^{\sigma,v} \geq \underline{\mathrm{DB}}(br_s^{\sigma,v}),\ pt_1^{\sigma,v} \leq \tau_s < pt_s^{\sigma,v},\ 1 \leq s \leq S^{\sigma,v},\ \text{if } \kappa_3^{\sigma,v} = 1, \tag{22}$$

$$\begin{aligned} &buf_{\tau_s}^{\rho} + \mathrm{SS}_s^{\sigma,v}(br_s^{\sigma,v}) \leq \overline{buf}^{\rho}, \\ &\quad pt_1^{\sigma,v} \leq \tau_s < pt_s^{\sigma,v},\quad 1 < s \leq S^{\sigma,v},\ \text{if } \kappa_3^{\sigma,v} = 1. \end{aligned} \tag{23}$$

Equation (8) allocates the required bandwidth for the video application in terms of the EV user's perceived video quality. Equations (9) and (10) delineate the upload and download durations for video conferencing/live video during each time slot. Equations (11) and (12) present the distinct quality metrics utilized to assess the effectiveness of video conferencing and live video, following the guidelines outlined in the cited reference [61]. Furthermore, Equation (13) defines the quality metric applicable to VoD, as referenced in the related source [62].

Equations (14)–(16) ensure that the video resolution remains equal to or exceeds the minimum resolution defined by the video service provider. Simultaneously, Equation (17) guarantees the complete pre-download of individual video segments before initiating their playback. Moreover, Equations (18) and (19) delineate the rebuffering time for each video segment.

Equations (20) and (21) establish the prerequisite that the base stations covering the road segments traversed by the electric vehicle (EV) must possess the necessary upload and download bandwidth to support video conferencing and live video, as required. Similarly, Equation (22) imposes an analogous requirement for VOD. Additionally, Equation (23) ensures that the buffer space within the EV is appropriately dimensioned to accommodate a video segment before its playback.

To be specific, the optimization problem outlined in Equation (8) is being approximated in this scenario by lowering the requested video resolution from the EV user to align with the minimum resolution established by the video service provider.

Step 5: Should the bandwidth obtained in the previous step continue to be inadequate for supporting the minimum resolution specified by the video service provider, the EV user will receive a notification.

Step 6: This module operates in the background execution mode. Following the time interval established by the system, the process returns to Step 2 to readjust the quality and bandwidth prerequisites of the video application, provided that the video application remains active.

2.3. Real-Time Bandwidth Allocation for RSUs

If the video bandwidth requirement of the EV user cannot be met when the EV arrives at the road section managed by the RSU, this module assists in allocating the bandwidth according to the characteristics of the video application. If the user's video application is video conferencing or live video, the bandwidth originally allocated for other VoD downloads in the same time slot will be reallocated to the required video conferencing or live video. Should the video bandwidth requirement remain unfulfilled, the EV will assume the role of the ultimate member within a fleet. Subsequently, the fleet will undergo expansion until the leading EV member enters a road segment where the base station's coverage can sufficiently accommodate the demanded video bandwidth. The bandwidth is then transferred from the leading EV member to the requesting EV through V2V communication.

In the case of VoD, if a video segment cannot be downloaded in time before the video segment is played, the RSU checks the routes of other EVs that arrive at the same road section at the same time as the EV playing the VoD to find out whether there is a base station that can provide bandwidth for pre-downloading the required video segment. Once a base station that can support bandwidth is found, the EV traveling through the road section covered by the base station will download the required video segment to the on-board storage of the EV. When the EV storing the downloaded video segment arrives at the same road section as the requesting EV, the video segment can be transmitted to the VoD user via V2V communication.

The steps of this module are described below.

Step 1: For VoD, proceed to Step 8.

Step 2: The RSU allocates the bandwidth of other VoD(s) in the same time slot to the demanding video conferencing/live video.

Step 3: If the bandwidth requirement for video conferencing/live video is satisfied, go to Step 6. Otherwise, proceed to the next step.

Step 4: In cases where the bandwidth necessary for video conferencing or live video is inadequate, the RSU initiates the process by including the EV as the final member of a fleet for bandwidth transfer and incorporating other EVs into a fleet. This fleet expansion persists until the foremost fleet member reaches a road segment where the base station's coverage can sufficiently supply the required bandwidth for the requesting EV. The establishment of the fleet must fulfill the subsequent requirement:

$$\underset{\mathrm{P}^{\sigma,v}}{\operatorname{argMin}} \sum_{1 \le i < \mathrm{P}^{\sigma,v}} \left| \left(x_{r_i^{\sigma,v}}, y_{r_i^{\sigma,v}} \right) - \left(x_{r_{i+1}^{\sigma,v}}, y_{r_{i+1}^{\sigma,v}} \right) \right|, \text{ if } \kappa_1^{\sigma,v} + \kappa_2^{\sigma,v} = 1, \tag{24}$$

subject to:

$$\mathcal{R}_t^{\sigma,v} = \left[\left(r_1^{\sigma,v}, r_2^{\sigma,v} \right), \cdots, \left(r_{i-1}^{\sigma,v}, r_i^{\sigma,v} \right), \cdots, \left(r_{i-1}^{\sigma,v}, r_{\mathrm{P}^{\sigma,v}}^{\sigma,v} \right) \right], \ t_0^{\sigma,v} \le t < t_e^{\sigma,v}, \\ \text{if } \kappa_1^{\sigma,v} + \kappa_2^{\sigma,v} = 1, \tag{25}$$

$$r^{\sigma,v}_{P^{\sigma,v}} = \sigma, \text{ if } \kappa_1^{\sigma,v} + \kappa_2^{\sigma,v} = 1, \tag{26}$$

$$at_{p_1^\sigma} \le t_0^{\sigma,v} \le t_e^{\sigma,v} \le at_{p^\sigma_{h^\sigma}}, \text{ if } \kappa_1^{\sigma,v} + \kappa_2^{\sigma,v} = 1, \tag{27}$$

$$\begin{gathered}\sum_\vartheta ub_t^{\vartheta,r_1^{\sigma,v}} \ge \underline{\mathrm{UR}}(uvr_t^{\sigma,v}),\ t_0^{\sigma,v} \le t < t_e^{\sigma,v},\\ \text{if } \kappa_1^{\sigma,v} = 1 \;\&\; \sum_\vartheta ub_t^{\vartheta,\sigma} < \underline{\mathrm{UR}}(uvr_t^{\sigma,v}),\end{gathered} \tag{28}$$

$$\begin{gathered}\sum_\vartheta db_t^{\vartheta,r_1^{\sigma,v}} \ge \underline{\mathrm{DR}}(dvr_t^{\sigma,v}),\\ t_0^{\sigma,v} \le t < t_e^{\sigma,v},\ \text{if } \kappa_1^{\sigma,v} + \kappa_2^{\sigma,v} = 1 \;\&\; \sum_\vartheta db_t^{\vartheta,\sigma} < \underline{\mathrm{DR}}(dvr_t^{\sigma,v}),\end{gathered} \tag{29}$$

$$\begin{gathered}ut_t^{r_{i+1}^{\sigma,v},r_i^{\sigma,v}} \ge \underline{\mathrm{UR}}(uvr_t^{\sigma,v}),\ t_0^{\sigma,v} \le t < t_e^{\sigma,v},\\ 1 \le i < \mathrm{P}^{\sigma,v} \text{ if } \kappa_1^{\sigma,v} = 1 \;\&\; \sum_\vartheta ub_t^{\vartheta,\sigma} < \underline{\mathrm{UR}}(uvr_t^{\sigma,v}),\end{gathered} \tag{30}$$

$$\begin{gathered}dt_t^{r_{i+1}^{\sigma,v},r_i^{\sigma,v}} \ge \underline{\mathrm{DR}}(uvr_t^{\sigma,v}),\ t_0^{\sigma,v} \le t < t_e^{\sigma,v}, 1 \le i < \mathrm{P}^{\sigma,v} \text{ if } \kappa_1^{\sigma,v} + \kappa_2^{\sigma,v}\\ = 1 \;\&\; \sum_\vartheta db_t^{\vartheta,\sigma} < \underline{\mathrm{DR}}(dvr_t^{\sigma,v}).\end{gathered} \tag{31}$$

Equation (24) serves as the foundation for configuring the fleet established during the transmission relay, optimized for the minimum fleet length. Building upon this, Equation (25) outlines the specific composition of the fleet, where Equation (26) identifies the designated EV as the terminal component of this fleet. To guarantee a harmonized sequence, Equation (27) specifies the scheduling of the initiation time for video conferencing or live video subsequent to the EV's departure, while also ensuring the timely conclusion of the application before the EV reaches its destination.

Equations (28) and (29) guarantee that the leading fleet member reaches a road segment where the base station's coverage can adequately provide the necessary bandwidth for the requesting EV. Additionally, Equations (30) and (31) establish that the V2V communication bandwidth between two successive fleet members must not fall below the relayed bandwidth acquired by the requesting EV.

To provide specific details, the process of forming the fleet as defined in Equation (24) entails a step-by-step assessment of all EVs that reach the same road segment as the requesting EV. Starting from the selected EV, the expansion of the fleet persists until the farthest member of the fleet enters a road segment where the coverage from the base station is proficient in supplying the required bandwidth for the EV that initiated the request.

Step 5: In instances where a certain VoD allocates bandwidth to video conferencing/live video, resulting in an inability to meet the minimum bandwidth requirement of the VoD due to reduced bandwidth availability, this module advances to the subsequent step in order to acquire the necessary bandwidth for the VoD. Conversely, the module concludes.

Step 6: To examine whether there are base stations in the coverage area of the road sections that can provide bandwidth for EVs to pre-download the required video segment(s):

$$\sum_\vartheta db_{\tau_s}^{\vartheta,\sigma} \ge \underline{\mathrm{DB}}(br_s^{\sigma,v}),\ \ pt_1^{\sigma,v} \le \tau_s < pt_s^{\sigma,v},\ 1 \le s \le S^{\sigma,v},\ \text{if } \kappa_3^{\sigma,v} = 1, \tag{32}$$

$$pt_s^{\sigma,v} \ge pt_{s-1}^{\sigma,v} + \frac{\mathrm{SS}_s^{\sigma,v}(br_s^{\sigma,v})}{\underline{\mathrm{DB}}(br_s^{\sigma,v})},\quad 1 < s \le S^{\sigma,v},\ \text{if } \kappa_3^{\sigma,v} = 1, \tag{33}$$

$$\begin{gathered}buf_{\tau_s}^\sigma + \mathrm{SS}_s^{\sigma,v}(br_s^{\sigma,v}) \le \overline{buf}^\sigma,\\ pt_1^{\sigma,v} \le \tau_s < pt_s^{\sigma,v},\quad 1 < s \le S^{\sigma,v},\ \text{if } \kappa_3^{\sigma,v} = 1.\end{gathered} \tag{34}$$

Step 7: If bandwidth is available in the base station coverage area of another road section, pre-download the video segment from the base station to the on-board storage of the EV while passing through the base station coverage area. Notify the managing RSU of the base station and this module ends. If not, proceed to the next step.

Step 8: Check if other EVs arrive at the same road section as EV σ at time $pt_s^{\sigma,v}$.

Step 9: If it is not possible to find an EV that arrives at the same time on the same road section as EV σ, the RSU notifies the EV user and this module ends. Instead, proceed to the next step.

Step 10: View the routes of the EVs arriving at the same section of the road at the same time as σ, and check if there is a base station that can provide bandwidth for the EV to pre-download VoD segment for σ:

$$\sum_{\vartheta} db_{\tau_{s'}}^{\vartheta,\rho} \geq \underline{\mathrm{DB}}(br_s^{\sigma,v}),\ \ pt_1^{\sigma,v} \leq \tau_{s'} < pt_s^{\sigma,v},\ 1 \leq s \leq S^{\sigma,v},\ \text{if } \kappa_3^{\sigma,v} = 1, \tag{35}$$

$$pt_s^{\sigma,v} \geq pt_{s-1}^{\sigma,v} + \frac{\mathrm{SS}_s^{\sigma,v}(br_s^{\sigma,v})}{\underline{\mathrm{DB}}(br_s^{\sigma,v})},\ \ 1 < s \leq S^{\sigma,v},\ \text{if } \kappa_3^{\sigma,v} = 1, \tag{36}$$

$$\begin{gathered} buf_{\tau_{s'}}^{\rho} + \mathrm{SS}_s^{\sigma,v}(br_s^{\sigma,v}) \leq buf^{\rho}, \\ pt_1^{\sigma,v} \leq \tau_s' < pt_s^{\sigma,v},\ 1 < s \leq S^{\sigma,v},\ \text{if } \kappa_3^{\sigma,v} = 1, \end{gathered} \tag{37}$$

$$dt_{\tau_{s'}}^{\sigma,\rho} \geq \underline{\mathrm{DB}}(br_s^{\sigma,v}),\ pt_1^{\sigma,v} \leq \tau_{s'} < pt_s^{\sigma,v},\ \ 1 < s \leq S^{\sigma,v},\ \text{if } \kappa_3^{\sigma,v} = 1. \tag{38}$$

Step 11: If EV ρ that supports pre-downloading is found, the VoD segment will be pre-downloaded by ρ, and then will be transferred to the requesting EV via V2V communication when two EVs meet. If not, the RSU notifies the EV user and this module ends.

3. Experimental Results and Discussion

In this paper, the proposed algorithm is simulated using a personal computer with an Intel Core i7 2.9 GHz CPU and 64 GB RAM. To assess the effectiveness of the proposed algorithm, two comparison targets are employed. The first target is the conventional first-come, first-served (FCFS) method. The second reference benchmark is a contemporary state-of-the-art algorithm that combines the DQN and AQCS [2]. Remarkably, taking into account the prevailing trend in recent literature that commonly employs the strategy of pre-downloading video segments at base stations and RSUs to enhance video service quality for EV users, this paper specifically opts to evaluate the latest DQN+AQC algorithm as the second comparative target.

The historical traffic data are obtained from a website of traffic volume counts for New York City [63]. Video applications are divided into three categories, namely, video conferencing, live video, and VoD. Video conferencing and live video utilize AVC [6]. VoD uses VVC [10] with a bit rate half that of HEVC. Each video application has different bandwidth requirements; this paper refers to the bandwidths suggested by Skype [64], and Table 1 shows the download and upload bandwidths required for video conferencing. In addition, Table 2 shows the bandwidth required for live video by referring to the literature on the use of AVC [65,66]. For VoD, the required bandwidth is shown in Table 3 by referring to the literature on the use of HEVC [66] and considering half of the bandwidth required for video using HEVC in the literature as the bandwidth required for video using VVC.

Table 1. Parameters of required bandwidth for video configuration with AVC.

Video Format	Download Required Bandwidth	Upload Required Bandwidth
Screen Sharing	300 Kbps	300 Kbps
High Quality	500 Kbps	500 Kbps
High Definition	1.5 Mbps	1.5 Mbps

Table 2. Parameters of required bandwidth for live video with AVC.

Video Format	Download Required Bandwidth
720 p 30 fps	3 Mbps
1080 p 60 fps	17 Mbps
2160 p 60 fps	60 Mbps

Table 3. Parameters of required bandwidth for VoD with VVC.

Video Format	Download Required Bandwidth
1080 p 60 fps	7.5 Mbps
2160 p 60 fps	20 Mbps
4320 p 60 fps	50 Mbps

Three types of base stations, including THz [28], mmWave [67], and sub-6 GHz [68], were set up in the simulated area to provide the bandwidth required for the video applications. Table 4 shows the bandwidth and transmission distance that can be provided by the three types of base stations. In addition, for V2V communication using VLC, this paper cites the relevant literature that involves simulations incorporating the use of car lights [45]. Additionally, Table 5 provides an overview of the variable bandwidths of VLC at different distances, distinguishing between daytime and nighttime circumstances.

Table 4. Parameters of bandwidth available for VLC in different situations.

Distance	Daytime Bandwidth	Nighttime Bandwidth
10 m	2790 Mbps	2810 Mbps
100 m	336 Mbps	362 Mbps

Table 5. Parameters of bandwidth available for THz, mmWave, and sub-6 GHz base stations.

Type of Base Station	Bandwidth	Transmission Distance
THz	54~24 Gbps	39 m
MmWave	2~1.8 Gbps	150 m
Sub-6 GHz	1~0.5 Gbps	622 m

Figure 3 shows the number of EVs traveling on the road during a day in our simulation. It can be observed that there is a spike in traffic during the morning and evening peak hours. The number of EVs starts to increase from 06:00 until the morning peak is reached at 09:00. After the morning peak, the traffic volume stays above 1500 vehicles. The number of EVs starts to increase slightly after 16:00, and then the traffic volume starts to decrease after the evening peak is reached at 19:00, with the lowest number of EVs in the early morning hours.

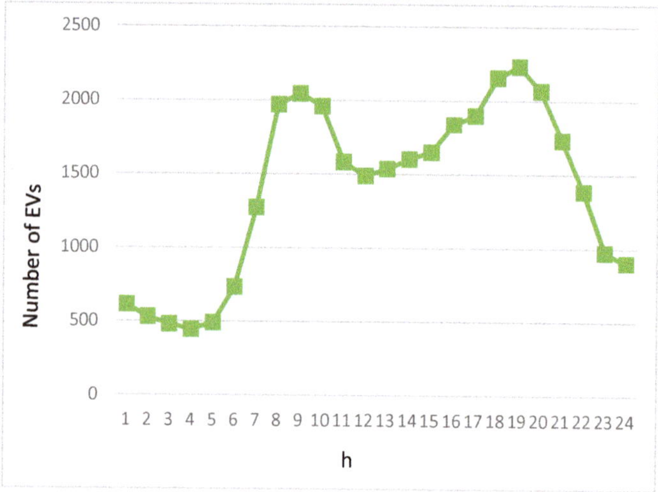

Figure 3. Volume of EVs within a day.

Figure 4 gives the user count of video applications within a day. As aforementioned, there are three types of video applications: video conferencing, live video, and VoD. It can be seen that the user counts of each application follow the same trend as the number of EVs. A small number of users still use video applications during the off-peak hours in the early hours of the morning, while the number of applications increases dramatically during the morning and evening peak traffic hours, when EV users have longer commute times during congestion and therefore have a greater demand for video applications. In addition, since VoD has the characteristics of flexible viewing time, repeatable viewing, and content diversity, it can be observed from Figure 4 that the number of users using VoD is higher than the number of users using video conferencing and live video in most of the time periods, no matter whether these are during peak or off-peak hours.

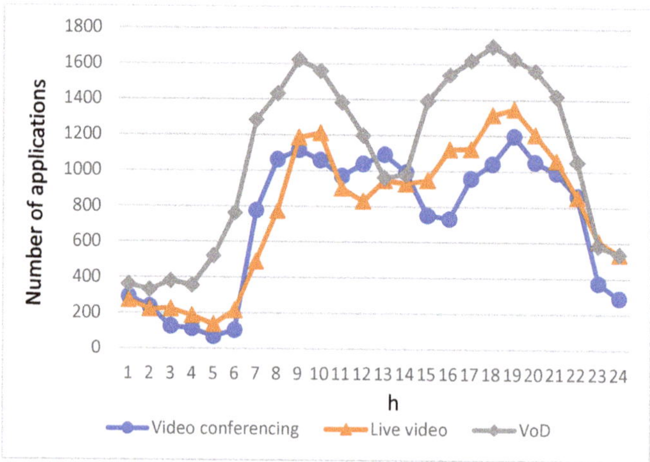

Figure 4. Number of users utilizing video conferencing, live video, and VoD within a day.

Figure 5 shows the video application bandwidth demand of users within a day. The bandwidth required for video conferencing is very low, as can be seen from Table 1. Even for high resolution, only 1.5 Mbps is required. Accordingly, even during the three peak hours of video conferencing users shown in Figure 4, the bandwidth required is still far less than the off-peak bandwidth demand of live video and VoD users. In addition, it can be seen from Figure 5 that the bandwidth demand for the use of live video and VoD is directly proportional to the number of their respective subscribers, with a significant increase in bandwidth demand during the morning and evening peak hours. Notably, although the number of users watching VoD is mostly higher than the number of users watching live video, as seen from Figure 4, the bandwidth requirement of VoD, as shown in Figure 5, is not much higher than that of live video, and even less than that of live video in several time slots.

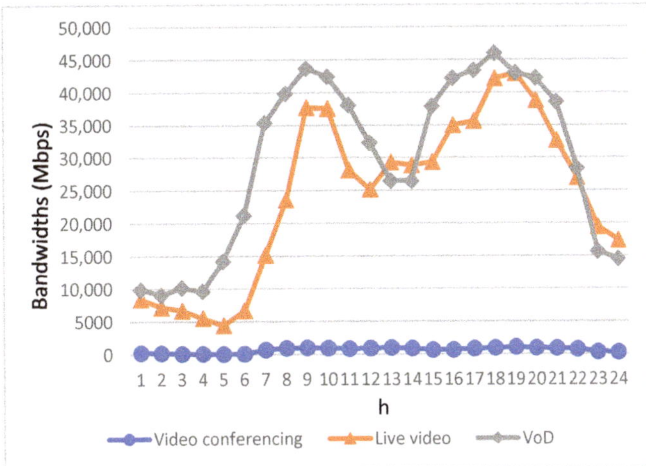

Figure 5. Bandwidth requirements for video conferencing, live video, and VoD within a day.

Figure 6 shows the bandwidth obtained by the users of each video application using the conventional FCFS method. Observing the data, it becomes evident that during the early morning hours when users' bandwidth requirements are minimal, the bandwidth accessible to EV users for each video application adheres to a direct proportionality with both their individual bandwidth demands and the sequence in which the users initiate their respective service applications. Due to the utilization of an FCFS strategy for bandwidth allocation by the base stations, applications initiated later in time struggled to secure sufficient bandwidth for real-time usage, particularly during peak periods. As a result, it is evident from the graph that, beginning at 08:00 and continuing until 22:00, the obtained bandwidth by EV users across different video applications does not align closely with the bandwidth requirements of these applications, as shown in Figure 5. Furthermore, the available bandwidth proves inadequate in fulfilling the requisite bandwidth for these applications, with the pronounced gap of insufficiency becoming notably conspicuous, particularly during the two peak periods of heightened user bandwidth demand.

Figure 7 illustrates the bandwidth allocation for each video application using the DQN+AQCS algorithm. As seen in Figure 7, during the early morning and midnight hours when users have lower bandwidth requirements, the allocated bandwidth for EV users aligns with their individual demands, similar to the FCFS approach. However, the DQN+AQCS algorithm consistently enhances the available bandwidth for EV users across various time intervals when compared to the FCFS algorithm. Notably, starting at 08:00 and continuing until 22:00, the DQN+AQCS algorithm leverages the video pre-download mechanisms at RSUs and base stations, resulting in a significant increase in the available bandwidth for EV users compared to the FCFS method. This improvement is particularly pronounced for VoD applications, where video segments can be pre-downloaded at RSUs and base stations before their scheduled playback time. Nevertheless, it is important to acknowledge that the provided bandwidth still falls short of meeting the demands of EV users during the peak demand period characterized by heightened bandwidth requirements, even though urgent applications were prioritized. Consequently, both users of live video and VoD applications experienced insufficient bandwidth during rush hours. The performance of live video applications was particularly affected because there is no provision for pre-downloading in this type of application.

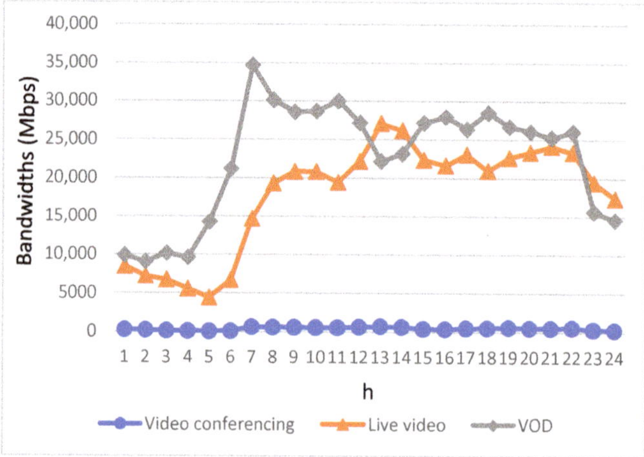

Figure 6. Bandwidth allocation for three types of video applications using the FCFS method.

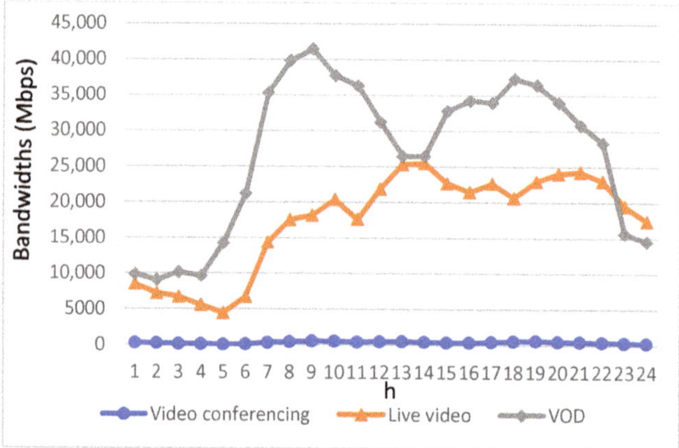

Figure 7. Bandwidth allocation for three types of video applications using the DQN+AQCS.

Figure 8 illustrates the bandwidth assigned to each video application after using the proposed algorithm. As expected, the bandwidth available to each video service application is proportional to their demanded bandwidth, and the available bandwidth is significantly increased during the peak of bandwidth demand after applying the proposed algorithm. To provide further clarity, in cases where the bandwidth demand of a video conferencing or live video application cannot be satisfied, the "Real-time Bandwidth Allocation" mechanism implemented at the RSU was triggered into action. This activation facilitates the acquisition of bandwidth for the specific video application in need. Initially, the bandwidth allocated for other VoD downloads within the same time slot is repurposed to cater to the demanding video's needs in a priority manner. In cases where the bandwidth requirement of the requesting video remains unfulfilled, there exists an alternative solution: bandwidth from base stations in other road sections can be reallocated to serve the requesting video's needs. This process is facilitated through the cooperative efforts of EVs that assist in transmitting bandwidth via V2V communications. Concerning VoD applications, the proposed algorithm engages in the process of identifying a suitable EV capable of pre-downloading the required video segment onto its onboard storage prior to its encounter with the requesting EV. Subsequently, when these two EVs cross paths on the same road section, the pre-downloaded video segment is then efficiently transmitted to the requesting EV through the utilization of V2V communication mechanisms.

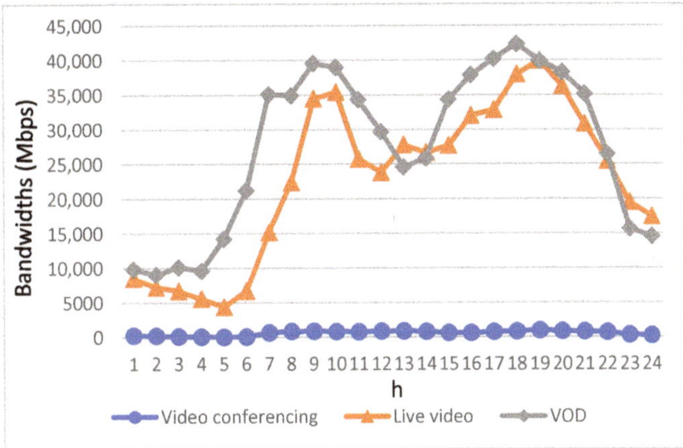

Figure 8. Bandwidth allocation for three types of video applications using the proposed algorithm.

Figure 9 illustrates various curves, each conveying distinct aspects of bandwidth requirements and allocation for video applications. The curves are color-coded as follows:

- The red curve represents the bandwidth requirements of video applications.
- The yellow curve depicts the bandwidth allocated to video applications using the conventional FCFS method.
- The green curve showcases the bandwidth allocated to video applications when employing the DQN+AQCS algorithm.
- Finally, the purple curve demonstrates the bandwidth allocation achieved by video applications when utilizing the proposed algorithm.

The yellow curve in Figure 9 vividly illustrates that video applications employing the FCFS method exhaust the available bandwidth provided by base stations in congested road segments entirely between 08:00 and 22:00. Consequently, there is an insufficient amount of bandwidth remaining to accommodate the bandwidth requests of applications initiated later during this congested timeframe. This limitation stems from the fact that applications launched earlier within this time frame have already consumed a significant portion of the available bandwidth. Comparatively, as indicated by the green curve in Figure 9, the DQN+AQCS algorithm offers a substantial improvement in the bandwidth accessible to EV users between 08:00 and 22:00. This enhancement can be attributed to the algorithm's implementation of a pre-downloading mechanism involving base stations and RSUs before the scheduled video segment playback.

As observed in the purple curve in Figure 9, when contrasting the proposed algorithm with the other two target algorithms, it consistently outperforms them during the high-bandwidth demand periods in the morning and evening. A significant portion of the required bandwidth for live video applications is efficiently delivered through real-time V2V communication. This process facilitates the seamless transfer of bandwidth from less congested road sections. The performance improvement of VoD applications involves the cooperative participation of other EVs in pre-downloading necessary video segments while passing by base stations with ample bandwidth. This adaptive approach proposed in the paper results in a considerable augmentation of the bandwidth allocated to each video application for EV users' requests, particularly during the two peak time periods. In sum, the proposed algorithm not only effectively addresses the bandwidth demands of most video applications, but also significantly enhances the utilization of the bandwidth within base stations located in less-congested road sections.

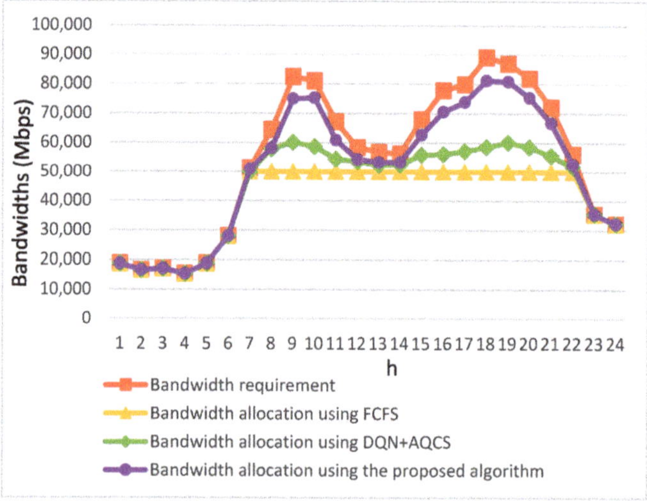

Figure 9. Comparative analysis of allocated bandwidth among the three approaches.

Figure 10 provides a visual comparison, offering insights into the normalized perceived video quality among EV users and enabling a clear assessment of the relative effectiveness of these approaches. Notably, during the time frame from 23:00 to 07:00 of the subsequent morning, a substantial portion of users' bandwidth demand is met. This occurs because EV users typically have reduced bandwidth requirements during late-night and early morning periods. Consequently, there is minimal impact on the perceived video quality for all three methods—FCFS, DQN+AQCS, and the proposed algorithm—during this time interval.

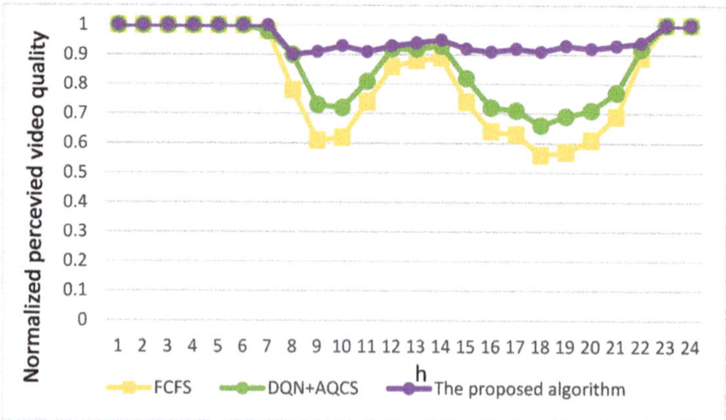

Figure 10. Comparative analysis of normalized perceived video quality among the three approaches.

However, as depicted by the yellow curve in Figure 10, the FCFS method performs poorly due to a significant bandwidth deficit during high-bandwidth-demand periods. Although the utilization of the DQN+AQCS algorithm, represented by the green curve in Figure 10, helps alleviate the reduction in perceived video quality for EV users during high-bandwidth-demand periods, it encounters a challenge of unstable perceived video quality between 08:00 and 22:00, especially during two peak time periods. Specifically, for the two compared approaches, the average bandwidth available to EV users decreases proportionally with the rise in video bandwidth demand within congested road sections, where the base station's bandwidth allocation has reached its upper limit. A significant drop in the quality of video usage by EV users is evident during peak hours. Only during the midday hours, when the video bandwidth scarcity is less severe, does the perceived video quality for EV users improve.

Conversely, as shown on the purple curve in Figure 10, the proposed algorithm effectively addresses these challenges by utilizing V2V communication to relay bandwidth from base stations with surplus capacity to the requesting live video applications and by efficiently pre-downloading video segments as requested by VoD applications. To offer a more detailed perspective, Figure 9 reveals that the proposed algorithm leads to only a slight decline in user-perceived video quality between 08:00 and 22:00. When compared to the FCFS method during the same time frame, it delivers a substantial average improvement of 30% in perceived video quality for EV users. Moreover, during the morning and evening peak hours, particularly at 09:00 and 19:00, this enhancement becomes even more significant, with improvements reaching 49% and 63%, respectively. In contrast to the DQN+AQCS approach, there is an average improvement of 17%. The most significant enhancements are observed at 10:00 and 18:00, where improvements reach 29% and 37%, respectively.

4. Conclusions

Although some studies in the literature have proposed bandwidth allocation algorithms to provide video services to EV users, there are no bandwidth allocation algorithms for video services that utilize emerging communication technologies for next-generation vehicular networks. In view of this, this paper integrates the THz base stations of 6G wireless networks with the existing mmWave and sub-6 GHz band base stations to provide the required bandwidth for EV users. A bandwidth management mechanism that suits the video quality requirements of different EV users is proposed to allocate the bandwidth from the three types of base stations mentioned above to the in-vehicle video services. Meanwhile, VLC technology is utilized in V2V communications to provide a high data rate and high security for inter-vehicle video transmissions.

A series of simulations is conducted to comprehensively compare the performance of the proposed algorithm against both the conventional FCFS method and a contemporary state-of-the-art approach known as DQN+AQCS. This rigorous evaluation process allowed for a thorough assessment of how the proposed algorithm performed in comparison to these two benchmark methods. The simulation results clearly demonstrated the efficacy of the proposed mechanisms, including video pre-downloading for EV users and bandwidth relay, in conjunction with the utilization of VLC for V2V communication. These mechanisms collectively enhanced the adaptability of base station bandwidth allocation, particularly during peak hours. This enhancement ensures that video conferencing, live video, and VoD applications for EV users can reliably access the necessary bandwidth support even in scenarios of bandwidth scarcity on congested road segments during peak times. Ultimately, the perceived video quality for EV users can be maintained at satisfactory levels.

In summary, the algorithm presented in this paper offers a dual advantage. Specifically, it effectively addresses the challenge of bandwidth contention among EV users, which stems from the constraints of base station bandwidth. Additionally, it represents a notable advancement over both the FCFS and DQN+AQCS approaches. To be more precise, it ensures access to sufficient bandwidth for real-time live video applications and guarantees uninterrupted playback of the vast majority of video applications even during periods of heightened demand. This enhancement significantly elevates the overall quality of the video viewing experience for EV users. Remarkably, even when the available bandwidth from congested base stations reaches its peak capacity, the proposed algorithm demonstrates a substantial improvement of 30% and 17% in average video quality compared to the FCFS approach and the DQN+AQCS method, respectively. This further solidifies the effectiveness of our proposed algorithm in providing satisfactory video quality even within the constraints of bandwidth limitations.

Author Contributions: Conceptualization and methodology, C.-J.H.; software, K.-W.H.; writing and editing, H.-W.C. All authors have read and agreed to the published version of the manuscript.

Funding: This research was funded by National Science and Technology Council, Taiwan, for financially supporting this research under Contract Number NSTC 112-2221-E-259-006.

Institutional Review Board Statement: Not applicable.

Informed Consent Statement: Not applicable.

Data Availability Statement: Not applicable.

Conflicts of Interest: The authors declare no conflict of interest.

Nomenclature

p_1^σ	Starting point of the driving route of EV σ
p_i^σ	The ith road section of the traveling route of EV σ
$p_{h^\sigma}^\sigma$	The final destination of the EV σ
$at_{p_i^\sigma}$	The time when EV σ reaches road section p_i^σ
$\mathrm{SVR}(\cdot)$	The predicted travel time of EV from p_i^σ to p_{i+1}^σ using SVR
$sp_{p_i^\sigma, p_{i+1}^\sigma}\left(at_{p_i^\sigma}\right)$	The average speed at the time of $at_{p_i^\sigma}$ arriving at the road section connecting p_i^σ and p_{i+1}^σ
$\rho_{p_i^\sigma, p_{i+1}^\sigma}\left(at_{p_i^\sigma}\right)$	The traffic volume passing through the road segment connecting p_i^σ and p_{i+1}^σ at the time of $at_{p_i^\sigma}$
$wd_{p_i^\rho, p_{i+1}^\rho}\left(at_{p_i^\sigma}\right)$	The day of the week on which the segment connecting p_i^σ and p_{i+1}^σ is traversed
$wt_{p_i^\rho, p_{i+1}^\rho}\left(at_{p_i^\sigma}\right)$	The weather condition at the time of $at_{p_i^\sigma}$
$\kappa_1^{\sigma,v}$	Whether the video application is a video conferencing application
$\kappa_2^{\sigma,v}$	Whether the video application is a live video application
$\kappa_3^{\sigma,v}$	Whether the video application is a VoD application
$ub_t^{\vartheta,\sigma}$	The upload bandwidth that base station ϑ allocates for EV σ at time slot t
$db_t^{\vartheta,\sigma}$	The download bandwidth that base station ϑ allocates for EV σ at time slot t
$\underline{\mathrm{UR}}(uvr_t^{\sigma,v})$	The minimum bandwidth requirement for uploading of video application v at time t
$\underline{\mathrm{DR}}(dvr_t^{\sigma,v})$	The minimum bandwidth requirement for downloading of video application v at time t
$uvr_t^{\sigma,v}$	The video resolution chosen by the EV user for uploading
$dvr_t^{\sigma,v}$	The video resolution chosen by the EV user for downloading
$\underline{vr}^v$	The minimum resolution for video conferencing or live video set by the software application provider
$\underline{br}^v$	The minimum resolution for VoD v set by the software application provider
$S^{\sigma,v}$	The number of selected video segment for VoD v
τ_s	The download time of selected video segment s
$pt_s^{\sigma,v}$	The playback time of selected video segment s
$\mathrm{SS}_s^{\sigma,v}(br_s^{\sigma,v})$	The length of VoD segment s when the bit rate is $br_s^{\sigma,v}$
$db_{\tau_s}^{\vartheta,\sigma}$	The bandwidth that base station ϑ allocates for the VoD segment in the time slot τ_s

$t_0^{\sigma,v}$	The time when the video conferencing/live video application starts
$t_e^{\sigma,v}$	The time when the video conferencing/live video application ends
$pt_1^{\sigma,v}$	The start time of the VoD application
$pt_{S^{\sigma,v}}^{\sigma,v}$	The end time of the VoD application
$buf_{\tau_s}^{\sigma}$	The remaining buffer of σ at τ_s
buf^{σ}	The buffer size of σ
$ult_t^{\sigma,v}$	The upload times of video conferencing/live video at time slot t
$dlt_t^{\sigma,v}$	The download times of video conferencing/live video at time slot t
$ud^{\sigma,v}$	The maximum upload delay limit for video conferencing/live video at time t
$dd^{\sigma,v}$	The maximum download delay limit for video conferencing/live video at time t
$\phi^{\sigma,v}$	The weight of the lower bound of delay for video transmission in Equation (8)
$\zeta^{\sigma,v}$	The weight of the video inter-frame transmission smoothness penalty in Equation (8)
$rbt_s^{\sigma,v}$	The rebuffering of the selected video segment of the video application
$\psi^{\sigma,v}$	The weight of rebuffering in Equation (8)
$sd^{\sigma,v}$	The starting delay time of the video application
$\omega^{\sigma,v}$	The weight of starting delay time of the video application in Equation (8)
$\beta^{\sigma,v}$	The weight of the inter-segment smoothness penalty of VoD v
$DS_t^{\sigma,v}(uvr_t^{\sigma,v})$	The video frame size for video conferencing/live video with uploading resolution $uvr_t^{\sigma,v}$
$DS_t^{\sigma,v}(dvr_t^{\sigma,v})$	The video frame size for video conferencing/live video with downloading resolution $dvr_t^{\sigma,v}$
$ur_t^{\sigma,v}$	The upload bit rate of v at time t
$dr_t^{\sigma,v}$	The download bit rate of v at time t
$\underline{DB}(br_s^{\sigma,v})$	The minimum bandwidth requirement of the video segment s of VoD v when the preset bit rate is $br_s^{\sigma,v}$
$dr_{\tau_s}^{\sigma,v}$	The actual bandwidth of VoD v downloaded at time τ_s
$buf_{\tau_s}^{\rho}$	The remaining buffer of EV ρ that downloads the segment for σ at τ_s
buf^{ρ}	The buffer size of EV ρ
$\mathcal{R}_t^{\sigma,v}$	The fleet formed at time t
$P^{\sigma,v}$	The number of EVs in the fleet
$x_{r_i^{\sigma,v}}$	The x coordinate of the ith EV in the fleet
$y_{r_i^{\sigma,v}}$	The y coordinate of the ith EV in the fleet
$ut_t^{r_{i+1}^{\sigma,v},r_i^{\sigma,v}}$	The bandwidth uploaded from $r_{i+1}^{\sigma,v}$ to $r_i^{\sigma,v}$ at time t
$dt_t^{r_{i+1}^{\sigma,v},r_i^{\sigma,v}}$	The bandwidth downloaded from $r_{i+1}^{\sigma,v}$ to $r_i^{\sigma,v}$ at time t
ρ	The EV that pre-downloads the VoD segment
$dt_{\tau_s'}^{\sigma,\rho}$	The bandwidth that EV σ downloads from EV ρ at time slot τ_s'

References

1. He, J.; Ammar, M.; Zegura, E. A Measurement-Derived Functional Model for the Interaction Between Congestion Control and QoE in Video Conferencing. In Proceedings of the International Conference on Passive and Active Network Measurement, Cottbus, Germany, 21–23 March 2023; pp. 129–159.
2. Dai, P.; Song, F.; Liu, K.; Dai, Y.; Zhou, P.; Guo, S. Edge Intelligence for Adaptive Multimedia Streaming in Heterogeneous Internet of Vehicles. *IEEE Trans. Mob. Comput.* **2021**, *71*, 1464–1478. [CrossRef]
3. Nguyen, T.V.; Nguyen, N.P.; Kim, C.; Dao, N.N. Intelligent Aerial Video Streaming: Achievements and Challenges. *J. Netw. Comput. Appl.* **2023**, *211*, 103564. [CrossRef]
4. Datta, S.; Ghosh, T. Linking Service Quality, Value, and Loyalty in the VoD Service Context Through a Review of Literature. *Cogent Bus. Manag.* **2022**, *9*, 2143311. [CrossRef]
5. Wang, Y.; Zhao, D.; Huang, C.; Yang, F.; Gao, T.; Zhou, A.; Zhang, H.; Ma, H.; Du, Y.; Chen, A. TrafAda: Cost-Aware Traffic Adaptation for Maximizing Bitrates in Live Streaming. *IEEE/ACM Trans. Netw.* **2023**. [CrossRef]
6. Wiegand, T.; Sullivan, G.J.; Bjontegaard, G.; Luthra, A. Overview of the H. 264/AVC Video Coding Standard. *IEEE Trans. Circuits Syst. Video Technol.* **2003**, *13*, 560–576. [CrossRef]
7. Om, K.; McGill, T.; Dixon, M.; Wong, K.W.; Koutsakis, P.H. 264 and H. 265 Video Traffic Modeling Using Neural Networks. *Comput. Commun.* **2022**, *184*, 149–159. [CrossRef]
8. Sullivan, G.J.; Ohm, J.R.; Han, W.J.; Wiegand, T. Overview of the High Efficiency Video Coding (HEVC) Standard. *IEEE Trans. Circuits Syst. Video Technol.* **2012**, *22*, 1649–1668. [CrossRef]

9. Bonnineau, C.; Hamidouche, W.; Fournier, J.; Sidaty, N.; Travers, J.F.; Déforges, O. Perceptual Quality Assessment of HEVC and VVC Standards for 8K Video. *IEEE Trans. Broadcast.* **2022**, *68*, 246–253. [CrossRef]
10. Bross, B.; Chen, J.; Ohm, J.R.; Sullivan, G.J.; Wang, Y.K. Developments in International Video Coding Standardization After AVC, With an Overview of Versatile Video Coding (VVC). *Proc. IEEE* **2021**, *109*, 1463–1493. [CrossRef]
11. Tsai, Y.H.; Lu, C.R.; Chen, M.J.; Hsieh, M.C.; Yang, C.M.; Yeh, C.H. Visual Perception Based Intra Coding Algorithm for H. 266/VVC. *Electronics* **2023**, *12*, 2079. [CrossRef]
12. Bross, B.; Wang, Y.K.; Ye, Y.; Liu, S.; Chen, J.; Sullivan, G.J.; Ohm, J.R. Overview of the Versatile Video Coding (VVC) Standard and Its Applications. *IEEE Trans. Circuits Syst. Video Technol.* **2021**, *31*, 3736–3764. [CrossRef]
13. Chen, M.J.; Lee, C.A.; Tsai, Y.H.; Yang, C.M.; Yeh, C.H.; Kau, L.J.; Chang, C.Y. Efficient Partition Decision Based on Visual Perception and Machine Learning for H. 266/Versatile Video Coding. *IEEE Access* **2022**, *10*, 42141–42150. [CrossRef]
14. Menasri, W.; Skoudarli, A. Performance Comparison of Throughput Between AVC, HEVC, and VVC Hardware CABAC Decoder. *J. Real-Time Image Process.* **2023**, *20*, 26. [CrossRef]
15. Choi, K. A Study on Fast and Low-Complexity Algorithms for Versatile Video Coding. *Sensors* **2022**, *22*, 8990. [CrossRef]
16. Belda, R.; Arce, P.; de Fez, I.; Guerri, J.C. Performance Evaluation and Testbed for Delivering SRT Live Content Using DASH Low Latency Streaming Systems. In Proceedings of the 19th ACM International Symposium on Performance Evaluation of Wireless Ad Hoc, Sensor, & Ubiquitous Networks, Montreal, QC, Canada, 24 October 2022; pp. 115–121.
17. Wang, C.X.; You, X.; Gao, X.; Zhu, X.; Li, Z.; Zhang, C.; Wang, H.; Huang, Y.; Chen, Y.; Haas, H.; et al. On the Road to 6G: Visions, Requirements, Key Technologies and Testbeds. *IEEE Commun. Surv. Tutor.* **2023**, *25*, 905–974. [CrossRef]
18. Li, Q.; Nayak, A.; Wang, X.; Wang, D.; Yu, F.R. A Collaborative Caching-Transmission Method for Heterogeneous Video Services in Cache-Enabled Terahertz Heterogeneous Networks. *IEEE Trans. Veh. Technol.* **2022**, *71*, 3187–3200. [CrossRef]
19. Liu, X.; Xu, B.; Zheng, K.; Zheng, H. Throughput Maximization of Wireless-Powered Communication Network with Mobile Access Points. *IEEE Trans. Wirel. Commun.* **2023**, *22*, 4401–4415. [CrossRef]
20. Zheng, K.; Jia, X.; Chi, K.; Liu, X. DDPG-Based Joint Time and Energy Management in Ambient Backscatter-Assisted Hybrid Underlay CRNs. *IEEE Trans. Commun.* **2022**, *71*, 441–456. [CrossRef]
21. Mei, J.; Dai, L.; Tong, Z.; Deng, X.; Li, K. Throughput-Aware Dynamic Task Offloading under Resource Constant for Mec with Energy Harvesting Devices. *IEEE Trans. Netw. Serv. Manag.* **2023**. [CrossRef]
22. Wang, Y.; Wong, V.W.; Wang, J. Flexible Rate-Splitting Multiple Access with Finite Blocklength. *IEEE J. Sel. Areas Commun.* **2023**, *41*, 1398–1412. [CrossRef]
23. Tang, F.; Chen, X.; Zhao, M.; Kato, N. The Roadmap of Communication and Networking in 6G for the Metaverse. *IEEE Wirel. Commun.* **2022**, 1–15. [CrossRef]
24. Pei, J.; Li, S.; Yu, Z.; Ho, L.; Liu, W.; Wang, L. Federated Learning Encounters 6G Wireless Communication in the Scenario of Internet of Things. *IEEE Commun. Stand. Mag.* **2023**, *7*, 94–100. [CrossRef]
25. Terahertz Spectrum in 6G: The Only Place to Go Is Up. Available online: https://rcrwireless.com/20220221/5g/terahertz-spectrum-in-6G-the-only-place-to-go-is-up (accessed on 2 July 2023).
26. Serghiou, D.; Khalily, M.; Brown, T.W.; Tafazolli, R. Terahertz Channel Propagation Phenomena, Measurement Techniques and Modeling for 6G Wireless Communication Applications: A Survey, Open Challenges and Future Research Directions. *IEEE Commun. Surv. Tutor.* **2022**, *24*, 1957–1996. [CrossRef]
27. Li, Y.; Chen, Y.; Yan, D.; Guan, K.; Han, C. Channel Characterization and Ray-Tracing Assisted Stochastic Modeling for Urban Vehicle-to-Infrastructure Terahertz Communications. *IEEE Trans. Veh. Technol.* **2022**, *72*, 2748–2763. [CrossRef]
28. Lin, Z.; Wang, L.; Ding, J.; Xu, Y.; Tan, B. Tracking and Transmission Design in Terahertz V2I Networks. *IEEE Trans. Wirel. Commun.* **2022**, *22*, 3586–3598. [CrossRef]
29. Moltchanov, D.; Beschastnyi, V.; Ostrikova, D.; Gaidamaka, Y.; Koucheryavy, Y.; Samouylov, K. Optimal Antenna Locations for Coverage Extension in Sub-Terahertz Vehicle-to-Vehicle Communications. *IEEE Trans. Wirel. Commun.* **2023**. [CrossRef]
30. Azari, M.M.; Solanki, S.; Chatzinotas, S.; Bennis, M. THz-Empowered UAVs in 6G: Opportunities, Challenges, and Trade-Offs. *IEEE Commun. Mag.* **2022**, *60*, 24–30. [CrossRef]
31. Ning, B.; Tian, Z.; Mei, W.; Chen, Z.; Han, C.; Li, S.; Yuan, J.; Zhang, R. Beamforming Technologies for Ultra-Massive MIMO in Terahertz Communications. *IEEE Open J. Commun. Soc.* **2023**, *4*, 614–658. [CrossRef]
32. Jiang, Y.; Li, G.; Ge, H.; Wang, F.; Li, L.; Chen, X.; Lu, M.; Zhang, Y. Machine Learning and Application in Terahertz Technology: A Review on Achievements and Future Challenges. *IEEE Access* **2022**, *10*, 53761–53776. [CrossRef]
33. Kouzayha, N.; Kishk, M.A.; Sarieddeen, H.; Alouini, M.S.; Al-Naffouri, T.Y. Coexisting Terahertz and RF Finite Wireless Networks: Coverage and Rate Analysis. *IEEE Trans. Wirel. Commun.* **2022**, *22*, 4873–4889. [CrossRef]
34. Wang, C.; Chun, Y.J. Stochastic Geometric Analysis of the Terahertz (THz)-mmWave Hybrid Network with Spatial Dependence. *IEEE Access* **2023**, *11*, 25063–25076. [CrossRef]
35. Moltchanov, D.; Sopin, E.; Begishev, V.; Samuylov, A.; Koucheryavy, Y.; Samouylov, K. A Tutorial on Mathematical Modeling of 5G/6G Millimeter Wave and Terahertz Cellular Systems. *IEEE Commun. Surv. Tutor.* **2022**, *24*, 1072–1116. [CrossRef]
36. Yin, F.; Zhang, Z.; Liu, D.; Li, P.; Zeng, M. Mobility-Based Proactive Cache and Transmission Strategy in Millimeter Wave Heterogeneous Networks. *IEEE Syst. J.* **2022**, *17*, 325–336. [CrossRef]

37. Sharda, P.; Reddy, G.S.; Bhatnagar, M.R.; Ghassemlooy, Z. A Comprehensive Modeling of Vehicle-to-Vehicle Based VLC System under Practical Considerations, an Investigation of Performance, and Diversity Property. *IEEE Trans. Commun.* **2022**, *70*, 3320–3332. [CrossRef]
38. Eldeeb, H.B.; Elamassie, M.; Sait, S.M.; Uysal, M. Infrastructure-to-Vehicle Visible Light Communications: Channel Modelling and Performance Analysis. *IEEE Trans. Veh. Technol.* **2022**, *71*, 2240–2250. [CrossRef]
39. Singh, G.; Srivastava, A.; Bohara, V.A.; Liu, Z.; Noor-A-Rahim, M.; Ghatak, G. Heterogeneous Visible Light and Radio Communication for Improving Safety Message Dissemination at Road Intersection. *IEEE Trans. Intell. Transp. Syst.* **2022**, *23*, 17607–17619. [CrossRef]
40. Tran, G.N.; Kim, S. Performance Analysis of Short Packets in NOMA VLC Systems. *IEEE Access* **2022**, *10*, 6505–6517. [CrossRef]
41. Kryszkiewicz, P.; Sroka, P.; Sybis, M.; Kliks, A. Path Loss and Shadowing Modeling for Vehicle-to-Vehicle Communications in Terrestrial TV Band. *IEEE Trans. Antennas Propag.* **2022**, *71*, 984–998. [CrossRef]
42. Aghaei, F.; Eldeeb, H.B.; Uysal, M. A Comparative Evaluation of Propagation Characteristics of Vehicular VLC and MMW Channels. *IEEE Trans. Veh. Technol.* **2023**, 1–10. [CrossRef]
43. Kamiya, S.; Tang, Z.; Yamazato, T.; Kinoshita, M.; Kamakura, K.; Arai, S.; Yendo, T.; Fujii, T. Achieving Successful VLC Signal Reception Using a Rolling Shutter Image Sensor While Driving at 40 km/h. *IEEE Photonics J.* **2023**, *15*, 7302811. [CrossRef]
44. Yang, Y.; Zhang, F.; Zeng, Z.; Cheng, J.; Guo, C. Visible Light Communication for Vehicular Applications: A Novel Architecture with Proof-of-Concept Prototype. *China Commun.* **2023**, *20*, 249–259. [CrossRef]
45. Li, G.; Niu, W.; Ha, Y.; Hu, F.; Wang, J.; Yu, X.; Jia, J.; Zou, P.; He, Z.; Yu, S.; et al. Position-Dependent MIMO De-multiplexing Strategy for High-Speed Visible Light Communication in Internet of Vehicles. *IEEE Internet Things J.* **2021**, *9*, 10833–10850. [CrossRef]
46. Wang, S.; Yang, J.; Bi, S. Adaptive Video Streaming in Multi-Tier Computing Networks: Joint Edge Transcoding and Client Enhancement. *IEEE Trans. Mob. Comput.* **2023**, 1–14. [CrossRef]
47. Liu, W.; Zhang, H.; Ding, H.; Yuan, D. Delay and Energy Minimization for Adaptive Video Streaming: A Joint Edge Caching, Computing and Power Allocation Approach. *IEEE Trans. Veh. Technol.* **2022**, *71*, 9602–9612. [CrossRef]
48. Jedari, B.; Premsankar, G.; Illahi, G.; Di Francesco, M.; Mehrabi, A.; Ylä-Jääski, A. Video Caching, Analytics, and Delivery at the Wireless Edge: A Survey and Future Directions. *IEEE Commun. Surv. Tutor.* **2020**, *23*, 431–471. [CrossRef]
49. Yuan, Y.; Wang, W.; Wang, Y.; Adhatarao, S.S.; Ren, B.; Zheng, K.; Fu, X. Vsim: Improving QoE Fairness for Video Streaming in Mobile Environments. In Proceedings of the IEEE INFOCOM 2022-IEEE Conference on Computer Communications, Online, 2–5 May 2022; pp. 1309–1318.
50. Liu, Y.; Wei, D.; Zhang, C.; Li, W. Distributed Bandwidth Allocation Strategy for QoE Fairness of Multiple Video Streams in Bottleneck Links. *Future Internet* **2022**, *14*, 152. [CrossRef]
51. Tung, T.Y.; Gündüz, D. DeepWiVe: Deep-Learning-Aided Wireless Video Transmission. *IEEE J. Sel. Areas Commun.* **2022**, *40*, 2570–2583. [CrossRef]
52. Chafii, M.; Bariah, L.; Muhaidat, S.; Debbah, M. Twelve scientific challenges for 6G: Rethinking the foundations of communications theory. *IEEE Commun. Surv. Tutor.* **2022**, *25*, 868–904. [CrossRef]
53. Xiong, R.; Zhang, C.; Zeng, H.; Yi, X.; Li, L.; Wang, P. Reducing Power Consumption for Autonomous Ground Vehicles via Resource Allocation Based on Road Segmentation in V2X-MEC with Resource Constraints. *IEEE Trans. Veh. Technol.* **2022**, *71*, 6397–6409. [CrossRef]
54. Zhang, Z.; Chang, Q.; Yang, S.; Xing, J. Sensing-Communication Bandwidth Allocation in Vehicular Links Based on Reinforcement Learning. *IEEE Wirel. Commun. Lett.* **2022**, *12*, 11–15. [CrossRef]
55. Yun, W.J.; Kwon, D.; Choi, M.; Kim, J.; Caire, G.; Molisch, A.F. Quality-Aware Deep Reinforcement Learning for Streaming in Infrastructure-Assisted Connected Vehicles. *IEEE Trans. Veh. Technol.* **2021**, *71*, 2002–2017. [CrossRef]
56. Cheng, K.; Fang, X. A Cost Efficient Edge Computing Scheme in Dual-band Cooperative Vehicular Network. In Proceedings of the 2023 IEEE Wireless Communications and Networking Conference (WCNC), Glasgow, UK, 26–29 March 2023; pp. 1–6.
57. Jiang, X.; Yu, F.R.; Song, T.; Leung, V.C. Resource Allocation of Video Streaming over Vehicular Networks: A Survey, Some Research Issues and Challenges. *IEEE Trans. Intell. Transp. Syst.* **2021**, *23*, 5955–5975. [CrossRef]
58. Katariya, V.; Baharani, M.; Morris, N.; Shoghli, O.; Tabkhi, H. DeepTrack: Lightweight Deep Learning for Vehicle Trajectory Prediction in Highways. *IEEE Trans. Intell. Transp. Syst.* **2022**, *23*, 18927–18936. [CrossRef]
59. Azadani, M.N.; Boukerche, A. A Novel Multimodal Vehicle Path Prediction Method Based on Temporal Convolutional Networks. *IEEE Trans. Intell. Transp. Syst.* **2022**, *23*, 25384–25395. [CrossRef]
60. Müller-Hannemann, M.; Rückert, R.; Schiewe, A.; Schöbel, A. Estimating the Robustness of Public Transport Schedules Using Machine Learning. *Transp. Res. Part C Emerg. Technol.* **2022**, *137*, 103566. [CrossRef]
61. Ma, X.; Li, Q.; Zou, L.; Peng, J.; Zhou, J.; Chai, J.; Jiang, Y.; Muntean, G.M. QAVA: QoE-Aware Adaptive Video Bitrate Aggregation for HTTP Live Streaming Based on Smart Edge Computing. *IEEE Trans. Broadcast.* **2022**, *68*, 661–676. [CrossRef]
62. Hewage, C.T.; Ahmad, A.; Mallikarachchi, T.; Barman, N.; Martini, M.G. Measuring, Modeling, and Integrating Time-Varying Video Quality in End-to-End Multimedia Service Delivery: A Review and Open Challenges. *IEEE Access* **2022**, *10*, 60267–60293. [CrossRef]
63. Traffic Volume Counts for New York City. Available online: https://data.cityofnewyork.us/Transportation/Automated-Traffic-Volume-Counts/7ym2-wayt (accessed on 20 July 2023).

64. How Much Bandwidth Does Skype Need. Available online: https://support.skype.com/en/faq/FA1417/how-much-bandwidth-does-skype-need (accessed on 14 July 2023).
65. Quinlan, J.J.; Zahran, A.H.; Sreenan, C.J. Datasets for AVC (H. 264) and HEVC (H. 265) Evaluation of Dynamic Adaptive Streaming over HTTP (DASH). In Proceedings of the 7th International Conference on Multimedia Systems, Klagenfurt, Austria, 10–13 May 2016; pp. 1–6.
66. Ichigaya, A.; Nishida, Y. Required Bit Rates Analysis for a New Broadcasting Service Using HEVC/H. 265. *IEEE Trans. Broadcast.* **2016**, *62*, 417–425. [CrossRef]
67. Wang, C.; Pang, M.; Zhong, D.; Cui, Y.; Wang, W. A MmWave Communication Testbed Based on IEEE 802.11 Ad with Scalable PtMP Configuration. *China Commun.* **2022**, *19*, 44–56. [CrossRef]
68. Zhang, Z.; Liu, Y.; Huang, J.; Zhang, J.; Li, J.; He, R. Channel Characterization and Modeling for 6G UAV-Assisted Emergency Communications in Complicated Mountainous Scenarios. *Sensors* **2022**, *23*, 4998. [CrossRef] [PubMed]

sensors

MDPI

Article

Performance Enhancement in Federated Learning by Reducing Class Imbalance of Non-IID Data

Mihye Seol and Taejoon Kim *

School of Information and Communication Engineering, Chungbuk National University, Chungju 28644, Republic of Korea
* Correspondence: ktjcc@chungbuk.ac.kr

Abstract: Due to the distributed data collection and learning in federated learnings, many clients conduct local training with non-independent and identically distributed (non-IID) datasets. Accordingly, the training from these datasets results in severe performance degradation. We propose an efficient algorithm for enhancing the performance of federated learning by overcoming the negative effects of non-IID datasets. First, the intra-client class imbalance is reduced by rendering the class distribution of clients close to Uniform distribution. Second, the clients to participate in federated learning are selected to make their integrated class distribution close to Uniform distribution for the purpose of mitigating the inter-client class imbalance, which represents the class distribution difference among clients. In addition, the amount of local training data for the selected clients is finely adjusted. Finally, in order to increase the efficiency of federated learning, the batch size and the learning rate of local training for the selected clients are dynamically controlled reflecting the effective size of the local dataset for each client. In the performance evaluation on CIFAR-10 and MNIST datasets, the proposed algorithm achieves 20% higher accuracy than existing federated learning algorithms. Moreover, in achieving this huge accuracy improvement, the proposed algorithm uses less computation and communication resources compared to existing algorithms in terms of the amount of data used and the number of clients joined in the training.

Keywords: federated learning; non-IID data; class imbalance mitigation

Citation: Seol, M.; Kim, T. Performance Enhancement in Federated Learning by Reducing Class Imbalance of Non-IID Data. *Sensors* **2023**, *23*, 1152. https://doi.org/10.3390/s23031152

Academic Editors: Yichuang Sun, Haeyoung Lee and Oluyomi Simpson

Received: 29 November 2022
Revised: 12 January 2023
Accepted: 16 January 2023
Published: 19 January 2023

1. Introduction

As the number of smartphones and Internet of Things (IoT) devices grows rapidly, the amount of data they are generating is growing explosively [1]. A mainstream in utilizing this large volume of data distributed over multiple devices is centralized data processing, i.e., transferring those devices' data to a server and training a machine learning model from it. However, transferring this huge amount of data to the processing server causes network overhead and increases communication costs. Additionally, data processing servers demand enormous storage and computing power, resulting in high maintenance costs. Federated learning (FL) has been proposed to solve these problems [2].

FL allows clients to cooperate to generate a global model without sharing the clients' data with a server. Federated Averaging (FedAVG) [3], a representative algorithm of FL, sends the local model parameters to a server after each device learns a local model using its own local dataset. The server configures a global model by aggregating the received local parameters. However, unlike central data processing, FL uses clients' resources to learn models, accordingly, the system heterogeneity (computing power, wireless channel environment, size of dataset, etc.) among clients has a significant impact on learning efficiency.

To resolve the problem of system heterogeneity among clients, a lot of research works were conducted to schedule devices on servers. In [4], the authors proposed a method of selecting clients based on the available amount of communication and computing resources with the goal of fast convergence and high accuracy of a global model. The methods of effectively utilizing communication resources were proposed in [5,6]. In [5], a method of controlling the frequency between a global aggregation and a local update was proposed. In [6], the authors proposed a method of applying hierarchical aggregation. In [7], the model parameters for servers and local clients were compressed for efficient use of communication resources. In addition, some research works were conducted to dynamically allocate batch size for clients based on the available amount of communication and computing resources [8–10].

One of the most important issues in FL is statistical heterogeneity, i.e., the negative effect of non-independent and identically distribution (non-IID) of the training dataset. The distribution of data generated by a client varies depending on the client's occupation, lifestyle, residential area, etc. As a result, the local data distribution of a client will be non-IID with a high probability. Accordingly, the class distribution of the client also has a class imbalance.

Class imbalance can be categorized into intra-client class imbalance and inter-client class imbalance. Intra-client class imbalance means that the distribution of data amount among classes, i.e., class distribution, in a client is different from Uniform distribution. The larger the distribution gap is, the more severe the imbalance is. Inter-client class imbalance means that the class distribution of each client is different from other clients' distribution. In [11,12], it was confirmed that the accuracy of FL was decreased when these intra- and inter-client class imbalances were considered.

Although a lot of research works have been conducted to prevent learning efficiency from decreasing when the class distribution of clients is imbalanced, to the best of our knowledge, an integrated research work incorporating three core components—(1) a method of reducing the intra-client class imbalance, (2) a method of reducing the inter-client class imbalance, and (3) a method of dynamic batch size allocation and learning rate control—has never been conducted.

We propose a novel algorithm that supports intra- and inter-client class imbalance mitigation and dynamic batch size allocation and learning rate control considering the amount of local dataset. First, the proposed algorithm performs data oversampling to make the class distribution of each client close to Uniform distribution. This oversampling scheme for FL, to the best of our knowledge, is the first approach incorporating an exponential decay factor, and it dynamically reflects the amount of oversampled data in the previous round. Second, to avoid performance degradation due to inter-client class imbalance, the clients to join FL are selected to balance the aggregated class distribution for each round, and the amount of data to be actually used for local learning is also adjusted by considering the class distributions of the selected clients. The combination of these two features in client selection is a unique contribution of this paper and shows significant performance improvement. Finally, the batch size and the learning rate of the selected clients are adjusted according to the amount of data for the clients. It is also the first approach presenting the dynamic batch size and learning rate adjustment assuming a common SGD update in an FL.

The performance of the proposed algorithm is validated over the CIFAR-10 [13] and MNIST [14] datasets in non-IID scenarios, and it is confirmed that the accuracy of the global model from the proposed algorithm achieves about 20% better performance than existing FL algorithms in non-IID situations. Moreover, despite this remarkable improvement in accuracy, the computing and communication resource usage in terms of the amount of data used for learning and the number of clients participating in learning are decreased compared to existing FL algorithms.

The main contributions of this paper are summarized as follows:

- To mitigate intra-client class imbalance, a novel data sampling to local datasets is introduced, which results in accuracy improvement in non-IID environments.
- An FL server intelligently selects clients and allocates the amount of data to be actually used in local learning by balancing the class distributions of the selected clients.

- The batch size and the learning rate of clients are dynamically controlled according to the amount of local dataset for each client.
- Performance evaluation in various non-IID scenarios confirms that the proposed algorithm achieves high accuracy and low usage of computing and communication resources compared to existing algorithms.

The remainder of this paper describes the following. Section 2 introduces the literature review, and Section 3 describes the overall system structure and defines the class distribution of clients. Section 4 describes the detailed procedure of the proposed algorithm, Section 5 shows the experimental results, and finally, Section 6 concludes the paper.

2. Related Works

In the literature, various research works were conducted to improve the performance of a global model under a non-IID dataset. In an intra-client class imbalance situation, in order to solve the learning efficiency reduction problem, there was an attempt to make the local class distribution of clients close to IID by sharing data among clients. In [12], a small IID dataset was created in a server by collecting data from clients to mitigate the negative effects of intra-client class imbalance. However, this approach does not meet the original purpose of FL because the clients' privacy is not protected by transmitting their data to the server to generate the small IID dataset.

In [15], both the statistical heterogeneity and the system heterogeneity were considered to prevent local models from deviating from a global model. Specifically, a proximal term was added to a loss function. Similarly, in [16], the elastic weight consolidation method was proposed to add a penalty term to a loss function to prevent the models of non-IID clients from drifting apart from each other.

Another research approach to alleviating inter-client class imbalance is to more intelligently select the clients to participate in an FL. The authors of [17] have improved the performance of FL by increasing the selection probability for the clients having a large gradient value. In [18], a scheme of group learning for clients with similar class distributions and merging the trained models into a global model was proposed. In [19], FL models could converge with fewer rounds through a hierarchical clustering of clients based on the similarity of local models of clients. In [20], the data augmentation scheme was proposed as a solution to a global imbalance situation in which the aggregated class data distribution of clients differs from Uniform distribution. In addition, mediator-based client rescheduling is introduced to alleviate local imbalance.

The level of IID for the local dataset was evaluated using weight divergence and multi-arm bandit [21]-based algorithms in [22] and [23], respectively. Moreover, in [23], the negative effects of local imbalance were reduced by increasing the selection probability of clients with high IID levels. The authors in [24] showed that, for performance improvement, the aggregation weights of local models should be finely adjusted considering the quality and quantity of local data, the number of classes, and the entropy of local data.

In FedNova [25], the performance degradation due to the differences in the number of local updates was reported, and this difference was from the heterogeneity in non-IID local datasets and computation resources. To solve this problem, a normalized model aggregation method was proposed. In [26], the performance degradation of stochastic gradient descent (SGD) method over non-IID data was mitigated by introducing a deep reinforcement learning-based client selection and client-specific batch size allocation scheme.

Although various studies have been conducted until now, research considering both intra- and inter-client class imbalance mitigation and dynamic batch size and learning rate adjustment considering the size of the dataset has not been conducted.

3. System Model and Data Distributions

3.1. System Model

An FL system for a multi-class classification task consists of a server to manage the global model and a set of clients $\mathcal{K} = \{1, 2, \ldots, K\}$. Each client has a local dataset, and client k's local dataset is denoted by $\mathcal{D}_k$. In rth round of the FL, client k, which is selected to participate in this learning, starts a local learning using its local dataset $\mathcal{D}_k$ with the initial global model vector w_r, received from the server. Client k makes up a mini batch set $\mathcal{B}_k$ from the local dataset $\mathcal{D}_k$ and proceeds with the local learning using an SGD optimizer. The update rule for the local learning is expressed as follows:

$$w_{k,r+1} \leftarrow w_{k,r} - \eta \frac{1}{|\mathcal{D}_k|} \sum_{x \in \mathcal{B}_k} \nabla f_k(w_{k,r}; x),\ \forall k \in \mathcal{K}, \tag{1}$$

where $|\mathcal{D}_k|$ denotes the cardinality of $\mathcal{D}_k$, $f_k(w_{k,r}; x)$ is a loss function for the local model vector $w_{k,r}$ and data x, and η is the learning rate. Each selected client trains the local model until a pre-determined local epoch and transmits the learned local vector to the server. The server updates the global model vector by aggregating the received local model vectors. In the aggregation process, a weight for each local model is required, and it is determined to be the amount of data used in each local training divided by the total amount of data for the entire clients participating in the learning. The aggregation with the weights is given by

$$w_{r+1} \leftarrow \sum_{k \in \mathcal{S}} \frac{|\mathcal{D}'_k|}{|\mathcal{D}|} w_{k,\ r+1}, \tag{2}$$

where $\mathcal{S}$ denotes the set of clients selected by the server to participate in the learning, $\mathcal{D} \triangleq \cup_{k \in \mathcal{S}} \mathcal{D}_k$. $\mathcal{D}'_k$ denotes data used by client k for local learning and has a relation of $\mathcal{D}'_k \subset \mathcal{D}_k$. This process is repeated until a specified round is reached. The main parameters of the system model are summarized in Table 1.

Table 1. Notation and definitions.

Notation	Definition
$\mathcal{K}$	Client index set
r	Round index
θ_{KLD}	Kullback–Leibler divergence threshold
h	Maximum number of selected clients
L	Number of classes
δ	Oversampling exponent
$b_{k,r}$	Batch size of client k at round r
$\eta_{k,r}$	Learning rate of client k at round r
w_r	Global model parameter at round r
$w_{k,r}$	Local model parameter of client k at round r
$\mathcal{D}_k$	Local dataset of client k
K	Number of clients
$\mathcal{B}_k$	Mini batch set for client k
$f_k(\cdot\,;\,\cdot)$	Local loss function of client k
α	Dirichlet distribution control parameter
n_k	Class data volume for client k
t_k	Average amount of class data for client k
$s_{k,r}$	Class training data volume for client k at round r
v_r	Class training data volume at round r
β	Number of SGD updates
$\eta_{\max}$	Maximum learning rate

3.2. Data Distributions

As shown in [27], the class distribution of a client is set using Dirichlet distribution. When a classification task has L classes to classify, it is assumed that all clients' local learning data are extracted according to a vector $\boldsymbol{q}$ ($q_\ell \geq 0$, $\ell \in [1, L]$ and $\| \boldsymbol{q} \|_1 = 1$), which corresponds to the class distribution. The class distribution of the clients is determined by $\boldsymbol{q} \sim \text{Dir}(\alpha \boldsymbol{p})$ of Dirichlet distribution, where $\boldsymbol{p}$ denotes the prior probability distribution and α is a parameter that adjusts the uniformity of the class distribution among the clients.

If $\alpha > 0$ and $\alpha \to \infty$, then the class distribution of all the clients approaches to Uniform distribution. Conversely, when α is close to 0, all the clients have only a single class of data, resulting in a non-IID class distribution.

4. Proposed Algorithm

4.1. Alleviating Intra-Client Class Imbalance

When the clients' class distribution is IID, the performance of the FL is very close to centralized learning methods. However, when the class distribution of the local dataset is non-IID, the accuracy of the FL decreases because the local model learned is biased to some class data. Hence, the performance of the FL model can be improved when the class distribution of each client is close to Uniform distribution. Consequently, an oversampling method of making the class distribution close to Uniform distribution without data sharing is proposed. The purpose of this scheme is to render the non-IID dataset to IID as close as possible.

Denote $\boldsymbol{n}_k = \left[n_k^1, n_k^2, \ldots, n_k^L\right]^T$ as class data distribution vector for client k, where n_k^j is the data amount of j-th class for client k. Then, the average amount of data over the classes is $t_k = \frac{1}{L}\sum_\ell^L n_k^\ell$. Client k randomly oversamples data elements in the classes having a smaller volume than t_k. This oversampling is conducted until the volume of each class reaches the average t_k ($n_k^\ell \geq t_k, \ell \in [1, L]$). One of the noteworthy features of this oversampling is that it reduces intra-client class imbalance without losing any of the data obtained with much effort. Due to this data reserving characteristics of the oversampling, it outperforms the other method of sampling at the level of an average over classes.

However, note that data oversampling can lead to overfitting, which reduces the generalizability of the model as the amount of local data increases. To avoid overfitting, the amount of oversampled data per round should be reduced as the round goes on. Accordingly, the training load on each client due to the oversampling is diminished rapidly. Specifically, in the proposed method, the amount of oversampling is exponentially reduced as $e^{-\delta r}$, where r is the round index and the exponent δ is increased when the amount of total oversampled data in the previous round is greater than the threshold θ_{over}. This scheme enables the early termination of the oversampling to prevent the local model from falling into overfitting. Figure 1 shows the local data sampling of a client, and the detailed process is represented in Algorithm 1.

Algorithm 1. Sampling. number of data per class is greater than or equal to the average

• **client executes:**
• **Input**
$\boldsymbol{n}_K$, r round index, δ oversampling exponent
• **Output**
$\boldsymbol{n}_K$
1: $t_k \leftarrow \frac{1}{L}\sum_\ell^L n_k^\ell \times e^{-\delta r}$
2: **repeat**
3: oversampling for $\mathcal{D}_k$
4: **until** $n_k^\ell \geq t_k,\ \ell \in [1, L]$
5: **return** $\boldsymbol{n}_K$
• **server executes:**
• **Input**
$\mathcal{S}$ selected client set
• **Output**
δ
1: **if** $\frac{\sum_{k\in\mathcal{S}}^k \left(\sum_\ell^L n_k^\ell - |\mathcal{D}_k|\right)}{\sum_{k\in\mathcal{S}}^k |\mathcal{D}_k|} > \theta_{\text{over}}$ //calculate oversample data rate
2: $\delta \leftarrow \delta + \Delta$
3: **return** δ

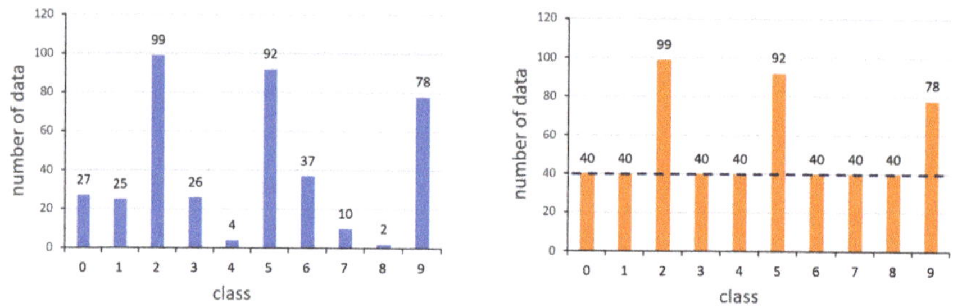

(**a**) Class distribution before oversampling (**b**) Class distribution after oversampling

Figure 1. Class data distribution through data oversampling, where the dotted line in (**b**) is the average amount of data per class.

4.2. Alleviating Inter-Client Class Imbalance

To alleviate the class imbalance among the clients, the server utilizes the sum of the class distributions in selecting the clients to participate in the FL. In addition to this client selection, the server determines the amount of data per class to learn for the selected clients. In each round, the negative effect of inter-client class imbalance can be alleviated by rendering the aggregated distribution of the total training data close to Uniform distribution.

Specifically, any client who wants to participate in rth round learning conducts data oversampling for intra-client class imbalance mitigation and transmits class data volume information $\boldsymbol{n}_k$ to the server. The client can transmit the information about the amount of data to contribute to the learning in this round by reflecting it in $\boldsymbol{n}_k$. This sharing of $\boldsymbol{n}_k$ may reveal the information of the clients to the server; however, note that not the content of data but only the class distribution information is transmitted. Moreover, $\boldsymbol{n}_k$ can be different from the actual class distribution of client k. The server manages the data amount information $\boldsymbol{s}_{k,r} = \left[s^1_{k,r}, s^2_{k,r}, \ldots, s^L_{k,r}\right]^T$ for each client k, where $s^\ell_{k,r}$, $\ell = 1, \ldots, L$ is the amount of ℓth class data for client k to learn in rth round. Considering all the selected clients, the server also manages the information of the amount of data per class required in the learning as $\boldsymbol{v}_r = \left[v^1_r, v^2_r, \ldots, v^L_r\right]^T$, where v^ℓ_r, $\ell = 1, \ldots, L$ is the total amount of ℓth class data to learn in rth round.

The server sorts the clients willing to participate in the learning in descending order of the amount of local dataset, i.e., $\sum_\ell^L n^\ell_k$. Let client k be on the top of the sorted client list. The server updates $\boldsymbol{v}_r$ and $\boldsymbol{s}_{k,r}$ with the data amount information $\boldsymbol{n}_k$. In $\boldsymbol{v}_r$, the server finds out the class with a maximum amount of data and the class with a minimum amount of data. Now, assume that the volume of the maximum amount class is denoted as m and the class index for the minimum amount of data is denoted as f. Through the sorted client list, the server searches for a client i having data to learn in class f. Similarly, $\boldsymbol{v}_r$ and $\boldsymbol{s}_{i,r}$ are updated using $\boldsymbol{n}_i$.

However, in this update process, we regulate the accumulated amount of data in each class to be equal to or smaller than the maximum value m so that this process does not break the balance among the classes. When $\boldsymbol{v}_r$ is updated, the uniformity of class distribution $\boldsymbol{v}_r$ is tested by calculating the Kullback–Leibler divergence (KLD) [28] between $\boldsymbol{v}_r$ and Uniform distribution. This client selection process terminates if the calculated KLD is below the threshold θ_{KLD}, or if the number of selected clients reaches the maximum number of clients h. Finally, the server informs the selected clients of the amount of data to learn by delivering $\boldsymbol{s}_{s,r}$ to client $s \in \mathcal{S}$. This process is expressed in Algorithm 2.

Algorithm 2. Client Selection. The server selects clients and adjusts client's training data

• **Input**
$\mathcal{K}$, $\mathcal{N} = \{n_1, n_{2,} \ldots, n_K\}$ client's class information set, θ_{KLD} KLD threshold,
h maximum number of selected client, L number of classes
• **Output**
selected client class information set $\mathcal{S}_{\text{info}} = \{s_{1,r}, s_{2,r}, \ldots, s_{h,r}\}^T$
initialize $S_{\text{info}} \leftarrow \varnothing$, data volume vector $v_r = [v_r^1, v_r^2, \ldots, v_r^L]^T$
Sort $\mathcal{N}$ in descending order by the amount of data $\sum_\ell^L n_{k'}^\ell$, $k \in \mathcal{N}$
repeat
    for each $n_c \in \mathcal{N}$ **do**
        if $\mathcal{S}_{\text{info}}$ is empty **then**
            for each ℓ, $\ell = 1, 2, \ldots, L$ **do**
                $v_r^\ell \leftarrow v_r^\ell + n_c^\ell$
                $s_{c,r}^\ell \leftarrow n_c^\ell$
            end for
            $m \leftarrow \max(v_r)$ //Maximum value among v_r
            add $s_{c,r}$ in $\mathcal{S}_{\text{info}}$
        else
            $f \leftarrow \text{argmin}_\ell\,(v_r)$
            if $n_c^f > 0$ **then**
                for each ℓ, $\ell = 1, 2, \ldots, L$ **do**
                    $v_r^\ell \leftarrow v_r^\ell + \min(m—v_r^\ell, n_c^\ell)$
                    $s_{c,r}^\ell \leftarrow \min(m—v_r^\ell, n_c^\ell)$
                end for
                add $s_{c,r}$ in $\mathcal{S}_{\text{info}}$
            end if
        end if
    end for
until $|\mathcal{S}_{\text{info}}| == h$ or $D_{KL}\,(P_{v_r}|P_{\text{uniform}}) < \theta_{KLD}$
return $\mathcal{S}_{\text{info}}$

4.3. Dynamic Batch Size and Learning Rate Control

In an FL, each client has a different amount of training data. Accordingly, each client needs to use different hyperparameters, e.g., batch size and learning rate. As shown in [26], under the non-IID dataset situation, if the batch size for local training is not properly adjusted, performance degradation is inevitable. Hence, the efficiency of the FL can be increased by dynamically controlling the batch size and the learning rate for each client by considering the amount of data $\sum_\ell^L s_{k,r}^\ell$ of the clients. By assuming a common number of SGD updates for the clients, an efficient batch size can be obtained. Specifically, in rth round of the proposed scheme, client k uses the value $\left\lfloor \frac{\sum_\ell^L s_{k,r}^\ell}{\beta} \right\rfloor$ as its batch size $b_{k,\,r}$, where β is the required number of SGD updates and $\lfloor\cdot\rfloor$ is the floor operator.

Note that the batch size is proportional to the amount of local dataset for each client, which leads to the improvement of the accuracy of the global model. As the clients learn using different batch sizes, it is necessary to control the learning rate for each client for the purpose of converging the global model. Note that when the batch size is small and the learning rate is too large, the loss function of the local model may diverge. Conversely, if the batch size is large and the learning rate is small, the convergence of local learning is too slow. It coincides with the relationship between the learning rate and the batch size shown in [29]. Therefore, the learning rate is adjusted to be proportional to the batch size. In addition, when the learning rate is too large, the model may not converge, accordingly, the learning rate is regulated with arctan function so that the learning rate does not exceed the maximum learning rate $\eta_{\max}$. This process is explained in Algorithm 3.

Algorithm 3. DynamicBL. dynamically allocate batch size and learning rate

• Input
$s_{k,r}$, $\eta_{\max}$ maximum learning rate, β number of SGD update
• Output
Batch size $b_{k,r}$, learning rate $\eta_{k,r}$
1: $b_{k,r} \leftarrow \left\lfloor \frac{\sum_{\ell}^{L} s_{k,r}^{\ell}}{\beta} \right\rfloor$
2: $\eta_{k,r} \leftarrow \eta_{\max} \times \arctan(b_{k,r})$
3: **return** $b_{k,r}$, $\eta_{k,r}$

4.4. Workflow

The training procedure of the proposed algorithm consists of local data sampling, client selection and training data allocation, and the control of dynamic batch size and learning rate. This procedure is followed by local training and local model aggregation. This overall process is shown in Figure 2.

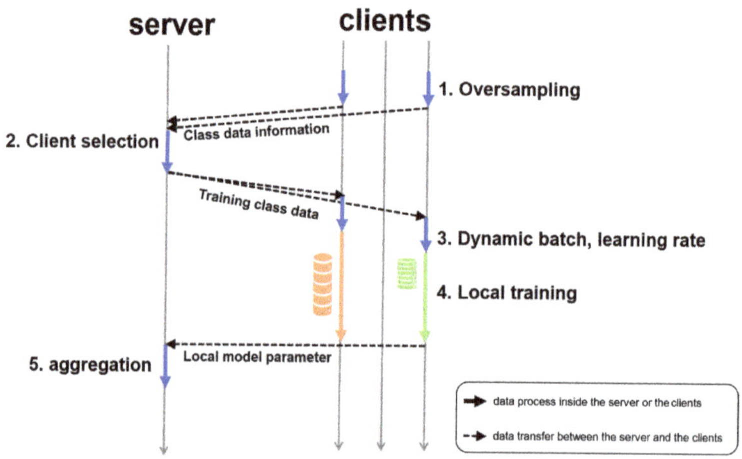

Figure 2. The workflow of the FL with the proposed algorithm, where the orange and green cylinders are the local datasets with different batch sizes. The solid arrow lines mean the data process inside the server or the clients, and the dotted lines mean the data transfer between the server and the clients.

- Local data sampling: a client who wants to participate in learning checks the class distribution of the local dataset and proceeds with oversampling, and then sends the data distribution information to the server.
- Client selection and allocation of training data: the server selects the clients to make the class distribution of learning data balanced for each round and delivers the information about the amount of training data to the selected clients.
- Dynamic batch and learning rate control: each client calculates the batch size and learning rate of local learning based on the amount of data it learns.
- Local training: Each client learns a local model using the amount of training data received from the server and the previously calculated batch size and learning rate. After learning, the client sends the local model parameters to the server.
- Aggregation: When the server receives all the selected clients' local model parameters, it updates the global model parameters using Equation (2). Then repeat until the final round.

5. Experiment Results

5.1. Experiment Setup

In the performance evaluation of the proposed algorithm, the representative dataset of CIFAR-10 and MNIST are adopted. The deep learning model for this classification task is the convolution neural network (CNN) with two 5×5 convolution layers for CIFAR-10, each with 64 and 128 filters. The two convolution layers are followed by a max pool layer, three fully connected layers, and a softmax layer. Then the classification probabilities are derived. We also perform experiments with a simple logistic regression classifier, which we train on the MNIST dataset.

In this evaluation, two baseline algorithms of FedAVG and FedNova [25] are used. FedAVG is a representative algorithm for FL, and FedNova reduces the negative effect of non-IID by normalizing the aggregation step of FedAVG with the number of local updates.

For the CIFAR-10 dataset, the range of the Dirichlet parameter which determines the data distribution of the clients is $\alpha \in [0,\ 0.2]$, and three test cases are considered according to the setting of Dirichlet distribution. In the case of the MNIST dataset, $\alpha = 0$ and all the clients have only a single class. The maximum number of clients participating in each round, h, is set to 10, and FedAVG and FedNova randomly select 10 clients in each round and conduct global aggregation. The initial oversampling decay exponent δ is set to 0.01, the oversampling decay exponent increment Δ is set to 0.1, and the KLD threshold θ_{KLD} is set to 0.1 which checks the similarity between the data distribution v_r and Uniform distribution.

In the setting for the local model training on CIFAR-10, FedAVG and FedNova set the local epoch, batch size, and learning rate to 5, 64, and 0.1, respectively. The local epoch is set as the same value in FedAVG [3]. For MNIST, local epoch, batch size, and learning rate are 5, 10, and 0.03, respectively. In the proposed algorithm, the number of SGD updates β is set to 3, 25, and 25 for Test Cases 1–3, respectively. The maximum learning rate $\eta_{\max}$ is set to 0.1. It is assumed that the computation capabilities of the clients are equal.

5.2. Results on Different Non-IID Data Distribution

As mentioned above, the three different non-IID scenarios of CIFAR-10 Cases 1–3 are considered. In Case 1, $\alpha = 0$ for all the 200 clients, where these clients have only a single class of data. In Case 2, $\alpha = 0$ for 180 out of 200 clients who have only a single class data, and $\alpha = 0.2$ for the remaining 20 clients. On average, a client with $\alpha = 0.2$ has 6 classes of data, where approximately 4 out of the 6 classes have 26–28% less amount of data than the average amount of data for each class. In Case 3, $\alpha = 0.2$ for all 100 clients, and it is the most balanced class distribution compared to other test cases. In Case 4, $\alpha = 0$ for 200 clients who have only a single class MNIST data. Table 2 shows the data distribution parameters and experimental parameters for all the test cases. Hyperparameter values are derived experimentally to obtain optimal results.

Table 2. Distribution setup and experiment parameters.

Distribution Setup				Experiment Parameter					
				Sampling	Client Selection	Dynamic Batch		Local Training	
Datasets	Case	K	α	θ_{over}	θ_{KLD}	$\eta_{\max}$	β	h	epoch
CIFAR-10	1	200	0	0.1	0.1	0.1	3	10	5
	2	200	0 or 0.2	0.1	0.1	0.1	25	10	5
	3	100	0.2	0.1	0.1	0.1	25	10	5
MNIST	4	200	0	0.1	0.1	0.1	25	10	5

In Figure 3, the average accuracy of the proposed global model is depicted for all the test cases. In obtaining the average accuracy, each algorithm is executed 10 times and averaged. As shown in (a)–(c) of this figure, the achieved accuracy of the proposed algorithm is highest when tested on Case 3 but lowest when tested on Case 1. Note that

Case 1 has both severe intra- and inter-client class imbalances. The proposed algorithm achieves 10.4% higher accuracy in Case 2 than in Case 1. As a result, the accuracy of FL decreases when intra-client class imbalance and inter-client class imbalance are very high; however, the accuracy can be improved even when the number of clients having a balanced intra-client class distribution is low. Since Case 3 has the lowest class imbalance, the proposed algorithm achieves the highest accuracy. Figure 3d shows the results of the MNIST dataset, where the proposed algorithm achieves higher accuracy than Case 1, but it has more fluctuation.

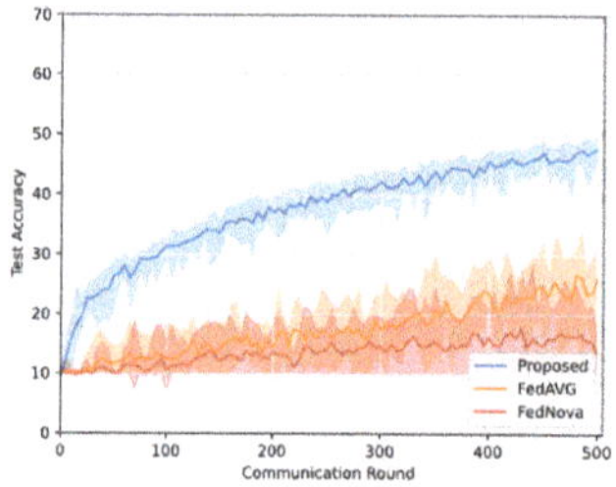

(**a**) Case 1 (α = 0)

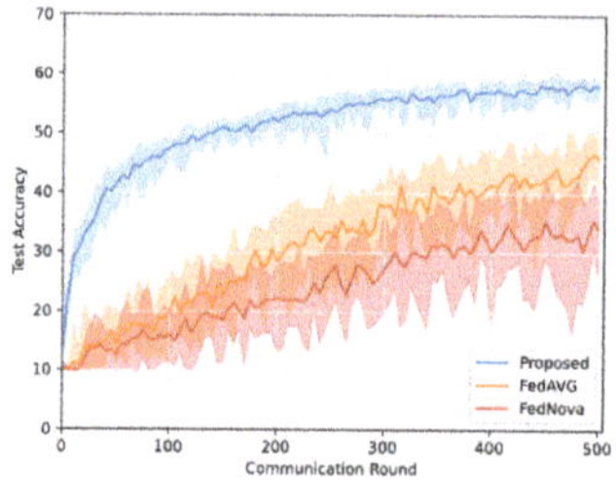

(**b**) Case 2 (α = 0 or 0.2)

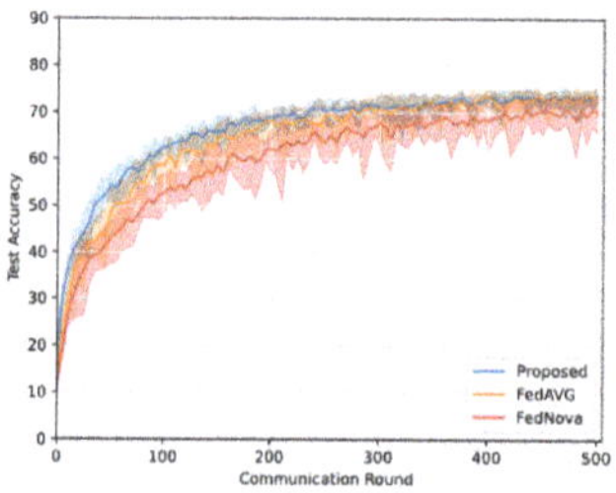

(**c**) Case 3 (α = 0.2)

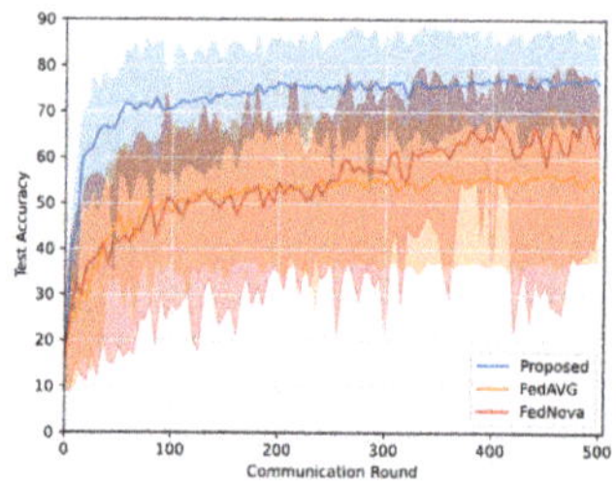

(**d**) Case 4 (α = 0)

Figure 3. Accuracy comparison between the proposed algorithm and two baseline algorithms (FedAVG, FedNova) on non-IID dataset, where the bold lines represent the average of the accuracy for 10 executions and the light area is the range for the maximum and minimum values. Cases 1–3 are the results of the CIFAR-10 dataset, and Case 4 is the result of the MNIST dataset.

Comparing the proposed algorithm with other baseline algorithms, as shown in Figure 3a, in Case 1, the accuracy of the proposed algorithm is improved by 21.8% and 34% compared to FedAVG and FedNova, respectively. Moreover, in this case, the baseline algorithms have a large fluctuation in accuracy from round to round, leading to poor training stability. In addition, when comparing the convergence of the three algorithms, the proposed algorithm converges at a faster rate than the baseline algorithms. In Case 1, since all the clients have only a single class of data, the intra-client class imbalance alleviation method in the proposed algorithm is skipped because this method generates duplicate data elements in non-empty classes. It is noteworthy that the proposed algorithm successfully improves the accuracy of the global model and reduces the variability of the global model without the intra-client class imbalance alleviation method. The baseline algorithms randomly select clients, hence, the sum of the class distributions of the selected clients is imbalanced. As a result, in the baseline algorithms, the accuracy decrement and the high fluctuation are inevitable. On the contrary, the proposed algorithm can improve the accuracy and stability of the global model by applying the inter-client class imbalance method and the dynamic batch size and learning rate control method. In Figure 3b, for Case 2 (α = 0 or 0.2), the proposed algorithm achieves 12.2% and 23.8% accuracy improvements over FedAVG and FedNova, respectively. Moreover, this higher accuracy is achieved within fewer communication rounds and with less fluctuation than FedAVG and FedNova.

In the early rounds in Figure 3b, a surge in the accuracy of the proposed algorithm is observed. This rapid accuracy increment in the early rounds is induced by the data oversampling method to mitigate intra-client class imbalance, which enables the learning with more data in the early stage of the learning. As shown in Figure 3c, for Case 3 (α = 0.2), the proposed algorithm does not improve accuracy significantly compared to FedAVG. Unlike Case 1 and Case 2, Case 3 has a balanced class distribution, accordingly, the performance of the proposed algorithm is similar to FedAVG. However, it still achieves about a 4% accuracy improvement over FedNova. In Figure 3d for Case 4, the proposed algorithm shows 21.1% and 11.4% improved accuracy over FedAVG and FedNova, respectively. FedNova performs better than FedAVG on MNIST, and vice versa on CIFAR-10.

5.3. Results on Class Imbalance Mitigation

Experiments are conducted to validate the effectiveness of the three core methods which constitute the proposed algorithm. For Cases 1–3 of the CIFAR-10 dataset, which showed the highest performance improvement, we compare it with the baseline algorithms. In the following experiments, the local data oversampling method to alleviate intra-client class imbalance is denoted as 'data sampling', the client selection and training data allocation method to alleviate inter-client class imbalance is denoted as 'client selection', and the dynamic batch size and learning rate control technique is expressed as 'dynamic batch'.

In Figure 4, the accuracy of 'client selection' method is compared with the proposed algorithm and two baseline algorithms on Case 1–3. As shown in Figure 4a, for Case 1, the 'client selection' method achieves a similar accuracy with the proposed algorithm, and the accuracy improvements of 'client selection' over FedAVG and FedNova are 22.5% and 34.6%, respectively. In Figure 4b of Case 2, 'client selection' lowers the fluctuation and achieves higher accuracy than FedAVG and FedNova by about 13% and 24.7%, respectively, but it is less accurate than the proposed algorithm. In Case 3, 'client selection' achieves a similar accuracy (about 73%) with FedAVG and a 4% higher accuracy than FedNova.

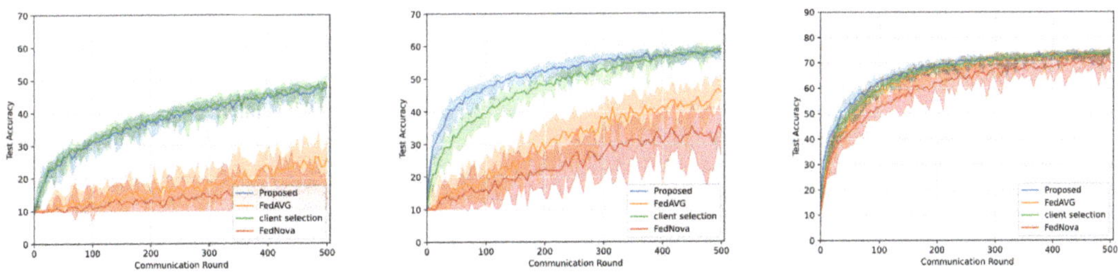

(**a**) Case 1 ($\alpha = 0$) (**b**) Case 2 ($\alpha = 0$ or 0.2) (**c**) Case 3 ($\alpha = 0.2$)

Figure 4. Accuracy comparison among the proposed algorithm, 'client selection', and two baseline algorithms on different non-IID situations.

In Figure 5, the accuracy of 'data sampling' is compared with the proposed algorithm and two baseline algorithms in Cases 1–3. In Case 1, 'data sampling' is not applied because all the clients have only a single class of data. However, it shows an accuracy improvement of 10.7% compared to FedNova. In Case 2, 'data sampling' achieves 4% and 16.4% improvement compared to FedAVG and FedNova, respectively. At the beginning of training, the training accuracy can be improved by increasing the amount of training data through 'data sampling'. In the latter part of training, the amount of oversampled data is reduced to avoid overfitting, hence, the improvement in accuracy gradually decreases compared to the early part of training. In Case 3, when only 'data sampling' is applied, the accuracy improvement is not noticeable because the effect of 'data sampling' is evident in scenarios having strong non-IID. Nevertheless, it shows an accuracy improvement of about 2% compared to FedNova.

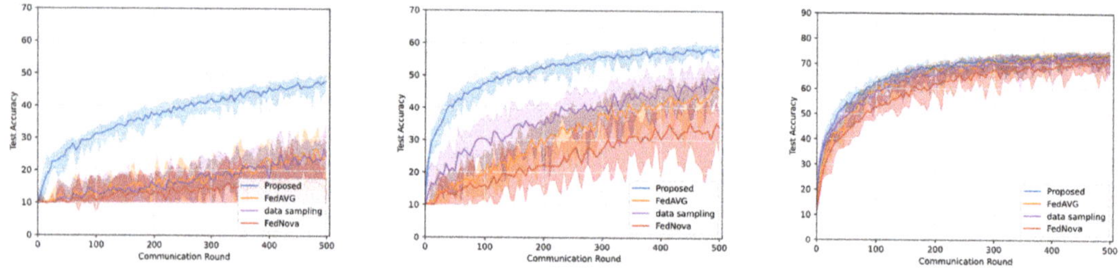

(**a**) Case 1 ($\alpha = 0$) (**b**) Case 2 ($\alpha = 0$ or 0.2) (**c**) Case 3 ($\alpha = 0.2$)

Figure 5. Accuracy comparison among the proposed algorithm, 'data sampling', and two baseline algorithms on different non-IID situations.

In Figure 6, the accuracy of the 'dynamic batch' is compared with the proposed algorithm and two baseline algorithms on Case 1–3. In Cases 1–3, the model fluctuations are similar to the baseline algorithms. In Case 1 and 3, the accuracy is similar to that of FedAVG; however, compared to FedNova, the proposed algorithm shows 11.2% and 4% improvement in Case 1 and Case 3, respectively. In Case 2, the accuracy is improved by about 4% and 16% compared to FedAVG and FedNova, respectively. Compared to Case 1, in Case 2, the clients have various class distributions, hence, if the batch size and the learning rate for each client are not properly adjusted, it is difficult to extract high performance, and it makes the effect of 'dynamic batch' conspicuous. In Case 3, the clients have more classes than in Case 2, accordingly, the contribution of the 'dynamic batch' in improving accuracy is relatively small.

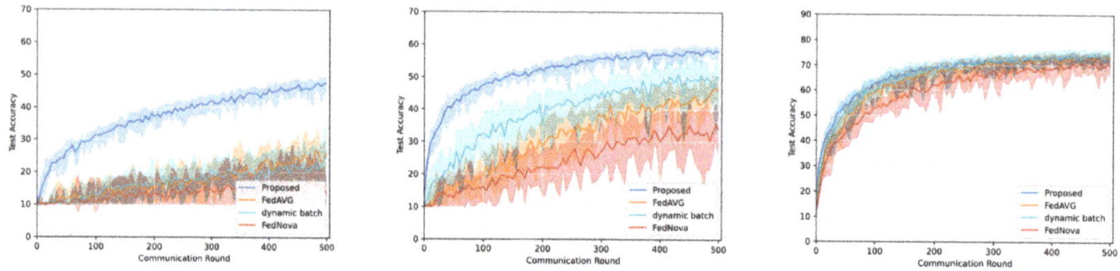

(**a**) Case 1 ($\alpha = 0$) (**b**) Case 2 ($\alpha = 0$ or 0.2) (**c**) Case 3 ($\alpha = 0.2$)

Figure 6. Accuracy comparison among the proposed algorithm, 'dynamic batch', and two baseline algorithms on different non-IID situations.

5.4. *Amount of Training Data*

Since the proposed algorithm determines the amount of local training data for the clients on each round, the clients can learn using only a subset of their local dataset. In Case 1–4, the total amount of data used for the proposed algorithm is compared to FedAVG and FedNova, and the results are shown in Figure 7.

As shown in Figure 7a, the amount of data samples used for the proposed algorithm is about 1,263,000, and it is roughly 1% more than the amount of data for FedAVG. In Case 1, since all the clients have only a single class, 'data sampling' is not applied, hence, the amount of training data does not increase. However, compared to FedAVG which randomly selects clients, the proposed algorithm is prone to select clients with more data to train. Thus, as shown in Figure 7a, the proposed uses 1% more data. In addition, in Case 1, most of the local datasets are used in training. For this reason, in Case 1, the number of training data is similar to FedAVG and FedNova, which uses all of the client's local data in training. However, it should be noted that even though the proposed algorithm uses a small amount of more data (about 1% more) than the baseline algorithms, the accuracy improvement is remarkably high by 21.8% and 34.4% compared to FedAVG and FedNova, respectively.

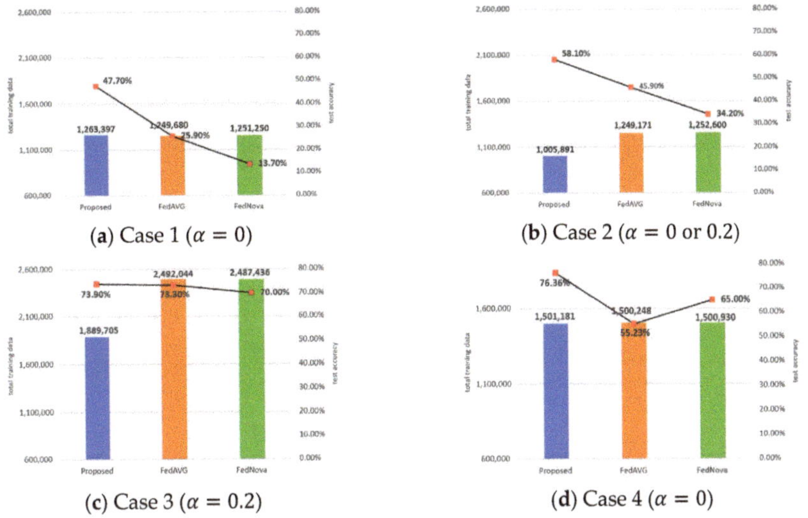

(a) Case 1 ($\alpha = 0$) (b) Case 2 ($\alpha = 0$ or 0.2) (c) Case 3 ($\alpha = 0.2$) (d) Case 4 ($\alpha = 0$)

Figure 7. Amount of training data (bar graph) and accuracy (red dot) comparison among the proposed algorithm, FedAVG, and FedNova on different non-IID situations. Cases 1–3 are the results of the CIFAR-10 dataset, and Case 4 is the result of the MNIST dataset.

In Figure 7b for Case 2, the proposed algorithm uses about 1,005,000 data samples, and the efficiency of the proposed algorithm is clearly shown in this figure. Specifically, the proposed algorithm uses 19% less data than the baseline algorithms, while the achieved accuracy is higher by 12.2% and 23.8% compared to FedAVG and FedNova, respectively. In Case 2, 'data sampling' is applied; however, the amount of oversampled data is quickly reduced to avoid overfitting. This mechanism also minimizes the potential burden of increasing the amount of data to train.

In Figure 7c for Case 3, the accuracy difference between the proposed algorithm and FedAVG is negligible at 0.6%, and between the proposed algorithm and FedNova, it is not high at 3.9%. However, the proposed algorithm uses 24% less amount of data than FedAVG and FedNova, and it is a huge gap.

In Figure 7d for Case 4, all three algorithms learn using a similar number of training data about 1,500,000. However, when comparing the test accuracy of the proposed algorithm with FedAVG and FedNova, it shows 21.1% and 11.4% improved results, respectively.

Figure 7 shows the adaptability of the proposed in improving accuracy and reducing the amount of training data. More specifically, in a severe non-IID situation like Case 1, the proposed algorithm mainly focuses on increasing the accuracy rather than reducing the used training data volume as shown in Figure 7a. When the level of non-IID is low like in Case 3, the proposed algorithm focuses on reducing the training data volume rather than increasing the accuracy as shown in Figure 7c. When the level of non-IID is medium like in Case 2, both the accuracy and the amount of training data are improved as shown in Figure 7b. Through the amount of training data used in Case 1–3, it is confirmed that, on average, the proposed algorithm achieves higher accuracy by using lower computing resources than FedAVG and FedNova.

5.5. Average Number of Clients

The average number of clients participating in the learning on each round is depicted in Figure 8. FedAVG and FedNova randomly select a fixed number of clients on every round, while the proposed algorithm can terminate the client selection process before the number of the selected client reaches the maximum h if the data information v_r for training becomes close enough to Uniform distribution. In Case 1 and Case 4, since all the clients have only a single class, the maximum number of clients must be selected to make v_r similar to Uniform distribution. In Cases 2 and 3, higher test accuracy is achieved even though fewer clients participate in the learning than FedAVG and FedNova. In FL, the reduced number of clients results in the reduced usage of communication resources. Therefore, it is confirmed that the proposed algorithm uses lower communication resources than the baseline algorithms.

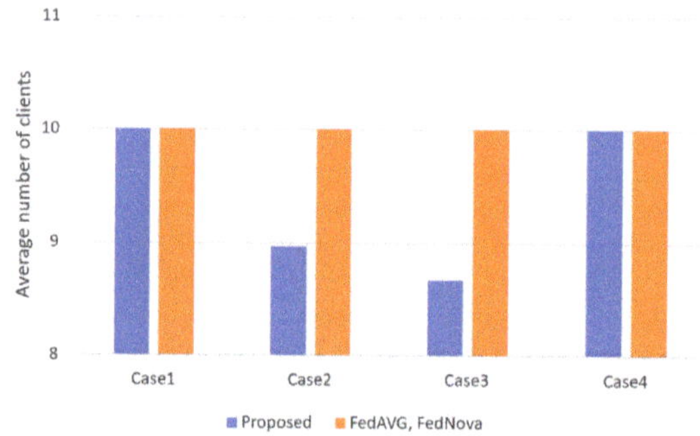

Figure 8. Average number of clients participating in the learning on different non-IID situations.

6. Conclusions

In FL, if the clients' local data distribution is non-IID, the accuracy and learning efficiency of the global model decreases. To solve this problem, the intra-client class imbalance is alleviated through local data sampling, and inter-client class imbalance is alleviated by selecting the clients and determining the amount of data to be used for training, which makes the aggregate of the training data class distributions balanced on every round. In addition, more efficient local learning is possible by dynamically determining the batch size and learning rate reflecting the amount of training data. The proposed algorithm achieves faster convergence speed and higher accuracy with lower computing and communication resource usage than existing algorithms in non-IID environments.

Author Contributions: Conceptualization, M.S. and T.K.; validation, M.S.; formal analysis, M.S. and T.K.; writing—original draft, M.S.; writing—review and editing, T.K. All authors have read and agreed to the published version of the manuscript.

Funding: This research is supported by the Basic Science Research Program through the National Research Foundation of Korea (NRF) funded by the Ministry of Education under Grant 2020R1I1A3068305.

Institutional Review Board Statement: Not applicable.

Informed Consent Statement: Not applicable.

Data Availability Statement: Publicly available datasets were analyzed in this study. This data can be found here: [https://www.cs.toronto.edu/~kriz/cifar.html, https://yann.lecun.com/exdb/mnist/].

Conflicts of Interest: The authors declare no conflict of interest.

References

1. Cisco. *Cisco Annual Internet Report (2018–2023)*; White Paper; Cisco: San Jose, CA, USA, 2020.
2. McMahan, B.; Ramage, D. Research Scientists. Federated Learning: Collaborative Machine Learning without Centralized Training Data. 2018. Available online: https://ai.googleblog.com/2017/04/federated-learning-collaborative.html (accessed on 12 June 2021).
3. McMahan, B.; Moore, E.; Ramage, D.; Hampson, S.; Arcas, B.A. Communication-efficient learning of deep networks from decentralized data. In Proceedings of the Artificial Intelligence and Statistics (AISTATS 2017), Fort Lauderdale, FL, USA, 20–22 April 2017; pp. 1273–1282.
4. Nishio, T.; Yonetani, R. Client selection for federated learning with heterogeneous resources in mobile edge. In Proceedings of the 2019 IEEE International Conference on Communications (ICC 2019), Shanghai, China, 20–24 May 2019; pp. 1–7. [CrossRef]
5. Wang, S.; Tuor, T.; Salonidis, T.; Leung, K.K.; Makaya, C.; He, T.; Chan, K. Adaptive Federated Learning in Resource Constrained Edge Computing Systems. *IEEE J. Sel. Areas Commun.* **2019**, *37*, 1205–1221. [CrossRef]
6. Wang, Z.; Xu, H.; Liu, J.; Huang, H.; Qiao, C.; Zhao, Y. Resource-Efficient Federated Learning with Hierarchical Aggregation in Edge Computing. In Proceedings of the IEEE INFOCOM 2021—IEEE Conference on Computer Communications, Vancouver, BC, Canada, 10–13 May 2021; pp. 1–10. [CrossRef]
7. Sattler, F.; Wiedemann, S.; Muller, K.-R.; Samek, W. Robust and Communication-Efficient Federated Learning from Non-i.i.d. Data. *IEEE Trans. Neural Netw. Learn. Syst.* **2020**, *31*, 3400–3413. [CrossRef] [PubMed]
8. Zhao, Y.; Gong, X. Quality-Aware Distributed Computation and User Selection for Cost-Effective Federated Learning. In Proceedings of the IEEE INFOCOM 2021—IEEE Conference on Computer Communications Workshops (INFOCOM WKSHPS), Vancouver, BC, Canada, 10–13 May 2021; pp. 1–6. [CrossRef]
9. Ma, Z.; Xu, Y.; Xu, H.; Meng, Z.; Huang, L.; Xue, Y. Adaptive Batch Size for Federated Learning in Resource-Constrained Edge Computing. *IEEE Trans. Mob. Comput.* **2021**, *22*, 37–53. [CrossRef]
10. Shi, D.; Li, L.; Wu, M.; Shu, M.; Yu, R.; Pan, M.; Han, Z. To Talk or to Work: Dynamic Batch Sizes Assisted Time Efficient Federated Learning over Future Mobile Edge Devices. *IEEE Trans. Wirel. Commun.* **2022**, *21*, 11038–11050. [CrossRef]
11. Li, Q.; Diao, Y.; Chen, Q.; He, B. Federated Learning on Non-Iid Data Silos: An Experimental Study. In Proceedings of the 2022 IEEE 38th International Conference on Data Engineering (ICDE), Kuala Lumpur, Malaysia, 9–12 May 2022; pp. 965–978. [CrossRef]
12. Zhao, Y.; Li, M.; Lai, L.; Suda, N.; Civin, D.; Chandra, V. Federated Learning with Non-IID Data. *arXiv* **2018**, arXiv:1806.00582. [CrossRef]
13. Krizhevsky, A.; Hinton, G. *Learning Multiple Layers of Features from Tiny Images*; Technical Report; University of Toronto: Toronto, ON, Canada, 2009.
14. Li, D. The MNIST Database of Handwritten Digit Images for Machine Learning Research [Best of the Web]. *IEEE Signal Process. Mag.* **2012**, *29*, 141–142.
15. Li, T.; Sahu, A.K.; Zaheer, M.; Sanjabi, M.; Talwalkar, A.; Smith, V. Federated Optimization in Heterogeneous Networks. *arXiv* **2018**, arXiv:1812.06127.
16. Shoham, N.; Avidor, T.; Keren, A.; Israel, N.; Benditkis, D.; Mor-Yosef, L.; Zeitak, I. Overcoming forgetting in federated learning on non-iid data. *arXiv* **2019**, arXiv:1910.07796.
17. Rizk, E.; Vlaski, S.; Sayed, A.H. Optimal Importance Sampling for Federated Learning. In Proceedings of the ICASSP 2021—2021 IEEE International Conference on Acoustics, Speech and Signal Processing (ICASSP), Toronto, ON, Canada, 6–11 June 2021; pp. 3095–3099. [CrossRef]
18. Kopparapu, K.; Lin, E. Fedfmc: Sequential efficient federated learning on non-iid data. *arXiv* **2020**, arXiv:2006.10937.
19. Briggs, C.; Fan, Z.; Andras, P. Federated Learning with Hierarchical Clustering of Local Updates to Improve Training on Non-IID Data. In Proceedings of the 2020 International Joint Conference on Neural Networks (IJCNN), Glasgow, UK, 19–24 July 2020; pp. 1–9. [CrossRef]
20. Duan, M.; Liu, D.; Chen, X.; Tan, Y.; Ren, J.; Qiao, L.; Liang, L. Astraea: Self-Balancing Federated Learning for Improving Classification Accuracy of Mobile Deep Learning Applications. In Proceedings of the 2019 IEEE 37th International Conference on Computer Design (ICCD), Abu Dhabi, United Arab Emirates, 17–20 November 2019; pp. 246–254. [CrossRef]
21. Auer, P.; Cesa-Bianchi, N.; Fischer, P. Finite-time analysis of the multiarmed bandit problem. *Mach. Learn.* **2002**, *47*, 235–256. [CrossRef]
22. Zhang, W.; Wang, X.; Zhou, P.; Wu, W.; Zhang, X. Client Selection for Federated Learning With Non-IID Data in Mobile Edge Computing. *IEEE Access* **2021**, *9*, 24462–24474. [CrossRef]
23. Yang, M.; Wang, X.; Zhu, H.; Wang, H.; Qian, H. Federated Learning with Class Imbalance Reduction. In Proceedings of the 2021 29th European Signal Processing Conference (EUSIPCO), Dublin, Ireland, 23–27 August 2021; pp. 2174–2178. [CrossRef]
24. Malandrino, F.; Chiasserini, C.F. Federated Learning at the Network Edge: When Not All Nodes Are Created Equal. *IEEE Commun. Mag.* **2021**, *59*, 68–73. [CrossRef]
25. Wang, J.; Liu, Q.; Liang, H.; Joshi, G.; Poor, H.V. Tackling the Objective Inconsistency Problem in Heterogeneous Federated Optimization. *Adv. Neural Inf. Process. Syst. (NeurIPS)* **2020**, *33*, 7611–7623.
26. Zhang, J.; Guo, S.; Qu, Z.; Zeng, D.; Zhan, Y.; Liu, Q.; Akerkar, R. Adaptive Federated Learning on Non-IID Data with Resource Constraint. *IEEE Trans. Comput.* **2022**, *71*, 1655–1667. [CrossRef]

27. Hsu, T.M.H.; Qi, H.; Brown, M. Measuring the effects of non-identical data distribution for federated visual classification. *arXiv* **2019**, arXiv:1909.06335.
28. Kullback, S.; Leibler, R.A. On Information and Sufficiency. *Ann. Math. Stat.* **1951**, *22*, 79–86. [CrossRef]
29. Goyal, P.; Dollár, P.; Girshick, R.B.; Noordhuis, P.; Wesolowski, L.; Kyrola, A.; Tulloch, A.; Jia, Y.; He, K. Accurate, Large Minibatch SGD: Training ImageNet in 1 h, CoRRabs/1706.02677. 2017. Available online: https://dblp.org/db/journals/corr/corr1706.html#GoyalDGNWKTJH17 (accessed on 30 April 2018).

sensors

MDPI

Article

Low-Latency Collaborative Predictive Maintenance: Over-the-Air Federated Learning in Noisy Industrial Environments

Ali Bemani *,† and Niclas Björsell †

Department of Electrical Engineering, Mathematics and Science, University of Gävle, 801 76 Gävle, Sweden; niclas.bjorsell@hig.se
* Correspondence: ali.bemani@hig.se
† These authors contributed equally to this work.

Abstract: The emergence of Industry 4.0 has revolutionized the industrial sector, enabling the development of compact, precise, and interconnected assets. This transformation has not only generated vast amounts of data but also facilitated the migration of learning and optimization processes to edge devices. Consequently, modern industries can effectively leverage this paradigm through distributed learning to define product quality and implement predictive maintenance (PM) strategies. While computing speeds continue to advance rapidly, the latency in communication has emerged as a bottleneck for fast edge learning, particularly in time-sensitive applications such as PM. To address this issue, we explore Federated Learning (FL), a privacy-preserving framework. FL entails updating a global AI model on a parameter server (PS) through aggregation of locally trained models from edge devices. We propose an innovative approach: analog aggregation over-the-air of updates transmitted concurrently over wireless channels. This leverages the waveform-superposition property in multi-access channels, significantly reducing communication latency compared to conventional methods. However, it is vulnerable to performance degradation due to channel properties like noise and fading. In this study, we introduce a method to mitigate the impact of channel noise in FL over-the-air communication and computation (FLOACC). We integrate a novel tracking-based stochastic approximation scheme into a standard federated stochastic variance reduced gradient (FSVRG). This effectively averages out channel noise's influence, ensuring robust FLOACC performance without increasing transmission power gain. Numerical results confirm our approach's superior communication efficiency and scalability in various FL scenarios, especially when dealing with noisy channels. Simulation experiments also highlight significant enhancements in prediction accuracy and loss function reduction for analog aggregation in over-the-air FL scenarios.

Keywords: predictive maintenance; over-the-air federated learning; analog aggregation; low latency; channel noise

Citation: Bemani, A.; Björsell, N. Low-Latency Collaborative Predictive Maintenance: Over-the-Air Federated Learning in Noisy Industrial Environments. *Sensors* **2023**, *23*, 7840. https://doi.org/10.3390/s23187840

Academic Editors: Yichuang Sun, Haeyoung Lee and Oluyomi Simpson

Received: 20 August 2023
Revised: 7 September 2023
Accepted: 8 September 2023
Published: 12 September 2023

1. Introduction

The pervasive connectivity of numerous devices and sensors, fueled by recent advancements in communication networks and Internet of Things (IoT) applications, has resulted in the generation of immense volumes of data. The copious volume of data at hand serves as the training datasets for machine learning (ML) algorithms that find utility across various applications in industry, such as process optimization, defining product quality, and PM, critical components within the purview of Industry 4.0 [1]. Conventionally, the training process for these ML algorithms has followed a centralized approach, wherein multiple devices transmit their raw and occasionally sensitive data to a PS possessing robust computational resources dedicated to training tasks. Nevertheless, the data required for training these ML algorithms are produced by numerous assets and devices in the industry, which often necessitate privacy-preserving measures to safeguard data integrity. Furthermore, the presence of bandwidth constraints poses a bottleneck when transmitting a massive amount of data from the devices to the PS [2]. For this reason, there has been considerable focus on federated and distributed learning algorithms, primarily due to their ability to train ML models in a fully decentralized manner.

FL algorithms have been proposed as an alternative scheme for privacy-preserving distributed ML, where each device participates in the training process using exclusively locally available data facilitated by a PS [3]. In the context of FL, devices communicate with the PS by exchanging model parameters and their respective local updates, while ensuring that the raw data remain localized on the devices. This approach offers not only privacy advantages but also proves to be a compelling solution for wireless edge devices, particularly when dealing with substantial dataset sizes. In the domain of PM applications, the volume of data obtained from online sensors is significant. Consequently, it becomes imperative to account for this substantial data volume when implementing federated learning techniques for PM. Additionally, time sensitivity and temporal awareness are essential attributes for the effective execution of PM activities. Therefore, time latency should be considered for PM applications.

The migration learning from centralized clouds to the edge enables edge servers to quickly obtain real-time data generated by edge devices, facilitating rapid AI model training. As a result, distributing these models from servers to devices in close proximity enables the devices to respond effectively to real-time events, making them well-suited for PM applications. Despite the rapid advancement of computing speeds, wireless transmission of large amounts of data by any device faces limitations due to limited radio resources and the adverse conditions of wireless channels. This creates a communication bottleneck that hinders fast-edge learning [4,5]. Hence, communication efficiency has been at the forefront of FL, and the paramount objective is to achieve high model accuracy while minimizing the number of communication rounds of resource usage. In the realm of cutting-edge research, the fusion of digital twins and multi-access edge computing has gained significant attention [6]. This innovative approach represents a crucial technology in the context of 6G networks, primarily serving as a fundamental enabler for the Industrial IoT. The primary objective of this research is to minimize the total task completion delay for IoT devices by optimizing various parameters.

Communication schemes for FL can be categorized as digital or analog. Digital communication, though burdensome for wireless networks, assigns communication resources to each client's ML model parameters. Analog communication reduces overhead by allowing shared resources for transmitting FL models. Early research aimed to reduce communication rounds or payload size. However, most FL literature assumes a perfect communication channel shifting focus to ML design. Recent research addresses this gap by emphasizing system design, especially for wireless FL [7–9]. Although early studies have delved deeply into optimizing communication design for FL, the crucial and practical matter of a noisy channel with over-the-air communication and computation (OACC) has not been thoroughly investigated. Incorporating the impact of noisy channels in OACC complicates convergence analysis due to noise propagation during each communication round. Moreover, the collective impact of these noisy communications in OACC on the final learning performance necessitates a comprehensive design and analysis approach.

This work addresses the impact of communication-induced noise during FL training on the convergence and accuracy performance of the ML model, and then proposes robust algorithms within the OACC framework that mitigate these effects and optimize client resources concurrently. Our focus lies in analog communication for model updates [10–12] and the exploration of new distributed algorithms capable of withstanding high levels of channel noise and low signal-to-noise ratio (SNR) in industrial environments, in accordance with PM applications.

1.1. Related Work

In recent years, substantial endeavors have been undertaken to implement the FL framework over wireless networks, with its origins traced back to McMahan's seminal works [13,14]. A comprehensive survey of FL has been conducted in [15], offering an overview of this paradigm where statistical models are trained on distributed networks at the edge. These studies explore the distinctive challenges and characteristics of FL compared to traditional approaches while identifying open problems that necessitate interdisciplinary research efforts. However, these works overlook the practical implications of channel effects and assume a seamless integration of FL algorithms into wireless networks. Recently, there has been significant attention given to exploring methods for mitigating the impact of these effects on FL algorithms [16–18].

This study primarily focuses on analog aggregation schemes for over-the-air transmission, which are driven by the inherent superposition property of signals in the wireless multiple-access channel. Analog over-the-air aggregation is a highly promising technique extensively used in Federated Averaging (FedAvg) due to the fact that the ps or clients require only the sum of their local gradients or model parameters. These schemes have been explored in various studies, including [2,19], and other relevant works.

In order to provide context and underscore the contributions of this article, we will now discuss previous studies that have investigated FL in the presence of imperfect or noisy communication. The main objective of [20] was to tackle the issue of noise in wireless communications for federated learning. The paper addressed this challenge by formulating the problem using an expectation-based model and a worst-case model. They introduced a sampling-based successive convex approximation algorithm for solving the problem. This approach successfully handles noise by incorporating it as a regularizer in the loss function during the training process. Simulation results showcased improved prediction accuracy and reduced loss function values, affirming the effectiveness of the proposed methods in mitigating the impact of noise. In [21], Amiri et al. examined the impact of a noisy channel on FedAvg. They determined that when dealing with noisy downlink transmission, the presence of noise cannot be neutralized through step-size design, resulting in an inability to ensure precise convergence. Consequently, addressing this fundamental problem has necessitated imposing strict demands on the estimation noise associated with the aggregated model weights.

In [22], researchers conducted an investigation into the influence of a noisy channel in uplink and downlink analog transmission on the training process. The findings of this analysis led the authors to conclude that for the FedAvg algorithm to converge, it is necessary for the noise variance to decrease at a rate of $\mathcal{O}(1/k^2)$, where k represents the communication round. Consequently, in order to effectively mitigate the impact of channel noise on the training procedure, the authors recommend two approaches. Firstly, it is advised to employ an increased transmission power gain of $\mathcal{O}(k)$ in both uplink and downlink transmissions. Alternatively, if the power gain remains fixed, extending the transmission time to $\mathcal{O}(k)$ is suggested. The implementation of either of these strategies is crucial for preserving the integrity of the training process when faced with channel noise.

Another critical issue is analog transmission, which is widely used in over-the-air aggregation within the wireless channel. It is extensively utilized and plays a crucial role in enhancing spectral efficiency and reducing multi-access latency. Gau et al. [23] developed a hardware transceiver and application software to train a real-world FL task using over-the-air analog aggregation. They focused on developing an over-the-air aggregation solution for wireless FL based on orthogonal frequency-division multiplexing (OFDM). The main challenge they faced was achieving perfect waveform superposition, which was complicated by the presence of frame timing offset and carrier frequency offset. To tackle these challenges, they proposed a two-stage waveform pre-equalization technique.

The focus of this study pertains to the exploration of algorithms aimed at resolving optimization problems through over-the-air analog aggregation. While numerous papers have discussed communication-efficient solutions for distributed learning problems, it is of paramount importance to thoroughly investigate the optimization of analog FL problems over noisy communication channels. This particular problem presents significant importance and complexity, necessitating meticulous attention and consideration.

1.2. Our Contributions

The major contributions of this work are summarized as follows.

- We propose a hierarchical approach to PM, building upon our prior work [1]. The key distinction lies in the utilization of OACC for the FL algorithm at the factory level. This choice is motivated by the benefits of low latency, making it suitable for PM applications while also improving spectral efficiency. At higher levels, such as fog and cloud servers, occasional requests are made to the factory level for the aggregate model, enabling averaging over multiple factory parameters, FedAvg. Our primary emphasis is on the factory level, specifically investigating FLOACC as the focal point of our research.
- We propose FSVRG-OACC as a distributed approach to solve the optimization problem for PM at the factory level based on OACC. FSVRG-OACC leverages analog over-the-air aggregation, which enables it to effectively handle highly noisy communication channels and allows for improved convergence in minimizing the cost function associated with the ML algorithm.
- FSVRG-OACC facilitates the transmission of local gradient updates by individual agents, capitalizing on the advantages of computation over the air. This algorithm effectively mitigates the impact of channel perturbations on convergence by incorporating the effects of the communication channel into the algorithm update process. The utilization of FSVRG-OACC ensures that convergence is not compromised, enabling efficient and robust optimization in the presence of varying channel conditions.
- The simulation results demonstrate the substantial reduction in convergence sensitivity to noise achieved by our proposed algorithm. This finding holds significant implications for the implementation of ML algorithms on analog over-the-air aggregation in highly noisy industrial environments.

The remainder of this paper is structured as follows. In Section 2, we introduce the system model, starting with the description of FL over-the-air analog aggregation and extending it to FLOACC. Section 3 presents our proposed algorithm, FSVRG-OACC, along with a comparison to other stochastic gradient descent algorithms. Following that, in Section 4, we evaluate the performance of FSVRG-OACC through a PM application and present the corresponding experimental results. Finally, in Section 5, we provide concluding remarks and discuss future research directions.

2. System Model

2.1. Federated Edge Learning System

We consider a distributed learning system specifically designed for a PM application at the factory level. This system comprises a single parameter server and K edge nodes, as depicted in Figure 1. Collaboratively, the edge nodes at each factory train a shared learning process involving the global model $\mathbf{w}$. Each machine collects a fraction of labeled training data via the interaction with a local dataset, denoted as $\mathcal{D}_1, \mathcal{D}_2, \ldots, \mathcal{D}_K$. The local loss function of the model vector $\mathbf{w}$ on $\mathcal{D}_k$ is given by

$$F_k(\mathbf{w}) = \frac{1}{\phi_k} \sum_{(\mathbf{x}_j, y_j) \in \mathcal{D}_k} f(\mathbf{w}, \mathbf{x}_j, y_j), \quad (1)$$

where ϕ_k is the number of data points stored on data partition $\mathcal{D}_k$ and $f(\mathbf{w}, \mathbf{x}_j, y_j)$ is the loss function quantifying the prediction error of the model $\mathbf{w}$ for each data sample j, which consists of the training sample $\mathbf{x}_j$ and its ground true label y_j. For convenience, we write $f(\mathbf{w}, \mathbf{x}_j, y_j)$ as $f_j(\mathbf{w})$ and assume uniform sizes for local datasets ϕ_k for all k. Table 1 presents the typical loss function used in FL for various applications. In the specific case of anomaly detection over edge nodes at the factory, the squared-SVM loss function has been employed in this study as a leading ML method that offers flexibility in modeling complex nonlinear boundaries between normal and abnormal data points. With these definitions, the global loss function on all the distributed datasets can be defined as

$$F(\mathbf{w}) = \frac{\sum_{j\in \cup_k \mathcal{D}_k} f_j(\mathbf{w})}{\cup_k \phi_k} = \frac{1}{K}\sum_{k=1}^{K} F_k(\mathbf{w}), \tag{2}$$

where $|.|$ denotes the size of the datasets and each dataset satisfies $\mathcal{D}_i \cap \mathcal{D}_j = \emptyset$ when $i \neq j$. The training target is to minimize the global loss function $F(\mathbf{w})$ according to the distributed process to find $\mathbf{w}^*$.

$$\mathbf{w}^* = \arg\min F(\mathbf{w}) \tag{3}$$

In this section, we begin by presenting the problem formulation of FL in the context of a hierarchical PM scenario. Subsequently, we delve into the specific case of FL-OACC, where all agents at the factory levels broadcast their updated models over the air, leveraging the advantages of computation through analog communication.

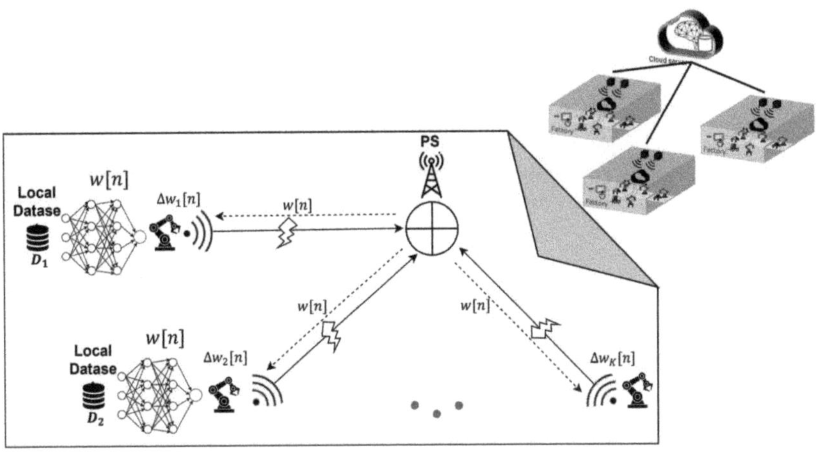

Figure 1. FL edge system model for a hierarchical PM scenario at the factory level.

Table 1. Several examples of loss functions.

Model	Loss Function
Linear regression	$\left\|1 - y_j\mathbf{w}^T x_j\right\|^2$
Logistic regression	$-\log\left(1+\exp\left(-y_j\mathbf{w}^T x_j\right)\right)$
K-means	$\min_l \left\|x_j - \mathbf{w}_l\right\|^2$
Cross-Entropy	$-\sum y_c p(y = c \mid x, \mathbf{w})$
Squared-SVM	$\lambda\|\mathbf{w}\|^2 + \max 0; 1 - y_j\mathbf{w}^T x_j$

To compute $F(\mathbf{w})$, one method involves directly uploading all local data, which raises privacy concerns. To address this issue, the FL framework is employed to solve the problem outlined in Equation (3) through a distributed approach. There are two approaches based on FedAvg for solving this distributed optimization problem:

1. Model averaging: In this approach, each agent minimizes its local loss function and transmits the model parameters to the PS for aggregation. In the second round of iteration, the agent receives the updated model from the PS.
2. Gradient averaging: In this approach, each agent calculates the gradient of its loss function and transmits the gradient to the PS for aggregation. In the second round of

iteration, the agent can update its model based on the gradient averaging received from the PS.

The agents generally employ a gradient descent (GD) algorithm or stochastic gradient descent (SGD) algorithm to minimize the loss function described in Equation (1). A single-step SGD for updating the model parameter for device K can be defined following the model and gradient averaging, respectively.

$$\mathbf{w}_k[n+1] = \mathbf{w}[n] - \alpha \nabla F_k(\mathbf{w}[n]), \tag{4}$$

$$\mathbf{w}_k[n+1] = \mathbf{w}_k[n] - \alpha \nabla F(\mathbf{w}[n]), \tag{5}$$

where α is the step size and ∇ is the gradient operator. Hence, the only operation performed by the PS is to compute the average of the model parameters or the gradients received from the agents.

$$\mathbf{w}[n+1] = \frac{1}{K} \sum_{k=1}^{K} \mathbf{w}_k[n+1], \tag{6}$$

$$\nabla F(\mathbf{w}[n]) = \frac{1}{K} \sum_{k=1}^{K} \nabla F_k(\mathbf{w}_k[n]). \tag{7}$$

The learning process entails iterating between Equations (4) and (6) or Equations (5) and (7) until the model converges. The averaging process on the PS serves as a motivation for the low-latency FL scheme, utilizing FLOACC. Further details about FLOACC are provided below.

2.2. Fl Over-the-Air Communication and Computation

In this section, we discuss the details of Federated Learning over-the-Air Communication and Computation, abbreviated as FLOACC. This method provides an efficient multi-access scheme in a low-latency scenario, which is crucial for applications like PM that require very fast and real-time task decision making. The idea of the over-the-air computation model for FL has been examined in several previous works, such as [8,16]. Their approach was inspired by the PS's lack of interest in individual model weight vectors. Instead, the server solely needs the average of the model weights, which is automatically provided by the wireless multiple-access channel in the form of their sum.

As depicted in Figure 2, the FLOACC enables simultaneous transmission of result vectors from all devices and assets at the factory level to the PS in an analog manner. Let $\mathbf{w}_k = \left[w_{k,1}, \cdots, w_{k,q}\right]^T$ denote $q \times 1$ local model parameter vector and $\nabla F_k(\mathbf{w}) = \left[\nabla F_{k,1}, \cdots, \nabla F_{k,q}\right]^T$ be the local gradient vector of the loss function from the k-th device. In FLOACC, it is assumed that the local gradient of each model is transferred over an analog medium. In this scenario, the transmitted symbols are denoted by $\{\tilde{\nabla} F_{k,i}\}$ and are normalized to have zero mean and unit variance $\mathrm{E}\left(\tilde{\nabla} F_{k,i} \tilde{\nabla} F_{k,i}^{*}\right) = 1$. By employing OFDM, it becomes feasible to allocate each element of the gradient vector to a distinct sub-carrier OFDM channel. This approach enables a significant reduction in the learning process latency for FLOACC.

During each round n, all the local devices at a factory simultaneously transmit their local gradient based on the distributed loss function as $h_k[n] p_k[n] \nabla F_k(\mathbf{w}[\mathbf{n}])$, where $h_k[n] \sim \mathcal{CN}(0,1)$ is the small-scale fading of the channel between the k-th device and the PS, and $p_k[n]$ is the allocated transmission power for each device k. In particular, the aggregated gradient in the n-th communication round, denoted by $y[n]$, is expressed as follows.

$$\hat{\nabla} F(\mathbf{w}[\mathbf{n}]) = \sum_{k=1}^{K} h_k[n] p_k[n] \nabla F_k(\mathbf{w}_k[n]) + Z[n], \tag{8}$$

where $Z[n] \sim \mathcal{CN}(0, \sigma_z^2 \mathbf{I})$ is additive white Gaussian noise. Note that in over-the-air analog aggregation, it is possible to transfer the model parameters as well. However, in FLOACC, we only consider the local gradient during aggregation. This is because the aggregated gradient is less sensitive in the optimization algorithms compared to the model parameters. We will discuss this issue further in the next section.

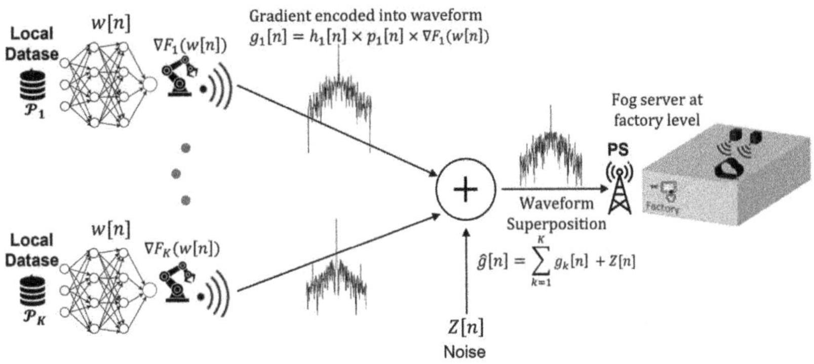

Figure 2. FLOACC scenario at the factory level.

We assumed that, similar to the LTE system, there are common downlink reference symbols in the radio resource block for the devices at the factory level to estimate the channel fading coefficient. Then, the local devices can transmit the signal with the specific power they calculated. The received signal at the PS has been depicted with gain blocks in Figure 3, so the PS can estimate the aggregated gradient as follows:

$$y[n] = \sum_{k=1}^{K} \frac{h_k[n] p_k[n] \nabla F_k(\mathbf{w}_k[n])}{\sqrt{\eta}} + \frac{Z[n]}{\sqrt{\eta}}, \quad (9)$$

where η is a receiver scaling factor. Owing to the imposed physical constraints, the transmission of each device is subject to a long-term transmission power constraint.

$$\mathrm{E}\left\| \left[\sum_{n=1}^{N} |p[n](h[n])|^2 \right] \right\| \leq P_0. \quad (10)$$

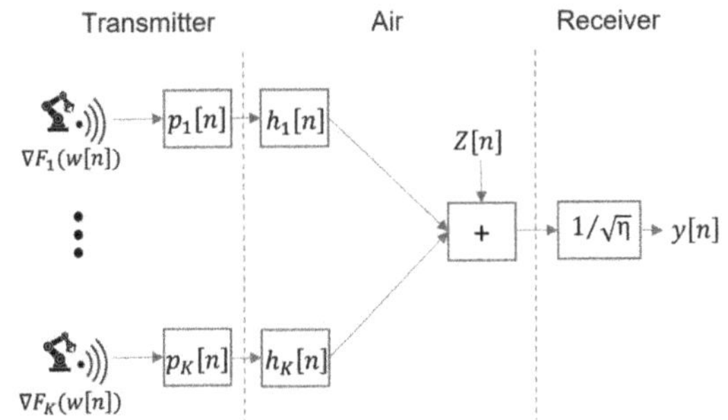

Figure 3. Illustration of the system design for FLOACC using analog transmission over-the-air aggregation.

2.3. Effective Noise and Definition of SNR in FLOACC

In order to keep the problem general, we adopt the effective noise model in analog aggregation. We assume accurate channel state information at the transmitters. To meet the aggregation requirement of FL, the local devices implement the channel inversion rule, which yields the instantaneous transmit power of user k during the communication round n for gradient aggregation.

$$p_k[n] = \frac{h_k[n]^H}{|h_k[n]|} \cdot \frac{\sqrt{\eta}}{|h_k[n]|}, \tag{11}$$

where H is Hermitian transpose. Based on this definition, the received signal can be formulated as follows.

$$y[n] = \sum_{k=1}^{K} \nabla F_k(\mathbf{w}_k[n]) + \frac{Z[n]}{\sqrt{\eta}}. \tag{12}$$

η represents a scalar that signifies the average transmission power, based on which the received SNR of the global gradient can be expressed as follows.

$$\mathrm{SNR}[n] = \mathbb{E}\left\| \frac{1}{K} \sum_{i=1}^{q} \frac{\sqrt{\eta} \sum_{k \in \mathcal{K}} \nabla F_k(\mathbf{w}_{k,i}[n])}{Z_i[n]} \right\|^2 = \frac{\eta \mathbb{E} \| \sum_{k \in \mathcal{K}} \nabla \mathbf{F_k}(\mathbf{w}_k[n]) \|^2}{\sigma_z^2 K^2}. \tag{13}$$

It is evident that the received SNR is consistent across all users due to devices with weaker channels compensating by transmitting at higher powers. In various studies, devices with significantly weak channels are often excluded from training due to their inability to pre-equalize their channels. Many research has been undertaken to optimize η to enhance model convergence over the wireless communication network. In [24,25], the optimal selection of p_k and η is determined by solving the corresponding optimization problem.

$$\mathrm{MSE} = \mathbb{E}\left\| y[n] - \sum_{k=1}^{K} \nabla F_k(\mathbf{w}_k[n]) \right\|^2 = \frac{1}{K^2} \sum_{k=1}^{K} \left(\frac{|h_k| p_k}{\sqrt{\eta}} - 1 \right)^2 + \frac{\sigma_z^2}{\eta K^2}, \tag{14}$$

$$\begin{aligned} &\min_{p_1, p_2, \dots, p_K, \eta} \frac{1}{K^2} \sum_{k=1}^{K} \left(\frac{|h_k| \sqrt{p_k}}{\sqrt{\eta}} - 1 \right)^2 + \frac{\sigma_z^2}{\eta K^2} \\ &\text{s.t. } p_k \le P_{\max}, \forall k \in \{1, 2, \dots, K\}. \end{aligned} \tag{15}$$

Extensive research has been conducted to optimize transmit power for the purpose of achieving model convergence and mitigating the effects of existing noise in the wireless transmission medium. However, to the best of our knowledge, no existing studies have investigated the comparison of convergence performance among different algorithms without any transmission power control. In the upcoming section, we delve into various gradient descent algorithms for ML optimization problems, considering the over-the-air scheme, and provide an analysis of which algorithm demonstrates superior convergence properties.

3. Proposed Adaptive FSVRG-OACC Algorithm

In this section, our specific focus lies on algorithms suitable for solving problem (3) within the context of over-the-air and analog aggregation. Firstly, in Section 3.1, we examine baseline algorithms that are compatible with distributed problems and analog gradient transmission (GT), while also highlighting the distinctions between model transmission (MT) and GT. Next, in Section 3.2, we delve into randomized methods that, in an initial approximation, integrate the advantages of cost-effective iterations from SGD with the rapid convergence of GD. Many of these methods can be categorized into one of two classes: dual methods of the randomized coordinate descent type and primal methods of the stochastic gradient descent with variance reduction type. Our emphasis lies on stochastic variance reduced gradient (SVRG), and we optimize this method for FL within the framework of analog aggregation while considering the presence of white Gaussian channel noise.

3.1. Baseline Algorithms

A fundamental approach for solving a distributed optimization problem with the structure (4) involves employing the GD algorithm, particularly when the functions f possess smoothness and convexity characteristics. In a distributed setting, there are two possible approaches: the GT method, where gradients are transmitted over the air and an aggregated gradient signal is received, which is subject to noise for model updating using GD, or the MT method, where the model is updated through GD iterations and then transmitted for model aggregation. In the presence of noise, the convergence deteriorates due to the elevated sensitivity of the loss function to the model parameters. Therefore, in over-the-air distributed algorithms, the GT approach is preferred. This preference stems from the lower sensitivity of the cost function to the aggregated gradient compared to the individual model parameters. GD demonstrates a fast convergence rate. However, each iteration has the potential to be computationally intensive on each local device. In contrast, SGD selects a random data label and performs the update, which offers a more efficient alternative.

3.2. Fsvrg-Oacc Algorithm

An additional algorithm within the SGD category is SVRG [26]. The SVRG algorithm operates through two nested loops. The outer loop involves calculating the full gradient of the entire function, $\nabla F(\mathbf{w}^t[n])$, which is typically a computationally expensive operation to be avoided whenever possible. In the inner loop, the update step is iteratively computed as follows.

$$\mathbf{w}[n+1] = \mathbf{w}[n] - \alpha\Big[\nabla F^i(\mathbf{w}[n]) - \nabla F^i(\mathbf{w}^t[n]) + \nabla F(\mathbf{w}^t[n])\Big], \tag{16}$$

where $\nabla F^i(\mathbf{w}[n])$ represents the stochastic gradient computed based on a randomly selected data label, $\nabla F^i(\mathbf{w}^t[n])$ denotes the stochastic gradient computed over the entire dataset, and α is stepsize. This iteration is specific to a single device, and its fundamental concept lies in utilizing stochastic gradients to estimate the gradient change from point $\mathbf{w}^t$ to $\mathbf{w}$, rather than directly estimating the gradient itself.

Indeed, this algorithm is naturally suited for centralized implementations since it necessitates computing the stochastic gradient over the complete dataset, thereby making it well-suited for centralized scenarios. But a notable contribution was made in [27], where they introduced FSVRG, which is particularly applicable in the context of distributed optimization. They demonstrated that existing SVRG algorithms are not suitable for distributed approaches and proposed the FSVRG algorithm, specifically designed for sparse distributed convex problems. The pseudocode for FSVRG is provided in Algorithm 1. This algorithm has been implemented and subjected to evaluation, and the results will be presented in the experimental section. The findings indicate that this algorithm does not demonstrate satisfactory convergence in the realm of FL for over-the-air analog aggregation, specifically in relation to the absence of power transfer control.

Let us now elucidate the motivation behind considering a different algorithm suitable for FL in the context of over-the-air analog aggregation. A crucial aspect that demands attention is the significant variation in the number of available data points among different devices, which may differ greatly from the average number of data points available to any single device. It looks like a similar issue to FSVR, but it should be noted that in our assumptions, analog communication is the sole communication type between local devices and the PS. As a result, the PS lacks information regarding the number of data points and the type of data distribution.

Additionally, this scenario frequently entails the local data being clustered around a specific pattern, which renders it unrepresentative of the overall distribution we aim to learn. Consequently, considering an aggregation on the entire gradient direction in each iteration could be a promising approach that could be undertaken in the concept of analog aggregation.

Algorithm 1: Federated SVRG

Parameters: ϕ = number of data points, ϕ_k = number of data points store on device k, α = stepsize, data partition $\{\mathcal{D}_k\}_{k=1}^{K}$.
for $n = 0, 1, \ldots$ **do**
  Compute $\nabla F(\mathbf{w}^t[n]) = \frac{1}{\phi}\sum_{i=1}^{\phi} \nabla F^i(\mathbf{w}^t[n])$ » Overall iterations
  for $k = 1$ *to* K ***do in parallel*** *over device* k **do**
    Initialize: $\mathbf{w}_k = \mathbf{w}^t$ and $\alpha_k = \alpha/\phi_k$ » Distributed loop
    Let $\{i_t\}_{t=1}^{\phi_k}$ be random permutation of $\mathcal{D}_k$
    for $t = 1, \ldots, \phi_k$ **do**
      $\mathbf{w}_k[n+1] = \mathbf{w}_k[n] - \alpha_k\left[\nabla F^{i_t}(\mathbf{w}_k[n]) - \nabla F^{i_t}(\mathbf{w}^t[n]) + \nabla F(\mathbf{w}^t[n])\right]$
    end
  end
  $\mathbf{w}^t[n] = \mathbf{w}^t[n] + \sum_{k=1}^{K} \frac{\phi_k}{\phi}(\mathbf{w}_k[n] - \mathbf{w}^t[n])$ » Model aggregation
end

From a practical perspective, in FSVRG-OACC, it is postulated that all devices possess a randomly allocated initialization value for the parameter vector, $\mathbf{w}_k$. This assumption holds significant importance in the practical execution of the algorithm. The proposed algorithm involves two communication rounds, which results in increased communication costs. However, it offers advantages in terms of the convergence algorithm. Algorithm 2 introduces the FSVRG-OACC, a modified FSVRG variant tailored for over-the-air analog aggregation.

During the initial communication round (distributed loop 1), all devices compute the complete gradient of the entire function and subsequently determine the internal gradient as $g_k = \left[\nabla F_k^{i_t}(\mathbf{w}^t[n]) - \nabla F_k(\mathbf{w}[n])\right]$, where i_t is sampled uniformly from the local dataset $\mathcal{D}_k$. These gradients are derived using SGD, rendering their computation relatively inexpensive. The computed internal gradient is then transmitted over the air to the PS, while the estimated aggregated internal gradient is sent back to the devices via the analog medium, as shown in Figure 2.

$$\hat{g} = \frac{1}{K}\sum_{k=1}^{K} \frac{h_k[n]p_k[n]g_k}{\sqrt{\eta}} + \frac{z[n]}{\sqrt{\eta}}. \tag{17}$$

The estimated aggregated internal gradient, denoted as $\hat{g}$, compels all devices in the second round of communication (Distributed loop 2) to move in the same direction. In this communication round, the updated gradient is denoted as $\hat{G}$, which was estimated in the first round as $\hat{G} = \nabla F^{i_t}(\mathbf{w_k}[n]) - \hat{g}$ for device k. Subsequently, the PS aggregates this gradient over the air from all devices and transmits it back to them for the remaining iterations. Each device k uploads the following gradient over the air for aggregation.

$$G_k = \nabla F^{i_t}(\mathbf{w_k}[n]) - \hat{g}, \tag{18}$$

$$G_k = \nabla F^{i_t}(\mathbf{w_k}[n]) - \left[\frac{1}{K}\sum_{k=1}^{K} \frac{h_k[n]p_k[n]\left[\nabla F_k^{i_t}(\mathbf{w}^t[n]) - \nabla F_k(\mathbf{w}[n])\right]}{\sqrt{\eta}} + \frac{z[n]}{\sqrt{\eta}}\right]. \tag{19}$$

By aggregating the gradient $\mathbf{G}_k$ at the PS and transmitting it back to each device, a uniform descent direction can be achieved across all devices.

$$\hat{G} = \frac{1}{K}\sum_{k=1}^{K} \frac{h'_k[n]p'_k[n]G_k}{\sqrt{\eta'}} + \frac{z'[n]}{\sqrt{\eta'}}, \tag{20}$$

$$\hat{G} = \frac{1}{K}\sum_{k=1}^{K} \frac{h'_k[n]p'_k[n]\left[\nabla F^{i_t}(\mathbf{w_k}[n]) - \left[\frac{1}{K}\sum_{k=1}^{K} \frac{h_k[n]p_k[n]\left[\nabla F_k^{i_t}(\mathbf{w}^t[n]) - \nabla F_k(\mathbf{w}[n])\right]}{\sqrt{\eta}} + \frac{z[n]}{\sqrt{\eta}}\right]\right]}{\sqrt{\eta'}} + \frac{z'[n]}{\sqrt{\eta'}}. \quad (21)$$

Algorithm 2: FSVRG With Over-the-Air Communication and Computation (FSVRG-OACC)

Parameters: ϕ = number of data points, ϕ_k = number of data points store on device k, α = stepsize, data partition $\{\mathcal{D}_k\}_{k=1}^{K}$, Randomly initialize $\mathbf{w}_k$ on each device.
for $n = 0, 1, \ldots$ **do**
» Overall iterations
  for $k = 1$ *to* K ***do in parallel*** *over device* k **do**
    1: Let $\{i_t\}_{t=1}^{\phi_k}$ be random permutation of $\mathcal{D}_k$ » Distributed loop **1**
    2: Compute $\nabla F_k^{i_t}(\mathbf{w}^t[n])$
    3: Compute $\nabla F_k(\mathbf{w}[n]) = \frac{1}{\phi_k}\sum_{i=1}^{\phi_k} \nabla F_k^{i_t}(\mathbf{w}^t[n])$
    4: Over-the-Air Gradient Aggregation: Each device k uploads
    $g_k = \left[\nabla F_k^{i_t}(\mathbf{w}^t[n]) - \nabla F_k(\mathbf{w}[n])\right]$ Over-the-Air.
  end
  Aggregated signal in PS (Server Side)
  $\hat{g} = \frac{1}{K}\sum_{k=1}^{K} \frac{h_k[n]p_k[n]g_k}{\sqrt{\eta}} + \frac{z[n]}{\sqrt{\eta}}$ » Model aggregation **1**
  Aggregated signal $\hat{g}$ is received by all edge devices
  for $k = 1$ *to* K ***do in parallel*** *over device* k **do**
    1: Initialize: $\alpha_k = \alpha/\phi_k$ » Distributed loop **2**
    2: Compute $\nabla F^{i_t}(\mathbf{w_k}[n])$
    3: Aggregated signal $\hat{G}$ is received by the edge device k
    $n = 0$ -> randomly initialize $\hat{G}$
    for $t = 1, \ldots, \phi_k$ **do**
      $\mathbf{w}_k[n+1] = \mathbf{w}_k[n] - \alpha_k \hat{G}$
    end
    4: Over-the-Air Gradient Aggregation: Each device k uploads
    $G_k = \left[\nabla F^{i_t}(\mathbf{w_k}[n]) - \hat{g}\right]$ Over-the-Air.
  end
  Aggregated signal in PS (Server Side)
  $\hat{G} = \frac{1}{K}\sum_{k=1}^{K} \frac{h'_k[n]p'_k[n]G_k}{\sqrt{\eta'}} + \frac{z'[n]}{\sqrt{\eta'}}$ » Model aggregation **2**
end

This approach is motivated by the primary goal of maintaining gradient step consistency among all clients by utilizing the aggregated gradient over-the-air signal in stochastic first-order methods. Therefore, the algorithm's complexity is exceptionally low due to the simplicity of first-order gradient calculations at each step. The only cost involved is the number of communication rounds.

In summary, the gradient update $\left[\nabla F^{i_t}(\mathbf{w_k}[n]) - \nabla F_k^{i_t}(\mathbf{w}^t[n]) + \nabla F_k(\mathbf{w}^t[n])\right]$ is divided into two parts for FSVRG-OACC. In the first distributed loop, the last two gradients $\nabla F_k^{i_t}(\mathbf{w}^t[n])$ and $\nabla F_k(\mathbf{w}^t[n])$ are calculated, and afterward, the distance between these two gradients is aggregated across all edge devices. In the second distributed loop, we can access the aggregated value of $\left[\nabla F_k^{i_t}(\mathbf{w}^t[n]) - \nabla F_k(\mathbf{w}^t[n])\right]$ for all devices. In the first iteration of this loop ($n = 1$), we calculate the stochastic gradient for each edge device and subtract it from the aggregated gradient obtained from the first distributed loop $\left[\nabla F^{i_t}(\mathbf{w_k}[n]) - \text{aggregation}(\nabla F_k^{i_t}(\mathbf{w}^t[n]) - \nabla F_k(\mathbf{w}^t[n]))\right]$. This gives us G_k, the overall gradient update for device k, which is then aggregated with all other gradients over the air.

Starting from the second iteration ($n = 2$) and beyond, we can utilize the whole aggregated gradient $\hat{G}$ to directly update the model parameters of each device.

4. Performance Evaluation of FSVRG-OACC

In this section, we evaluate the performance of the proposed algorithm on the task of anomaly detection for a PM application in an FL manner with over-the-air analog aggregation. It aims to investigate the convergence characteristics of four optimization algorithms, GD, SGD, FSVRG, and FSVRG-OACC, in the context of over-the-air analog aggregation. The primary focus is to analyze the performance of these algorithms in terms of both model accuracy and convergence time, which are crucial performance metrics in the field. Similar to prior research in the domain of Federated Machine Learning Algorithm for Collaborative Predictive Maintenance [1], we employ the widely recognized and benchmarked CMAPSS [28] dataset in this study.

The C-MAPSS model, developed by NASA and implemented in MATLAB/Simulink, represents a nonlinear dynamic model of a commercial turbofan engine. By manipulating the input parameters of this simulation model, it becomes possible to simulate diverse degradation patterns under various engine conditions. In order to generate datasets reflecting different fault modes, four distinct time series (FD001, FD002, FD003, FD004) were produced using these simulation tools. These datasets comprise multivariate time series, each of which is further divided into training and testing subsets. For the purpose of this study, we selected and analyzed two datasets: FD003 and FD004. The FD003 dataset comprises 100 test trajectories and 100 train trajectories, focusing on a single fault mode and two types of degradation: high-pressure compressor degradation and fan degradation. On the other hand, the FD004 dataset includes 248 test trajectories and 249 train trajectories, covering six fault modes and two types of degradation. The time series data in each dataset include 21 sensor observations, three operating settings, a trajectory ID, and a cycle count. The Remaining Useful Life (RUL) of an engine is estimated based on the number of operation cycles remaining before the engine fails.

4.1. Algorithm Implementation

In this study, a federated SVM model was employed to predict the RUL using the provided time series data. Specifically, FD003 and FD004 datasets were distributed among ten devices located at the factory site, with the purpose of participating in a collaborative PM scenario. It is crucial to emphasize that the data distribution was non-independent and not identically distributed (non-IID). This implies that the instances of failures were not randomly distributed among the edge devices, and the data distribution was not evenly spread across all edge devices. Instead, specific edge devices were linked to particular types of failures, while others experienced different types. We intentionally opted for this non-IID distribution as a worst-case scenario to thoroughly assess the effectiveness of our proposed method. This collaboration aimed to achieve low latency through over-the-air analog aggregation. During all the simulations, the optimizer parameters were configured to achieve optimal performance. In particular, the learning is kept fixed at 0.01, and the ℓ_2-regularization parameter λ is set to $\lambda = 0.1$. The momentum factor is kept equal to 0.9 and the number of local epochs is set to 1. The definition of the loss function for the federated SVM is as follows.

$$\begin{aligned} f(\mathbf{w}) &= \frac{1}{\phi_k} \sum_{j \in \mathcal{D}_k} f_j(\mathbf{w}) + \lambda \|\mathbf{w}\|_2^2 \\ f_j(\mathbf{w}) &= \max(0, 1 - y_j \mathbf{w}^T x_j), \end{aligned} \tag{22}$$

The input data, denoted as x_j, are structured in a 2D format resembling an image, where one dimension corresponds to the sequence length, and the other dimension represents the number of sensor measurements. The output variable, y_j, indicates the condition of the engine. Specifically, if the input data are associated with a particular condition whose

RUL is below a specified threshold, the output value is assigned as $\{-1\}$, indicating the detection of an anomaly. Conversely, if the RUL exceeds the threshold, the output value is assigned as $\{1\}$. In all the conducted experiments, the reported convergence and accuracy averaged over five independent runs were calculated over 500 communication rounds.

4.2. *Performance of FSVRG-OACC*

4.2.1. General Performance

We first conduct a comparison between FSVRG-OACC and standard FSRVG, SGD, and GD methods to illustrate the notable enhancements in convergence rate and accuracy achieved in the context of over-the-air analog aggregation with a noisy environment. Throughout the experiments, we assessed the performance of each method and observed the significant gains achieved by FSVRG-OACC. We assume the environment noise variance to be $\sigma_z^2 = 1$ for $\mathbf{Z}_k \sim \mathcal{CN}(0, \sigma_z^2 \mathbf{I})$, and the probability of noise presence as $p = 1$. In this study, our primary focus is not on the transmission policy. Therefore, we make the assumption that the channel inversion policy is employed for estimation. Additionally, we assume that each component of $Z[n]$ has a zero mean and a variance of σ_z^2.

The performance is demonstrated on the FD003 and FD004 datasets. In this experiment, for all of $K = 10$ devices at the factory site, this implies that the transmitting power is set to $P = 300$ mW and the receiving scale factor, $\sqrt{\eta}$, is set to 0.1. The accuracy formula used is as follows:

$$\text{accuracy} = \frac{\mu_P + \mu_N}{\mu_P + \mu_N + \Gamma_P + \Gamma_N}, \quad (23)$$

where μ_P represents the number of true positives, Γ_P represents false positives, μ_N represents true negatives, and Γ_N represents false negatives. The results have been plotted in Figure 4 for FD003 and Figure 5 for FD004. We observed the following results.

1. In the case of FD003, it is evident that both the GD-MT and FSVRG algorithms fail to converge under the given noise environment and transmission power settings. Consequently, these algorithms are excluded from the accuracy analysis. On the other hand, the GD-GT and SGD-GT algorithms demonstrate convergence, but they exhibit significant fluctuations during the convergence phase. In contrast, our proposed FSVRG-OACC method shows excellent convergence performance under the same environmental conditions and noise levels.
   As depicted in the accuracy plot for FD003, both SGD-DT and GD-GT algorithms experience a considerable drop in accuracy. However, our proposed algorithm achieves an average accuracy of 91% and demonstrates higher stability compared to the other algorithms.
2. In the case of FD004, we observed similar results, although this dataset presents greater challenges due to its inclusion of six fault modes, making the prediction algorithm significantly more complex compared to other CMAPSS datasets. Despite these difficulties, our proposed FSVRG-OACC algorithm demonstrates robust convergence, with only minimal fluctuations occurring after the convergence stage. These fluctuations can be attributed to the estimation anomalies present in the most challenging instances of the dataset.
   In the accuracy plot, our proposed method achieved a commendable accuracy of 61% on the model. In contrast, the average accuracy of the other two converged methods is lower than that of the proposed method, and their accuracy values exhibited less fluctuation during the communication rounds.

In conclusion, we provide the average local training runtime and accuracy of FSVRG-OACC, SGD-GT, and GD-GT at the client level, as summarized in Table 2. It is noteworthy to mention that the computational runtime is primarily governed by the computation of gradients in all 500 iterations. Additionally, we assume that the communication round is negligible in comparison to the gradient computation. As is evident, the runtime in FSVRG-OACC exceeds that of the other algorithms. However, this method, employed in

FLOACC, demonstrates substantial reduction in communication latency while achieving commendable convergence and consistent accuracy.

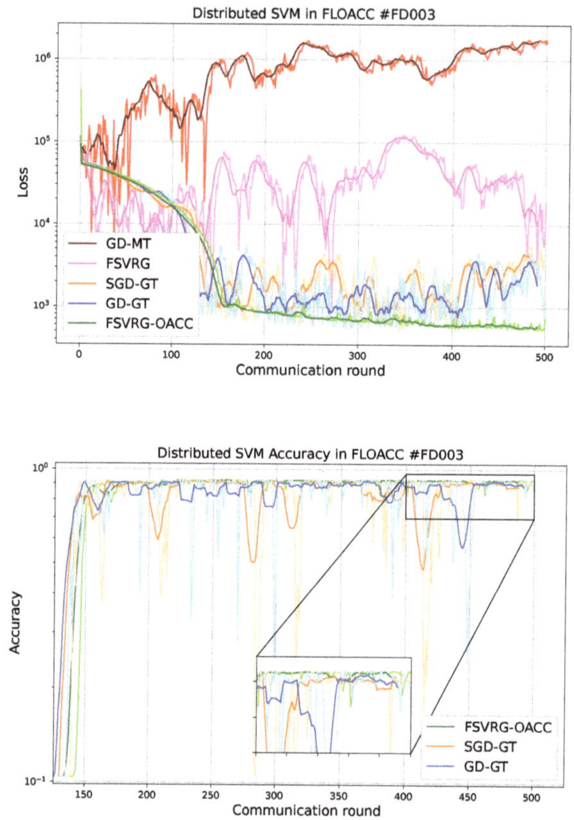

Figure 4. Convergence and accuracy versus communication round for different algorithms in Over-the-Air analog aggregation ($\sigma_z^2 = 1$, $p = 1$, and dataset: FD003). The pale graphs represent the signal, while the bold graphs depict the windowed average of the signal.

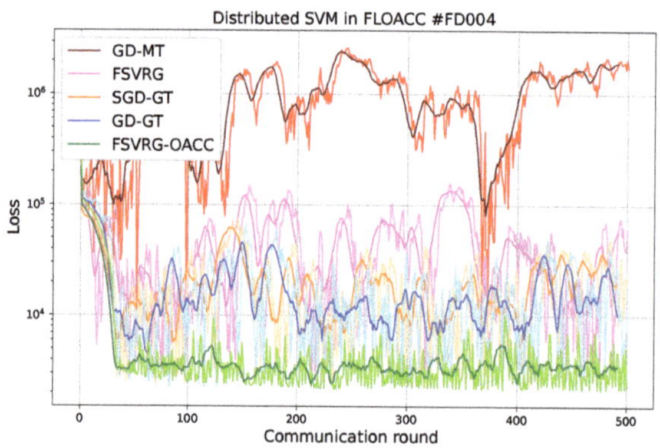

Figure 5. *Cont.*

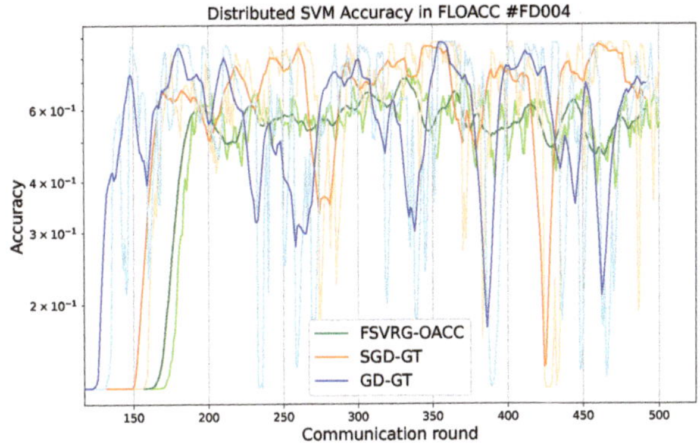

Figure 5. Convergence and accuracy versus communication round for different algorithms in Over-the-Air analog aggregation ($\sigma_z^2 = 1$, $p = 1$, and dataset: FD004). The pale graphs represent the signal, while the bold graphs depict the windowed average of the signal.

Table 2. Performance analysis of different algorithms on FLOACC.

Dataset	Evaluation Metrics	Optimizer		
		GD-GT	SGD-GT	FSVRG-OACC
FD003	Runtime Final accuracy	45.8 s 76%	21.3 s 61%	66.7 s 91%
FD004	Runtime Final accuracy	70 s 42%	20 s 41%	105 s 61%

We compared our results with a scenario where there is no communication noise. In a previous study, we looked at how well the FL algorithm works for PM applications when there is no communication interference. We summarized those results in Table 3. As evident from the results, there is a significant disparity in accuracy between GD-GT and SGD-GT when applied to FLOACC and noiseless communication settings. This demonstrates that employing a power transmission policy can enhance the accuracy of both GD and SGD in the context of FLOACC, bringing them closer to achieving results similar to those in noiseless communication scenarios. It is worth noting that, for the FSVRG-OACC method and FSVRG applied to noiseless communication, the difference in accuracy remains minimal. This observation is particularly noteworthy in the case of FD003, where the RUL prediction task is less challenging than FD004. These findings underscore the robustness and effectiveness of our proposed approach, which achieves these results without the need for any power transmission policies.

Table 3. Performance analysis of FL algorithm on noiseless communication channel [1].

Dataset	Evaluation Metrics	Optimizer		
		GD-GT	SGD-GT	FSVRG
FD003	Runtime Final accuracy	69 s 94.2%	20.5 s 92.2%	161 s 91.7%
FD004	Runtime Final accuracy	143.5 s 78.4%	45.5 s 71.7%	337 s 86.6%

4.2.2. Performance of FSVRG-OACC with Varying Noise Level

In this section, we conduct an analysis to assess the robustness of the proposed algorithms for Over-The-Air analog aggregation in the presence of varying levels of noise.

The performance evaluation is conducted using a part of the FD003 dataset while systematically varying the noise level σ_z. Specifically, we consider noise levels σ_z from the set $\{0.25, 0.75, 1.5, 2.5\}$ to assess the algorithm's performance under different noise conditions. For the total of 10 local devices, the receiver scaling factor, denoted as $\sqrt{\eta}$, is varied across different values $\{0.4, 0.13, 0.06, 0.04\}$. To investigate the impact of noise, we manipulate the probability of noise presence, denoted as p, throughout the experiments. The set $\{0.1, 0.25, 0.5, 1\}$ is utilized to vary the probability of noise. The results for GD-GT and SGD-GT have been plotted in Figures 6 and 7.

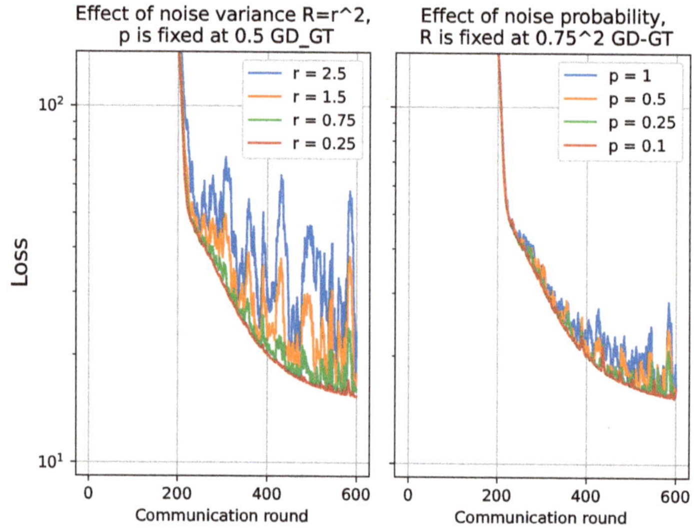

Figure 6. Convergence analysis for GD-GT by varying noise level $r \equiv \sigma_z = \{0.25, 0.75, 1.5, 2.5\}$, and the probability of noise presence $p = \{0.1, 0.25, 0.5, 1\}$.

For both GD-GT and SGD-GT, as the noise level increases, there is a significant degradation in performance, accompanied by amplified fluctuations in the loss function. Additionally, we observe a parallel trend in the cost function when the probability of noise presence is elevated. Furthermore, we conducted the same analysis on the FSVRG-OACC method, and the corresponding results are presented in Figure 8. Notably, the proposed algorithm demonstrates significantly enhanced robustness against higher noise levels and increased probability of noise presence.

All of these analyses provide evidence supporting the suitability of the proposed FSVRG-OACC method for analog aggregation and FL over the air. It is important to note that all of these analyses were conducted under the same transmission power conditions. However, we anticipate that employing a power control policy in communication with this method would yield even more accurate results in convergence and model accuracy.

On the other hand, implementing our proposed approach in real-world industrial settings has practical implications and challenges. A significant part of this approach is adjusting the local gradients in large edge devices using an analog waveform and sending them through the same wireless channels. Our main challenge is achieving perfect waveform superposition, which is crucial for our algorithm's success. This task is complicated because of frame timing offset and carrier frequency offset. To overcome these issues, we require high-performance devices with substantial computational capabilities. Another important challenge emerges when we only have access to partial information, resulting in some edge devices being unable to effectively join the FLOACC aggregation process. This led to a deviation in the gradient descent regime. This deviation highlights the necessity for robust device selection and strategies to manage situations where certain devices may have limited participation capabilities.

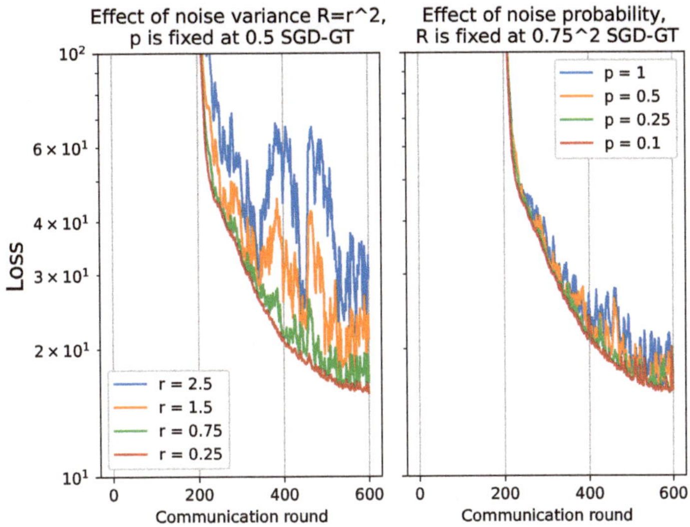

Figure 7. Convergence analysis for SGD-GT by varying noise level $r \equiv \sigma_z = \{0.25, 0.75, 1.5, 2.5\}$, and the probability of noise presence $p = \{0.1, 0.25, 0.5, 1\}$.

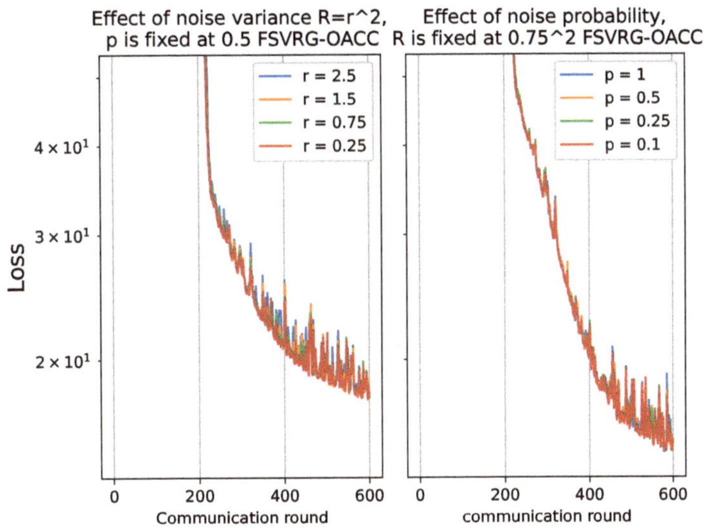

Figure 8. Convergence analysis for FSVRG-OACC by varying noise level $r \equiv \sigma_z = \{0.25, 0.75, 1.5, 2.5\}$, and the probability of noise presence $p = \{0.1, 0.25, 0.5, 1\}$.

5. Conclusions

In this paper, our focus was on FL scenarios that leverage wireless transmission channels for both communication and computation. The design presented capitalizes on the waveform superposition property inherent in a multi-access channel, which enables efficient update aggregation, optimizing the communication process. We specifically emphasized the potential of over-the-air analog aggregation in FL for hierarchical PM scenarios. Our study focused on the factory site as the lower hierarchical component, recognizing the criticality of meeting low-latency requirements for time-sensitive applications. We have demonstrated that this challenge can be effectively addressed through the utilization of analog aggregation.

Throughout our investigation, we thoroughly examined the impact of practical channel effects, including noise and fading, on the learning algorithm's performance. To enhance the robustness of learning algorithms against channel noise effects, we proposed a novel algorithm called FSVRG-OACC. The effectiveness and superiority of the proposed algorithm were demonstrated through the application of a distributed SVM for anomaly detection using the CMPASS dataset. The simulation results validate the effectiveness of FSVRG-OACC in reducing aggregation distortion while operating at the same transmission power level. Furthermore, the learning behavior of the proposed algorithm can be further enhanced by incorporating power control and device selection policies into its design.

Author Contributions: Conceptualization, N.B. and A.B.; methodology, A.B.; software, A.B.; validation, N.B. and A.B.; formal analysis, N.B. and A.B.; investigation, N.B. and A.B.; resources, N.B.; data curation, A.B.; writing—original draft preparation, A.B.; writing—review and editing, N.B. and A.B.; visualization, N.B. and A.B.; supervision, N.B.; project administration, N.B.; funding acquisition, N.B. All authors have read and agreed to the published version of the manuscript.

Funding: The research was funded by the European Commission within the European Regional Development Fund, Swedish Agency for Economic and Regional Growth, Region Gävleborg, and the University of Gävle.

Institutional Review Board Statement: Not applicable.

Informed Consent Statement: Not applicable.

Data Availability Statement: The CMAPSS dataset is available at https://data.nasa.gov/dataset/C-MAPSS-Aircraft-Engine-Simulator-Data/xaut-bemq (accessed on 19 August 2023).

Conflicts of Interest: The authors declare no conflict of interest.

Abbreviations

The following abbreviations are used in this manuscript:

IoT	Internet of Things
ML	Machine Learning
PM	Predictive Maintenance
PS	Parameter Server
FL	Federated Learning
OACC	Over-the-Air Communication and Computation
SNR	Signal-to-Noise Ratio
FedAvg	Federated Averaging
OFDM	Orthogonal Frequency-Division Multiplexing
GD	Gradient Descent
SGD	Stochastic Gradient Descent
non-IID	non-independent and not identically distributed
FSVRG	Federated Stochastic Variance Reduced Gradient
FLOACC	Federated Learning Over-the-Air Communication and Computation
FSVRG-OACC	Federated Stochastic Variance Reduced Gradient-Over-the-Air Communication and Computation

References

1. Bemani, A.; Björsell, N. Aggregation Strategy on Federated Machine Learning Algorithm for Collaborative Predictive Maintenance. *Sensors* **2022**, *22*, 6252. [CrossRef] [PubMed]
2. Amiri, M.M.; Gündüz, D. Machine Learning at the Wireless Edge: Distributed Stochastic Gradient Descent Over-the-Air. *IEEE Trans. Signal Process.* **2020**, *68*, 2155–2169. [CrossRef]

3. Konečný, J.; McMahan, H.B.; Yu, F.X.; Richtárik, P.; Suresh, A.T.; Bacon, D. Federated learning: Strategies for improving communication efficiency. *arXiv* **2016**, arXiv:1610.05492.
4. Peng, C.; Hu, Q.; Wang, Z.; Liu, R.W.; Xiong, Z. Online-Learning-Based Fast-Convergent and Energy-Efficient Device Selection in Federated Edge Learning. *IEEE Internet Things J.* **2023**, *10*, 5571–5582. [CrossRef]
5. Liu, S.; Yu, J.; Deng, X.; Wan, S. FedCPF: An Efficient-Communication Federated Learning Approach for Vehicular Edge Computing in 6G Communication Networks. *IEEE Trans. Intell. Transp. Syst.* **2022**, *23*, 1616–1629. [CrossRef]
6. Zhang, L.; Wang, H.; Xue, H.; Zhang, H.; Liu, Q.; Niyato, D.; Han, Z. Digital twin-assisted edge computation offloading in industrial Internet of Things with NOMA. *IEEE Trans. Veh. Technol.* **2022**. [CrossRef]
7. Zheng, S.; Shen, C.; Chen, X. Design and Analysis of Uplink and Downlink Communications for Federated Learning. *IEEE J. Sel. Areas Commun.* **2021**, *39*, 2150–2167. [CrossRef]
8. Amiri, M.M.; Gündüz, D. Federated Learning Over Wireless Fading Channels. *IEEE Trans. Wirel. Commun.* **2020**, *19*, 3546–3557. [CrossRef]
9. Krouka, M.; Elgabli, A.; Ben Issaid, C.; Bennis, M. Communication-Efficient Federated Learning: A Second Order Newton-Type Method With Analog Over-the-Air Aggregation. *IEEE Trans. Green Commun. Netw.* **2022**, *6*, 1862–1874. [CrossRef]
10. Jing, S.; Xiao, C. Federated Learning via Over-the-Air Computation With Statistical Channel State Information. *IEEE Trans. Wirel. Commun.* **2022**, *21*, 9351–9365. [CrossRef]
11. Yang, P.; Jiang, Y.; Wang, T.; Zhou, Y.; Shi, Y.; Jones, C.N. Over-the-Air Federated Learning via Second-Order Optimization. *IEEE Trans. Wirel. Commun.* **2022**, *21*, 10560–10575. [CrossRef]
12. Du, J.; Jiang, B.; Jiang, C.; Shi, Y.; Han, Z. Gradient and Channel Aware Dynamic Scheduling for Over-the-Air Computation in Federated Edge Learning Systems. *IEEE J. Sel. Areas Commun.* **2023**, *41*, 1035–1050. [CrossRef]
13. Konečný, J.; McMahan, B.; Ramage, D. Federated optimization: Distributed optimization beyond the datacenter. *arXiv* **2015**, arXiv:1511.03575.
14. McMahan, B.; Moore, E.; Ramage, D.; Hampson, S.; Arcas, B.A.Y. Communication-efficient learning of deep networks from decentralized data. *Proc. Artif. Intell. Statist.* **2017**, *54*, 1273–1282.
15. Li, T.; Sahu, A.K.; Talwalkar, A.; Smith, V. Federated Learning: Challenges, Methods, and Future Directions. *IEEE Signal Process. Mag.* **2020**, *37*, 50–60. [CrossRef]
16. Zhu, G.; Wang, Y.; Huang, K. Broadband Analog Aggregation for Low-Latency Federated Edge Learning. *IEEE Trans. Wirel. Commun.* **2020**, *19*, 491–506. [CrossRef]
17. Chen, M.; Yang, Z.; Saad, W.; Yin, C.; Poor, H.V.; Cui, S. A Joint Learning and Communications Framework for Federated Learning Over Wireless Networks. *IEEE Trans. Wirel. Commun.* **2021**, *20*, 269–283. [CrossRef]
18. Yang, H.H.; Liu, Z.; Quek, T.Q.; Poor, H.V. Scheduling Policies for Federated Learning in Wireless Networks. *IEEE Trans. Commun.* **2020**, *68*, 317–333. [CrossRef]
19. Krouka, M.; Elgabli, A.; Issaid, C.B.; Bennis, M. Communication-Efficient and Federated Multi-Agent Reinforcement Learning. *IEEE Trans. Cogn. Commun. Netw.* **2022**, *8*, 311–320. [CrossRef]
20. Ang, F.; Chen, L.; Zhao, N.; Chen, Y.; Wang, W.; Yu, F.R. Robust Federated Learning With Noisy Communication. *IEEE Trans. Commun.* **2020**, *68*, 3452–3464. [CrossRef]
21. Amiri, M.M.; Gündüz, D.; Kulkarni, S.R.; Poor, H.V. Convergence of Federated Learning Over a Noisy Downlink. *IEEE Trans. Wirel. Commun.* **2022**, *21*, 1422–1437. [CrossRef]
22. Wei, X.; Shen, C. Federated Learning Over Noisy Channels: Convergence Analysis and Design Examples. *IEEE Trans. Cogn. Commun. Netw.* **2022**, *8*, 1253–1268. [CrossRef]
23. H. Guo et al., Over-the-Air Aggregation for Federated Learning: Waveform Superposition and Prototype Validation. *J. Commun. Inf. Netw.* **2021**, *4*, 429–442. [CrossRef]
24. Cao, X.; Zhu, G.; Xu, J.; Huang, K. Optimized Power Control for Over-the-Air Computation in Fading Channels. *IEEE Trans. Wirel. Commun.* **2020**, *19*, 7498–7513. [CrossRef]
25. Liu, W.; Zang, X.; Li, Y.; Vucetic, B. Over-the-Air Computation Systems: Optimization, Analysis and Scaling Laws. *IEEE Trans. Wirel. Commun.* **2020**, *19*, 5488–5502. [CrossRef]
26. Xiao, L.; Zhang, T. A proximal stochastic gradient method with progressive variance reduction. *SIAM J. Optim.* **2014**, *24*, 2057–2075. [CrossRef]
27. Konečný, J.; McMahan, H.B.; Ramage, D.; Richtárik, P. Federated optimization: Distributed machine learning for on-device intelligence. *arXiv* **2016**, arXiv:1610.02527.
28. Li, X.; Ding, Q.; Sun, J.Q. Remaining useful life estimation in prognostics using deep convolution neural networks. *Reliab. Eng. Syst. Saf.* **2018**, *172*, 1–11. [CrossRef]

MDPI
St. Alban-Anlage 66
4052 Basel
Switzerland
www.mdpi.com

MDPI Books Editorial Office
E-mail: books@mdpi.com
www.mdpi.com/books

www.ingramcontent.com/pod-product-compliance
Lightning Source LLC
Chambersburg PA
CBHW061549220926
51850CB00005B/293
* 9 7 8 3 7 2 5 8 0 7 2 5 3 *